Regenerative Medicine, Artificial Cells and Nanomedicine – Vol. 3

SELECTED TOPICS IN NANOMEDICINE

Regenerative Medicine, Artificial Cells and Nanomedicine
ISSN: 2010-2836

Series Editor: Thomas Ming Swi Chang

Published

Vol. 1: Artificial Cells: Biotechnology, Nanomedicine, Regenerative Medicine, Blood Substitutes, Bioencapsulation, and Cell/Stem Cell Therapy
by T. M. S. Chang

Vol. 2: Present and Future Therapies for End-Stage Renal Disease
edited by E. A. Friedman and M. C. Mallappallil

Vol. 3: Selected Topics in Nanomedicine
edited by T. M. S. Chang

Regenerative Medicine, Artificial Cells and Nanomedicine – Vol. 3

SELECTED TOPICS IN
NANOMEDICINE

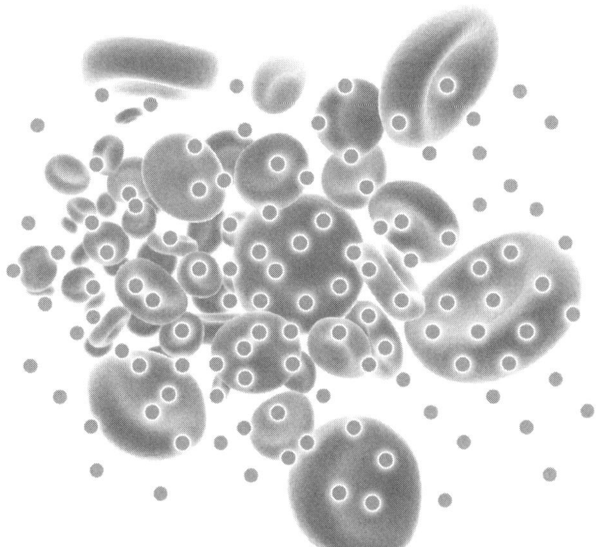

Edited by

Thomas Ming Swi Chang

McGill University, Canada

World Scientific

NEW JERSEY • LONDON • SINGAPORE • BEIJING • SHANGHAI • HONG KONG • TAIPEI • CHENNAI

Published by

World Scientific Publishing Co. Pte. Ltd.
5 Toh Tuck Link, Singapore 596224
USA office: 27 Warren Street, Suite 401-402, Hackensack, NJ 07601
UK office: 57 Shelton Street, Covent Garden, London WC2H 9HE

Library of Congress Cataloging-in-Publication Data
Selected topics in nanomedicine / [edited by] Thomas Ming Swi Chang.
p. ; cm. -- (Regenerative medicine, artificial cells and nanomedicine ; vol. 3)
Includes bibliographical references and index.
ISBN 978-9814472852 (hardcover : alk. paper)
I. Chang, Thomas Ming Swi, editor of compilation. II. Series: Regenerative medicine, artificial cells and nanomedicine ; v. 3.
[DNLM: 1. Nanomedicine--methods. 2. Nanostructures--therapeutic use. 3. Artificial Cells. 4. Blood Substitutes. QT 36.5]
RS201.N35
615.3'99--dc23

2013032339

British Library Cataloguing-in-Publication Data
A catalogue record for this book is available from the British Library.

Typeset by Stallion Press
Email: enquiries@stallionpress.com

Printed by FuIsland Offset Printing (S) Pte Ltd Singapore

Contents

Chapter 1

Artificial Cells: The Beginning of Nanomedicine

Thomas Ming Swi Chang

A. Introduction

1. *Nanomedicine*

Nanomedicine is the use of nanotechnology and nanobiotechnology in medicine. There are different ways to define these features, but the most common ones include:

i. Nano diameter
ii. Nanodimension thickness membrane
iii. Nanobiotechnological complexes
iv. Others

Artificial cells have all the features for nanotechnology and nanobiotechnology. What are artificial cells and how do they initiate a number of areas in nanomedicine?

Director, Artificial Cells and Organs Research Centre, Departments of Physiology, Medicine and Biomedical Engineering, Faculty of Medicine, McGill University, Montreal, Quebec, Canada H3G1Y6, www.artcell.mcgill.ca artcell.med@mcgill.ca

2. *The "What," "When," "How," and "Why" of Artificial Cells*

What and when: The first artificial cells were prepared in 1957 while the author was an undergraduate at McGill University (Chang, 1957) (Fig. 1). His further research shows the potential of artificial cells in basic research and applications (Chang, 1964, 1965, 1966, 1971a, 1971b, 1972; Chang *et al.*, 1966; Chang and Poznansky, 1968) (Fig. 1). This initial research forms the basic principle that has been greatly extended into different areas, including nanomedicine.

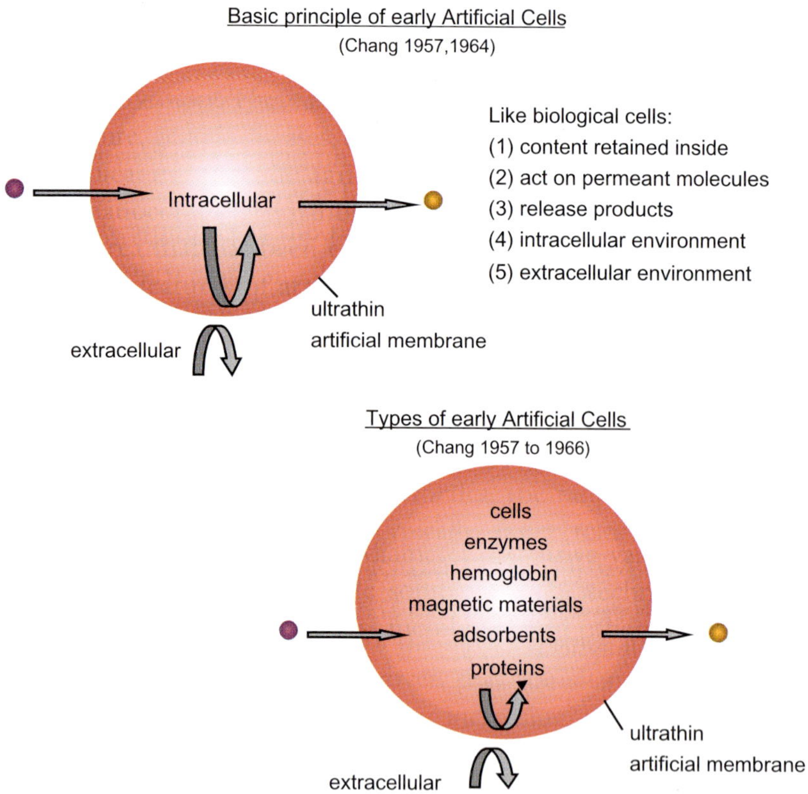

Fig. 1. Upper: Basic principles of early artificial cells. Lower: Different types of early artificial cells based on these basic principles. (*With copyright permission from Chang 2007 Monograph on Artificial Cells.*)

How does it develop: Thus, major progress in other areas has led to stepwise progress in artificial cells. First, there is the coming of age of polymer chemistry and biomaterials. Then there is the increasing interest in biotechnology, molecular biology, genomics and nanobiotechnology. Examples of the ongoing development and extension of the basic features of artificial cells include microcapsules, bioencapsulation, nanocapsules, nanoparticles, polymersomes, nanosensors, macroenapsulation, red blood cell mimics, polyhemoglobin, conjugated hemoglobin, nanotubules, lipid vesicles, liposomes, polymer-tethered lipids, synthetic cells and others (Chang, 2005, 2007, 2013) (Fig. 2).

Uses for artificial cells: The principle of artificial cells is already in routine patient use as a miniature device for the treatment of severe acute poisoning; approved for routine use in patients for blood transfusion in South Africa to avoid HIV-contaminated blood and in Russia to boost blood supply; in a number of drug delivery systems. It is also being extensively developed around the world for many other

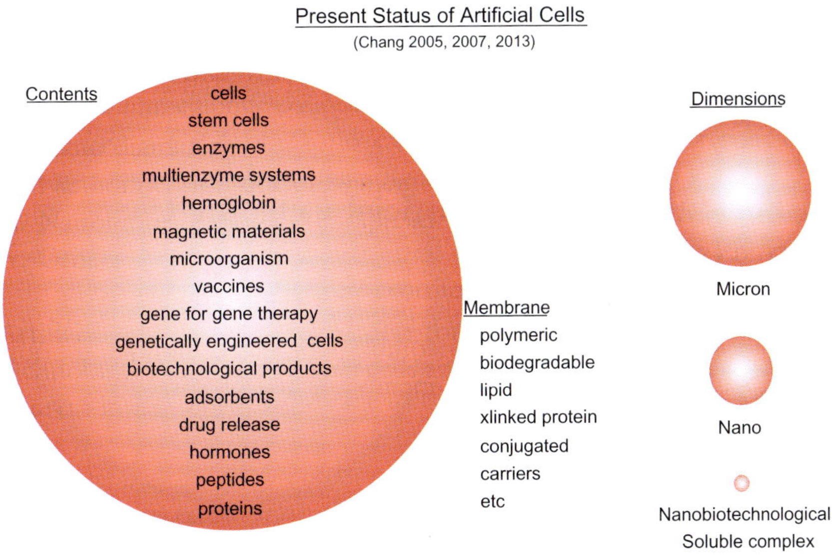

Fig. 2. Present status of artificial cells with wide variations in contents, membrane material and dimensions. (*With copyright permission from Chang 2007 Monograph on Artificial Cells.*)

ARTIFICIAL CELL: APPLICATIONS

Microdevice and nanodevice
Drug delivery
Blood Substitutes
Enzyme and gene therapy
Cell & Stem Cell Therapy
Biotechnology & Nanobiotechnology
Nanomedicine
Regenerative medicine
Agriculture, Industry, Aquatic culture
Nanocomputers and nanorobatics
Nanosensors etc

Fig. 3. Some areas of use for artificial cells.

areas of application. These include treatment for diabetes, liver failure, kidney failure, genetic diseases, endocrine diseases, cancer, biosensors, etc. Non-medical uses include agriculture, industry, food science, aquatic culture, nanocomputers, nanorobatics and others (Fig. 3).

One area that has been extensively developed is the use of some of the basic features of artificial cells in nanomedicine. This chapter contains a brief summary. Detailed papers on selected areas of nanomedicine will follow in this book.

B. Basic Features of Artificial Cells for Use in Nanomedicine

These include nanodimension diameter, extensive variations in contents, large number of possible soluble nanobiotechnological complexes, nanodimension thickness membrane and others.

1. *Nanodimension Diameter*

The original method of emulsion followed by evaporation (Fig. 4) (Chang, 1957, 1964) has been extended to the use of emulsion followed by interfacial polymerization (Chang, 1964; Chang *et al.*, 1966). This basic principle has been extended even further to many variations

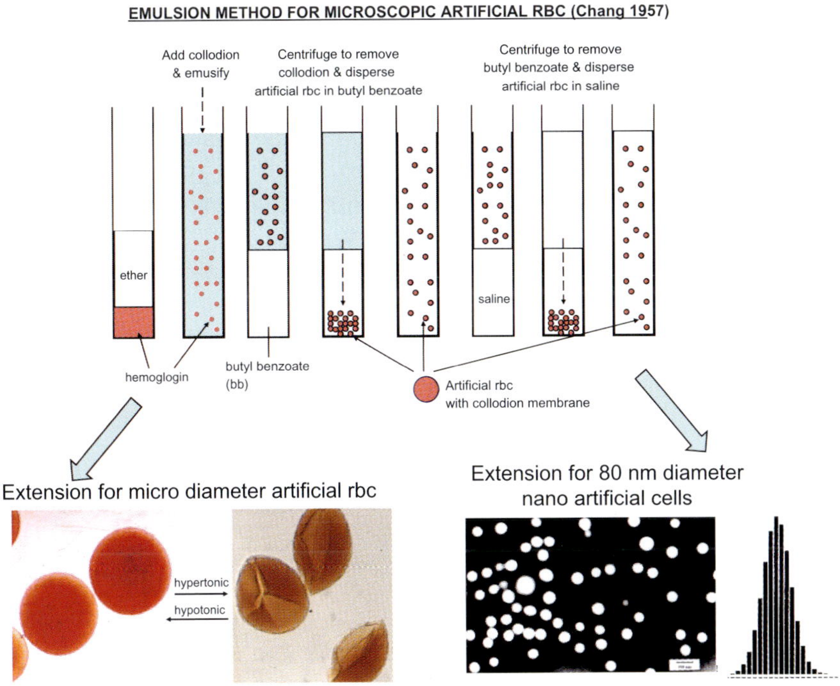

Fig. 4. (Upper): Original (1957) emulsion phase separation method of preparing artificial cells. This principle has since been extended to methods for the preparation of microscopic or nanodimension artificial cells that are also called microcapsules, nanocapsules, lipid membrane vesicles, microparticles, nanoparticles, polymersomes, etc. (Lower left): Micro dimension artificial rbc with ultrathin nylon-protein membrane. (Lower right): E/M of nanodimension (80 nanometer) artificial red blood cells containing hemoglobin and enzymes. (*With copyright permission from Chang 2007 Monograph on Artificial Cells.*)

to prepare nanodimension diameters material for use in nanomedicine. For example, decreasing the emulsion diameter to nanodimension range by different means would result in nanodimension systems (Fig. 4).

2. *Variations in Contents*

Artificial cells can contain the same biological material as biological cells, including hemoglobin and all red blood cell enzymes, microsomes, cytosol, polymerases, ribosomes and transcription/translation system. What is even more useful in nanomedicine is that the content could be from almost anything. Some examples include adsorbents, magnetic materials, drugs, enzymes, multienzyme systems, multi-compartment systems, hemoglobin, vaccines, genes, hormones, peptides, magnetic material and many other materials, which can be included separately or in combination (Fig. 2).

3. *Soluble Nanobiotechnological Complexes*

The first nanobiotechnology approach reported is the crosslinking of hemoglobin into ultrathin polyhemoglobin (PolyHb) membrane with nanodimension thickness (Chang, 1964, 1972a) (Fig. 3). This is used to form the membrane of artificial red blood cells (Chang, 1964, 1972a). If the emulsion is made very small, then the whole submicron artificial cells can be crosslinked into PolyHb of nanodimension. Glutaraldehyde can crosslink hemoglobin into soluble PolyHb each consisting of an assembly of 4–5 hemoglobin molecules (Chang, 1971b) (Fig. 5). Sebacyl chloride crosslinks hemoglobin and diamine to form polyamide conjugated hemoglobin (Chang, 1964, 1972a) (Fig. 5). An extension of this is the crosslinking of single protein molecule to soluble polymers (Wong *et al.*, 1968; Park *et al.*, 1981) (Fig. 5).

4. *Nanodimension Thickness Membrane with Possible Variations in Properties*

The artificial cell membranes, especially ultrathin nylon membrane (Chang, 1964), can be ultrathin and yet strong (Fig. 6).

The membrane of an artificial cell separates its contents from the outside, but at the same time the membrane can be prepared to selectively allow different types of molecules to cross. For example, the membrane material includes polymer, biodegradable polymer, lipid, crosslinked protein, lipid-polymer complex, lipid-protein complex and membrane with transport carriers. This way, one can prepare artificial cell membranes that selectively allow the movement of molecules according to molecular size, lipid solubility, affinity to carrier mechanisms, etc. (Fig. 7).

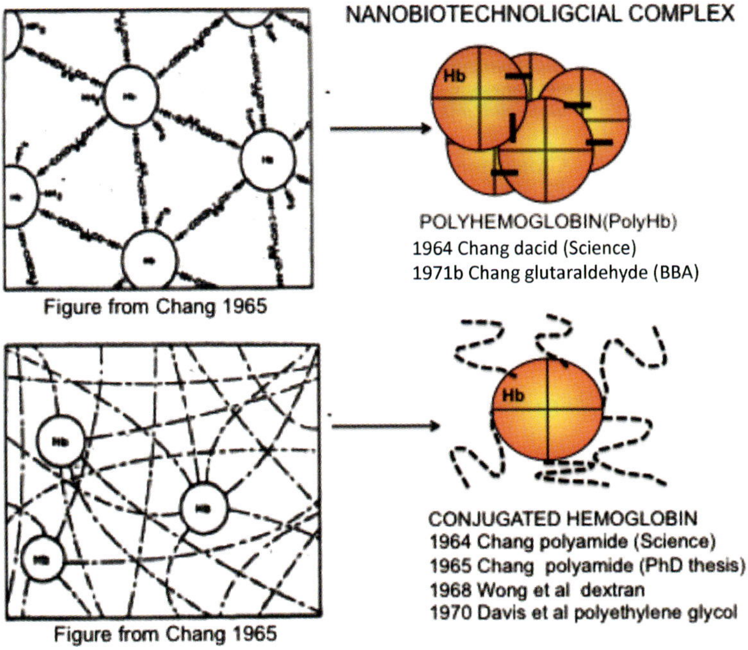

Fig. 5. (Upper left): Basic method of using bifunctional agents to assemble and crosslink hemoglobin (Hb) into PolyHb. (Upper right): Soluble complex of polyhemoglobin. (Lower left): Basic method of conjugating hemoglobin to polymer. (Lower right): Conjugation of hemoglobin to soluble dextran or polyethylene glycol. (*With copyright permission from Chang 2007 Monograph on Artificial Cells.*)

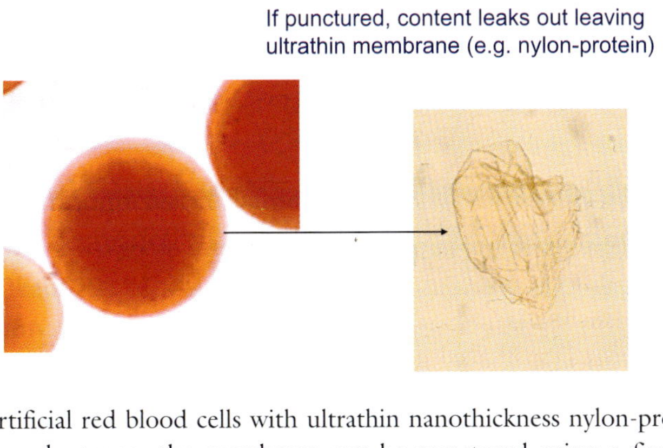

Fig. 6. Artificial red blood cells with ultrathin nanothickness nylon-protein membrane. Although strong, the membrane can be punctured using a fine needle to release the content, leaving behind the flexible ultrathin membrane. (*With copyright permission from Chang 2007 Monograph on Artificial Cells.*)

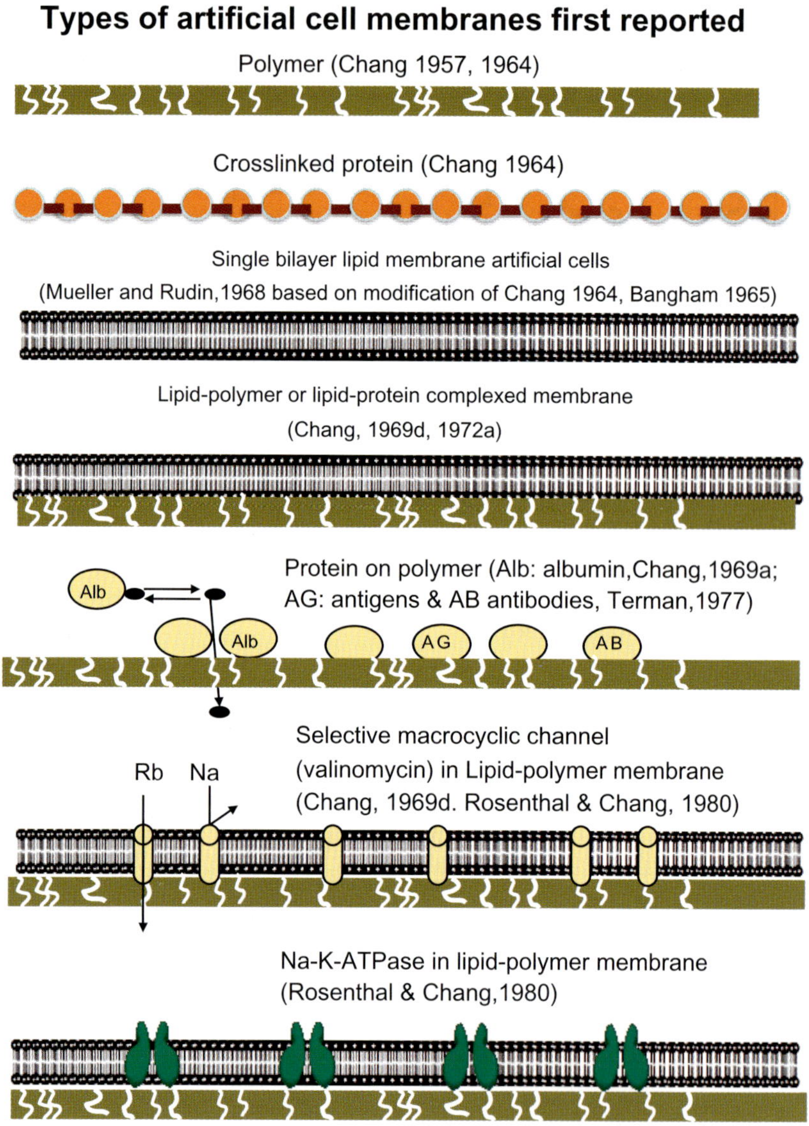

Fig. 7. Schematic representation of examples of ultrathin artificial cell membranes: polymer membrane, crosslinked protein membrane, lipid membrane, lipid-polymer membrane, membrane with albumin or other proteins, lipid-polymer membrane with cyclic channel carrier, lipid-polymer membrane with Na-K-ATPase transport carriers. (*With copyright permission from Chang 2007 Monograph on Artificial Cells.*)

Surface properties of artificial cell membranes can be varied by (1) incorporation of negative or positive charge (Chang, 1972a); (2) incorporation of albumin to increase blood compatibility (Chang, 1969a, 1972a); (3) incorporation of antigens to bind antibodies or antibodies to bind antigen (Terman *et al.*, 1977); (4) incorporation of polysaccharide like heparin (Chang, 1972a) or polyethylene glycol (PEG) (Chang, 2005; Torchilin, 2005) to increase compatibility or retention time in circulation (Fig. 8).

The ultrathin membranes of 100 to 200 nanometer and the micro/nano dimeters of artificial cells result in a large surface-to-volume relationship. This way a small volume of artificial cells of 90 micron diameter can have a potential mass transfer that is more than 100 times that of a large artificial kidney (Fig. 9). For nano dimension artificial cells, the mass transfer is even more amazing,

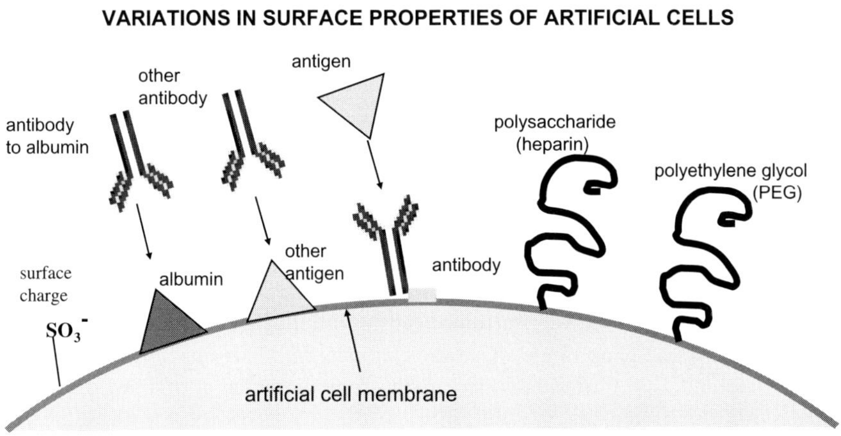

Fig. 8. Surface properties of artificial cell membranes can be varied by (1) incorporation of negative or positive charge; (2) incorporation of albumin to increase blood compatibility; (3) incorporation of antigens to bind antibodies or antibodies to bind antigen; (4) incorporation of polysaccharides like heparin or polyethylene glycol (PEG) to increase compatibility or retention time in circulation. (*With copyright permission from Chang 2007 Monograph on Artificial Cells.*)

	ARTIFICIAL KIDNEY 1 large machine dialysis membrane	ARTIFICIAL CELLS 60 ml in suspension 90 micron diameter
Membrane thickness (d)	>1 micron	100 to 200 nm
Total membrane area (A)	1 square meter	4 square meters
MASS TRANSFER (A/d):	**1**	**200 to 400 times higher**

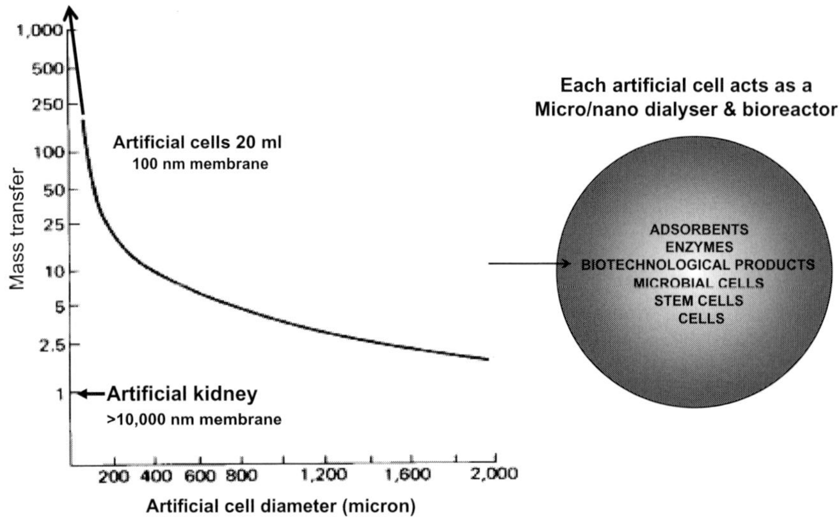

Fig. 9. (Upper): Mass transfer of 60 mL of 90 micron diameter artificial cells with 100 to 200 nanometer thick membrane is 200 to 400 times that of an artificial kidney machine. (Lower left): Theoretical analysis of mass transfer with that of an artificial kidney machine as 1. This is compared to ratio of mass transfer of 20 ml of 100 nm membrane thickness artificial cells of different diameters. With decreasing diameter to the nano range, this ratio can increase to an amazing 1,000 times. (Lower right): Thus, artificial cells of different diameter containing different bioactive material can become micro/nano dialyser/bioreactor with unlimited possibilities. (*With copyright permission from Chang 2007 Monograph on Artificial Cells.*)

around 1,000 times that of an artificial kidney. By combining this high mass transfer rate with unlimited contents possible, one can have unlimited feasibilities for micro/nano dialyser/bioreactor in many areas of application (Fig. 9).

C. Some Specific Examples of Uses in Nanomedicine

1. *Artificial Cells with Ultrathin Membranes for Miniaturization of Medical Devices*

The principle of artificial cells has been successfully used to miniaturize large artificial kidney machines into a 70 gram column of artificial cells. This has been a routine treatment for patients with acute poisoning around the world for many years (Chang *et al* 1973; Chang 1975; Winchester 1988; Lin *et al* 2004; Singh *et al* 2004) (Fig. 10). As mentioned above, the ultrathin membranes of 100 to 200 nanometer and the small diameters of artificial cells result in a large surface-to-volume relationship. This way a small volume of artificial cells can have

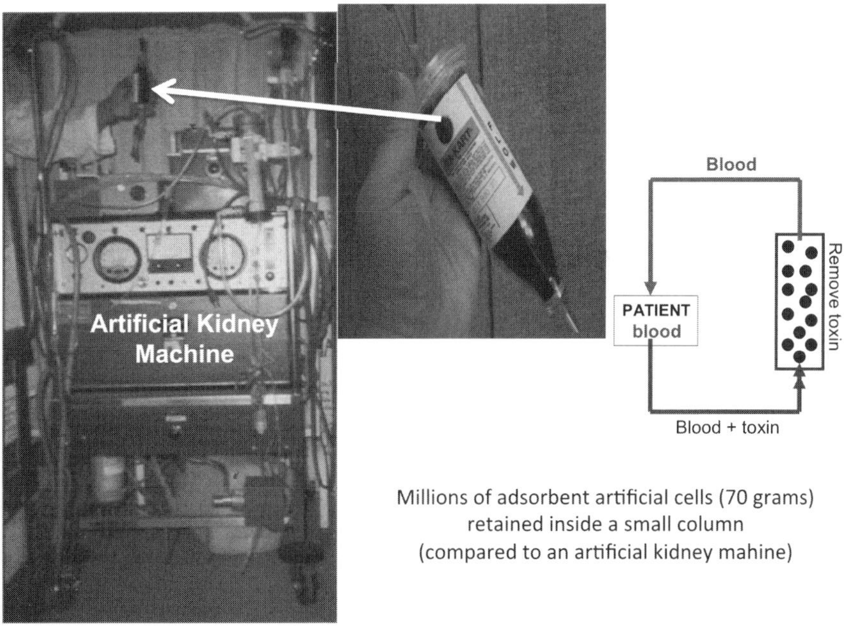

Fig. 10. A hemoperfusion device containing 70 grams of ultrathin membrane artificial cells containing activated charcoal on the right is compared to the same one held in the hand just above an artificial kidney machine. (*With copyright permission from Chang 2007 Monograph on Artificial Cells.*)

a potential mass transfer that is more than 100 times that of a large artificial kidney (Figs. 9 and 10). This allows artificial cells containing bioadsorbents or other bioreactants to be much more effective when compared to standard haemodialysis in removing toxins and drugs from the blood of patients (Chang, 1966, 1969a, 1975) (Fig. 10).

The most common routine use of this approach is the use of ultrathin polymeric artificial cells containing activated charcoal (Chang, 1969a, 1975; Chang *et al.*, 1973) (Figs. 10 and 11). The coating membrane prevents the major problems of release of embolizing particles and damage to blood cells in contact with blood (Fig. 11). This was first successfully used in suicidal overdose patients (Chang *et al.*, 1973) (Fig. 12). Since then, this has been become a routine treatment worldwide for acute poisoning in adults and children, especially in suicidal overdose (Chang, 1975; Winchester, 1988; Singh *et al.*, 2004; Lin *et al.*, 2004; Lopez *et al.*, 2002; Kawasahi *et al.*, 2000). This is particularly useful in places where dialysis machines are not easily available.

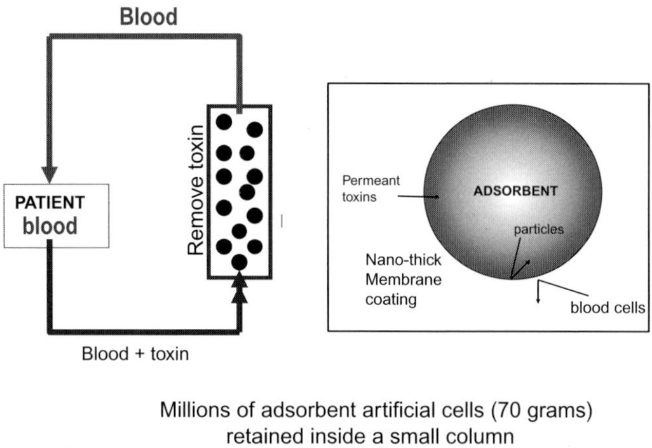

Fig. 11. (Right): The ultrathin membrane coating prevents the adsorbents from releasing harmful embolizing particles. It also prevents any damaging effects on blood cells. (Left): Toxins from circulation blood of patients can be rapidly removed from blood circulating through the column. This is now a routine method for the treatment of acute poisoning. (*With copyright permission from Chang 2007 Monograph on Artificial Cells.*)

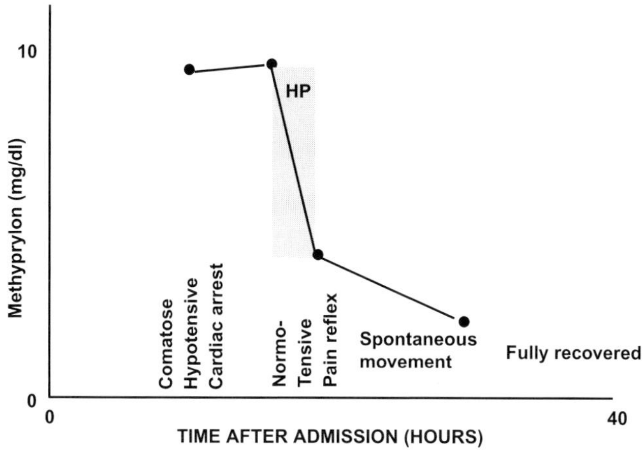

Fig. 12. One of the many examples of clinical and laboratory results of hemoperfusion in patients. This is a suicidal methryprylon overdose patient who ingested 3 times the lethal dose (Chang *et al* 1973). Hemoperfusion was carried out when standard hemodialysis at that time was not able to lower the drug level. (*With copyright permission from Chang 2007 Monograph on Artificial Cells.*)

This approach is effective in removing toxic products in kidney failure patients (uremia), resulting in the relief of uremic symptoms (Chang, 1972a, 1975; Winchester, 1988). Components for the removal of other uremic wastes need to be developed. This can also remove toxic molecules in patients with severe liver failure, resulting in the recovery of consciousness of grade 4 hepatic coma patients (Chang, 1972b, 1975; Gazzard *et al.*, 1974). Detoxification is only one of the functions of the liver, and this approach is being used as the detoxification component of hybrid liver support systems that are being developed.

The success in the clinical uses of artificial cells containing bioadsorbents has led to artificial cells containing immuoadsorbents (Chang, 1980). Terman *et al.* 1977 found that the albumin coating method (Chang 1969a) (Figs. 7 and 8) can be used to remove albumin antibodies from the circulating blood (Fig. 8). This has led to the use of hemoperfusion system with surface containing other antigen or antibody for the treatment of immunological diseases (Terman, 1977; Hakim *et al.*, 1990; Yang *et al.*, 2004). This includes clinical trials on lupus and other immunological diseases.

2. *Soluble Nanobiotechnological Complexes for Blood Substitutes and Oxygen Therapeutics*

Nanobiotechnology to assemble hemoglobin into polyhemoglobin for use as oxygen carrier for transfusion

The 1971 basic method of glutaraldehyde crosslinked polyhemoglobin (PolyHb) (Chang, 1971b) (Fig. 5) has been developed for clinical use as follows (Fig. 13). One is glutaraldehyde human PolyHb (Gould *et al.*, 2002; Moore *et al.*, 2009). Their clinical trial shows that this can replace blood lost in trauma surgery by keeping the blood hemoglobin at an acceptable level (Gould *et al.*, 2002). Their more recent clinical trials giving this in the ambulance to about 700 hemorrhagic shock patients show that one infusion of PolyHb can

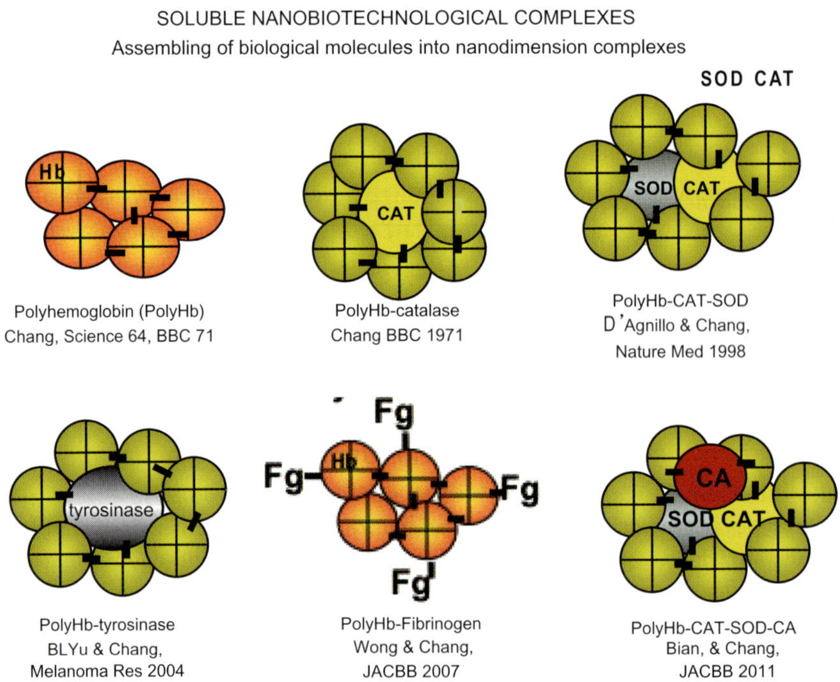

Fig. 13. Examples of different types of nanobiotechnology-based polyHb-enzyme systems. (*Updated with copyright permission from Chang 2007 Monograph on Artificial Cells.*)

maintain the patient for 12 hr after reaching the hospital. In the control group, most of the patients needed blood transfusion on reaching the hospital. In addition to hemoglobin from outdated human donor blood, other sources of hemoglobin have also been used for PolyHb. These included, for example, bovine Hb (Pearce and Gawryl, 2006), porcine Hb (Zhu *et al.*, 2007; Zhu and Chen, 2013) and Hb from human placental blood (Li *et al.*, 2006). Recombinant hemoglobin (Hoffman *et al.*, 1990), Arenicola marina Hb (Rousselot *et al.*, 2006) and others are also potential sources. One type of glutaraldehyde-crosslinked bovine PolyHb, first developed by Wong BL and C Rausch, has been studied extensively in clinical trials (Jahr *et al.*, 2008; Greenburg *et al.*, 2008; Greenburg, 2013). Russia and South Africa have approved this for routine clinical use in patients.

Unlike red blood cells, there is no blood group, thus PolyHb can be given on the spot without waiting for typing and cross-matching in the hospital. It is also free from infective agents such as HIV, hepatitis C, bacteria, parasites and so on. Whereas donor blood has to be stored at 4°C and is only good for 42 days, PolyHb can be stored at room temperature for more than one year. Thus, PolyHb can have important uses in a number of clinical conditions. Furthermore, in heart attack or stroke due to arterial obstruction, red blood cells, being 7 micron in diameter, cannot go through the obstruction (Fig. 14). PolyHb, being in solution, can more readily go through the obstruction to supply the needed oxygen (Fig. 14). However, this has to be done early since if there is much delay, PolyHb alone might result in the production of oxygen radicals that can cause tissue injury (ischemia reperfusion injuries). Even antioxidant enzymes normally present in red blood cells are not enough to prevent this problem in severe sustained hemorrhagic shock.

Nanobiotechnology and the assembling of hemoglobin with enzymes that remove oxygen radicals

We use glutaraldehyde crosslinking to assemble a nanobiotechnology complex of PolyHb-SOD-CAT by crosslinking hemoglobin, superoxide dismutase and catalase (D'Agnillo and Chang, 1998) (Fig. 14). This way, one can increase the antioxidant enzymes to a much higher level than those in red blood cells.

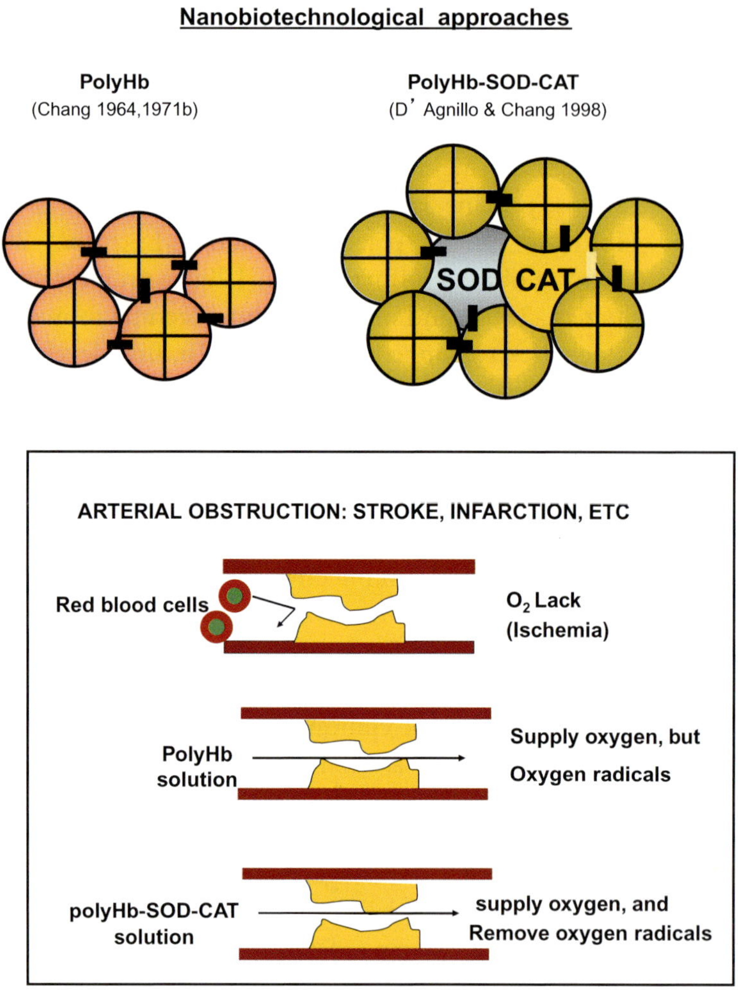

Fig. 14. Arterial obstruction from the narrowing of the artery can result in stroke and heart attack. Red blood cells, being 7 to 8 micron in diameter, have difficulty flowing through obstructed vessels to supply the needed oxygen. PolyHb, being a solution, can perfuse through to supply the needed oxygen. However, if oxygen lack is prolonged, reperfusion with an oxygen carrier can release damaging oxygen radicals. One possible solution is to use PolyHb-SOD-CAT that has the dual function of being an oxygen carrier and having the ability to remove oxygen radicals. (*With copyright permission from Chang 2007 Monograph on Artificial Cells.*)

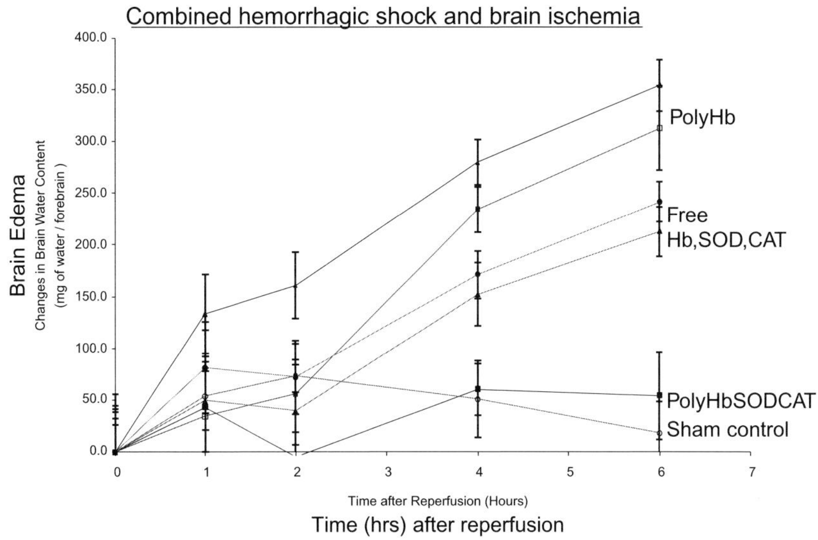

Fig. 15. Different fluids were infused after 90 min of combined hemorrhagic shock and brain ischemia. Brain edema was followed. PolyHb-SOD-CAT treated animals did not have any brain edema. On the other hand, PolyHb or free solution of Hb, SOD and CAT caused significant increases in brain edema. (*With copyright permission from Artificial Cells, Blood Substitutes and Biotechnology.*)

Thus in a rat stroke model, after 60 min of ischemia, reperfusion with PolyHb resulted in significant increase in the breakdown of the blood-brain barrier and an increase in brain water (brain edema) (Powanda and Chang, 2002). On the other hand, polyHb-SOD-CAT did not result in these adverse changes (Powanda and Chang, 2002) (Fig. 15). Hsia extended the PolyHb-SOD_CAT approach to prepare a hemoglobin with synthetic antioxidant based on the convalent binding of nitroxides (Buehler *et al.*, 2004; Ma and Hsia, 2013). A Hb-haptoglobin complex can also be used to protect against oxidative stress (Jia and Alayash, 2013).

Nanobiotechnology to assemble hemoglobin with enzymes resulting in a carrier for both oxygen and carbon dioxide with enhanced antioxidant activity

PolyHb-SOD-CAT is an oxygen carrier with enhanced antioxidant properties. Since red blood cells also carry out the important function of transport of carbon dioxide from the tissues to the lung for excretion, do we need this component? Sims *et al.* (2001) used a novel microelectrode to measure tissue pCO_2 in animal model of severe hemorrhagic shock. He shows that mortality is related to the elevation of tissue pCO_2. The enzyme carbonic anhydrase (CA) in red blood cells is the major means for the transport of tissue CO2 to the lung. We therefore use the nanobiotechnological method to assemble this enzyme with hemoglobin and antioxidant enzymes to form PolyHb-SOD-CAT-CA (Bian *et al.*, 2011). Our recent study in a rat hemorrhagic shock model shows that this is more efficient than red blood cells in lowering the elevated pCO_2 level in the tissue (Bian *et al.*, 2013) (Fig. 16).

Temperature stability of PolyHb-enzymes

How stable are this type of PolyHb-enzyme complexes especially with storage at different temperatures? Our study shows that the freeze-dried powder preparation is very stable with very long-term storage at −80°C and 4°C. It is also much more stable than the

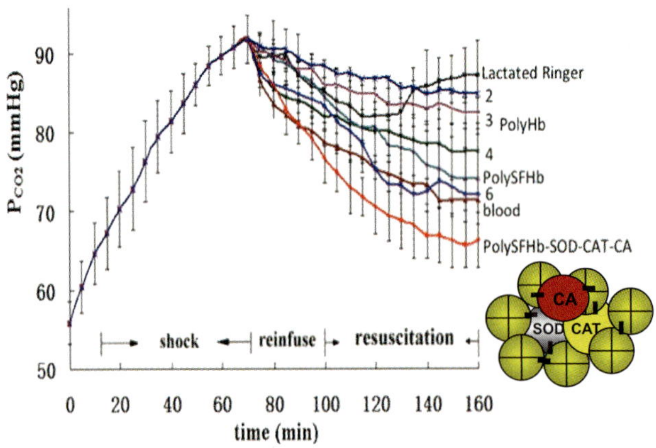

Fig. 16. Hemorrhagic shock rats maintained at a mean arterial blood pressure of 30 mm Hg. The tissue pCO_2 increases steady with time. Reinfusion of different fluids shows that lactated Ringer salt solution or PolyHb did not lower the tissue pCO_2 as much as blood or PolyHb-SOD-CAT-CA. The latter is even more effective than blood. (*From Bian and Chang 2013 with copyright permission from J Artificial Cells, Nanomedicine and Biotechnology.*)

solution at room temperature and even at 37°C (Bian, Yang and Chang, 2013). The freeze-dried powder is much easier to store since it takes up little space. Being a light compact powder it also much easier than the solution for transportation, especially during emergencies, disasters or war.

Nanobiotechnology to assemble hemoglobin and fibrinogen into an oxygen carrier with platelet-like activity

PolyHb can replace the hemoglobin level in very severe hemorrhage, but platelets also need to be replaced (Gould *et al.*, 2002). We studied this in a rat model and found that replacing more than 80% of the total blood volume with PolyHb leads to defects in blood clotting (Wong and Chang, 2007) (Fig. 17). We use nanobiotechnology to assemble hemoglobin with fibrinogen to form PolyHb-fibrinogen (Wong and Chang, 2007) (Figs. 13 and 17). Using this, we can replace up to 98% of the total blood volume without causing any clotting problems (Wong and Chang, 2007) (Fig. 17).

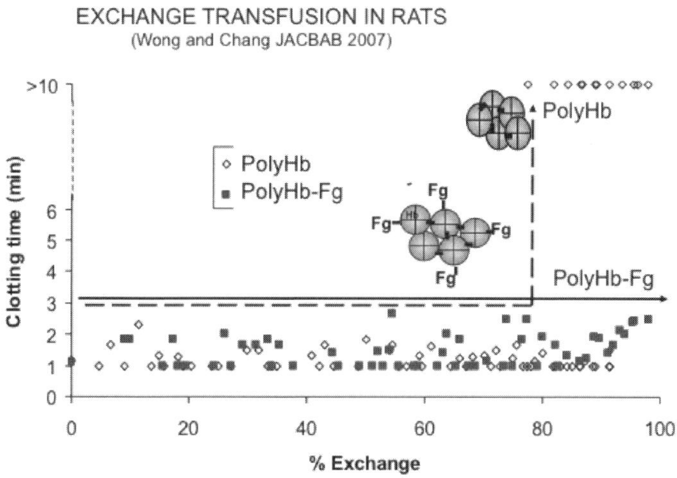

Fig. 17. Exchange transfusion in rats. There is a clotting problem when more than 80% of blood has been exchanged with PolyHb. There is no problem with clotting when 98% of the blood is replaced with PolyHb-fibrinogen with platelet-like activity. (*From Wong and Chang, 2007 with copyright permission from J Artificial Cells, Blood Substitutes and Biotechnology.*)

Nanobiotechnology for the assembling of hemoglobin with other enzymes

Abnormal microcirculation in tumor leads to decrease in perfusion by oxygen carrying red blood cells. PolyHb can more easily perfuse the abnormal microcirculation of tumors to supply oxygen needed for chemotherapy or radiation therapy. This way, PolyHb decreases the growth of the tumor and increases the lifespan in a rat model of gliosarcoma brain tumor (Pearce and Gawryl, 1998). We have recently crosslinked tyrosinase with hemoglobin to form a soluble PolyHb-tyrosinase complex (BL Yu and Chang, 2004) (Figs. 13 and 18). This has the dual function of supplying the needed oxygen and at the same time lowering the systemic levels of tyrosine needed for the growth of melanoma. Intravenous injections delayed the growth of the melanoma without causing adverse effects in the treated animals (Yu and Chang, 2004) (Fig. 18).

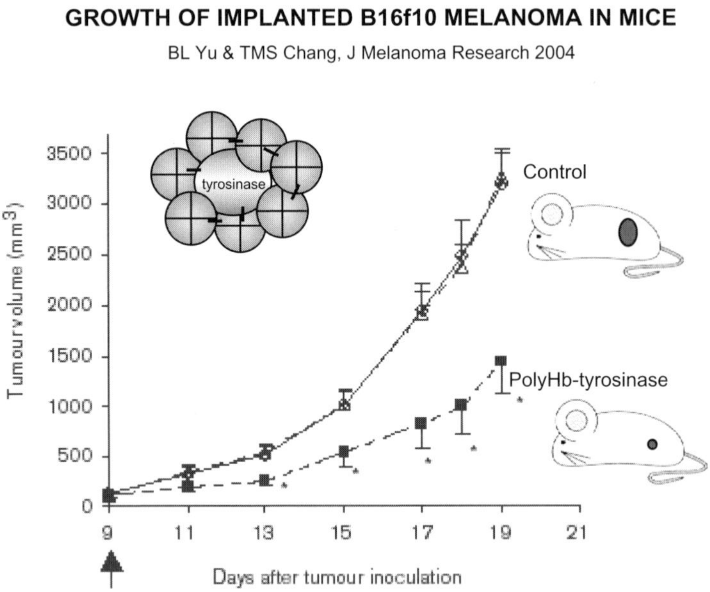

Fig. 18. Effects of daily intravenous injection of PolyHb-tyrosinase on tumor growth of B16F10 melanoma in mice. (i) sham control: no intravenous injection; (ii) saline control; (iii) PolyHb-tyrosinase group. (*With copyright permission from Chang 2007 Monograph on Artificial Cells.*)

Conjugated hemoglobin

In the presence of diamine, sebacyl chloride crosslinks hemoglobin with polyamide to form conjugated hemoglobin (Chang, 1964, 1965) (Fig. 5). An extension of this is the crosslinking of a single hemoglobin molecule to soluble polymers like dextran (Wong *et al.*, 1988) or PEG (Yabuki *et al.*, 1990; Wislow 2006; Liu and Xia, 2008) (Fig. 5). PEG-Hb shares many of the advantages of PolyHb as described above. More details are available in a later chapter (Seetharama *et al.*, 2013). Clinical trials are ongoing for two types of PEG-Hb (Winslow, 2006; Liu and Xia, 2008). Later phases of clinical trials on PEG-Hb have not yet been published, thus we cannot give any detailed description here. PEG-Hb also shows promising results for cancer therapy (Shorr *et al.*, 1966; Han *et al.*, 2012). Hsia has applied his nitroxide to conjuated hemoglobin as described in a later chaper (Ma and Hsia, 2013).

Stem cells for producing blood substitutes?

Do we need nanobiotechnology to prepare blood substitutes when stem cells can be used to produce blood cells? After much research, it is still not possible to scale this up sufficiently for the large volume of rbc needed. When scaling up becomes a reality, this will be an important source of rbc for many clinical conditions. However, these cells will still have many of the same problems of rbc. These include:

- Red blood cells require refrigeration at 4°C but still have a short storage time of less than 42 days. PolyHb can be stored at room temperature for more than a year. Freeze-dried powder of PolyHb and PolyHb-enzyme is even more stable.
- Unlike red blood cells, blood substitutes can be freeze-dried into powder form. In this form, they are light and compact with ease of transport especially in case of emergency, major disaster or war.
- Unlike rbc, blood substitutes can better perfuse obstructed microcirculation as in stroke, heart attack, ischemic limbs, sickle cell anemia and other conditions. They can also better perfuse disturbed microcirculations as in tumor, hemorrhagic shock and other conditions.

- Unlike rbc, blood substitutes can be enhanced with higher enzyme levels than rbc to be more effective than rbc against severe ischemia-reperfusion injury, fatal elevation of tissue pCO2 and other conditions.
- Nanobiotechnology can combine PolyHb with other enzymes and other bioreactants for specially designed oxygen therapeutics.

General discussion

There has been much progress in the area of blood substitutes and oxygen therapeutics (Liu and Xia, 2008; Mozzarelli, 2010; Zapol, 2012). In addition to PolyHb and conjugated Hb, there are other ways of modifying hemoglobin to prevent adverse effects of unmodified hemoglobin. Even PolyHb needs to be carefully prepared since those that contain high levels of uncrosslinked hemoglobin or low-molecular-weight PolyHb could also have adverse effects (Chang, 2007; Bucci, 2011, 2013). There are also other factors, including pathological characteristics of patients, like endothelial dysfunction (Yu *et al.*, 2010). However, vasoconstriction can be avoided if nitric oxide removal is prevented by a specially designed recombinant Hb (Doherty *et al.*, 1998) or a modified form of stabilized intramolecularly crosslinked Hb (Wong *et al.*, 2011) or by the administration of nitric oxide (Yu *et al.*, 2010; Zapol, 2012).

Thus, one should not attempt to combine the clinical trial results of different types of hemoglobin-based blood substitute and different clinical conditions into a single meta-analysis as has been done (Natanson *et al.*, 2008).

3. *Enzyme and Gene Therapy*

Enzymes inside artificial cells would not leak out, but can act on external permeant substrates (Chang, 1964, 1972a) (Fig. 19). This would avoid protein sensitization, anaphylactic reaction, or antibody production with repeated injection (Fig. 19).

Implanted urease artificial cells convert systemic urea into ammonia (Chang, 1964, 1966). Implanting artificial cells containing catalase replaces the defective enzyme in mice with a congenital defect in

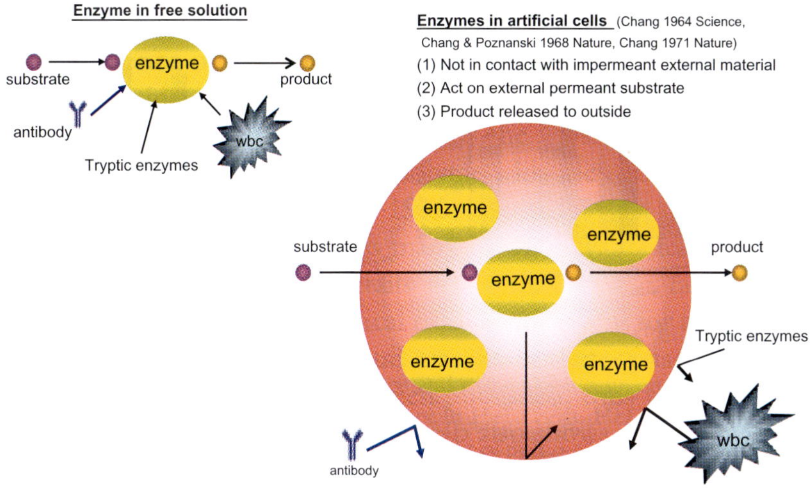

Fig. 19. (Left): Problems related to injection of enzymes in free solution. (Right): Enzymes inside artificial cells no longer have these problems. (*With copyright permission from Chang 2007 Monograph on Artificial Cells.*)

catalase — acatalasemia (Chang and Poznanski, 1968) (Fig. 20). Artificial cells containing asparaginase implanted into mice delay the onset and growth of lymphosarcoma (Chang, 1971a) (Fig. 20). Davis' group (Park *et al.*, 1981) extended this using PEG conjugated asparaginase that is now being studied by other groups in clinical trials for leukemia with promising results (Wetzler *et al.*, 2007). As mentioned above, PolyHb-tyrosinase suppresses the growth of the skin cancer, melanoma (Yu and Chang, 2004). Biodegradable PEG-PLA nano artificial cells containing PolyHb-tyrosinase are also being studied (Fustier and Chang, 2012; Wang and Chang, 2012). Implantation of artificial cells is useful for these shorter-term uses. However, for hereditary diseases where treatment is needed for the duration of the patient's life, we have to consider the potential accumulation of injected material over a long period of time.

Giving enzyme artificial cells by mouth avoids the need for repeated injection (Fig. 21). For example, artificial cells containing urease and ammonia adsorbent can lower the systemic urea level (Chang, 1972a). In Lesch–Nyhan Disease, artificial cells containing

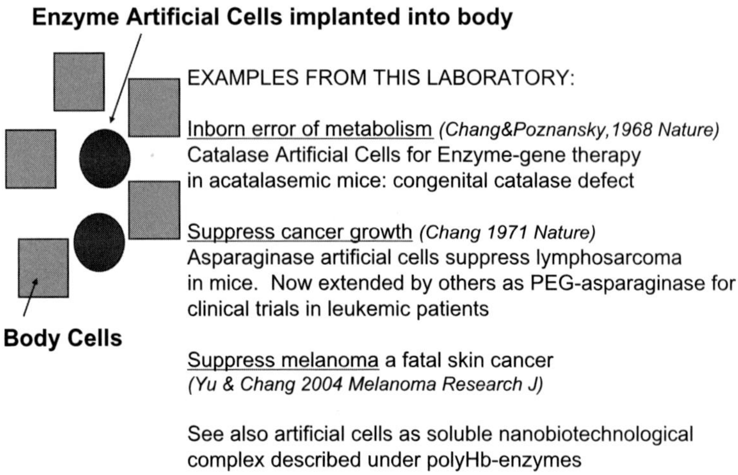

Fig. 20. Examples of the effect of implantation of artificial cells containing different enzymes for enzyme therapy. (*With copyright permission from Chang 2007 Monograph on Artificial Cells.*)

xanthine oxidase lower the toxic systemic hypoxanthine levels in an infant with this disease (Chang, 1989; Palmour *et al.*, 1989). Artificial cells containing phenylalanine ammonia lyase given by mouth lower the systemic phenylalanine levels in phenylketonuria (PKU) rats (Bourget and Chang, 1986; Chang, 2007). This leads to investigation into recombinant sources of this enzyme (Sarkissian *et al.*, 1999; Liu *et al.*, 2002). In the same way, oral artificial cells containing tyrosinase when given orally lowered the systemic tyrosine level (Chang, 2007).

There has been much recent extension of research using polymeric artificial cells of micro or nano dimensions for enzyme therapy (De Vocht *et al.*, 2009; Garnacho *et al.*, 2008; Haider *et al.*, 2008; Kaminsky and Kosenko, 2012).

Multienzyme systems with cofactor recycling

Most enzymes in the body function as multienzyme systems with cofactor recycling. After basic research on artificial cells containing multienzyme systems (Chang, 1985) we look into their possible use. Thus, artificial cells containing three different enzymes can convert

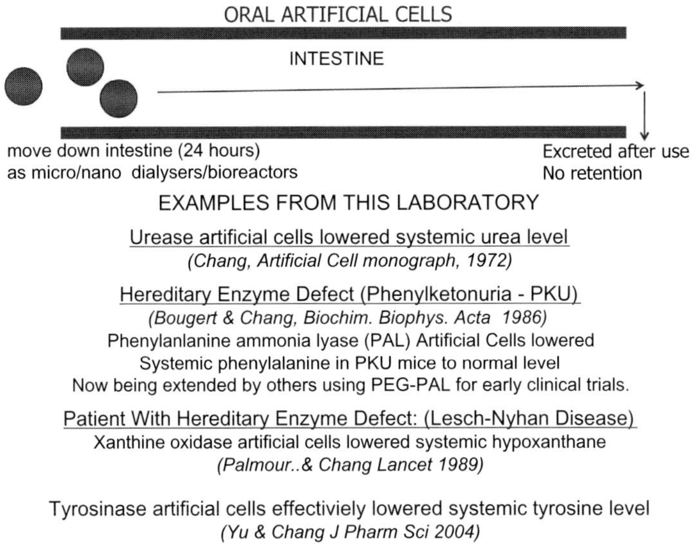

Fig. 21. Examples of artificial cells containing different enzymes given orally to avoid accumulation in long-term enzyme therapy. (*With copyright permission from Chang 2007 Monograph on Artificial Cells.*)

metabolic waste like urea and ammonia into essntial amino acids (Gu and Chang, 1990) (Fig. 22). The needed cofactor, NADH, can be recycled and retained inside the artificial cell by crosslinking to dextran or by the use of lipid–polymer membrane. All the multienzyme system in red blood cells can be incuded inside nanodimension artificial red blood cells (Chang *et al.*, 2003) (Fig. 22).

4. *Artificial Cells with Intracellular Compartments*

Biological cells contain organelles that allow separate compartments inside the cell to carry out specific functions more effectively. Artificial cells are also prepared to have intracellular compartments (Chang, 1965, 1972a; Chang *et al.*, 1966) (Fig. 23). Specific enzyme systems or other biologically active systems can be enclosed separately or in combination into each of these intracellular compartments to allow for more efficient stepwise functions. This principle has been extended

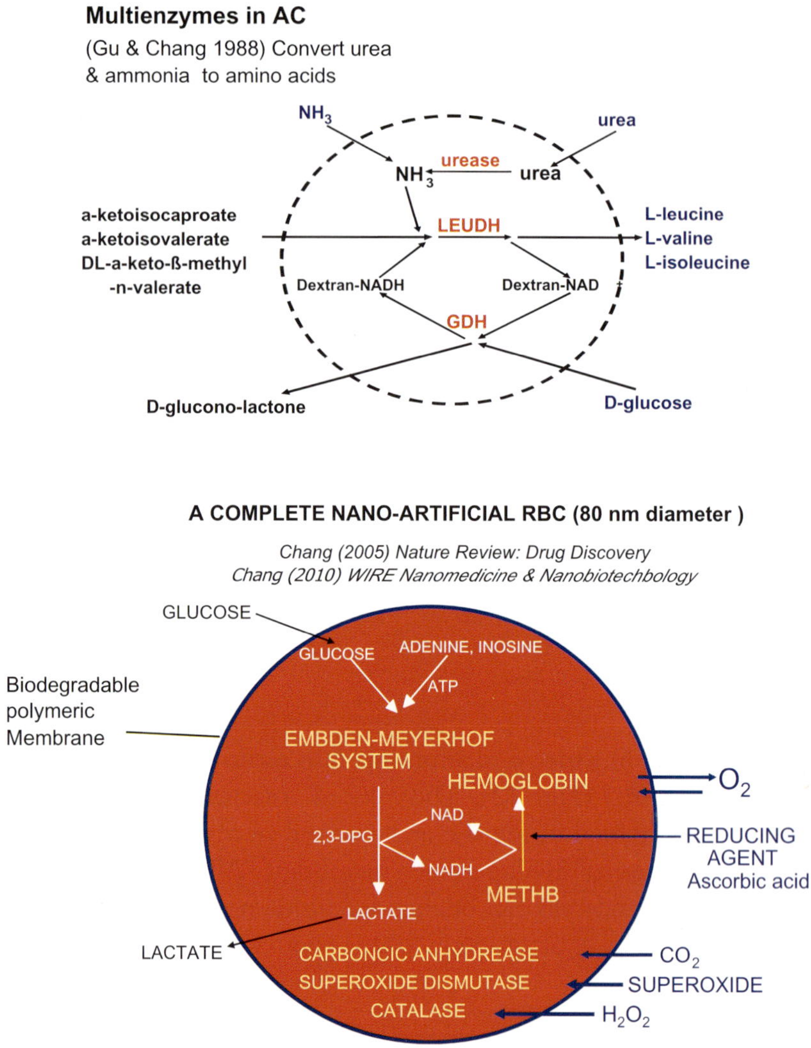

Fig. 22. (Upper): Artificial cells containing multienzyme systems to convert the waste, urea, into useful essential L-branched-chain amino acids, L-leucine, L-valine and L-isoleucine (Gu and Chang, 1988a). (Lower): Nano artificial red cells contain all the enzymes of red blood cells (Chang *et al.*, 2003). (*With copyright permission from Chang 2007 Monograph on Artificial Cells.*)

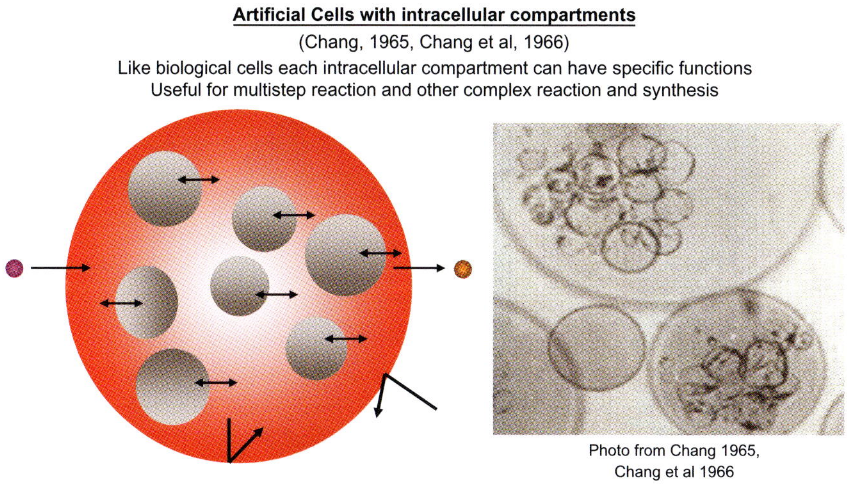

Fig. 23. Artificial cells can be prepared with intracellular multicompartments (Chang, 1972a). (*With copyright permission from Chang 2007 Monograph on Artificial Cells.*)

into potential uses for therapy as reveiwed in a later chapter (Hosta-Rigau and Städler, 2013).

5. *Nanodimension Artificial Cells with Biodegradable Membranes*

The use of artificial cells in enzyme therapy is a form of drug delivery. However, like biological cells, enzymes are retained inside the artificial cells at all times, and act on substrates diffusing into the artificial cells. Artificial cells in different modified forms have also been used in drug delivery in a different way. Here, they are used to separate the drug from the external environment, and to release the drug at a specific site and a specific rate when and where it is needed (Fig. 24). Drug delivery is an extremely large and wide area, and many excellent reviews and books are available.

Luzzi (1970) reported the use of artificial cells with nano thickness nylon membrane prepared as reported by Chang (1964). He microencapsulated drugs for slow release in oral administration.

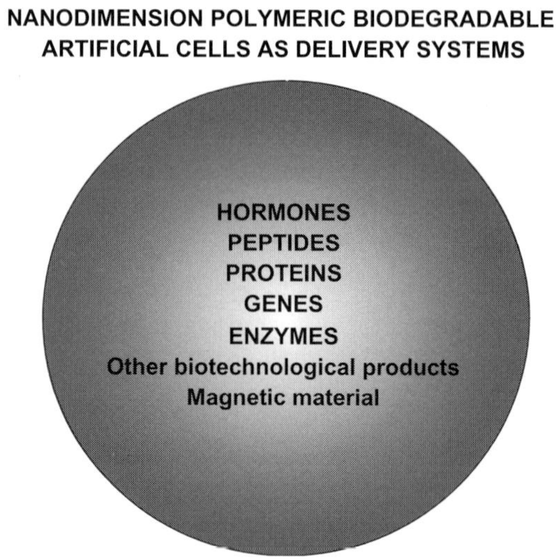

Fig. 24. Examples of biologically active material inside ultrathin polymeric membrane artificial cell carriers of micro/nano dimensions.

Others have also extended this approach. The modern approaches in drug delivery systems are based on injectable biodegradable systems of biodegradable polymer or bilayer lipid.

Biodegradable polymeric artificial cells

This was reported first in the form of biodegradable polylactide membrane artificial cells containing enzymes, hormones, vaccines and other biologicals (Chang, 1976). The polylactide polymer can degrade in the body into lactic acid, and finally into water and carbon dioxide. Variations in preparation can result in artificial cells that release insulin at different rates (Chang, 1976) (Fig. 25).

We have also used these for the slow release of prostaglandin E2 and ciprofloxacin. Biodegradable drug delivery systems are now used widely in different forms, ranging from microscopic to nanodimensions. They are also known as nanoparticles, nanocapsules, polymersomes, nanotubules, etc. Copolymers of polyethylene glycol (PEG) and polylactic acid (PLA) have been used to increase the circulation time of nanodimensional artificial cells. This also forms the basis for preparing nanodimension PEG–PLA membrane artificial red blood

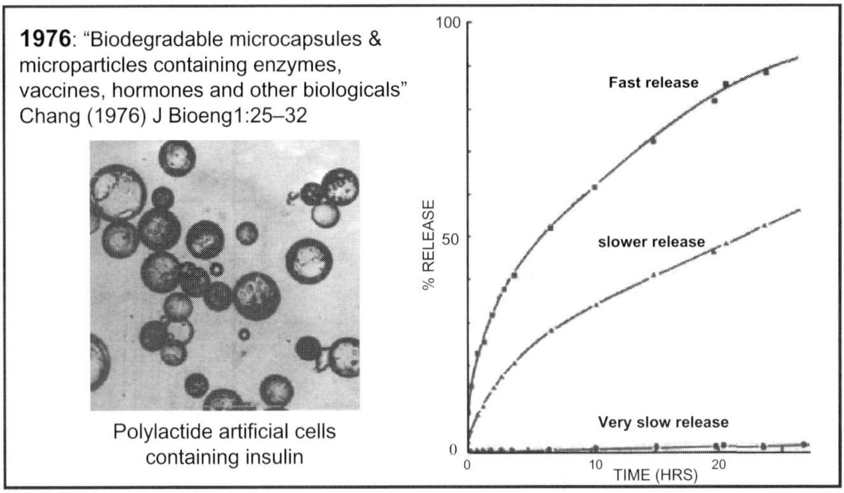

2013
Biodegradable polymers:
Microcapsules, Microparticles
Nanocapsules, Nanoparticles
Polymersomes, etc

Fig. 25. Biodegradable membrane artificial cells have been prepared to contain enzymes, hormones, vaccines and other biologicals (Chang, 1976). Variations in the molecular weight of polylactide and thickness of the membrane can result in artificial cells that release insulin at different rates. (*Updated with copyright permission from Chang 2007 Monograph on Artificial Cells.*)

cells (Chang *et al.*, 2003) and PEG-PLA nanoencapsulated polyHb-tyrosinase (Fustier and Chang, 2012; Wang and Chang, 2012).

Bilayer lipid membrane artificial cells

Bangham (1965) reported the preparation of liposomes each consisting of microspheres of onion-like concentric lipid bilayers — multi-lamellar. These were initially used as membrane models in basic research. After his Ph.D. at McGill, Gregoriadis went to the U.K. to continue his research and first reported the use of liposomes as drug delivery systems (Gregoriadis, 1976). However, the large amount of lipid in the multi-lamellar liposome limits the amount of water-soluble drugs that can be enclosed. Thus, the basic principle and method of preparing artificial cells using ether as the dispersing phase (Chang, 1957, 1964) (Fig. 4) was extended by researchers into what they call an "ether evaporation method" to form single bilayer (unilamellar) lipid membrane liposomes (Deamer and Bangham, 1976). These lipid membrane artificial cells have since been extensively studied for use as drug delivery systems (Torchilin, 2005). This is also being investigated as a red blood cell substitute, as will be discussed in a later chapter (Sakai, 2013). The modern single bilayer lipid vesicles are actually single bilayer lipid membrane artificial cells.

Membrane of artificial cells started with polymer, then lipid bilayer, then lipid-polymer and back to polymer

Surface charges and polysaccharides have been incorporated into both polymeric artificial cells (Chang, 1964, 1972a) and also into liposomes for possible targeting of drug. Polyethylene glycol (PEG) has also been incorporated to both types of surface to result in longer circulation time. Thus, PEG-lipid vesicles are becoming more like the lipid–polymer membrane artificial cells (Chang, 1972a) (Fig. 7), and are no longer pure lipid vesicles. Further development led to the incorporation of antibodies into the polymeric or lipid membrane of artificial cells (Fig. 8), to allow for targeting to cells with the corresponding antigens. Discher's group (Photos *et al.*, 2003) tried to increase the strength of the PEG–lipid membrane artificial cells by using self-assembling of block copolymers. Poly(ethylene glycol) (PEG) was the hydrophilic block and polyethylene or polybutadiene (PB) was the hydrophobic block. A later chapter by Nallani's group (De Hoog *et al.*, 2013) in this book gives a detailed overview of this area. This significantly increased the strength when compared to PEG–lipid membrane artificial cells. Thus, polymeric membrane artificial cells have branched off into multi-lamellar liposomes that have then evolved into lipid membrane artificial cells, then polymer(PEG)–lipid membrane artificial cells, and finally back to the original polymeric membrane artificial cells (Chang, 1964) that are called by different names including polymersomes, nanocapsules, nanoparticles and others.

Later chapters in this book give a few examples of the extension and variations possible (Chandra *et al.*, 2013; Siu *et al.*, 2013; Shao *et al.*, 2013; Nimesh and Chandra, 2013; Wohl *et al.*, 2013; Mobed-Miremadi, 2013; Eldridge *et al.*, 2013; Qasaimeh, 2013).

The principle of loading magnetic particles into artificial cells (Chang, 1966) has also been used for loading into both polymeric and lipid membrane artificial cells allowing for magnetic targeting and other applications (Fig. 26).

6. *Artificial Cells Containing Magnetic Materials*

Artificial cells containing biological materials and magnetic materials have been prepared to have useful functions (Chang, 1966) (Fig. 26). This way, external magnetic fields can direct their movement; remove or separate them from a mixture; retain them at specific site of action; stir or agitate them as in bioreactors, and other possibilities. This principle is now being used very extensively in bioreactors; in removing specific materials from a mixture as in diagnostics kits; in drug delivery systems; for locating radioactive material or chemotherapeutic agents

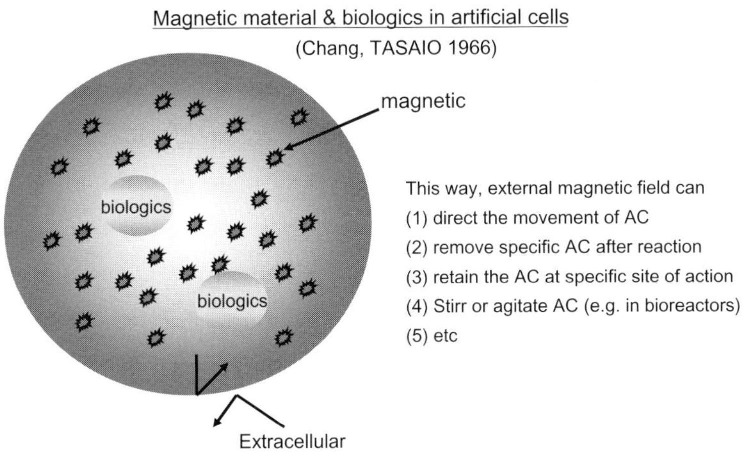

Fig. 26. Artificial cells containing biologics can also contain magnetic material. (Chang, 1966) (*With copyright permission from Chang 2007 Monograph on Artificial Cells.*)

at site of tumor and other areas of application. This is now an explosive area, as reviewed by two later chapters in this book (Qin *et al.*, 2013; Peng *et al.*, 2013).

7. *Artificial Cells Containing Biological Cells*

The first artificial cells containing intact biological cells were reported in 1964 based on a drop method (Chang, 1964), and it was proposed that "*protected from immunological process, encapsulated endocrine cells might survive and maintain an effective supply of hormone*" (Chang, 1965, 1972a; Chang *et al.*, 1966) (Fig. 27). This basic principle has been extensively developed around the world.

Although a nano thickness membrane can be used, a somewhat thicker membrane is more common used. Thus, only a very brief summary will be given here.

This has been investigated by many centers for use in cell therapy (Lim and Sun, 1980; Achischer *et al.*, 1996; De Vos *et al.*, 2002; Orive *et al.*, 2003; Hunkeler *et al.*, 2003; Chang, 2007). Examples include artificial cells containing endocrine tissues for instance, islets for diabetes. Another extensively investigated area is artificial cells containing genetically engineered cells for many types of diseases. The results in animals have been promising. However, one injection can only function for less than one year, and this is not practical for long-term illness like diabetes. Repeated injection would have retention problems. Ongoing research is being carried out by many groups around the world on improving the longer term biocompatibility for the membrane. We are looking at another two approaches. As in the case of artificial cells for enzyme therapy described above, we have investigated the oral administration of artificial cells containing cells (Prakash and Chang, 1996). Another of our recent approaches to this problem is to use artificial cells containing bone marrow stem cells to help liver regeneration in severe acute liver failure, since the liver can regenerate and does not require long-term treatment. In these studies, rats with 90% of the liver removed did not recover, On the other hand, most of the rats that received one injection of artificial cells containing stem cells recovered (Liu and Chang, 2012).

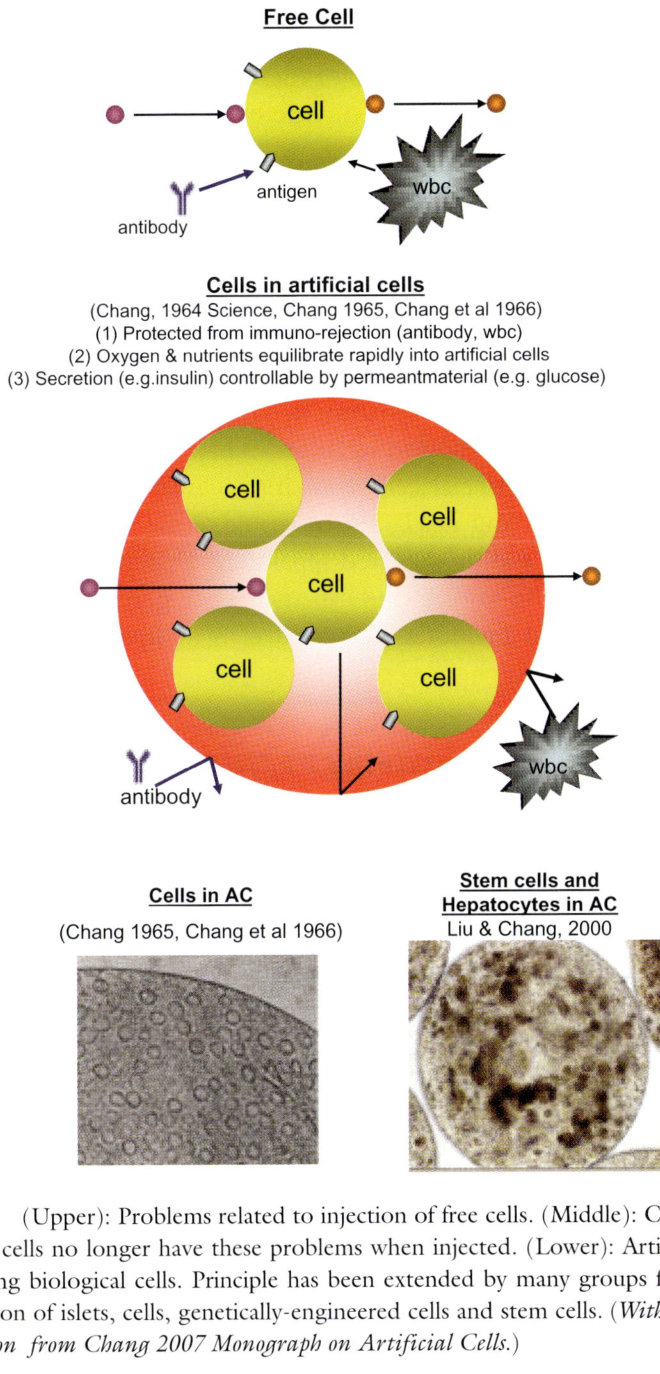

Fig. 27. (Upper): Problems related to injection of free cells. (Middle): Cells inside artificial cells no longer have these problems when injected. (Lower): Artificial cells containing biological cells. Principle has been extended by many groups for bioencapsulation of islets, cells, genetically-engineered cells and stem cells. (*With copyright permission from Chang 2007 Monograph on Artificial Cells.*)

8. *Nanobiosensors*

Nanobiosensors is an area that is of increasing interest, and different approaches are being investigated. One of the many approaches is to use a biosensor where a lipid bilayer is "tethered" on ultrathin polymeric support to form a lipid–polymer complex based on Chang (1969d, 1972a). In this form, different "channels" can also be inserted into the membrane to allow for selective movement of specific solute for detection (Phung *et al.*, 2011a). This principle is based on the basic study reported earlier (Chang, 1972a) (Fig. 7). Another approach that is also possible is to encapsulate enzymes inside artificial cells of microscopic or nanodimension. This way, the product of enzymatic reaction can be followed by fluorescence or other methods. The ability to prepare artificial cells with intracellular compartmentation (Chang, 1972a; Chang *et al.*, 1966) would allow multi-step enzyme reaction to occur and be detected separately. Depending on the type of reaction being followed, one can use either polymeric membrane artificial cells, lipid membrane artificial cells or lipid–polymer membrane artificial cells.

9. *Artificial Cells Containing Radioisotopes or Radio-Opaque Material*

The general principle of artificial cells could be explored in many other areas. Thus, artificial cells containing radioactive isotopes or antimetabolites might be used for intra-arterial injection into tumors. In this case, some of them might lodge at the tumor site, while others would be carried by lymphatic channels to regional lymph nodes. Artificial cells containing radio-opaque material could provide a contrast medium. Nanodimension PEG-polymeric or PEG–lipid membrane artificial cells can circulate readily in the bloodstream, and may form the basis of vehicles for contrast materials in angiography. More details on the use of magnetic nanoparticles will be discussed in later chapters (Boridy *et al.*, 2013; Zhang *et al.*, 2013).

10. *Nonmedical Uses of Artificial Cells*

There are many developments and uses of the principle of artificial cells for agriculture, bioreactors, cosmetics, food production and aquatic culture (Poncelet and Neufeld, 2012). Others include, for instance, the use of artificial cells for energy production by artificial photosynthesis (Hammarstrom and Styring, 2008). Another large area is the use of artificial cells in nanorobatics and nanocomputers that in 2004 forms the European Commission sponsored Programmable Artificial Cell Evolution (PACE) and in 2008 becomes the European Centre for Living Technology (Bedau *et al.*, 2012). There are some centers that are attempting to prepare "living artificial cells" (Murtas, 2009). However, these are not within the scope of this book on nanomedicine.

D. The Future of Artificial Cells in Nanomedicine

The 1972 monograph Artificial Cells (Chang, 1972a) predicted that: "Artificial Cell is not a specific physical entity... It is clear that different approaches can be used to demonstrate this idea." In the last 55 years (Chang, 1957), artificial cells have progressed way beyond this 1972 prediction. Artificial cells can now be of micro-, nano-, and molecular dimensions. There are also unlimited possibilities in variations for artificial cell membranes and contents (Figs. 1 to 9). The other 22 chapters in this contain a few selected topics from the very vast area of nanomedicine. We have just touched the surface of the potential of nanomedicine.

Acknowledgments

TMSC acknowledges the support of this research continuously since 1966 by the Medical Research Council of Canada (MRC), which is now called the Canadian Institutes of Health Research (CIHR); also the Quebec Virage Centre of Excellence in high technology. There are also a number of shorter-term supports, in particular, the Quebec Hemovigillance and Transfusion Medicine Program under FRSQ.

References

Abuchowski A, Es TV, Palczuk NC, Davis FF. (1977) Alteration of immunological properties of bovine serm alnmi by convalent attachment of polyethylene glycol. *J Biol Chem* **252:** 2578–3581.

Acharya SA, Tsai AG, Intaglietta M. (2013) EAF PEG hemoglobins: novel class of non-hypertensive resuscitation fluids: simplicity and advantages of extension arm chemistry for PEGylation. In: Chang TMS (ed), *Selected Topics in Nanomedicine*, World Science Publisher/Imperial College Press.

Aebischer P, *et al.* (1996) Gene therapy for amyotrophic lateral sclerosis (ALS) using a polymer encapsulated xenogenic cell line engineered to secrete hCNTF. *Hum Gene Ther* **1:** 851–860.

Alayash AI. (2004) Oxygen therapeutics: can we tame Hb. *Nature Rev Drug Discov* **3:** 152–159.

Amberson WR. (1937) Blood substitute. *Biol Rev* **12:** 48.

Bangham AD, *et al.* (1965) Diffusion of univalent ions across the lamellae of swollen phospholipids. *J Mol Biol* **13:** 238–252.

Bedau MA, McCaskill JS, Packard NH, Rasmussen S. (2010) Living technology: exploiting life's principles in technology. *Artificial Life* **16:** 89–97. doi:10.1162/artl.2009.16.1.16103

Bian Y, Rong Z, Chang TMS. (2011) Polyhemoglobin-superoxide dismutase-catalase-carbonic anhydrase: a novel biotechnology-based blood substitute that transports both oxygen and carbon dioxide and also acts as an antioxidant. *Artificial Cells, Blood Substitutes & Biotechnology* **39:** 127–136.

Bian YZ, Wei G, Chang TMS. (2013) Lowering of elevated tissue PCO_2 in a hemorrhagic shock rat model after reinfusion of a novel nanobiotechnological polyhemoglobin-superoxide dismutase-catalase-carbonic anhydrase that is an oxygen and carbon dioxide carrier with enhanced antioxidant properties. *Artificial Cells, Nanomedicine and Biotechnology, an international journal* **41:** 60–68.

Bian YZ, Yang L, Chang TMS. (2013) Freeze dried polyhemoglobin-superoxide dismutase-catalase-carbonic anhydrase powder maintains long term stability at 37C, 20C, 4C and –70C (to be submitted).

Boridy S, Maysinger D. (2013) Polymeric and metallic nanoparticles for imaging and treating inflammation in the nervous system. In: Chang TMS (ed), *Selected Topics in Nanomedicine*, World Science Publisher/Imperial College Press.

Bourget L, Chang TMS. (1986) Phenylalanine ammonia-lyase immobilized in microcapsules for the depleture of phenylalanine in plasma in phenylketonuric rat model. *Biochim Biophys Acta* **883:** 432–438.

Bucci E. (2011) Basic science offers a challenge for developing hemoglobin based oxygen carriers into therapeutic agent. *Artificial Cells, Blood Substitutes, and Biotechnology* **39:** 206–213.

Bucci E. (2013) Evidence regarding some pharmacologic characteristics of haemoglobin based oxygen carriers. In: Chang TMS (ed), *Selected Topics in Nanomedicine*, World Science Publisher/Imperial College Press.

Buehler PW, Haney CR, Gulati A, *et al.* (2004) Polynitroxyl hemoglobin: a pharmacokinetic study of covalently bound nitroxides to hemoglobin platforms. *Free Radical Biology & Medicine* **37(1):** 124–135.

Bunn HF, Jandl JH. (1968) The renal handling of hemoglobin. *Trans Assoc Am Physicians* **81:** 147.

Chandra R, Madan J, Singh P, *et al.* (2013) Noscapines: novel carrier systems to target the tumour cells. In: Chang TMS (ed), *Selected Topics in Nanomedicine*, World Science Publisher/Imperial College Press.

Chang TMS. (1957) Hemoglobin corpuscles. Report of a research project for Honours Physiology, Medical Library, McGill University. Also reprinted in full in Chang's 2007 Monograph on "Artificial Cells."

Chang TMS. (1964) Semipermeable microcapsules. *Science* **146:** 524–525.

Chang TMS. (1965) Semipermeable aqueous microcapsules. Ph.D. Thesis, McGill University.

Chang TMS. (1966) Semipermeable aqueous microcapsules ("artificial cells"): with emphasis on experiments in an extracorporeal shunt system. *Trans Am Soc Artif Intern Organs* **12:** 13–19.

Chang TMS, Macintosh FC, Mason SG. (1966) Semipermeable aqueous microcapsules: I. Preparation and properties. *Can J Physiol Pharmacol* **44:** 115–128.

Chang TMS, Poznansky MJ. (1968) Semipermeable microcapsules containing catalase for enzyme replacement in acatalsaemic mice. *Nature* **218:** 242–245.

Chang TMS. (1969a) Removal of endogenous and exogenous toxins by a microencapsulated absorbent. *Can J Physiol Pharmacol* **47:** 1043–1045.

Chang, TMS. (1969d) Lipid-coated spherical ultrathin membranes of polymer or cross-linked protein as possible cell membrane models. *Fed Proc* **28**:461

Chang TMS. (1971a) The *in vivo* effects of semipermeable microcapsules containing L asparaginase on 6C3HED lymphosarcoma. *Nature* **229:** 117–118.

Chang TMS. (1971b) Stabilisation of enzymes by microencapsulation with a concentrated protein solution or by microencapsulation followed by cross-linking with glutaraldehyde. *Biochem Biophys Res Commun* **44:** 1531–1536.

Chang TMS. (1972a) *Artificial Cells.* Charles C. Thomas, Springfield, IL (out of print but available for free online viewing/download at www.artcell.mcgill.ca).

Chang TMS. (1972b) Haemoperfusions over microencapsulated adsorbent in a patient with hepatic coma. *Lancet* **2:** 1371–1372.

Chang TMS, Coffey JF, Barre P, *et al.* (1973) Microcapsule artificial kidney: treatment of patients with acute drug intoxication. *Can Med Assoc J* **108:** 429–433.

Chang TMS. (1975) Microencapsulated adsorbent hemoperfusion for uremia, intoxication and hepatic failure. *Kidney Int* **7:** S387–S392.

Chang TMS. (1976) Biodegradable semipermeable microcapsules containing enzymes, hormones, vaccines, and other biologicals. *J Bioengineering* **1:** 25–32.

Chang TMS. (1980) Blood compatible coating of synthetic immunoadsorbents. *Trans Am Soc Artif Intern Organs* **26:** 546–549.

Chang TMS. (1985) Artificial cells with regenerating multienzyme systems. *Meth Enzymol* **112:** 195–203.

Chang TMS. (1989) Preparation and characterization of xanthine oxidase immobilized by microencapsulation in artificial cells for the removal of hypoxanthine. *Biomater Artif Cells Artif Organs* **17:** 611–616.

Chang TMS. (1997) *Red Blood Cell Substitutes: Principles, Methods, Products and Clinical Trials*, Vol. I (Monograph). Karger/ Landes Systems, Basel, Switzerland (available for free online viewing at www.artcell.mcgill.ca or www.artificialcell.info).

Chang TMS, Powanda D, Yu WP. (2003) Analysis of polyethyleneglycolpolylactide nano-dimension artificial red blood cells in maintaining systemic hemoglobin levels and prevention of methemoglobin formation. *Artif Cells Blood Substit Biotechnol* **31:** 231–248.

Chang TMS. (2005). Therapeutic applications of polymeric artificial cells. *Nature Review: Drug Discovery* **4:** 221–235.

Chang TMS. (2007) Monograph on "ARTIFICIAL CELLS: biotechnology, nanotechnology, blood substitutes, regenerative medicine, bioencapsulation, cell/stem cell therapy," World Scientific Publisher/Imperial College Press, 435 pages. (available for free online viewing/download at www.artcell.mcgill.ca)

Chang TMS. (2009) Nanobiotechnology for hemoglobin based blood substitutes. *Critical Care Cinics* **25:** 373–382.

Chang TMS. (2010) Blood replacement with engineered hemoglobin and hemoglobin nanocapsules. *Wiley Interdisciplinary Reviews Nanomedicine and Nanobiotechnology* **2:** 418–430.

Chang TMS. (2011) Nanobiotechnology-based Blood Substitutes and the Cardiovascular Systems in Transfusion Medicine Chapter in *Nanomedicine and Cardiovascular System*, pp. 77–98.

Chang TMS. (2012) Foreword to *Theory, Techniques and Applications of Nanotechnology in Gene Silencing*, Nimesh S and Chandra R (eds), pp. v–vi.

Chang TMS. (2013) Artificial Cells, Blood Substitutes and Nanomedicine website containing reviews and monographs for free online viewing or download: www.artcell.mcgill.ca

D'Agnillo F, Chang TMS. (1998) Polyhemoglobin-superoxide dismutasecatalase as a blood substitute with antioxidant properties. *Nature Biotechnol* **16:** 667–671.

Deamer DW, Bangham AD. (1976) Large-volume liposomes by an ether vaporization method. *Biochim Biophys Acta* **443:** 629–634.

De Hoog HPM, Liedberg B, Nallani M. (2013). Bio-functional polymer vesicles for applications in nanomedicine. In: Chang TMS (ed), *Selected Topics in Nanomedicine*, World Science Publisher/Imperial College Press.

De Vos P, Hamel AF, Tatarkiewicz K. (2002) Considerations for successful transplantation of encapsulated pancreatic islets. *Diabetologia* **45:** 159–173.

De Vocht C, Ranquin A, Willaert R, *et al.* (2009) Assessment of stability, toxicity and immunogenicity of new polymeric nanoreactors for use in enzyme replacement therapy of MNGIE. *Journal of Controlled Release* **137(3):** 246–254.

Djordjevich L, Miller IF. (1980) Synthetic erythrocytes from lipid encapsulated hemoglobin. *Exp Hematol* **8:** 584.

Eldridge J, Colby AH, Willmott G, *et al.* (2013) Use of tunable pores for accurate characterization of micro- & nanoparticle systems in nanomedicine. In: Chang TMS (ed), *Selected Topics in Nanomedicine*, World Science Publisher/Imperial College Press.

Fustier C, Chang TMS. (2012 online) PEG-PLA nanocapsules containing a nanobiotechnological complex of polyhemoglobin-tyrosinase for the depletion of tyrosine in melanoma: preparation and *in vitro* characterisation. *Journal Nanomedicine and Biotherapeutics Discovery* **2012, 2:** 1.

Garnacho C, Dhami R, Simone E, *et al.* (2008) Delivery of acid sphingomyelinase in normal and Niemann–Pick disease mice using intercellular adhesion molecule-1-targeted polymer nanocarriers. *Journal of Pharmacology and Experimental Therapeutics* **325(2):** 400–408.

Gazzard BG, Portmann BA, Weston MJ, *et al.* (1974) Charcoal haemoperfusion in the treatment of fulminant hepatic failure. *Lancet* **1:** 1301–1307.

Gould SA, *et al.* (2002) The life-sustaining capacity of human polymerized hemoglobin when red cells might be unavailable. *J Am Coll Surg* **195:** 445–452.

Gregoriadis G (ed). (1976) *Drug Carriers in Biology and Medicine*. Academic Press, Inc., New York.

Greenburg AG, Pitman A, Pearce LB, Kim HW. (2008) Clinical contextualization and the assessment of adverse events in HBOC trials. *Artificial Cells, Blood Substitutes, and Biotechnology* **36:** 477–486.

Greenburg AG. (2013) The Clinical Development of Hemopure®: A Selective Commentary. In: Chang TMS (ed), *Selected Topics in Nanomedicine*, World Science Publisher/Imperial College Press.

Gu KF, Chang TMS. (1990) Production of essential L-branched-chained amino acids, in bioreactors containing artificial cells immobilized multienzyme systems and dextran-NAD+. *Appl Biochem Biotechnol* **26:** 263–269.

Haider T, Husain Q. (2008) Concanavalin a layered calcium alginate–starch beads immobilized β galactosidase as a therapeutic agent for lactose intolerant patients. *International Journal of Pharmaceutics* **359(1):** 1–6.

Hakim RM, Milford E, Himmelfarb J, *et al.* (1990) Extracorporeal removal of antiHLA antibodies in transplant candidates. *Am J Kidney Dis* **16:** 423.

Hammarstrom L, Styring S. (2008) Coupled electron transfers in artificial photosynthesis. *Philosophical Transactions of the Royal Society of London. Series B, Biological Sciences* **363:** 1283–1291; discussion 1291, doi:10.1098/rstb.2007.2225

Han JQ, Yu MH, Dai M, *et al.* (2012) Decreased expression of MDR1 in PEG-conjugated hemoglobin solution combined cisplatin treatment in a tumor xenograft model. *Artificial Cells, Blood Substitutes and Biotechnology* **40:** 239–244.

Hoffman SJ, Looker DL, Roehrich JM, *et al.* (1990) Expression of fully functional tetrameric human hemoglobin in *Escherichia coli. Proc Natl Acad Sci USA* **87:** 8521–8525.

Hosta-Rigau L, Städler B. (2013) Subcompartmetalized systems toward therapeutic cell mimics. In: Chang TMS (ed), *Selected Topics in Nanomedicine*, World Science Publisher/Imperial College Press.

Hunkeler Dl. (2003) Bioartificial organ grafts: a view at the beginning of the third millennium. *Artif Cells Blood Substit Immobil Biotechnol* **31:** 365–382.

Jia YP, Alayash AI. (2013). Molecular Basis of Haptoglobin and Hemoglobin Complex Formation and Protection against Oxidative Stress and Damage. In: Chang TMS (ed), *Selected Topics in Nanomedicine*, World Science Publisher/ Imperial College Press.

Jahr JS, Mackenzie C, Pearce LB, *et al.* (2008) HBOC-201 as an alternative to blood transfusion: efficacy and safety evaluation in a multicenter phase III trial in elective orthopaedic surgery. *J Trauma* **64:** 1484–1497.

Kaminsky YG, Kosenko EA. (2012) Argocytes Containing Enzyme Nanoparticles. Reduce Toxic Concentrations of Arginine in the Blood. *Bulletin of Experimental Biology and Medicine*, 1–3.

Kawasaki C, Nishi R, Uekihara S, *et al.* (2000) Charcoal hemoperfusion in the treatment of phenytoin overdose. *Am J Kidney Dis* **35:** 323–326.

Li T, Yu R, Zhang HH, *et al.* (2006) A method for purification and viral inactivation of human placenta hemoglobin. *Artf Cells Blood Sub Biotech* 2006 (in press).

Lim F, Sun AM. (1980) Microencapsulated islets as bioartificial endocrine pancreas. *Science* **210:** 908–909.

Lin CC, Chan TY, Deng JF. (2004) Clinical features and management of herb induced aconitine poisoning. *Ann Emerg Med* **43:** 574–579.

Liu Q, Xiu RJ. (2008) Proceedings of the XI ISBS Symposium at Peking Union Medical College, Chinese Academy of Medical Sciences, *Artificial Cells, Blood Substitutes and Biotechnology, an international journal.* **36(3)**: 169–293.

Liu ZC, Chang TMS. (2008) Long term effects on the histology and function of livers and spleens in rats after 33% toploading of PEG-PLA-nano artificial red blood cells. *Artificial Cells, Blood Substitutes & Biotechnology* **36:** 513–524.

Liu J, *et al.* (2002) Study on a novel strategy to treatment of phenylketonuria. *Artif Cells Blood Substit Immobil Biotechnol* **30:** 243–258.

Liu ZC, Chang TMS (2012) Intrasplenic transplantation of bioencapsulated mesenchymal stem cells enhances the regeneration of remnant livers and improves the recovery rates of 90% partial hepatectomized rats. *Stem Cells International* Volume 2012 (2012), Article ID 697094, 6 pages doi:10.1155/2012/697094

Lopez Lago AM, *et al.* (2002) Paraquat poisoning and hemoperfusion with activated charcoal. *Ann Intern Med* **19:** 310–312.

Ma L, Hsia CJC. (2013) Polynitroxylated Hemoglobin as a multifunctional therapeutic for critical care and transfusion medicine. In: Chang TMS (ed), *Selected Topics in Nanomedicine*, World Science Publisher/Imperial College Press.

Mobed-Miremadi M. (2013) High throughput methods for miniaturization of implantable artificial cells. In: Chang TMS, *Selected Topics in Nanomedicine*, World Science Publisher/Imperial College Press.

Moore E, Moore FA, Fabian TC, *et al.* (2009) Human polymerized hemoglobin for the treatment of hemorrhagic shock when blood is unavailable: the USA Multicenter Trial. *J Am Coll Surg* **208:** 1–13.

Mozzarelli A. (2010) XII ISBS Symposium Proceeding. *Artificial Cells, Blood Substitutes and Biotechnology, an international journal* **38(6)**: 295–342.

Murtas G. Artificial assembly of a minimal cell. *Molecular BioSystems* **5:** 1292–1297, doi:10.1039/b906541e (2009).

Natanson C, Kern SJ, Lurie P, *et al.* (2008) Cell-free hemoglobin-based blood substitutes and risk of myocardial infarction and death: a meta-analysis. *JAMA* **299:** 2304–2312.

Nimesh S, Chandra R. (2013) Application of polyethylenimine based nanoparticles for RNA therapeutics. In: Chang TMS (ed), *Selected Topics in Nanomedicine*, World Science Publisher/Imperial College Press.

Nimesh S, Chandra R. (2013) Chitosan based nanocarriers for efficient and targeted siRNA delivery. In: Chang TMS (ed), *Selected Topics in Nanomedicine*, World Science Publisher/Imperial College Press.

Orive G, Hernandez RM, Gascon AR, *et al.* (2003) Cell encapsulation: promise and progress. *Nat Med* **9:** 104–107.

Palmour RM, Goodyer P, Reade T, Chang TMS. (1989) Microencapsulated xanthine oxidase as experimental therapy in Lesch–Nyhan disease. *Lancet* **2:** 687–688.

Park YK, Abuchowski A, Davis S, Davis F. (1981) Pharmacology of Escherichia coli-L-asparaginase polyethylene glycol adduct. *Anticancer Res* **1:** 373–376.

Pearce LB, Gawryl MS. (1998) Overview of preclinical and clinical efficacy of Biopure's HBOCs. In: Chang TMS (ed), *Blood Substitutes: Principles, Methods, Products and Clinical Trials*, Vol. 2, pp. 82–98. Karger, Basel.

Peng M, Chao X, Chen C, Cui Y. (2013) GoldMag composite particles and their applications in nanomedicine. In: Chang TMS (ed), *Selected Topics in Nanomedicine*, World Science Publisher/Imperial College Press.

Photos PJ, Bacakova L, Discher B, *et al.* (2003) Polymer vesicles *in vivo*: correlations with PEG molecular weight. *J Control Release* **90:** 323–334.

Phung T, Zhang Y, Dunlop J, Dalziel J. (2011) Bilayer lipid membranes supported on Teflon filters: a functional environment for ion channels. *Biosensors and Bioelectronics* **26:** 3127–3135, doi:10.1016/j.bios.2010.12.013

Poncelet D, Neufeld R. (2012). 20th Bioencapsulation Conference.

Powanda D, Chang TMS. (2002) Cross-linked polyhemoglobin-superoxide dismutase-catalase supplies oxygen without causing blood brain barrier disruption or brain edema in a rat model of transient global brain ischemia-reperfusion. *Artif Cells Blood Substit Immobil Biotechnol* **30:** 25–42.

Poznansky MJ, Chang TMS. (1974) Comparison of the enzyme kinetics and immunological properties of catalase immobilized by microencapsulation and catalase in free solution for enzyme replacement. *Biochim Biophys Acta* **334:** 103–115.

Prakash S, Chang TMS. (1996) Microencapsulated genetically engineered live E. coli DH5 cells administered orally to maintain normal plasma urea level in uremic rats. *Nature Med* **2:** 883–887.

Qasaimeh MA, Sebastien Ricoult S, Juncker D. (2013) Microfluidic probes to process surfaces, cells, and tissues. In: Chang TMS (ed), *Selected Topics in Nanomedicine*, World Science Publisher/Imperial College Press.

Qin GW, Li S, Ren YP, *et al.* (2013) Soft magnetic nanomaterials towards nanomedicine: magnetism principles, preparation, characterization and potential applications. In: Chang TMS (ed), *Selected Topics in Nanomedicine*, World Science Publisher/Imperial College Press.

Rousselot M, Delpy E, Drieu La Rochelle C, *et al.* (2006) Arenicola marina extracellular hemoglobin: a new promising blood substitute. *Biotechnol J* **1(3):** 333–345.

Sakai H. (2013) Biocompatibility of a highly concentrated fluid of hemoglobin-vesicles as a transfusion alternative. In: Chang TMS (ed), *Selected Topics in Nanomedicine*, World Science Publisher/Imperial College Press.

Sarkissian CN, Shao Z, Blain F, Scriver C. (1999) A different approach to treatment of phenylketonuria: phenylalanine degradation with recombinant phenylalanine ammonia lyase. *Proc Natl Acad Sci* **96:** 2339–2344.

Shao W, Paul A, Prakash S. (2013) Carbon nanotubes in cancer and stem cell therapeutics. In: Chang TMS (ed), *Selected Topics in Nanomedicine*, World Science Publisher/Imperial College Press.

Shorr RG, Viau AT, Abuchowski A. (1996) Phase 1B safety evaluation of PEG hemoglobin as an adjuvant to radiation therapy in human cancer patients. *Artificial Cells, Blood Substitutes and Immobilization Biotechnology, an international journal* **24:** (abstracts issue) 407.

Sims C, Seigne P, Menconi M, *et al.* (2001). Skeletal muscle acidosis correlates with the severity of blood volume loss during shock and resuscitation. *J Trauma* **51:** 1137–1146.

Singh SM, McCormick BB, Mustata S, *et al.* (2004) Extracorporeal management of valproic acid overdose: a large regional experience. *J Nephrol* **17:** 43–49.

Siu YS, Ho KM, Li P. (2013) Core-shell nanoparticles containing hairy polyethyleneimine shells as effective nanocarriers for gene delivery. In: Chang TMS (ed), *Selected Topics in Nanomedicine*, World Science Publisher/Imperial College Press.

Terman DS, Tavel T, Petty D, *et al.* (1977) Specific removal of antibody by extracorporeal circulation over antigen immobilized in colodion charcoal. *Clin Exp Immunol* **28:** 180.

Torchilin VP. (2005) Recent advances with liposomes as pharmaceutical carriers. *Nat Rev Drug Discov* **4:** 145–160.

Tsuchida E, Sakai H, Horinouchi H, Kobayashi K. (2006). *Artificial Cells, Blood Substitutes and Biotechnology, an international journal* **34:** 581–588.

Wang Y, Chang TMS. (2012) Nanobiotechnological nanocapsules containing polyhemoglobin-tyrosinase: effects on murine B16F10 melanoma cell proliferation and attachment. *Journal of Skin Cancer*, special issue on 'melanoma' http://www.hindawi.com/journals/jsc/2012/673291/

Wei G, Bian YZ, Chang TMS. (2013) PLA-PEG nanoencapsulated nano artificial red blood cells that act as O_2 and CO_2 carrier with enhanced antioxidant activity: polyhemoglobin-superoxide dismutase-catalase-carbonic anhydrase. *Journal Artificial Cells, Nanomedicine and Biotechnology* (in press).

Wetzler M, Sanford BL, Kurtzberg J, *et al.* (2007) Effective Asparagine Depletion with Pegylated. Asparaginase Results in Improved Outcomes in Adult Acute Lymphoblastic Leukemia: Cancer and Leukemia Group B Study 9511. *Blood* **109.10:** 4164–4167.

Winslow RM (ed). 2006. *Blood Substitutes.* Academic Press, Amsterdam.

Winchester JF (ed). (1996) *Replacement of Renal Function by Dialysis*, 4th ed. Kluwer Academic Publishers, Boston.

Wohl BM, Fejerskov B, Chong SF, Zelikin AN. (2013) Surface mediated drug delivery. In: *Selected Topics in Nanomedicine*, World Science Publisher/Imperial College Press.

Wong BL, Kwok SY, Lau SH. (2011) Treatment methods using a heat stable oxygen carrier-containing pharmaceutical composition. United States Patent 8,048,856.

Wong JT. (1988) Rightshifted dextran-hemoglobin as blood substitute. *Biomater Artif Cells Artif Organs* **16:** 237–245.

Wong N, Chang TMS. (2007) Polyhemoglobin-fibrinogen: a novel blood substitutes with platelet-like activity for extreme hemodilution. *Artificial Cells, Blood Substitutes and Biotechnology* **35:** 481–489.

Yabuki A, Yamaji K, Ohki H, Iwashita Y. (1990) Characterization of a pyridoxylated hemoglobin-polyoxyethylene conjugate as a physiologic oxygen carrier. *Transfusion* **30:** 516–520.

Yang L, Cheng Y, Yan WR, Yu YT. (2004) Extracorporeal whole blood immunoadsorption of autoimmune myasthenia gravis by cellulose tryptophan adsorbent. *Artif Cells Blood Substit Biotechnol* **32:** 519–528.

Yu BL, Chang TMS. (2004) *In vitro* and *in vivo* effects of polyhemoglobintyrosinase on murine B16F10 melanoma. *Melanoma Res J* **14:** 197–202.

Yu B, M Shahid, EM Egorina, *et al.* (2010) Endothelial dysfunction enhances vasoconstriction due to scavenging of nitric oxide by a hemoglobin-based oxygen carrier. *Anesthesiology* **112:** 586–594.

Zapol W. (2012) Proceedings of the XIII ISBS Symposium at Mass General Hospital of Harvard Medical School. *Artificial Cells, Blood Substitutes and Biotechnology, an International Journal* **40:** Issue 3.

Zhang Y, Li Y, Budak G, *et al.* (2013) Nanoparticles for imaging and therapy — functionalization, endocytosis and characterization. In: Chang TMS (ed), *Selected Topics in Nanomedicine*, World Science Publisher/Imperial College Press.

Zhu XL, Chu W, Wang T, *et al.* (2007). Variations in dominant antigen determinants of glutaraldehyde polymerized human, bovine and porcine hemoglobin. *Artificial Cells, Blood Substitutes and Biotechnology* **35:** 518–532.

Zhu YJW. Chen C. (2013) Glutaraldehyde polymerized porcine haemoglobin. In: Chang TMS (ed), *Selected Topics in Nanomedicine*, World Science Publisher/ Imperial College Press.

Chapter 2

The Clinical Development of Hemopure®: A Selective Commentary

A Gerson Greenburg

Introduction

Hemopure® (HBOC-201, hemoglobin glutamer-250 (bovine)) (OPK Biosciences, Cambridge, MA) is an oxygen therapeutic, the class often mislabeled "blood substitutes." Oxygen therapeutics are *not* blood substitutes, for they lack coagulation and immune system factors and other essential properties of blood. The development of Hemopure® began in the early 1980s, when the safety of the blood supply became a significant concern. This was, after all, the beginning of the AIDs era, and with HIV transmission linked to blood transfusions, decreasing the risk with an alternative oxygen-carrying solution seemed reasonable. There emerged a sense of urgency to develop alternatives for red blood cell transfusions as well as a questioning of transfusion practices, including the elusive "transfusion trigger." Prior to this time, the concept of a "blood substitute" primarily for use in trauma was being pursued. Today, the blood supply in the developed nations is considered reasonably safe and abundant. However, periodic shortages still occur, seasonally or in association with natural and man-made disasters, elective surgery is cancelled, and stockpiles of

---

Professor of Surgery, Emeritus, Alpert School of Medicine, Brown University Consultant, OPK Biosciences, Cambridge MA.

blood or oxygen-carrying solutions for use in acute disasters do not exist. In populations in the developing world, where blood banks are rare or do not exist, children with malaria or women with obstetrical hemorrhage often succumb for lack of an oxygen-carrying solution to deal with the acute anemia. Moreover, it is not clear that the blood supply in the developed nations is truly safe given the emergence of new transfusion-transmitted pathogens. Most recently Babesiosis, a tick-borne agent, has been identified as one such pathogen placing certain populations at increased risk.[1] There is ongoing concern about emerging strains of hepatitis.[2] There are other threats to the safety of the blood supply that should be of concern in the near future and beyond.[3,4]

The search for a "blood substitute" has century-long historical roots, with the major developments linked to an increased understanding of the physiology of human aerobic dependency, macro- and microcirculation, hemoglobin structure and function as well as significant advances in technology. A safe, shelf storable, compatible without typing and crossmatching ***oxygen carrying solution*** for use when red blood cells are neither an option nor available is a desirable therapeutic agent. A brief historical review of the field was provided by Winslow,[5] and Chen has provided a more recent overview.[6]

The formulation and physical–chemical properties of Hemopure® (HBOC-201) have evolved since the early 1980s. Inquiry to the FDA was initiated in 1986, and with the current formulation "finalized" by the early 1990s, discussions with FDA regarding clinical trials were initiated. The physical–chemical properties and other parameters of interest for the current formulation of HBOC-201 are shown in Table 1.

It must be noted that OPK Biosciences, formerly Biopure, also provides a fully approved bovine-hemoglobin formulation for veterinary use, Oxyglobin®. Its composition is slightly different, more tetrameric hemoglobin and less polymerized molecule; it is the only approved veterinary oxygen therapeutic. It is worth noting that this formulation has been used by many investigators in pre-clinical studies referring to it as "bovine hemoglobin" without specifying it as the veterinary product; there is tacit implication that this is the human

Table 1. Properties of Hemopure®

Hemoglobin concentration	13 g/dL
Average molecular weight	250 kD
Tetramer content	< 3%
Plasma half-life	19 hr
p50	40 mm Hg
pH	7.4–8.0
Colloid (COP)	26 mmHg
Viscosity	2.4 centipoise
Volume	250 mL
Buffer	Modified lactated Ringer's solution
Blood compatibility	Universal — no blood typing
Infectivity risk	Viral inactivation/removal
Shelf life	three years, room temperature (2–30°C)

product, implying the results would be the same as with HBOC-201. When reviewing pre-clinical studies, this fact must be considered.

This chapter will emphasize some aspects of HBOC-201 clinical development. Some very recent, specific and relevant preclinical studies are included as they support the safety and efficacy of the product in contradistinction to the view held by the FDA. Indeed, some of these studies were requested by the FDA to support an IND application for trauma clinical trials. A relatively complete review of the early and significant preclinical development, through late 2005, is available.[7]

At this writing, the literature reflects more than 279 preclinical studies, pharmacology and toxicity and efficacy models in whole animals, organ perfusion and tissue culture. Physiologic effects, models of trauma, reanimation, ischemic rescue, biochemistry and metabolism have also been explored. The veterinary product, Oxyglobin®, has also undergone extensive testing in a variety of species before it was *licensed* as a therapeutic, where it has been used in many thousands of applications. The safety and toxicity of Hemopure® has been evaluated in at least

16 species including dog, cat, rat, guinea pig, rabbit, hamster, miniature pig, sheep, two species of monkey, baboon, river otter, horse, miniature horse and calf. Constraints of space prohibit a formal exploration and review of all of these studies and their impact on the clinical development program. The data generated by many of these studies was sufficient to permit FDA approval of the formulation to proceed to Phase I clinical trials in 1992.

The Clinical Trials

A general summary of the 16 Phase I, I/II and II studies is available.[8] There were four Phase I trials and nine Phase I/II studies. Areas of investigation included normal volunteer safety and pharmacokinetic studies followed by protocols in a variety of surgical patients, e.g. aortic aneurysm, liver resection, orthopedic surgery, prostatectomy, gynecologic surgery and general surgery. Many of these were not randomized 1:1 and control solutions varied to include lactated Ringer's solution and hydroxyethyl starch. Doses ranged up to 245 g in various delivery rate and sequence patterns. This information is made explicit because it is critical to appreciate this level of variation in dose and dosing algorithms, randomization paradigms, control solutions and clinical settings with a single product when reviewing a most negative and poorly constructed meta-analysis of the HBOC field in general.[9] In that analysis, all of the Hemopure® clinical studies, including the Phase III trials, were pooled and analyzed as a single study, contrary to the conventions in place for an analysis of type.

Nine Phase II trials were completed including cardiac surgery, aortic aneurysm, general surgery, orthopedic surgery and some non-surgical studies in sickle cell anemia both in and out of crisis, weaning from ventilation and as adjunctive to radiation therapy for glioblastoma. Again, a variety of control solutions were employed including red blood cells, saline and lactated Ringer's. Most of the surgical studies used RBCs as controls, the objective to establish the ability to avoid RBCs. They were often randomized to 1:1 design as the target endpoint was avoidance of blood transfusion. Doses varied but the maximum was 10 units of HBOC-201 with various dosing schedules.

As stated by Pearce *et al.*[8] there were 16 Phase I, I/II, II and III studies completed with a total of 806 patients receiving HBOC-201 with maximum doses from 7 to 1230 g of hemoglobin and 661 controls.

Phase III Trials

HBOC-201 has been evaluated in six Phase III clinical trials including the major orthopedic surgery trial, HEM-0115, a general surgical trial, HEM-0114 and two trials in patients undergoing cardiac revascularization. The largest trial, HEM-0115, was in orthopedic surgery. In it, 688 patients were randomized to packed red blood cells (PRBC) or HBOC-201 ***when a transfusion was indicated***.[10]

This was a multicenter, single blind, randomized red cell-controlled, parallel-group trial. Male and female patients (non-pregnant/non-lactating), 18 years or older were the primary population. Patients were expected to require at least two units of PRBC before midnight of postoperative day 3. Patients were randomized to HBOC-201 or PRBC at the first transfusion decision. This decision was based on the investigator's assessment of transfusion need and total hemoglobin of less than 10.5 g/dL. Subsequent transfusion required a total hemoglobin of less than 10 g/dL and at least one clinical sign of inadequate tissue perfusion or ongoing blood loss. Decreased urine output (less than 0.5 mL/kg within 2 hr), hypotension (systolic blood pressure less than 90 mmHg), pulse rate of 100 beats per minute or more, ECG evidence of myocardial ischemia or base deficit of 4 or worse. In retrospect the hemoglobin level criteria were considerably more liberal than the "transfusion triggers" of the modern era but reflected the guidelines practiced by anesthesiologists and surgeons at the time. For those patients randomized to HBOC-201 a loading dose of 2 units, 65 g of hemoglobin in 500 mL was given. Additional dosing was used, as indicated to a maximum of 10 units, 2500 mL or 325 grams of product. After the protocol volume was exceeded or six days elapsed, any additional need to increase oxygen-carrying capacity was met with

PRBCs. Patients could also be crossed over if the investigator felt it necessary. Bear in mind that this was a multi-institutional international study and control over the local standards of care was not possible. This resulted in significant variance in the data due to different practices and philosophies. However, of particular importance was the absence of a matching crossover group in the PRBC arm which significantly confounded the data. Unfortunately, the protocol did not consider this subset in the data analysis. Further details can be found in the paper by Jahr *et al.*[10]

The patients in this crossover subset have subsequently been shown to be the major contributors to the adverse safety profile used by the FDA used to deny Hemopure® a license. In essence, the safety profile was compromised by increases in cardiovascular events, myocardial infarctions, congestive heart failure, pulmonary edema and cerebral vascular events. Further *post hoc* analysis, not part of the original protocol, shows this subset of patients to be responsible for the majority of adverse events. These events have subsequently been explained using *clinical contextualization*, a detailed "root cause analysis," a clinically directed exploration of each patient's course.[11] As a scientific principle, it is necessary to explain variances and deviations in the data and observations and to use any and all known analytical tools as needed. While "root-cause analysis" is not generally accepted to explain the results of clinical trials, it is a valid tool to explain the deviations and differences, not to rationalize them. While the principle of randomized trials is generally accepted, the need to explain deviations in the data is another critical scientific principle. Einstein is noted to have said, "Not everything that can be counted, counts and not everything that counts can be counted." Most trials are not of sufficient size to randomize all the variables.

With respect to efficacy, the projected endpoint of 35% transfusion avoidance for the HBOC-201 group was exceeded with an overall PRBC transfusion avoidance of 59%. Given concerns about one-unit blood transfusions, an adjustment was made in the distribution for this factor, removing the patients receiving one unit of PRBCs, matching them to patients in the HBOC-201 arm with 50% of patients in the HBOC-201 arm avoiding PRBC transfusion.

Transfusion avoidance is one of many mechanisms for management of the blood supply. An earlier trial in cardiac surgery patients with another product[12] demonstrated the potential for significant avoidance and savings of bank blood. An increase in the supply of PRBCs would be expected if this paradigm of HBOC-201 use became a standard of practice along with other conservation strategies. Transfusion triggers have evolved, actually decreased, over time, and the use of transfusion in many surgical procedures including major orthopedic surgery have also decreased. Some would suggest that no transfusion today would be triggered at the levels predefined in this study. The transfusion trigger is a dynamic and actually very individually determined point of reference; for an individual patient it may be higher or lower than that used in this study.

The Safety Profile

The FDA, in denying the Hemopure® BLA application, stressed the known vasoactivity of HBOCs *generally* as the basis for the observed cardiovascular adverse events, the key factor responsible for the negative safety profile. The actual increases in systolic blood pressure for the population was 10 to 15 mmHg above that seen with the red blood cell arm and was most often associated with the two unit loading dose; possibly just a volume loading effect. There is the suggestion that the hemoglobin binding of nitric oxide is the vasoactivity mechanism and thus responsible for the adverse effects. This hypothesis has led to a cottage industry exploring this mechanism and ways to modulate it. It was also suggested that there is a direct cardiac toxicity associated with HBOC-201 which was not demonstrated in pre-clinical models with the product. Most of the severe adverse events observed in the trial in the HBOC-201 arm, after careful review of each case, clinical contextualization, were clearly not the result of the HBOC-201 toxicity nor could they be related to the volume of product infused. Rather, the adverse outcomes in most of these cases arose from significant deviations in clinical care and patient management and not the study solution.[11] Additionally, further studies have demonstrated that neither the vasoactivity nor a direct cardiac toxicity can be invoked as causative of the adverse event profile shown in the formal intent-to-treat analysis.

In the HEM-0115 trial the intent-to-treat analysis was unfavorable with respect to safety as noted by the safety committee and the incidence of adverse and severe adverse events reported by the investigators. There was clearly an imbalance in outcomes between the two randomized groups. Again, this form of the analysis did not take into account the cross-over group and thus it became necessary to do additional analysis to determine what, if anything, could account for the differences.

To make sense of the data the groups were compared by separating them into high and low needs for oxygen-carrying capacity augmentation. Recall, in this trial, as in most of the HBOC trials, these are clinical decisions and not based on acquired physiologic parameters related to global oxygen delivery and consumption which, while ideal, are difficult to work into a protocol of this magnitude. The PRBC subset was divided into two groups, those requiring three or fewer units and those requiring more than three. Similarly, the HBOC-201 group was divided into two sets, one where there was success, i.e. avoidance of PRBCs, and a second where there was failure. These sets are well described in the Jahr reference.[10] Dose in this context serves as a surrogate for "need," and any attempt to define a dose-response curve is clearly obviated. The increase in cardiac and central nervous system events was primarily in the higher-needs group of HBOC-201 subjects. The clinical contextualization of these patients revealed age, volume of fluid overload and under treatment of oxygen deficit to be critical in provoking the adverse events. Patients over 80 years of age contributed over 40% of the cardiac severe adverse events and 50% of the deaths to the study. Of note, only 9% of the patients in the HBOC-201 arm of the study were in this age group. A third of the patients were older and labile; older patients do not do well with acute anemia. An additional one-third of the patients were significantly over hydrated or undertreated, fluid overload and pulmonary edema or delay in crossing over to PRBC for significant acute anemia.

Using a logistic regression model in the HEM-0115 data designed to identify factors related to the adverse cardiac events, age, a history of cardiac disease and calculated hemoglobin deficit significantly correlated to the events while the use of Hemopure® was not statistically significant. Hemoglobin deficit is a defined measure of the magnitude and duration of anemia, a data-construct useful in identification of total hemoglobin deficit over time relative to a specific "transfusion trigger." In the case of the HEM-0115 data this point was set at 9 g/dL, for the time of the trial a reasonable level. Today, this point might be set lower, say 8 g/dL — possibly adjusted for each patient by age and other risk factors. It is a surrogate measure of oxygen delivery deficits and should be considered one of a series of composite end-point measures in future trials.

To address the issue of cardiac toxicity, specific animal and human studies have been undertaken. In one study, the hypothesis was tested using intracoronary infusion of pre-oxygenated HBOC-201 during brief, total coronary artery occlusion to demonstrate preservation of left ventricular function.[13] The authors concluded: "Intracoronary infusion of oxygenated HBOC-201 is capable of preserving left ventricular function, likely through maintenance of myocardial oxygenation. It is hypothesized that, in an acute setting, HBOC-201 could serve as an oxygen bridge to reperfusion by PCI extending the "golden time period during which permanent myocardial damage is unlikely." A clear demonstration of efficacy and no acute cardiac toxicity was observed. This demonstration of perfusion of an occluded coronary vessel with oxygenated HBOC-201 without significant functional hemodynamic or electrocardiographic alterations is a valid observation regarding the effects of HBOC-201 on cardiac function. This study was undertaken after a preclinical demonstration of a dose-response potential beneficial effect was realized[14] concluding "... preoxygenated HBOC-201 can match the oxygen delivery role of endogenous blood in the heart on an almost equivalent-volume basis."

Given the concern about vasoconstriction, the issue was addressed in a secondary analysis of the HEM-0115 trial data by Freilich *et al.*[15] Clearly the HEM-0115 trial showed that "... potentially vasoactivity-related adverse safety signals were more frequent with HBOC-201 than RBC in ***older patients*** undergoing orthopedic surgery..."

However, "... incidences of these safety signals were generally lower and group differences narrowed in subpopulations with stable trauma, hypotension, and younger age..." which "... suggests an acceptable safety profile in younger acute trauma populations, especially in settings where rapid access to safe blood transfusions is unavailable..." This hypothesis remains untested as the regulatory agency continues to deny approval of a trauma trial with HBOC-201 putting forth the vasoactivity as toxicity argument. This is of course the argument used by Natanson[9] to brand all HBOCs the same and thus not to be used in any clinical study independent of potential benefit to individuals.

The trauma situation, civilian or military, often involves significant blood loss and hypotension. What level of hypotension deserves treatment has been a vexing question and how to treat that hypotension an open issue. A recent reference by Eastridge[16] demonstrates in a large population a clear relationship between systolic BP and mortality that correlates closely with the relationship of base deficit, a marker of hypoperfusion, and mortality. The critical point is that the moderate vasoactivity associated with HBOC-201 could be beneficial in the trauma situation as proposed studies could explore. Increasing the perfusion pressure while increasing the systemic oxygen delivery may be beneficial with demonstrated clinical efficacy.

Regarding the issue of vasoconstriction generally or specifically cardiac or cerebral vascular, a number of studies have been reported. They fail to demonstrate the vasoconstriction purported to be the offending physiologic event responsible for the emergence of the significant adverse events. It should be noted that these studies were undertaken at the request of the FDA as they continued to deny approval of protocols for trauma resuscitation studies based in large part on their closely held hypothesis of vasoconstriction, resulting from nitric oxide binding by hemoglobin, being *important* in the evolution of the observed adverse events.

To address the FDA concerns about vasoactivity being the critical factor in the emergence of adverse events, two specifically requested studies were designed and implemented. Both studies failed to support the hypothesis proposed for the toxicity of HBOC-201. The study by Mongan[17] explored the changes in regional blood flow associated with serial normovolemic exchange. It demonstrated skeletal muscle as the only organ with a significant decrease in flow with HBOC-210 hemodilution. There were no differences in global oxygen consumption suggesting a coupling of oxygen delivery and consumption "unimpaired by local vasoconstriction." These observations, in detail were in significant contrast to the albumin control animals. Despite mild to moderate increases in systemic and pulmonary blood pressure these changes were "without consequence on regional organ blood flow."

In another critical study, Muir *et al.*[18] demonstrated "... the administration of HBOC-201 maintained tPO2 in three vital organs after profound hemodilution." When compared to albumin hemodilution. This is useful information as it shows critical tissues, those of concern with the emergence of adverse events, are likely not responding to product-induced vasoconstriction. There have also been demonstrations of significant brain oxygenation in complex trauma models as well as in clinical situations.[19] A clinical trial proposed by Proctor *et al.* based on his work with animal models of polytrauma would be a very reasonable undertaking.[20] A more detailed description of the challenge to the vasoactivity argument can be found in the transcript of an FDA/NIH Workshop on the Safety of Hemoglobin Based Oxygen Carriers (www.fda.gov/BiologicsBloodVaccines/News Events/Work-shopsMeetingsConferences/ucm091975.htm).

The impact of the Natanson meta-analysis[9] has been significant. This flawed analysis, failing to observe many of the key guiding principles of constructing such an analysis, has left its mark on the field. Based in the premise that vasoconstriction is the offending event and thus responsible for the deaths, myocardial infarctions and most of the other adverse events, it fails to recognize the significant differences in the solutions, e.g. hemoglobin concentration, percent of tetramer, degree of polymerization, trial designs, control solutions and randomization schemes that were in the trials highly varied. Moreover, it treats all of the clinical trials with HBOC-201 as a single trial, not appropriate for the analysis. Upon close examination of the data used to construct the meta-analysis using the appropriate techniques and analysis inclusion criteria, the entire variance in the analysis, the negative result, rests with the results of a single trauma trial.[21] If this study with its 111 patients is removed from the meta-analysis, there is no significance to the tested hypothesis and the conclusions stated are without merit. How the flaws in the analysis survived the scrutiny of reviewers, statistical consultants and the editorial process is a mystery. It should be noted that the product used in the DCLHb trial was primarily a tetrameric hemoglobin with well-documented vasoactive properties, more so than most of the other solutions included.

The authors of the meta-analysis, convinced of their conclusions, notified regulatory agencies throughout the world of their findings, requesting a stop to all clinical studies, and in the case of Hemopure® cancellation of the license in South Africa. The license was indeed withdrawn in response. However, this action was challenged, new data was provided for review, an independent critique of the meta-analysis was obtained and the regulatory agency undertook its own analysis of the data and critique of the errant publication. In the end, the license was restored and physicians and patients in South Africa once again have access to this therapeutic agent. A listing of current HBOC-201, Hemopure®, approvals is noted in Table 2.

Additional Clinical Information

Compassionate use of Hemopure® has been documented in a number of publications. A report on the treatment of 54 patients indicates that with significant anemia, i.e. hemoglobin concentrations of around 3 g/dL, about half of the patients in this mixed diagnosis population survived.[22] Reviewing data from these patients permitted the development of principles of management. Regarding the adverse event "hypertension," elevations in blood pressure, generally 10–20% in systolic, were noted, usually with the first infusion. The occasional greater increase was noted and in those instances interventions, standard care in intensive care units was implemented. Control of blood pressure elevations was obtained, when indicated, with nitrates, beta-blockers, or calcium channel blockers with the expected outcome. There was a change from bolus or rapid infusion, e.g. two to three hours per unit to continuous infusion and the initial elevations in

Table 2. Current Regulatory Approvals for Hemopure®

Product	Country/Region	Regulatory Authority	Application/Registration #
Hemoglobin glutamer 200 — (bovine) Oxyglobin	Active Substance — Certificate of suitability (CEP) for TSE	European Directorate for the Quality of Medicines (EDQM)	R1-CEP 2001-134-Rev 00
Hemoglobin glutamer 250 — (bovine) Hemopure	Certificate of suitability (CEP) for TSE	European Directorate for the Quality of Medicines (EDQM)	R1-CEP 2002-239-Rev 00
Oxyglobin®	European Commission — Centralized	European Medicines Agency	EU/2/99/015/001-002
Oxyglobin®	USA	FDA/CVM	NADA 141-067
Hemopure®	Russian Federation	Roszdravnadzor/Minzdrav	JIJI 0000 1 1
Hemopure®	South Africa	Medicines Control Council	34/30.4/0261

Oxyglobin and Hemopure are the trade names of the products produced by OPK Biosciences. TSE is transmissible spongiform encephalopathy, a required quality assurance test.

blood pressure were essentially eliminated. The current recommendation is for constant infusion, meeting the oxygen delivery needs of the patient if there is no active significant bleeding. Increases in met-hemoglobin have been observed, higher levels with lower native hemtocrit. This non-oxygen-carrying molecule effectively decreases oxygen-carrying capacity, and while levels up to 10% in patients with normal levels of hemoglobin may not have a clinical impact, when dealing with hemoglobin levels of 5 or 6 g/dL this could be important. The addition of ascorbic acid, 1–2 g per day, was effective in lowering the met-hemoglobin levels or modulating the rate of increase. Acute treatment with the reducing agent methylene blue was used for rapid rises. Because many of these patients were treated with significant volumes of intravenous saline and/or plasma expanders prior to the onset of HBOC-201 infusion, the judicious use of diuretics during the infusion of the solution was advocated to prevent volume overload. As a group, these patients did not exhibit increases in adverse events especially myocardial infarction, other cardiovascular events or cerebral vascular accidents. Indeed, some published cases report and demonstrate electrocardiographic evidence of reversal of cardiac ischemia with treatment further supporting the concept of efficacy and not the hypothetical toxicity of hemoglobin solutions.[23–26] In early 2008, FDA approval for compassionate use of Hemopure® in a 21-year-old woman with acute hemolytic anemia was granted. Over the ensuing 19 days this patient received 54 units of Hemopure® while her native hemoglobin level hovered at 1 gm/dL. The total dose of hemoglobin, a supposedly "toxic substance," was 1.73 kg. While she did not survive her acute illness, there was no evidence of tissue toxicity from this volume and quantity of HBOC-201. She was maintained on ventilator support and heavy sedation to minimize oxygen demands, and when the sedation was eased the nurses reported she had intact brain function. Clearly this case represents evidence of the potential for a life-saving oxygen-carrying solution. It is recognized that the compassionate-use cases do not constitute a basis for regulatory approval. However, they clearly demonstrate efficacy, perhaps arguable given the relationship of hemoglobin concentration and mortality in specific situations is not resolved. What is known, however, is that the relationship does exist; mortality decreases as a function of hemoglobin concentration, "anemia" for many diseases and clinical situations.

One could argue that the transfusion trigger used in the HEM-0115 trial of less than 10.5 g/dL hemoglobin is not operative today and thus the data is suspect for application. Would there be the same results if the transfusion trigger was lowered to 8 gm/dL? Would there be more adverse events in the control group as a result? Would the argument of a hemoglobin deficit associated cardiac ischemic event disappear? These are all good questions in search of answers and worthy of exploration. What is the ideal model for evaluation of HBOC-201 in the trauma situation? I would argue that urban trauma is not be the best model, rather a model with long transit times, nearer to 90–120 min, longer than the "golden hour," would be more revealing.

HBOC-201 can be seen as adjunctive to the management of blood bank PRBCs if used as a hemodilution agent with intra-operative autologous transfusion as well as for immediate use when blood is neither an option nor available. The safety profile is not without risk, as with any therapeutic, and much of the risk is related to the population being treated. Older patients with underlying cardiac disease and acute anemia do not do as well as other populations. As with any therapeutic, there are "side-effects" which are not toxicity that can be managed clinically. There are potential benefits for specific at-risk populations for whom the risk may indeed outweigh the potential for survival, whether in as-yet unproven in the clinical trauma trials or as an agent in the management of spontaneous or iatrogenic anemias. The potential of new infectious agents affecting the blood supply, an increasing incidence of shortages and no stockpile of an oxygen-carrying solution known or on the horizon all point to the ongoing evaluation of a hemoglobin-based oxygen carrier. It would seem prudent and obvious that the continued development and testing of oxygen therapeutics and their evolving derivatives be supported.

References

1. Johnson ST, Cable RG, Leiby DA. (2011) Lookback investigations of Babesia microti-seropositive blood donors: seven-year experience in a Babesia-endemic area. *Transfusion* 2011, Sep 12.
2. Beale MA, Tettmar K, Szypulska R, *et al.* (2011) Is there evidence of recent hepatitis E virus infection in English and North Welsh blood donors? *Vox Sang* **100:** 340–342.
3. Grimshaw K, Sahler J, Spinelli SL, *et al.* (2011) New frontiers in transfusion biology: identification and significance of mediators of morbidity and mortality in stored red blood cells. *Transfusion* **51:** 874–880.
4. Dodd RY. (2010) Emerging pathogens in transfusion medicine. *Clin Lab Med* **30:** 499–509.
5. Winslow RM. (2006) Historical background. In: Winslow RM (ed), *Blood Substitutes*, Academic Press, pp. 5–16.
6. Chen JY, Scerbo M, Kramer G. (2009) A review of blood substitutes: examining the history, clinical trial results, and ethics of hemoglobin-based oxygen carriers. *Clinics (Sao Paulo)* **64:** 803–813.
7. Rentko VT, Pearce B, Moon-Massat PF, Gawryl M. (2006) Hemopure® (HBOC-201, Hemoglobin Glutamer-250 (Bovine)): Preclinical Studies. In: Winslow RM (ed), *Blood Substitutes*, Academic Press, pp. 424–436.
8. Pearce LB, Gawryl MS, Rentko VT, *et al.* (2006) HBOC-201 Hemoglobin Glutamer-250 (Bovine), Hemopure: Clinical Studies. In: Winslow RM (ed), *Blood Substitutes*, Academic Press, pp. 437–450.
9. Natanson C, Kern SJ, Lurie P, *et al.* (2008) Cell-free hemoglobin-based blood substitutes and risk of myocardial infarction and death: a meta-analysis. *JAMA* **299:** 2304–2312.
10. Jahr JS, Mackenzie C, Pearce LB, *et al.* (2008) HBOC-201 as an alternative to blood transfusion: efficacy and safety evaluation in a multicenter phase III trial in elective orthopedic surgery. *J Trauma* **64:** 1484–1497.
11. Greenburg AG, Pitman A, Pearce LB, Kim HW. (2008) Clinical contextualization and the assessment of adverse events in HBOC trials. *Art Cells, Blood Substitutes, and Biotechnology* **36:** 477–486.
12. Greenburg AG, Kim HW, Hemolink Study Group (2004) Use of an oxygen therapeutic as an adjunct to intraoperative autologous donation to reduce transfusion requirements in patients undergoing coronary artery bypass graft surgery. *J Am Coll Surg* **198:** 373–383.
13. Meliga E, Vranckx P, Regar E, *et al.* (2008) Proof-of-concept trial to evaluate haemoglobin based oxygen therapeutics in elective percutaneous coronary revascularization. *EuroInterv* **4:** 99–107.
14. Te Lintel Hekkert M, Dubé GP, Regar E, *et al.* (2010) Preoxygenated hemoglobin-based oxygen carrier HBOC-201 annihilates myocardial ischemia during

brief coronary artery occlusion in pigs. *Am J Physiol Heart Circ Physiol* **298:** H1103–H1113.
15. Freilich D, Pearce LB, Pitman A, *et al.* (2009) HBOC-201 vasoactivity in a phase III clinical trial in orthopedic surgery subjects — extrapolation of potential risk for acute trauma trials. *J Trauma* **66:** 365–376.
16. Eastridge BJ, Slainas J, Wade CE, Blackbourne LH. (2011) Hypotension is 100 mm Hg on the battlefield. *Am J Surgery* **202:** 404–408.
17. Mongan PD, Moon-Massat PF, Rentko V, *et al.* (2009) Regional blood flow after serial normovolemic exchange transfusion with HBOC-201 (Hemopure®) in anesthetized swine. *Trauma* **67:** 51–60.
18. Muir WW, Ilangovan G, Zweier JL, *et al.* (2011) Vital organ tissue oxygenation after serial normovolemic exchange transfusion with HBOC-201 in anesthetized swine. *Shock* **35:** 597–603.
19. Marinaro J, Smith J, Tawil I, *et al.* (2009) HBOC-201 use in traumatic brain injury: case report and review of literature. *Transfusion* **49:** 2054–2059.
20. Dudkiewicz M, Harpaul TA, Proctor KG. (2008) Hemoglobin-based oxygen carrying compound-201 as salvage therapy for severe neuro- and polytrauma (Injury Severity Score = 27–41). *Crit Care Med* **36:** 2838–2848.
21. Sloan EP, Koenigsberg M, Gens D, *et al.* (1999) Diaspirin cross-linked hemoglobin (DCLHb) in the treatment of severe traumatic shock: a randomized controlled efficacy trial. *JAMA* **282:** 1857–1864.
22. Mackenzie CF, Moon-Massat PF, Shander A, *et al.* (2011) When blood is not an option: factors affecting survival after the use of a hemoglobin-based oxygen carrier in 54 patients with life-threatening anemia. *Anesth Analg* **110:** 685–693.
23. Mullon J, Giacoppe G, Clagett C, *et al.* (2000) Transfusions of polymerized bovine hemoglobin in a patient with severe autoimmune hemolytic anemia. *N Engl J Med* **342:** 1638–1643.
24. Fitzgerald MC, Chan JY, Ross AW, *et al.* (2011) A synthetic haemoglobin-based oxygen carrier and the reversal of cardiac hypoxia secondary to severe anaemia following trauma. *Med J Aust* **194:** 471–473.
25. Donahue LL, Shapira I, Shander A, *et al.* (2010) Management of acute anemia in a Jehovah's Witness patient with acute lymphoblastic leukemia with polymerized bovine hemoglobin-based oxygen carrier: a case report and review of literature. *Transfusion* **50:** 1561–1567.
26. Pachinburavan M, Marik PE. (2008) Bovine blood and neuromuscular paralysis as a bridge to recovery in a patient with severe autoimmune hemolytic anemia. *Clin Transl Sci* **1:** 172–173.

Chapter 3

Glutaraldehyde Polymerized Porcine Hemoglobin: Preparation, Safety Evaluation and Pharmacodynamics Study

Chao Chen*,†, Hongli Zhu*,‡ and Kunping Yan*,‡

Abstract

In an attempt to avoid many of the liabilities of existing hemoglobin-based oxygen carriers (HBOCs), Lifegen Ltd. is developing a glutaraldehyde polymerized porcine hemoglobin product (pPolyHb). In this article we review the preclinical characteristics of pPolyHb that suggest its advantages over other HBOCs. pPolyHb was shown to efficiently supply oxygen to tissues and maintain hemodynamic stability in rats in both the exchange transfusion model and the hemorrhagic shock and resuscitation model. pPolyHb was also able to effectively reverse anaerobic metabolism and repay the oxygen debt as measured by lactic acid accumulation and base excess maintenance. The safety profile of pPolyHb was established by acute and chronic

* College of Life Science, Northwest University, Xi'an, P. R. China.
†National Engineering Research Center for Miniaturized Detection Systems, Northwest University, Xi'an, P. R. China.
‡Shaanxi Lifegen Co. Ltd., Xi'an, P. R. China.

toxicology evaluation, as well as by immune safety studies. No significant elevation of blood pressure or pro-inflammatory effects were observed. Furthermore, *in vitro* studies demonstrated the antioxidative properties of pPolyHb in inhibiting H_2O_2-mediated endothelial cell damage and apoptosis. Taken together, these preclinical studies predict both the clinical efficacy and safety of pPolyHb for use in a range of physiologic and pathologic conditions where blood or red blood cells (RBC) are not available.

Introduction

Hemoglobin-based therapeutics have a wide range of potential clinical applications. They can be used as oxygen carriers in the prevention and treatment of disorders or pathological conditions associated with a deficiency in tissue oxygen delivery, such as severe blood loss, hemorrhagic shock, trauma, and cardiac events.[1] These therapeutics can also be used for the temporary maintenance of patients with rare blood types. The potential benefits of hemoglobin-based oxygen carriers (HBOCs) include universal compatibility, immediate availability, absence of infectious agents, and long-term storage. Moreover, certain characteristics of HBOCs, such as viscosity, molecular size, oncotic pressure and oxygen delivery kinetics, can be altered according to specific clinical objectives.

So far, four main types of HBOCs have been developed and tested in clinical trials: crosslinked tetrameric hemoglobin, recombinant human hemoglobin, polymerized hemoglobin (PolyHb), and conjugated hemoglobin.[2]

Baxter's HemAssist® (DCLHb), an example of intramolecularly crosslinked tetrameric hemoglobin (Hb), was the first HBOC to reach clinical trials. HemAssist is a human hemoglobin tetramer stabilized by a diaspirin linkage.[3] This approach prevents dimer formation and improves P_{50}. However, HemAssist was shown to result in intensified vasoconstriction, possibly because it scavenges nitric oxide (NO) produced by the vascular endothelium and thus prevents smooth muscle relaxation. In Phase III clinical trials in trauma, surgery and acute ischemic stroke, excess mortality was observed among patients receiving HemAssist compared to those receiving banked red blood cells.[4,5] In elective surgery patients, HemAssist treatment did allow a significant reduction in the need for blood transfusion.[4,6] However, serious adverse events occurred, including acute respiratory distress syndrome, multi-organ failure and cardiac lesions, and the study had to be suspended due to safety concerns.[6]

The clinical failure of HemAssist® spurred further research aimed at solving the problem of vasoconstriction. One of the strategies was to introduce mutations into the heme pocket that would decrease the rate of NO binding to hemoglobin, while still permitting O_2 transport.[7] Recombinant human Hb from *E. coli* was thus produced based on the above rationale. In another approach, recombinant human Hb was formed by the fusion of the two α subunits of each Hb molecule to prevent breakdown into dimers. This product has advanced to clinical trials, but was discontinued due to adverse effects.[8] A second-generation product with reduced vasoactivity has been developed and tested in preclinical studies.[9]

Another approach to reduce interference with NO signaling was to limit extravasation of HBOCs by increasing the size of the hemoglobin molecule, using either polymerization or conjugation techniques. Polymerized hemoglobin (PolyHb) has been prepared by intermolecular crosslinking of stabilized Hbs using bifunctional agents, such as glutaraldehyde and O-raffinose. Examples of PolyHb products include Hemopure® (Biopure Corporation, Cambridge, MA), PolyHeme® (Northfield Laboratories, Inc., Evanston, IL), and HemoLink® (Hemosol Corporation, Mississauga, Canada).

Hemopure (HBOC-201), a bovine Hb polymerized with glutaraldehyde, has been examined as an alternative to red blood cell transfusion in various clinical trials. Most of these trials were elective surgical cases, including cardiac, orthopedic, hepatic, vascular and general surgical procedures.[10, 11] Phase I studies in normal volunteers indicated that Hemopure was well tolerated over a range of doses and regiments, with minor incidents of elevated blood pressure and gastrointestinal discomfort as well as some elevations of liver enzymes and lipase. Phase II trials in cardiac surgery[10] and abdominal aortic surgery,[12] and Phase III studies in elective orthopedic surgery, demonstrated that Hemopure significantly reduced the use of allogeneic RBC transfusion.[13] However, certain adverse events, including anemia, tachycardia, vomiting, jaundice, lipase increase, oliguria and hypertension were observed. In 2001, Hemopure was approved in South Africa for treatment of adult surgical patients who are acutely anemic. Recently, Hemopure (OPK Biotech LLC, Cambridge, Massachusetts, USA, originally Biopure Corporation) has been approved for human use in Russia.[14]

PolyHeme has also been developed as an alternative to banked red blood cells in surgery and trauma.[15] This product was produced by crosslinking stroma-free hemoglobin from outdated human blood with pyridoxylate to form intramolecular crosslinks, followed by intermolecular polymerization with glutaraldehyde. PolyHeme has finished US multicenter Phase III clinical trials[16] at urban Level I trauma centers. There was no significant difference between day 30 mortality in the as-randomized (13.4% PolyHeme vs. 9.6% control) or per-protocol (11.1% PolyHeme vs. 9.3% control) cohorts. Allogeneic blood use, however, was lower in the PolyHeme group (68% vs. 50% in the first 12 hr). Patients resuscitated with PolyHeme, without stored blood for up to 6U in 12 hr post injury, had outcomes comparable with those for the standard of care. Although there were more adverse events, such as anemia, hypertension, fever, and electrolyte imbalance, in the PolyHeme group, the benefit-to-risk ratio of PolyHeme was considered favorable in situations where blood is needed but not available. However, in May 2009, the FDA refused to approve PolyHeme.[17]

Another PolyHb product, Hemolink, was prepared by polymerizing human hemoglobin using an oxidized trisaccharide, O-raffinose, followed by a reduction step. This product was studied in Phase II clinical trials in dialysis and as an oxygen-carrying replacement fluid in acute normovolemic hemodilution. Hemolink was noted to have a mild systemic pressor effect.[18] In Phase II and III studies in cardiac surgery, patients receiving up to four units of Hemolink reduced their transfusions of banked red blood cells for up to five days after surgery, compared to controls receiving pentastarch.[19,20] However, there were safety concerns, including myocardial infarction, which led to early termination of a Phase III trial of Hemolink.[21]

An example of a conjugated Hb is Hemospan® (Sangart Inc., San Diego, California, USA). Hemospan was prepared by conjugating polyethylene glycol (PEG) to human hemoglobin. This product was designed to have a low P_{50}, a large molecular diameter, and a high colloid osmotic pressure (COP) and viscosity.[22] It was formulated to provide a balance between oxygen capacity and oncotic pressure and was not expected to be a complete replacement for blood. Preclinical studies demonstrated that Hemospan could improve microcirculatory blood flow and tissue oxygenation. Moreover, it was free of any significant toxicity in doses exceeding those proposed for human use (estimated to be 250–500 mg/kg body weight, or a maximum of about 1 L per patient). The clinical development plan for Hemospan focused on its unique properties of tissue oxygenation. Phase I and II trials have been completed in Europe[23] and another Phase II trial in elective prostatectomy has been initiated in the United States.[24]

Generally, among the different types of HBOCs, glutaraldehyde polymerized products have made the most progress towards commercial development. Studies have shown that these products could reduce the use of allogeneic RBC transfusion, maintain hemodynamic stability, deliver oxygen to tissues effectively and improve survival. Nonetheless, some adverse effects were also observed during the pre-clinical and clinical studies of glutaraldehyde polymerized products, including hypertension, cardiac events and oxidative stress.

Lifegen Ltd., China, is developing a new HBOC product, glutaraldehyde polymerized porcine hemoglobin (pPolyHb), in an attempt to avoid the adverse effects that have hindered the development of HBOCs. Preliminary studies show that pPolyHb has good oxygen capacity and volume expansion ability, as well as antioxidative activity. Its favorable safety profile has also been established in rats. So far, no significant vasoactivity or elevated blood pressure has been observed in animal models. pPolyHb is currently at the pre-clinical stage of development.

Preparation and Physiochemical Properties

Porcine blood was freshly collected and stored at 4°C with a citric acid/sodium citrate buffer as the anticoagulant. Erythrocytes were thoroughly washed with normal saline and carefully lysed using hypotonic phosphate buffer. Cell stroma and hemoglobin were separated by ultrafiltration to obtain stroma-free hemoglobin (SFHb), then purified by specific steps. After concentration using a 10 kDa hollow fiber column, purified hemoglobin was obtained. This purification technique allowed the retention of superoxide dismutase (SOD) and catalase (CAT) activities.

pPolyHb was prepared by polymerization of the purified hemoglobin with glutaraldehyde, and the reaction was quenched with dimethylaminoborane (DMAB). The polymerized hemoglobin was purified by 100 kDa tangential flow filtration (TFF), wherein small molecules, including excess glutaraldehyde and unpolymerized tetramer-hemoglobin, were removed. pPolyHb was formulated at a concentration of 11 g/dL in normal saline at a pH between 7.3 and 7.5. The physiochemical characteristics and structural properties of pPolyHb are listed in Table 1.[25] The protein sample was stored at 4°C under nitrogen gas until use.

Preclinical Studies

It has been reported by other groups that the administration of some types of HBOCs may induce elevation of blood pressure, oxidative stress and pro-inflammatory effects. In order to evaluate the safety profile and pharmacodynamic properties of pPolyHb, various preclinical studies were done in anticipation of clinical trials.

Exchange Transfusion in Rats

Rats were heparinized before exchange transfusion through the venous catheter at 60 units/100 g body weight. Blood was removed from the femoral artery and exchange fluid (either pPolyHb or Hetastarch (HES)) was simultaneously transfused via the femoral

Table 1. Physiochemical and Structural Characteristics of pPolyHb[25]

Parameter	Results
Polymerized porcine hemoglobin	10.5±0.5 g/dL
Methemoglobin	<5%
Endotoxin	<1.0 EU/mL
Osmolality	300–330 mOsm
pH	7.4±0.05
P_{50}	28±3 mmHg
Average molecular weight of pPolyHb	600±50 kD
64 kD tetramer	<2%
Na^+	135–155 mmol/L
K^+	3.0–5.0 mmol/L
Ca^{2+}	1–3 mmol/L
Cl^-	140–160 mmol/L

vein.[26] Exchange transfusions were done at a rate of 0.3 mL/min to a total volume of solution that equaled 120–140% of estimated total blood volume (70 mL/kg). Residual erythrocyte hemoglobin was less than 2g/dL.

Relative to HES, exchange transfused animals that received pPolyHb showed a significantly higher rate of survival (p <0.05).[26] At the present stage in the optimization of the product, pPolyHb exchange transfusion is able to maintain 100% survival of rats, whereas the average survival time of the HES group is only 10 hr (unpublished data).

Mean arterial blood pressure (MAP) and heart rate (HR) are indexes used for demonstrating the function of the cardiovascular system. The results demonstrated that MAP and HR, as well as systolic blood pressure (SP) and diastolic blood pressure (DP), of rats in the pPolyHb group were stable throughout the whole process of blood exchange. Notably, no significant elevation of blood pressure was observed when pPolyHb was transfused (Fig. 1). These results

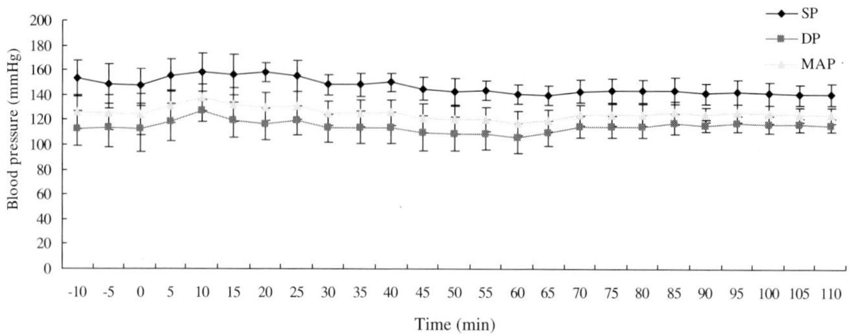

(a)

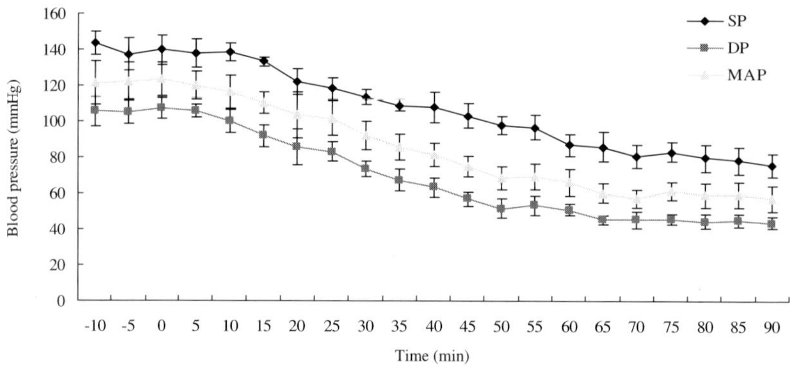

(b)

Fig. 1. Change of blood pressure in rat exchange transfusion model. (Modified from Zhu HL *et al.*, 2011.) (a) Changes of MAP, SP, DP in pPolyHb group; (b) Changes of MAP, SP, DP in Hetastarch group.
Mean arterial blood pressure (MAP), systolic blood pressure (SP) and diastolic blood pressure (DP) were monitored every 5 min throughout the experiment.

demonstrate that pPolyHb has excellent volume expansion capability and is also able to maintain hemodynamic stability. In contrast, the animals treated with Hetastarch displayed a significant decrease in the values of each of these hemodynamic parameters.[26]

Lactate, base excess (BE), and HCO_3^- are markers of anaerobic metabolism. pPolyHb was shown to effectively reverse anaerobic metabolism caused by large volume exchange and erythrocyte hemoglobin loss (Fig. 2). This appears to be due to its superior oxygen-carrying and releasing capability, which helps avoid metabolic acidosis and allows survival. This conclusion was also supported by assessing pH and PCO_2 trends in circulating blood. Specifically, both the pH and PCO_2 values decreased markedly in the HES group, whereas these parameters remained stable in the pPolyHb group when erythrocyte hemoglobin was under 2 g/dL.[26]

Hemorrhage Shock and Resuscitation in Rats

While the exchange transfusions performed with rats demonstrated that pPolyHb allowed rat survival even when erythrocyte hemoglobin was under 2 g/dL, it is important to evaluate its ability to do so under conditions of hemorrhagic shock. Several animal models were used to simulate the clinical situation of hemorrhagic shock and resuscitation.

Fifty Per Cent Controlled Hemorrhage and Resuscitation in Rats

Hemorrhagic shock was induced in rats by removal of 50±5% of the total blood volume via the femoral artery at a rate of 0.5 mL/min, over an average time of 90±5 min, in a two-step process. Initial removal of 20±3% of the total blood volume was followed by a 10 min rest period, and then the second stage involved the removal of 30±2% of the total blood volume. Throughout the bleeding process, the blood pressure was maintained between 40–60 mmHg. No intervention was made if the MAP decreased below 40 mmHg during the shock period. At the conclusion of the 90 min shock period, rats were randomized to receive one of four resuscitation fluids: lactated Ringer's (LR) solution equivalent to three times the volume of shed blood; 6% Hetastarch 200–0.5 (HES) in sodium chloride solution equivalent to the volume of shed blood; 2 g/kg pPolyHb plus saline equivalent to two times the volume of shed blood; or reinfusion of the rat's own shed whole blood. The control group received no resuscitation fluid, in order to establish and quantify the severity of the injury with this model. Fluid administration was continued over a one-hour period at a rate of 0.5 mL/min.

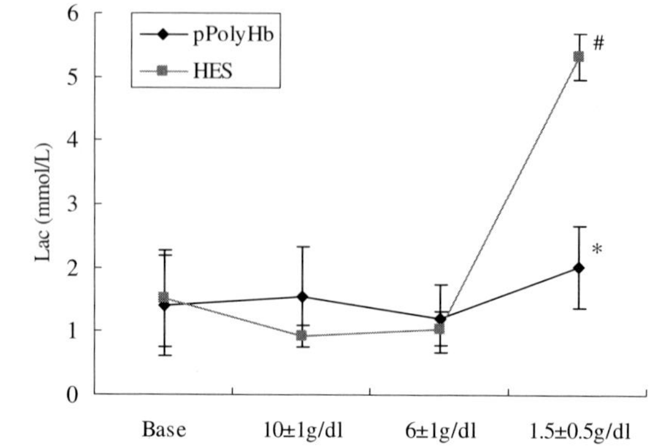

(a)

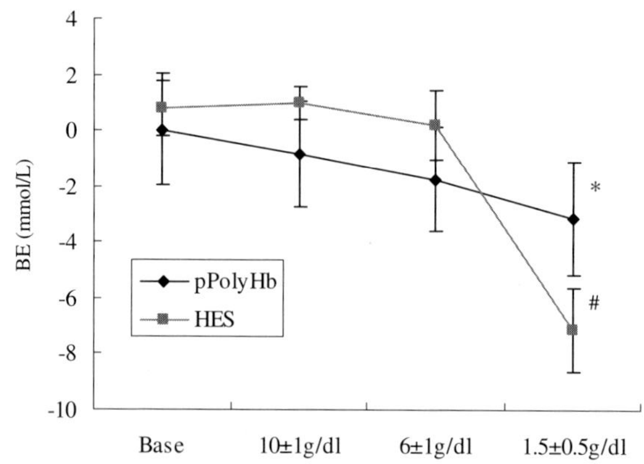

(b)

Fig. 2. Lactate, base excess (BE) and HCO_3^- levels in the rat exchange transfusion model. (Zhu HL *et al.*, 2011.) (a) Comparison of lactate levels in the pPolyHb group and the HES group; (b) Comparison of BE levels in the pPolyHb group and the HES group; (c) Comparison of HCO_3^- levels in the pPolyHb group and the HES group. Blood samples were drawn before the start of blood exchange (baseline) and at different stages of exchange (i.e. when erythrocyte hemoglobin was 10±1 g/dL, 6±1g/dL, and 1.5±0.5g/dL) and the levels of lactate, BE and HCO_3^- were tested. Blood gas analysis was performed on an ABL 725 (Radiometer, Copenhagen, Denmark). $*P < 0.05$ in comparison to hetastarch group; $\#P < 0.05$ in comparison to baseline.

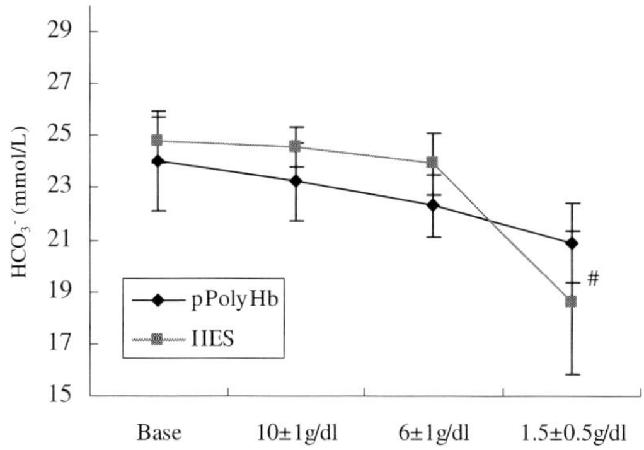

(c)

Fig. 2. (*Continued*)

Hemodynamic parameters were monitored every 5 min throughout the experiment. Blood was sampled serially during the experiment at the following time points: at baseline; at 90 min (completion) of shock; and at 45 and 180 min after resuscitation.

The results showed that resuscitation with pPolyHb could restore both systemic and pulmonary mean pressures more effectively than either HES or LR. With pPolyHb, these parameters returned to baseline levels at 45 min after resuscitation and remained stable until the end of the experiment. Additionally, no significant elevation of blood pressure was observed in this hemorrhage and resuscitation model (unpublished data). These results demonstrated that pPolyHb was able to efficiently maintain hemodynamic stability. pPolyHb resuscitation also restored the baseline level of lactate and base excess (BE), as well as HCO_3^-, whereas lactic acidosis developed in the HES and LR groups (unpublished data). These results demonstrated the effectiveness of pPolyHb at reversing anaerobic metabolism and compensating for the acid-base state, and were consistent with the observation that all animals resuscitated with pPolyHb survived.

Sixty Per Cent Controlled Hemorrhage and Resuscitation in Rats

Similar procedures were followed to induce hemorrhagic shock, except that 60±5% of the total blood volume was removed from the femoral artery, with removal of 25±3% of total blood volume in the first step, and 35±2% in the second step. The resuscitation procedure was as described above. This protocol allowed examination of the time-course of the metabolic changes that accompany hemorrhage, as well as assessment of the overall length of time of animal survival.

The survival rate for the pPolyHb group was 100%, which was same as that for the group reinfused with shed blood. This survival rate was much higher than for either the HES group (56%) or the LR group (37%) (unpublished data). Rats resuscitated with pPolyHb were also able to maintain better oxygen delivery and extraction than rats receiving only HES or LR. This finding is important, because in hemorrhagic shock, oxygen extraction exceeds oxygen delivery, resulting in lactic acidosis if not corrected. Repayment of the oxygen debt is considered a necessary endpoint of successful resuscitation.

Perioperative Hemorrhagic Shock in Canine Model

The infusion of pPolyHb in a canine model of perioperative hemorrhagic shock was also investigated. Blood was withdrawn in three steps until the endohemoglobin concentration dropped to 5 g/dL. Isovolumic pPolyHb was infused after each withdrawal until the total hemoglobin concentration was restored to 10 g/dL. Electrocardiogram (ECG), central venous pressure (CVP), MAP, HR, oxygen saturation (SaO_2) and cardiac output (CO) were measured, and none of these parameters showed a significant change over the course of the study. Lactate and BE levels, as well as oxygen consumption, were in the normal range. The dogs survived the study (unpublished data).

In Vitro Antioxidation Study

A significant impediment to the development of HBOCs has been redox-related safety issues. Most HBOC products have been shown to be cytotoxic to cells through a mechanism that involves the oxidant activity of H_2O_2.[27] Hb or HBOC preparations in their reduced forms are not, by themselves, cytotoxic to cells.[28] However, when oxidized by H_2O_2, the accumulation of highly reactive Fe^{4+} can induce significant cytotoxicity through a mechanism that involves the production of toxic oxygen intermediates such as ferryl-Hb ($HbFe^{4+} = O$) and a globin tyrosyl-based radical ($\bullet HbFe^{4+} = O$). •OH and $HbFe^{4+} = O$ can initiate lipid peroxidation, while $\bullet HbFe^{4+} = O$ is an effective modifier of low-density lipoproteins. Ferryl-Hb species may also contribute to many other pathological events, including damage to large molecules and apoptosis.[29]

A novel function demonstrated for pPolyHb is in protecting a human umbilical vein endothelial cell line from H_2O_2-induced cytotoxicity. Several lines of evidence support this conclusion. First, an H_2O_2 consumption assay showed that pPolyHb was able to neutralize exogenously added H_2O_2, indicating that pPolyHb could decrease H_2O_2-mediated cytotoxicity.[25] Second, ferrylhemoglobin formation was not observed as a result of incubation of pPolyHb and H_2O_2, suggesting a lack of toxic product formation by pPolyHb (Fig. 3). Third, MTT and cell morphology assays showed that cell death caused by H_2O_2 could be prevented by the presence of pPolyHb (Fig. 4). Fourth, flow cytometry and immunostaining with an apoptotic marker revealed that pPolyHb inhibited apoptosis induced by H_2O_2 (Fig. 5). Consistent with the involvement of pPolyHb in preventing H_2O_2-induced cell cytotoxicity, the SOD and CAT activities of pPolyHb were much higher than those of purified Hb.[25] These results provide the first evidence that pPolyHb has a protective function against H_2O_2-induced cell toxicity.

Pro-inflammatory Study

Although various physiological properties of pPolyHb and other HBOCs, including efficacy and toxicity, have been extensively studied, it is also critical to establish the effects of pPolyHb on the host inflammatory response.

We assessed the levels of three indicators of inflammation, complement C3a, IL-6, and TNF-α, when pPolyHb was incubated with cultured cells or administered to healthy rats. As shown in Fig. 6, in cultured cells, all three inflammatory markers remained at a low level in the presence of pPolyHb, at a concentration similar to the negative controls (nonspecific protein (BSA) or medium (PBS)). Moreover, these levels were not significantly changed from the baseline level in healthy rats after pPolyHb administration, demonstrating that pPolyHb did not induce pro-inflammatory activation.[30]

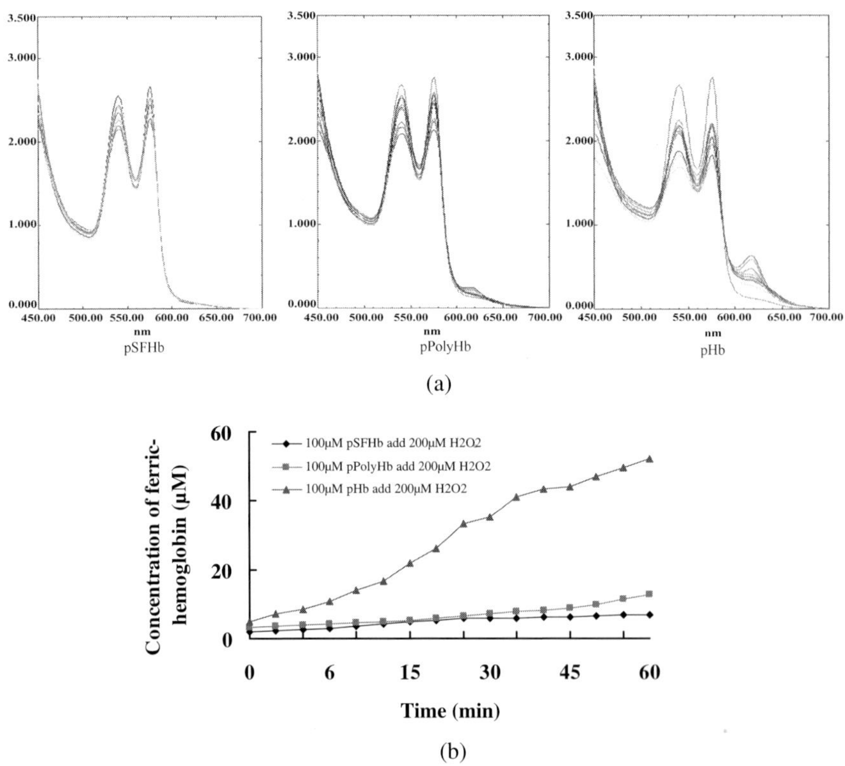

Fig. 3. pPolyHb inhibition of ferrichemoglobin or ferrylhemoglobin formation. (Zhang W *et al.*, 2012.) pSFHb (porcine stoma-free hemoglobin), pPolyHb or pHb (purified porcine hemoglobin), solutions were incubated with H_2O_2 at room temperature for 60 min. (a) Spectral analysis of hemoglobin with a SPD-10AVP plus UV-VIS spectrophotometer; (b) and (c) Ferrichemoglobin and ferrylhemoglobin formation. Each curve represents measurements taken every 2 min for the first 10 min, followed by every 5 min for the next 50 min.

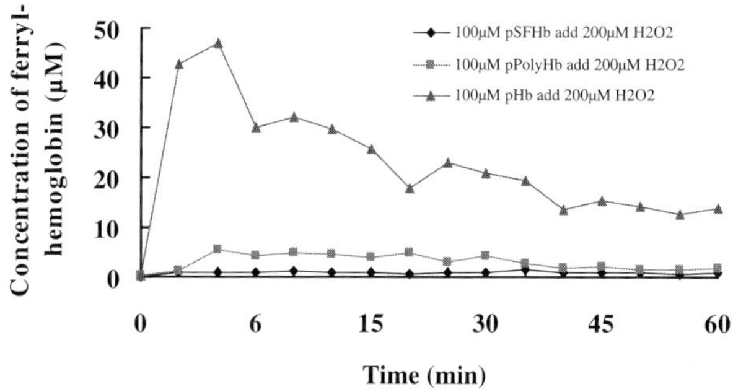

(c)

Fig. 3. (*Continued*)

We further investigated pPolyHb's immune effects in a controlled severe hemorrhagic shock and resuscitation model. The course of hemorrhagic shock (HS) results in a possible challenge to the immune system, which may contribute to multiorgan failure (MOF) or death.[31] Current emergency resuscitation fluids and blood products possess potential intrinsic immunomodulating properties.[32] Thus it was necessary to investigate whether pPolyHb invokes an innate immune response in such critical circumstance. The results showed that in untreated animals, C3a, IL-6 and TNF-α rose to their highest levels at the end of the shock process. This may be due to the infiltration and activation of large numbers of leukocytes that release inflammatory mediators and initiate inflammatory responses in tissues. However, pPolyHb resuscitation was able to reverse inflammation and restore the levels of the three inflammatory mediators to baseline, suggesting that infusion of pPolyHb does not induce an inflammatory response[30] and may in fact help to attenuate the existing inflammatory response. Similar results were also observed for the bovine polyHb product, HBOC-201. In these studies, in swine models of moderately severe HS and severe controlled HS, HBOC-201 had no significant effects, either adverse or beneficial, on immune function.[33, 34]

Thus, both *in vitro* and *in vivo* results showed that there was no inflammatory reaction caused by addition or infusion of pPolyHb.

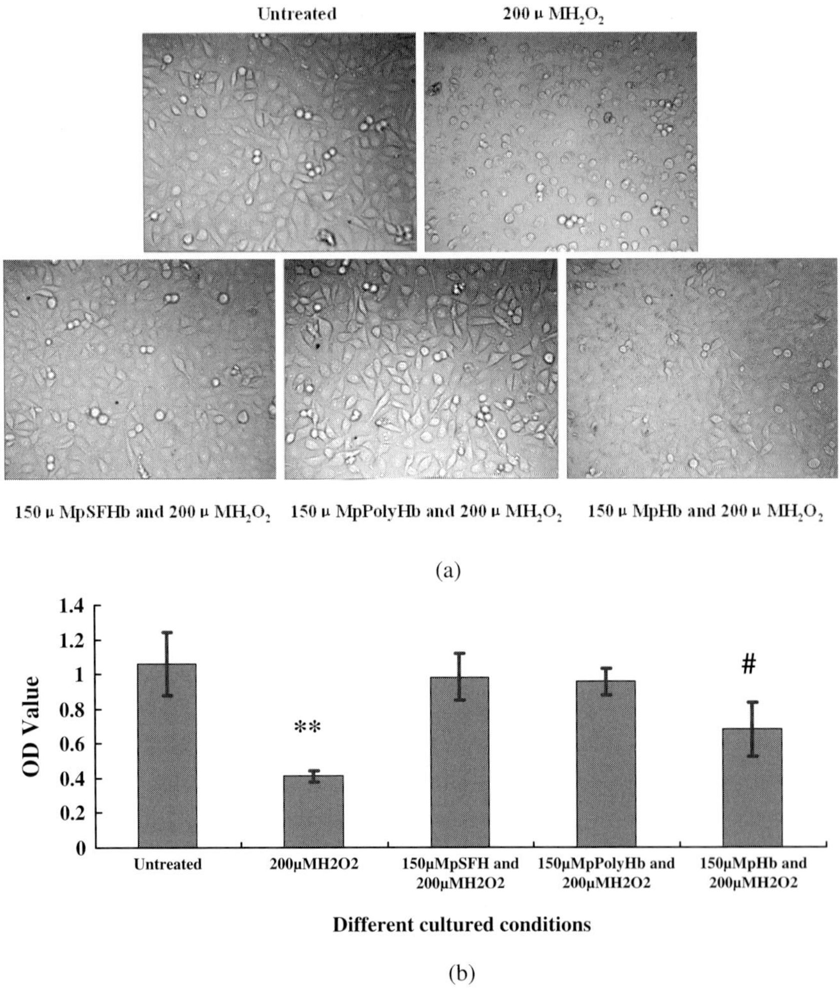

Fig. 4. pPolyHb inhibition of H_2O_2-induced cell damage. (Zhang W *et al.*, 2012.) EVC-304 cells were incubated with FBS-free medium, H_2O_2, pSFHb, pPolyHb, or pHb combined with H_2O_2 as indicated. (a) Cell morphology, as detected by phase contrast microscopy; (b) MTT assay analysis of living cells. Statistical significance indicated by $^{**}P < 0.01$ for pPolyHb vs. H_2O_2; $\#P < 0.05$ for pPolyHb vs. pHb.

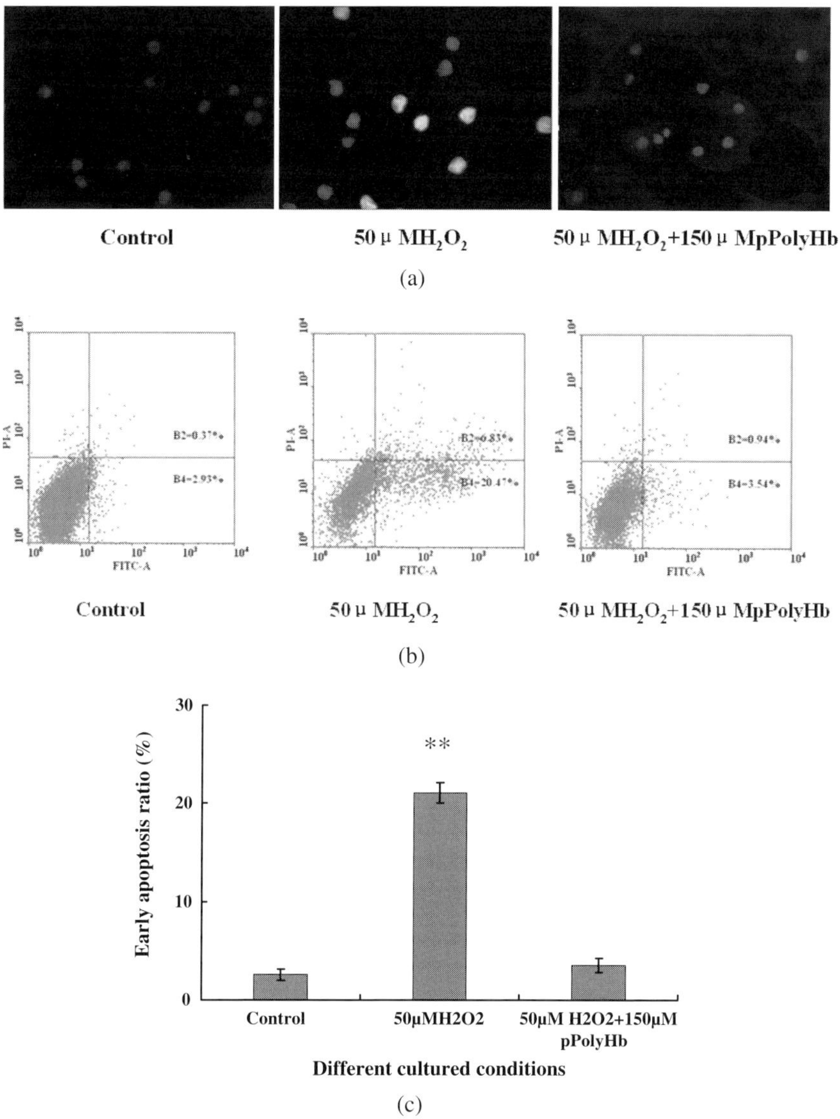

Fig. 5. pPolyHb inhibition of H_2O_2-induced cell apoptosis. (Zhang W *et al.*, 2012) EVC-304 cells were incubated with FBS-free medium, H_2O_2, or pPolyHb combined with H_2O_2, as indicated. (a) Fluorescence microscopic examination of nuclear staining with Hoechst 33258; (b) flow cytometric analysis of annexin V-stained cells; (c) quantitative analysis of data in (b). Statistical significance indicated by $^{**}P < 0.01$.

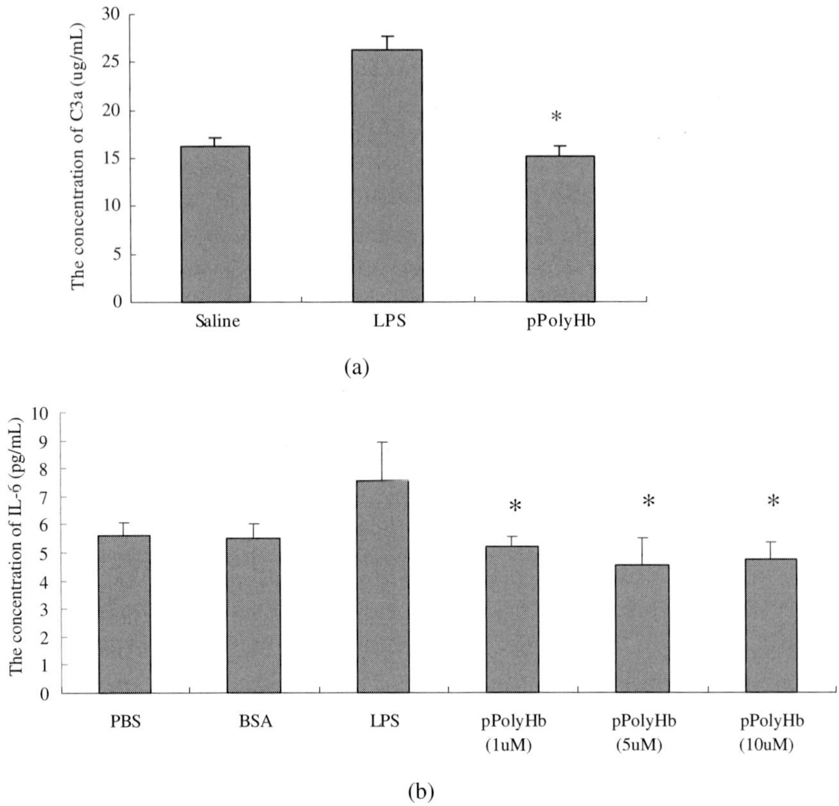

Fig. 6. Concentrations of inflammatory mediators upon treatment with different agents *in vitro*. (Zhu HL *et al.*, 2011.) (a) Comparison of complement C3a levels. Plasma from Sprague Dawley rats was mixed with saline (negative control), LPS (positive control) or pPolyHb. 1hr later, complement C3a levels were determined using the Rat Complement C3a Elisa kit (R&D Systems) according to the manufacturer's protocol. $^*P < 0.05$ in comparison to LPS group. (b) Comparison of IL-6 levels. Raw264.7 cells were treated with PBS, BSA (negative control), LPS (positive control), or various concentrations of pPolyHb, for 12–18 hr. Supernatant was collected and IL-6 levels detected using the Mouse IL-6 ELISA kit (R&D Systems) according to the manufacturer's protocol. $^*P < 0.05$ in comparison to LPS group. (C) Comparison of TNF-α levels. Raw264.7 cells were treated as in (b), and TNF-α levels detected using the Mouse TNF-α ELISA kit (R&D Systems) according to the manufacturer's protocol $^*P < 0.05$ in comparison to LPS group.

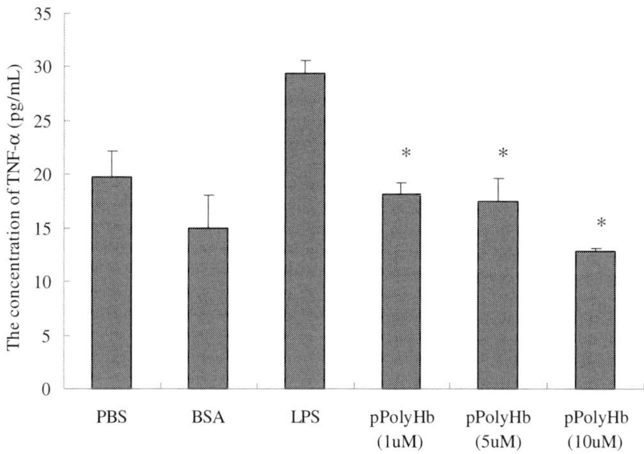

(c)

Fig. 6. (*Continued*)

These studies of immune function add to the emerging body of evidence that predicts that pPolyHb will be a safe resuscitative fluid.

Safety Evaluation

Acute and Chronic Toxicology Evaluation

In an acute toxicological study, rats were administrated escalating doses of pPolyHb via the tail vein. The endpoints were survival, clinical observations, clinical chemistry, clinical pathology, hematology, and histopathology, measured 1, 3 and 14 days after dosing with pPolyHb, or the control fluids, HES or LR. No significant differences were found in regard to body weight gain and food consumption at maximal feasible dose (MFD) or less. In rats receiving HES or LR, various abnormalities were noted, including thinness, mild depression and rough hair, but these abnormalities were not seen in the pPolyHb group. There were no significant differences with regard to clinical chemistry or hematology among different groups. The results indicated that infusion of pPolyHb did not cause adverse consequences (unpublished data).

In a chronic toxicological study, rats received escalating doses of pPolyHb via the tail vein once a day for two weeks. No significant differences were found, relative to controls, in regard to body weight gain or food consumption at any dose of pPolyHb. Reversible elevations in some liver enzymes were noted in some rats, but there were not dose-related. At days 7 and 14 after dosing, eight rats in each group were euthanatized. Tissues were harvested, fixed and subjected to blinded histologic examination. These studies showed no signs of abnormal transudates, no edema of the lungs or other organs, or any other significant findings. In samples of heart, kidneys, lung, pancreas, spleen, and liver, no pathological abnormalities were found in any organ that could be attributed to pPolyHb. All rats that received pPolyHb survived if not euthanatized (unpublished data).

Safety observations were also made in exchange transfusion studies performed with 30% of blood replacement. The results indicated that there were no significant abnormalities in regard to clinical observations and clinical chemistry at 3, 7 or 14 days after transfusion. Additionally, no pathologic abnormalities were observed in organ samples including brain, heart, liver, lung, and kidneys. All rats that received pPolyHb survived if not euthanatized. The safety profiles of pPolyHb in rat models of hemorrhage and resuscitation were also evaluated. Tissues were harvested 7 and 14 days after resuscitation and subjected to blinded histologic examination. No pathological lesions or significant findings were observed in the samples of brain, heart, liver, lung or kidneys. Rats that received pPolyHb recovered better from hemorrhage stress compared to HES controls, and could turn over and move freely shortly after the operation (unpublished data).

Discussion

For the last 20 years or so, therapeutic development of HBOCs, using hemoglobin from different sources and produced via different methods, has been vigorously pursued.[35] Human and bovine hemoglobin have been the most widely used raw materials for manufacturing of HBOCs, although both have limitations. While human hemoglobin is a good source for HBOC preparation in respect to immune response and other safety considerations, outdated or banked human RBC is nearly unavailable in many countries, and particularly in developing countries. Additionally, the possible threat of human blood-transmitted diseases such as hepatitis and HIV has hindered the use of human Hb as a candidate source for HBOCs in developing countries. As an alternative, some companies have used bovine hemoglobin to prepare HBOCs. However, due to the possibility of cross-species transmission of mad cow disease, bovine Hb is not an optimal source for HBOC manufacture. In light of these considerations, porcine Hb has been added to the portfolio of potential HBOC raw materials.[36] Like bovine hemoglobin, porcine hemoglobin is highly similar in amino acid sequence to human hemoglobin. Specifically, porcine hemoglobin is 85% identical to human Hb,[37] with 22 amino acid substitutions in the α-subunit (141 amino acid residues) and 21 substitutions in β-subunit (146 amino acid residues), whereas bovine hemoglobin is 87% identical with human hemoglobin, with 17 amino acid substitutions in the α-subunit and 23 substitutions in β-subunit. However, unlike bovine hemoglobin, porcine hemoglobin presents no known problems of cross-species transmission of disease.

Although HBOCs, with their capacity for delivering oxygen, could potentially function as red blood cell substitutes or primary resuscitation solutions, unfortunately, progress to date has not fulfilled early expectations because of unanticipated and unwanted adverse clinical effects. Among these adverse effects, vasoactivity and oxidative stress are of the most concern.

Some HBOCs under development have shown a systemic pressor effect and in some cases a pulmonary pressor effect as well.[38] One of the hypotheses for the observed vasopressor effects is that HBOCs may trigger vasoconstriction by scavenging the NO produced by the vascular endothelium. NO is continuously formed in the vascular endothelium and rapidly diffuses to adjacent smooth muscle beds where it mediates vascular relaxation by activation of the heme-dependent enzyme guanylate cyclase. The intercellular junctions of the endothelial lining of the vascular wall allow molecular dimension Hb to enter into the interstitial space, binding and removing nitric oxide needed for maintaining the normal tone of smooth muscles.[2] This results in constriction of the blood vessels and other smooth muscles, which causes a decrease in heart rate, cardiac stroke volume, cardiac output as well as diminished perfusion of vital organ systems. Some HBOCs have been shown to produce dose-dependent vasoconstriction in isolated blood vessels.[39] The large size of pPolyHb may limit its passage into the interstitial space where hemoglobin binds and removes NO needed for maintaining the normal tones of smooth muscles. Also, the low tetrameric hemoglobin content (less than 2%) of pPolyHb ensures that this product does not induce vasoconstriction.

Another hypothesis for the vasoconstriction observed with some HBOCs is that ferryl-Hb, a toxic oxygen intermediate of Hb oxidation, mediates the production of angiotensin II[40] and 8-iso prostaglandin F2α,[41] a vasoconstrictive isoprostane, through the peroxidation of arachidonic acid. Our *in vitro* studies have provided evidence for the antioxidant activity of pPolyHb. No ferrylhemoglobin formation was observed as a result of incubation of pPolyHb and H_2O_2, and therefore, the possibility of vasoconstriction caused by ferryl-Hb mediated formation of 8-iso prostaglandin F2α and angiotensin II with this product is unlikely.

Another significant impediment to progress in the development of HBOCs has been their oxygen binding affinity and cytotoxicity.[42] H_2O_2 can convert ferrous hemoglobin (containing bound Fe^{2+}) into ferric Hb (Fe^{3+}), which cannot deliver O_2 and is toxic to cells.[43] Under some conditions, H_2O_2 even reacts with ferric Hb as well as other heme proteins to produce an even higher oxidation state of the iron (ferryl, Fe^{4+}), which can cause lipid peroxidation, carbohydrate degradation and protein crosslinking.[29,44] An additional oxidative pathway that can contribute to the toxicity of HBOCs is the formation of heme degradation products during the autoxidation of $HbFe^{2+}$, particularly under oxidative stress conditions. The autoxidation of $HbFe^{2+}$ produces $HbFe^{4+}$, and O_2 , ·OH as well in the heme pocket, which results in heme degradation.[45] The low level of circulating heme can then produce pronounced vascular pathologies.[46]

In order to suppress or control the oxidative toxicity induced by Hb, the technique to prepare pPolyHb was designed to crosslink antioxidant enzymes, such as SOD and CAT. Our *in vitro* studies have shown that pPolyHb can significantly inhibit H_2O_2-mediated endothelial cell damage and apoptosis, providing evidence that pPolyHb has a protective function against H_2O_2-induced cell toxicity.

Summary

pPolyHb (glutaraldehyde polymerized porcine hemoglobin) was designed as a new blood substitute that retains certain properties of red blood cells, such as oxygen capacity, volume expansion and antioxidative capability, while avoiding many of the safety liabilities of other HBOCs. Preclinical studies have established the fundamental pharmacological properties of pPolyHb. Exchange transfusion experiments in rats demonstrated that pPolyHb retains hemodynamic stability, exhibits superior volume expansion capability, and effectively reverses anaerobic metabolism caused by extensive blood volume exchange. pPolyHb also maintains a suitable level of oxygen delivery to tissues during transfusion, as measured by lactic acid levels and base excess. It was also found that resuscitation with pPolyHb in the hemorrhagic shock rat model and the perioperative hemorrhagic shock canine model restored baseline levels of lactate acid and base excess, showing the effectiveness of pPolyHb at reversing anaerobic metabolism in clinically relevant shock models. The hemodynamic and blood gas parameters were also restored to baseline, whereas lactic acidosis developed in control animals. All these properties contribute to the effectiveness of pPolyHb in promoting animal survival. Moreover, acute and chronic toxicology evaluations, as well as immunological studies, suggest the safety of pPolyHb administration. Specifically, there were no significant abnormalities found in regard to clinical observations, clinical chemistry, hematology or histopathology. Additionally, no elevation of blood pressure or pro-inflammatory toxicities were observed in rats. Furthermore, *in vitro* studies indicated that pPolyHb can inhibit H_2O_2-mediated endothelial cell damage and apoptosis. Thus, both the pharmacologic and safety profile of pPolyHb indicate the potential of pPolyHb as a new hemoglobin-based oxygen carrier. This product will be tested in future clinical trials.

Acknowledgments

The authors are grateful to the publishers Wiley Blackwell and Informa Healthcare for their permission to use certain figures and content in preparing this chapter. The authors also acknowledge grant support from the following: National High-Tech R&D Program (863 Program) (Program Number: 2012AA021902), National Natural Science Foundation of China (Program Number: 81102367), China Scholarship Council of the Ministry of Education, the "Science and Technology Research and Development Project" of Shaanxi Province (Grant No: 2011KTCL03-23 and 2011K12-03-10), and Xi'an Science and Technology Bureau (Grant No: CXY1131). The authors would like to thank Professor TMS Chang for his constructive suggestions in preparing this manuscript.

References

1. Stollings JL, Oyen LJ. (2006) Oxygen therapeutics: oxygen delivery without blood. *Pharmacotherapy* **26:** 1453–1464.
2. Chang TMS. (2007) *Artificial Cells: Biotechnology, Nanomedicine, Regenerative Medicine, Blood Substitutes, Bioencapsulation, and Cell/Stem Cell Therapy, Vol. 1 of Regenerative Medicine, Artificial Cells and Nanomedicine*. World Scientific, Singapore.
3. Chatterjee R, Welty EV, Walder TY, *et al.* (1986) Isolation and characterization of a new hemoglobin derivative cross-linked between the α chains (Lysine 99α1-99α2). *J Biol Chem* **26:** 9929–9937.
4. Saxena R, Wijnhoud AD, Carton H, *et al.* (1999) Controlled safety study of a hemoglobin-based oxygen carrier, DCLHb, in acute ischemic stroke. *Stroke* **30:** 993–996.
5. Sloan EP, Koenigsberg M, Gens D, *et al.* (1999) Diaspirin cross-linked hemoglobin (DCHLb) in the treatment of severe traumatic hemorrhagic shock. A randomized controlled efficacy trial. *JAMA* **282:** 1857–1864.

6. Schubert A, Przybelski RJ, Eidt JF, *et al.* (2003) Diaspirin crosslinked hemoglobin reduces blood transfusion in noncardiac surgery: a multicenter, randomized, controlled, double-blinded trial. *Anesth Analg* **97:** 323–332.
7. Olson JS, Foley EW, Rogge C, *et al.* (2004) No scavenging and the hypertensive effect of hemoglobin-based blood substitutes. *Free Rad Biol Med* **36:** 685–697.
8. Hayes JK, Stanley TH, Lind GH, *et al.* (2001) A double-blind study to evaluate the safety of recombinant human hemoglobin in surgical patients during general anesthesia. *J Cardiothorac Vasc Anesth* **15:** 593–602.
9. Burhop KE, Doyle M, Shick M, Matthews M. (2003) The development and preclinical testing of a novel second generation recombinant hemoglobin solution. Abstract S-I-6, Ninth International Symposium on Blood Substitutes, Tokyo, Japan, March 3–5.
10. Levy JH, Goodnough LT, Grelich PE, *et al.* (2002) Polymerized bovine hemoglobin solution as a replacement for allogeneic red blood cell transfusion after cardiac surgery: results of a randomized, double-blind trial. *J Thorac Cardiovasc Surg* **124:** 35–42.
11. Sprung J, Kindscher JD, Wahr JA, *et al.* (2002) The use of bovine hemoglobin glutamer-250 (Hemopure) in surgical patients: results of a multicenter, randomized, single-blinded trial. *Anesth Analg* **94:** 799–808.
12. Lamuraglia GM, O'Hara PJ, Baker WH, *et al.* (2000) The reduction of the allogenic transfusion requirement in aortic surgery with a hemoglobin-based solution. *J Vasc Surg* **31:** 299–308.
13. Jahr JS. (2002) A novel blood substitute: use of HBOC-201 (Hemopure) to decrease need for RBC: result of pivotal trial in orthopedic surgery patients. *Crit Care Med* **29:** S243.
14. http://www.prweb.com/releases/2011/7/prweb8630451.htm
15. Gould SA, Moore EE, Hoyt DB, *et al.* (1998) The first randomized trial of human polymerized hemoglobin as a blood substitute in acute trauma and emergent surgery. *J Am Coll Surg* **187:** 113–120.
16. Ernest EM, Frederick AM, Timothy CF, *et al.* (2009) Human polymerized hemoglobin for the treatment of hemorrhagic shock when blood is unavailable: The USA multicenter trial. *J Am Coll Surg* **208:** 1–13.
17. Chen JY, Michelle S, George K. (2009) A review of blood substitutes: examining the history, clinical trial results, and ethics of hemoglobin-based oxygen carriers. *Clinics* **64:** 803–813.
18. Carmichael FJL, Ali ACY, Campbell JA, *et al.* (2000) A phase I study of oxidized raffinose cross-linked human hemoglobin. *Crit Care Med* **28:** 2283–2292.
19. Cheng DCH. (2001) Safety and efficacy of O-raffinose cross-linked human hemoglobin (Hemolink) in cardiac surgery. *Can J Anesth* **48:** S41–S48.
20. Greenburg AG, Kim HW, Hemolink Study Group. (2004) Use of an oxygen therapeutic as an adjunct to intraoperative autologous donation to reduce

transfusion requirements in patients undergoing coronary artery bypass graft surgery. *J Am Coll Surg* **198:** 373–383.
21. http://www.hemosol.com/presspop.cfm?newsID=2063
22. Winslow RM. (2005) Targeted O_2 delivery by low-P_{50} hemoglobin: a new basis for hemoglobin-based oxygen carriers. *Artif Cells Blood Substit Immobil Biotechnol* **33:** 1–12.
23. Bjorkholm M, Fagrell B, Przybelski R, *et al.* (2005) A phase I single blind clinical trial of a new oxygen transport agent (MP4), human hemoglobin modified with maleimide-activated polyethylene glycol. *Haematologica* **90:** 505–515.
24. http://www.sangart.com/press/?pID=1045727680
25. Zhang W, Yan KP, Dai PG, *et al.* (2012) A novel hemoglobin based oxygen carrier (HBOC), porcine polymerized hemoglobin (pPolyHb), inhibits H_2O_2-induced cytotoxicity of endothelial cells. *Artificial Organ* **36:** 151–160.
26. Zhu HL, Dang XD, Yan KP, *et al.* (2011) Pharmacodynamic study of polymerized porcine hemoglobin (pPolyHb) in a rat model of exchange transfusion. *Artificial Cells, Blood Substitutes and Biotechnology* **39:** 119–126.
27. Baldwin AL. (2005) Blood substitutes and redox responses in the microcirculation. *Antioxid Redox Signal* **6:** 1019–1030.
28. D'Agnillo F, Alayash AI. (2000) Interactions of hemoglobin with hydrogen peroxide alters thiol levels and course of endothelial cell death. *Am J Physiol Heart Circ Physiol* **279:** H1880–H1889.
29. Goldman DW, Breyer RJ 3rd, Yeh D, *et al.* (1998) Acellular hemoglobin-mediated oxidative stress toward endothelium: a role for ferryl iron. *Am J Physiol* **275:** H1046–H1053.
30. Zhu HL, Yan KP, Dang XD, *et al.* (2011) Immune safety evaluation of polymerized porcine hemoglobin (pPolyHb) — a potential red blood cell substitute. *Artificial Cells, Blood Substitutes and Biotechnology* **39:** 398–405.
31. Mannick JA, Rodrick ML, Lederer JA. (2001). The immunologic response to injury. *J Am Coll Surg* **193:** 237–244.
32. Rhee P, Wang D, Ruff P, *et al.* (2000). Human neutrophil activation and increased adhesion by various resuscitation fluids. *Crit Care Med* **28:** 74–78.
33. Dong F, Hall CH, Golech SA, *et al.* (2006) Immune effects of resuscitation with HBOC-201, a hemoglobin-based oxygen carrier, in swine with moderately severe hemorrhagic shock from controlled hemorrhage. *Shock* **25:** 50–55.
34. Hall CH, Malkevich N, Handrigan M, *et al.* (2007) Innate immune responses in swine resuscitated from severe traumatic hemorrhagic shock with hemoglobin-based oxygen carrier-201. *Artificial Cells, Blood Substitutes and Biotechnology* **35:** 259–274.
35. Stowell CP. (2005) What happened to blood substitutes? *Transfus Clin Biol* **12:** 374–379.
36. Zhu XL, Chu W, Wang T, *et al.* (2007) Variations in dominant antigen determinants of glutaraldehyde polymerized human, bovine and porcine hemoglobin. *Artif Cells Blood Substit Immobil Biotechnol* **35:** 518–532.

37. Shikama K, Matsuoka A. (2003) Human haemoglobin: a new paradigm for oxygen binding involving two types of ab contacts. *Eur J Biochem* **270:** 4041–4051.
38. Przybelski RJ, Dailey EK, Birnbaum ML. (1997) The pressor effect of hemoglobin — good or bad? In: Winslow RM, Vandegriff KD, Intaglietta M (eds), *Advances in Blood Substitutes: Industrial Opportunities and Medical Challenges.* Birkhäuser, Boston, pp. 71–85.
39. Hart J, Ledvina M, Muldoon S. (1997). Actions of diaspirin crosslinked hemoglobin on isolated rat and dog vessels. *J Lab Clin Med* **129:** 356–363.
40. Simoni J, Simoni G, Moeller JP, Wesson DE. (2007) Evaluation of angiotensin converting enzyme (ACE)-like activity of acellular hemoglobin. *Artif Cells Blood Substit Immobil Biotechnol* **35:** 191–210.
41. Simoni J, Simoni G, Wesson DE, *et al.* (2001) Free hemoglobin (Hb) mediates the synthesis of 8-iso prostaglandin F2a, a vasoconstrictive isoprostane. *Artif Cells Blood Substit Immobil Biotechnol* **29:** 162.
42. Winslow RM. (2006) Current status of oxygen carriers ('blood substitutes'): 2006. *Vox Sanguinis* **91:** 102–110.
43. Giulivi C, Cadenas E. (1998) Heme protein radicals: formation, fate, and biological consequences. *Free Radic Biol Med* **24:** 269–279.
44. Giulivi C, Davies KJA. (1994) Hydrogen peroxide-mediated ferrylhemoglobin generation *in vitro* and in red blood cells. *Methods Enzymol* **231:** 490–496.
45. Nagababu E, Rifkind JM. (2000) Reaction of hydrogen peroxide with ferrylhemoglobin: superoxide production and heme degradation. *Biochemistry* **39:** 12503–12511.
46. Jeney V, Balla J, Yachie A, *et al.* (2002) Pro-oxidant and cytotoxic effects of circulating heme. *Blood* **100:** 879–887.

Chapter 4

Evidence Regarding Some Pharmacologic Characteristics of Hemoglobin-Based Oxygen Carriers

Enrico Bucci

Abstract

We have a library of chemically treated and recombinant HBOCs candidate for clinical use. The problem is how and when HBOC solutions could be beneficial, non-toxic or, at least, with manageable toxicity. Investigations of microcirculatory effects of HBOCs confirm very good peripheral oxygen delivery by all of the tested HBOCs. Main toxic events appear to be extravasation and hyperoxygenation. Brain and heart infarcts protection by carbon monoxide infused with carbonmonoxy HBOCs is under scrutiny.

Introduction

We have a library of chemically treated and recombinant HBOC candidates for clinical use. Most promising are HBOCs either intramolecular

---

Department of Biochemistry and Molecular Biology, University of Maryland Medical School, Baltimore MD 21201.

crosslinked, or polymerized with either glutaraldehyde or Z-link procedures. Also promising are HBOCs either decorated with polyoxiethylene glycol or obtained with recombinant procedures.[1–3]

The meta-analysis of Natanson *et al.*[4] regarding the use of glutaraldehide polymerized HBOCs for resuscitation purposes indicates that none of the new procedures was more efficient than classical procedures. Notably there were common severe adverse events, myocardial ischemias and infarctions.

Interpretation as to the cause of these events was very difficult. After discussing the possible relevance of protocols, of fluid composition, of the oxygen affinities of blood and of the carriers, and other minor details, the causes remained undetermined.

Oxygen toxicity was not considered.

This paper presents a brief survey of a vast literature, focusing on data considered very relevant for development and use of HBOCs, namely toxicity linked to hyperoxygenation and beneficial effects of CO infusions (not inhaled).

Oxygen Distribution *In Vivo* by Red Cells and HBOCs

Oxygen is transported in plasma in two ways. Gaseous oxygen in the lungs dissolves in plasma as free oxygen with PO_2 about 120 mmHg. It has a low water solubility, thus its concentration in plasma remains in the range of 10^{-5} M. Red cells absorb oxygen on their hemoglobin content. The amount absorbed is about 98% of total hemoglobin saturation.

Normal subjects metabolism requires oxygen at the rate of about 1 L min^{-1}. As described in previous papers,[5] at the capillaries this amount is matched by oxygen released from the red cells, which replaces oxygen consumed by metabolism. The release also buffers the partial pressure of oxygen (PO_2) to near 30 mmHg, i.e. at the P_{50} of the red cells. Thus, with an hematocrit of 40–50%, a large quantity of free oxygen is released to the system at low PO_2.

After the failure of the trials reported by Natanson *et al.*,[4] research is now focused on the polymeric form of hemoglobin obtained either by chemical treatments[1] or recombinant procedures.[3] Various forms of PEG decorated hemoglobins are still being developed.[2] These treatments allow the formation of large polymeric molecules with NMW between 500 kDA and 25 mDA, which do not extravasate. PEG decorated HBOCs also do not extravasate. These HBOCs do not increase MAP after infusion, because there is no scavenging of NO from the vessels muscles. Polymers have much higher oxygen affinities than red cells.[1,3] PEG decorated Hbs also have higher affinity than the red cells.[2] It was shown that high affinity does not prevent oxygen delivery to tissues, probably because they also facilitate oxygen diffusion.[6] Facilitated oxygen diffusion is due to the Brownian motion of cell free HBOCs in plasma, which add to their bound oxygen a virtual increased diffusion rate.[7]

While oxygen transport and delivery by red cells, as briefly described above, is under strict physiologic control, and succeeds in offering to metabolism large amounts of oxygen at minimal oxygen tension, facilitated diffusion on the other hand makes oxygen delivery by HBOCs not under physiologic control, regulated only by the oxygen binding characteristics of the carrier. Thus delivery may exceed oxygen consumption increasing the risk hyperoxygenation and oxygen radicals formation, as reported below.

Oxygen Toxicity

Due to formation of oxygen radicals, oxygen is a very toxic gas, still indispensable for metabolism. These radicals are oxygen molecules with single unpaired electrons, the most important of which are superoxide (O_2^-) and the hydroxyl radical ($HO^\bullet$). Hydrogen peroxide (H_2O_2) is an intermediate between the two, when O_2^- reacts with water (H_2O). They are very strong oxidative molecules which react at random with tissue components including membrane phospholipids, thiol proteins, glutathione, ascorbate. Blocking aconitase, O_2^- interferes with the Krebs cycle. In combination with Fe^{3+} O_2^- oxidizes DNA molecules producing mutations (damaging mutations are rapidly eliminated, favorable ones are retained and transmitted to progeny — Darwinian evolution?). Only 0.1% of metabolized oxygen is transformed in oxygen radicals, but they are devastating. Senescence may be in part due to the prolonged toxic actions of oxygen radicals.[6,7]

Oxygen radicals are mostly a byproduct of mitochondrial respiration. Physiology is well equipped for neutralizing oxygen radicals. Catalase and peroxydase enzymes are available in solution and membrane bound. Most important is superoxide dismutase, which reduces the lifetime of O_2^- by a factor of 10^9. There are antioxidant reagents like glutathione, thioredoxin, ascorbic acid. Also, the physiology of oxygen transport and delivery, as briefly described above and better discussed in other papers,[5] succeeds in offering to metabolism large amounts of oxygen at minimal oxygen tension. In fact, low oxygen tension decreases the risk of formation of oxygen radicals.

These activities reduce to a minimum but do not completely eliminate oxygen radicals, which stimulate protein synthesis, DNA replication and kill bacteria.[8,9]

Thus, an equilibrium is established between production and reduction of radicals. Excessive amounts of oxygen alter the equilibrium in favor of radicals formation, producing a very toxic oxygen stress (OS).[9]

Hyperoxygenation

Hyperoxygenation is a major cause of OS conditions. The extra oxygen introduced in the blood stream by HBOCs and oxygen-facilitated diffusion increase the risk of hyperoxygenation.

Besides the natural defenses against oxygen radicals, discussed above, the vascular system responds to hyperoxygenation with a vasoconstriction, which reduces blood flow and consequently oxygen supply. It should be stressed that this vasoconstriction is under physiologic control, is local and tends to produce normoxia. This is at variance with the vasoconstriction produced by NO scavenging by hemoglobin extravasation, which is systemic, eludes physiology and results in anoxia.

Experimentally, hyperoxygenation is detected as a vasoconstriction not dependent on NO scavenging. Matheson *et al.*[1] have shown that infusion of a non-extravasating polymer (not detectable in the lymph) produced vasoconstriction, not dependent on NO scavenging; in fact it persisted when NO production was prevented by inhibition of NO synthase. Even the vasoconstriction produced by extravasating HBOCs revealed a hyperoxygenation component when NO production was inhibited.[10]

It should be mentioned that those experimental animals were bled-transfused using HBOC fluids as blood replacement; half of the bled hemoglobin was replaced. No calibration was attempted for monitoring the dependence of vasoconstriction on the amount of infused HBOCs.

These experiments suggest that HBOCs are oxygen-delivery drugs, to be used carefully in order to avoid hyperoxygenation. It should also be stressed that hyperoxygenation is particularly dangerous for coronary systems, where the immediate proximity to heart allows an intravascular PO_2 near and above 100 mmHg, rather than the milder peripheral 30 mmHg, adding to the risk of radical formation. The resulting OS would damage endothelial cells resulting in exaggerated vasoconstriction and ischemia. Extravasation would worsen the damage.

Whether OS is at the basis of the common adverse effects found in the clinical trials reported by Natanson *et al.*,[4] where HBOCs were profusely used as blood replacement, remains to be considered.

Effects of Hemoglobin-Carbon Monoxide

There are numerous reports that inhaled CO has beneficial effects in a variety of clinical events involving transplants, post-hemorrhagic resuscitation, heart diseases, reperfusion injuries of lungs and myocardial ischemia.[11] Usage of this technology is limited by the toxic effect of CO when inhaled.

While gaseous CO is toxic, Hb-bound CO has no toxicity. This is evident in resuscitation experiments performed with either bloods with the red cells saturated with CO[12] or suspension of vesicles encapsulating carbonmonoxy-Hb.[13] No adverse toxic events were present in treated animals. Instead it was noted that liver and lungs were less damaged than after resuscitation with oxygenated vesicles suspensions.

Endogenous CO produced by degradation of heme by hemoxigenase HO-1 and HO-2 is known to regulate many physiological processes involving nervous, cardiovascular, muscle and skin systems, apparently inhibiting toxic effects by oxygen radicals.[14]

Chemicals have been devised which bind and release CO both *in vitro* and *in vivo*. These CO release molecules (CORM-) are based on metals like Mn and Ru, or on borocarboxilate. CORM-3(Manganese) and CORM-A1(Boron), are water soluble. Infused in mice (less than 0.1 mg/mouse), they protected the heart from ischemia and infarcts. Hemoglobin inside red cells was not involved; no COHb was detected.[15,16]

Recent reports indicate that infusion of CO-saturated PEGHb protects heart[17] and brain from infarcts in the rat.[18] There was a reduction close to 80% of the size of brain infarcts when reperfusion was performed after a 10 mL top load infusion of a 4% solution of CO-PEGHb.[18] Reperfusion of ischemic heart with 4% solutions of CO-PEGHb reduced infarcts size by about 50%; the oxy derivative did not provide cardioprotection.[17]

The brain and heart protection against ischemic infarcts by carboxygenated HBOCs is in line with the protection provided by CO-RMS. Thus, the beneficial effects are in all probability produced by the CO introduced into the systems with CO-HBOCs.

In this scenario, this writer misses experiments of resuscitations and/or infarcts protection with either saline or albumin solutions saturated with CO.

Infused CO-RMS and CO-HBOCs find in plasma a zero PCO. Thus the release of carried CO is almost immediate, and free CO rapidly diffuses within and outside the vessels. The small concentrations of oxygen radicals and diffused CO in blood and tissues make reciprocal interferences impossible. Thus the hypothesis is formulated that there is a CO-receptor, which signals to a cascade of favorable events.

The variety of tissues where the protective action of CO is produced suggests that there is a common ubiquitous receptor for the CO moiety. Mitochondria are ubiquitous and are the natural focus of present research. Detailed and competent discussions of this problem are presented in papers by Bilban *et al.*[19] and Motterlini.[20] At present, the receptor of CO which signals for its effects remains undetermined.

Therapeutic stimulation of endogenous CO is limited by the biochemistry and physiology of heme oxygenases. Therapeutic use of CO-RMS is very promising although still in its infancy. *The effect of CO-HBOCs on ischemic infarcts may suggest a simple means for adding the protective action of CO to a variety of clinical situations.*

References

1. Matheson B, Kwansa HE, Bucci E, *et al.* (2002) Vascular response to infusions of a nonextravasating hemoglobin polymer. *J Appl Physiol* **93:** 1479–1486.
2. Vandegriff KD, Malavalli A, Wooldridge J, *et al.* (2003). MP4, a new nonvasoactive PEG-Hb conjugate. *Transfusion* **43:** 509–516.
3. Fronticelli C, Koehler C. (2009) Design of recombinant hemoglobins for use in transfusion fluids. *Crit Care Clin* **25:** 357–371.
4. Natanson C, Kern SJ, Lurie P, *et al.* (1998) Cell free hemoglobin-based blood substitutes and risk of myocardial infarctions and death. *JAMA* **299:** 2304–2312.
5. Bucci E. (2011) Basic science offers a challenge for developing hemoglobin based oxygen carriers into therapeutic agent. *Artificial Cells, Blood Substitutes, and Biotechnology* **39:** 206–213.
6. Koehler RC, Fronticelli C, Bucci E. (2008) Independence of oxygen transport on oxygen affinity of hemoglobin-based oxygen carriers. *Biochem Biophys Acta* **1784:** 1387–1394.
7. Wittenberg JW. (1966) The molecular mechanism of hemoglobin-facilitated oxygen diffusion. *J Biol Chem* **241:** 104–114.
8. Fridovich I. (1998) Oxygen toxicity: a radical explanation. *J Exp Biol* **201:** 1203–1209.
9. Lane N. (2009) *Oxygen.* Oxford University Press, Oxford NY.
10. Rebel A, Ulatowski JA, Kwansa H, *et al.* (2003) Cerebrovascular response to decreased hematocrit: effect of cell-free hemoglobin, plasma viscosity, and CO_2. *Am J Physiol Heart Circ Physiol* **285:** H1600–H1608.
11. Zuckerbraun BS, McCloskey CA, Gallo D, *et al.* (2005) Carbon monoxide prevents multiple organ injury in a model of hemorrhagic shock and resuscitation. *Shock* **23:** 527–532.
12. Cabrales P, Tsai AG, Intaglietta M. (2007) Hemorrhagic shock resuscitation with carbon monoxide saturated blood. *Resuscitation* **71:** 306–318.

13. Sakai H, Hirohisa H, Tsuchida E, Kobayashi K. (2009) Hemoglobin vesicles and red blood cells as carriers of carbon monoxide prior to oxygen for resuscitation after hemorrhagic shock in a rat model. *Shock* **31:** 507–514.
14. Slebos DJ, Stefan W, Ryter M, *et al.* (2003) Heme oxygenase-1 and carbon monoxide in pulmonary medicine. *Resp Res* **4:** 7.
15. Clark JE, Naughton P, Shurey S, *et al.* (2003) Cardioprotective actions by a water-soluble carbon monoxide releasing molecule. *Circ Res* **93:** e2–e8.
16. Guo Y, Adam B, Stein A, *et al.* (2004) Administration of a CO-releasing molecule at the time of reperfusion reduces infarct size *in vivo. AJP — Heart* **286:** H1649–H1653.
17. Vandegriff KD, Young V, Lohman J, *et al.* (2008) CO-MP4, a polyethylene glycol-conjugated haemoglobin derivative and carbon monoxide carrier that reduces myocardial infarct size in rats. *Br J Pharmacol* **154:** 1649–1661.
18. Klaus JA, Kibler KK, Abuchowski A, Koehelr RC. (2010) Early treatment of transient focal cerebral ischemia with bovine pegylated carboxy hemoglobin transfusion. *Artificial Cells, Blood Substitutes, and Biotechnology* **38:** 223–229.
19. Bilban M, Haschemi A, Wegiel B, *et al.* (2008) Heme oxygenase and carbon monoxide initiate homeostatic signaling. *J Mol Med* **86:** 267–279.
20. Motterlini R, Sawle P, *et al.* (2005) CORM-A1: a new pharmacological active carbon monoxide releasing molecule. *FASEB J* **19:** 284–286.

Chapter 5

EAF PEG Hemoglobins: Novel Class of Nonhypertensive Resuscitation Fluids: Simplicity and Advantages of Extension Arm Chemistry for PEGylation

Seetharama A. Acharya*, Amy G. Tsai† and Marcos Intaglietta†

Introduction

Chemically modified derivatives of hemoglobin (Hb), particularly intramolecularly crosslinked Hb, have been developed as potential blood substitutes with the aim of overcoming dissociation of Hb into dimers and to have oxygen affinity in the same range as that of Hb inside red blood cells (RBCs). Most of these chemical modifications neutralize nephrotoxicity of acellular Hbs by minimizing glomerular filtration of αβ-dimers of Hb, however, almost all products cause hypertension.[1] This common feature of almost all of the derivatives of Hb was considered the major obstacle in the realization of the use of Hb as blood substitutes. The molecular basis of the hypertensive activity of Hb and its derivatives is attributed to the intrinsic nitric oxide (NO) scavenging activity of Hb. Accordingly the current focus

* Departments of Medicine and of Physiology & Biophysics, Albert Einstein College of Medicine, Bronx, NY.
†Department of Bioengineering, University of California, San Diego, CA.

in developing Hb-based blood substitutes is to generate mutant Hbs through site-directed mutagenesis that exhibit reduced NO scavenging activity and with oxygen affinity comparable to that of Hb in RBCs.[1] Oligomerization of Hb and mutant Hbs are additional design strategies for producing nonhypertensive Hbs.[1] Though significant success has been achieved in realizing these goals, the realization of the goal of using Hb derivatives has not been yet accomplished.

High Oxygen Affinity Hb Derivatives are Vasoinactive

The high oxygen affinity derivatives of Hb are non-hypertensive, leading Winslow and his colleagues to propose that preferential unloading of oxygen by acellular Hb on the arterial side of circulation activates autoregulatory mechanisms leading to the vasoconstrictive activity of Hb. Accordingly they have advanced the concept that all Hb derivatives designed earlier to endow an oxygen affinity to Hb comparable to that of RBCs are destined to fail as blood substitutes. The high oxygen affinity was proposed as the uniquely required Hb property for developing blood substitutes.[2]

High Oxygen Affinity PEGylated Hbs are also Vasoinactive

DecaPEGylated bovine Hb, with 10 copies of 5 KDa polyethylene glycol (PEG) chains conjugated to uncrosslinked bovine Hb through a urethane linkage (P5K10 Bv Hb), was developed by Enzon as a potential blood substitue. The extensive studies of Intaglietta and his colleagues and Winslow and his colleagues have identified this PEG-Hb as representing a new class of Hb derivatives that are vasoinactive *in vivo* when used to formulate blood substitutes. As this decaPEGylation of Bv Hb generates species of PEG-Hbs with high oxygen affinity, the absence of vasoactivity with P5K10 Bv Hb is consistent with the activation of the autoregulatory principles by the high oxygen affinity species of Hb advanced by Winslow and his colleagues.[2]

P5K10 Bv Hb is also distinct from other Hb derivatives since a 4 gm% solution has higher viscosity and colloidal osmotic pressure than other Hb derivatives previously designed as blood substitutes. Additionally it also exhibits a very large hydrodynamic volume, with a molecular radius of approximately 15 nm. These are the solution properties of plasma expanders. Accordingly Intaglietta and his team have advanced the concept of engineering plasma expander like properties to Hb by PEGylation; this process is coupled to an increase in the oxygen affinity of bovine Hb. Accordingly this is consistent with both concepts of inducing plasma expander like properties to Hb as well as designing Hb derivative a high oxygen affinity species to overcome the vasoconstrictive activity of acellular Hb.[3]

PEGylated Hb of Ajinomoto is a pyridoxylated and PEGylated Hb.[4] This PEG-Hb carries 10 copies of PEG 3.4 K; some of the PEG chains are involved in intra and/or inter molecular crosslinks, but unlike other products discussed above, it exhibits a low oxygen affinity. However this PEG-Hb exhibited only transient vasoinactivity due to its low oxygen affinity by comparison with PEGylated bovine Hb. This suggests that the optimum total mass of PEG to be conjugated to Hb to generate a nonhypertensive molecule may be at least a total mass of 50 K (10 copies of PEG 5K) to ensure the absence of hypertensive responses.

Extension Arm Facilitated (EAF) PEGylation of Hb

A new PEGylation platform has been developed at Albert Einstein College of Medicine and is referred to as Extension Arm facilitated PEGylation.[5,6] This approach has built in it some of the elements of click chemistry, and is a very flexible modular protocol of simple chemistry and with excellent yields of reaction products. The conceptualization of Extension Arm chemistry facilitated PEGylation of proteins is represented pictorially in Fig. 1. The approach first introduces

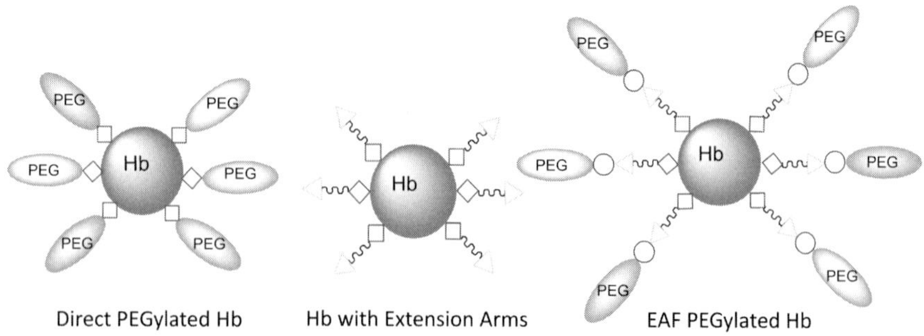

Fig. 1. Pictorial comparison of directly PEGylated vs. Extension Arm Facilitated PEGylated Hb: In directly PEGylated Hb, the functional groups of the protein side chains on the surface form covalent linkage with the functional group of the PEG chains. Accordingly the PEG chains are very close to the surface of the protein and are likely to be involved in extensive noncovalent interactions with the protein surface. EAF PEGylation is, in principle, a two-step process. In the first step, protein (Hb) is modified monofunctionally using a heterobifunctional reagent. The second functional group of the bifunctional reagent (the targeted functional group for the modification by a PEG reagent with complimentary functional) is either protected and released on completion of the reaction of the first functional group with the protein or generated *in situ* on reaction of the functional group targeted for the protein modification. In the second stage, the protein modified with extension arms carrying new function group targeted for modification with PEG reagent is subjected to PEGylation. The functional group at the distal end of the extension arms and the complementary functional chemistry for the PEG reagent can be appropriately designed to make the PEGylation reaction nearly quantitative. The maleimide reaction that exhibits the highest site selectivity has been chosen for the initial studies. The extension arms increase the accessibility of the functional groups to the macromolecular PEG reagents further facilitating the quantitative aspects of the PEGylation reaction.

short aliphatic chains (~1 nm is length, this could be any desired length) on the surface functional groups protein and that carries a reactive functional group at the distal end, and these new functional groups are the targeted sites to conjugate fictionalized PEG reagents. EAF PEGylation was designed at Einstein to have a better control on the PEG conjugation reaction and minimize the influence of PEG shell on the protein core. The EAF PEGylated protein will hence have a central protein core of high packing density and an outer PEG shell of lower packing density, and sandwiched in between is a zone of extension arms. The extension arms are located in the protein hydration layer and facilitate placing the PEG chains into the bulk water (or solution) phase, thereby decreasing the potential interactions between the protein surface and PEG chains on the one hand and attenuating the perturbation of the protein hydration layer generally associated with direct PEGylation on the other.

The initial EAF PEGylation protocol for protein was developed using the thiol maleimide chemistry for targeting the PEGylation reaction.[5] A unique hetero bifunctional reagent 2-iminothiolane that generates the second functional group of the reagent, in this case a thiol, *in situ* on reaction of the reagent at the amino groups of the protein has been used as the molecule to introduce a thiol at the distal end of the extension arm. This is presented schematically in Fig. 2. The thiol group is the targeted site for conjugation of PEG chains using maleimide PEG. The thiol maleimide reaction is one of the most quantitative reactions in protein chemistry; hence this reaction can be optimized for maximum efficacy in terms of using PEG reagents. Maleimido phenyl urethane of PEG 5K, which is generated in a modular synthesis involving the reaction p-maleimido phenyl isocyanate with the hydroxyl function of monomethoxy PEG 5K, was used as the functionalized PEG to conjugate PEG to the thiol group of the extension arm.[5]

Thus, new EAF PEGylation chemistry developed here introduces a spacer arm (small carbon chain), (δ-thio succinimido phenyl) butirimidyl chain, between the protein and the urethane linkage present in the Enzon PEG Hb, generated using the direct PEGylation protocol and using succinimidyl carbonate chemistry (Fig. 3). The spacer arm between the protein and the PEG chain is an integrated unit formed by condensation of the extension arm on the protein with the liker chemistry (PEG functionalization chemistry). EAF hexa PEGylated Hb, $[\delta\text{-(thiosuccinimido phenyl)butirimidyl PEG-5K}]_6$ Hb prepared using this protocol, was found to be vasoinactive just as the PEGylated bovine Hb of Enzon even though it carries only six copies of PEG K chains. The total molecular mass of PEG conjugated to Hb in this EAF PEGylated Hb was only 30 K, and is significantly

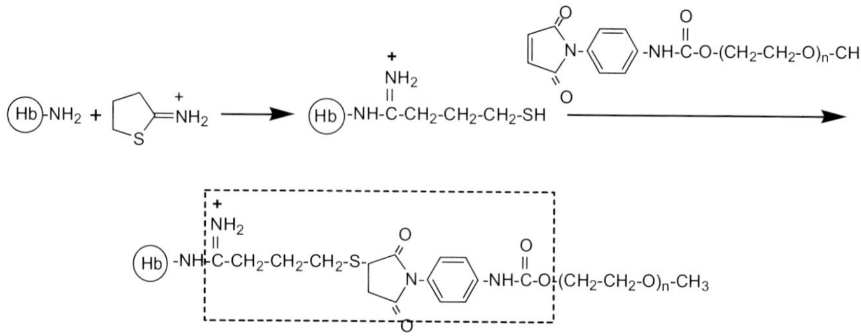

Thiosuccinimido phenyl linkage

Fig. 2. Schematic Representation of the thiol maleimide chemistry based EAF PEGylation of Hb. 2-iminothiolane (2-IT); the cyclic form of the heterobifunctional reagent, δ-mercapto butirimidate is used for the thiolation of Hb. On the reaction of 2-IT with the amino groups of Hb, an amidine linkage is formed between the bifunctional reagent and the amino groups Lys residues with the release of thiol group at the distal end of the extension arm. Since the derivatization the ε-amino group of Lys as a substituted amidine linkage conserves the positive charge of the amino group under the physiological conditions, this approach is referred to as conservative EAF PEGylation. The thiol group at the distal end of the extension arm is the targeted site for the modification by maleimido phenyl urethane of PEG. By controlling the concentration of 2-IT, the number of the extension arms introduced on the protein and hence the numbers of PEG-chains conjugated onto the protein could be controlled. Besides, since with 2-IT the thiol groups are generated only after the reaction of the reagent with the protein, the protein could be incubated with 2-IT and the PEG reagent simultaneously without the danger of the reagent consuming the maleimide PEG reagent.

lower than the 50 K mass of the PEG conjugated in the Enzon bovine PEG. In this EAF hexaPEGylated Hb, the two Cys-93(β) of Hb are PEGylated at the thiol group of Cys-93(β) without the extension arm by maleimido phenyl urethane of PEG-5K, and accordingly this PEG-Hb is a high oxygen affinity derivative of Hb. Consistent with this, the oxygen affinity of EAF hexaPEGylated Hb is slightly higher than that of Enzon PEGylated bovine Hb.[7] The relative roles PEGylation of Cys-93(β) of Hb in EAF PEG Hb and the absence of PEGylation on the corresponding Cys of bovine Hb in PEGylated

A

B

C

D

E

F

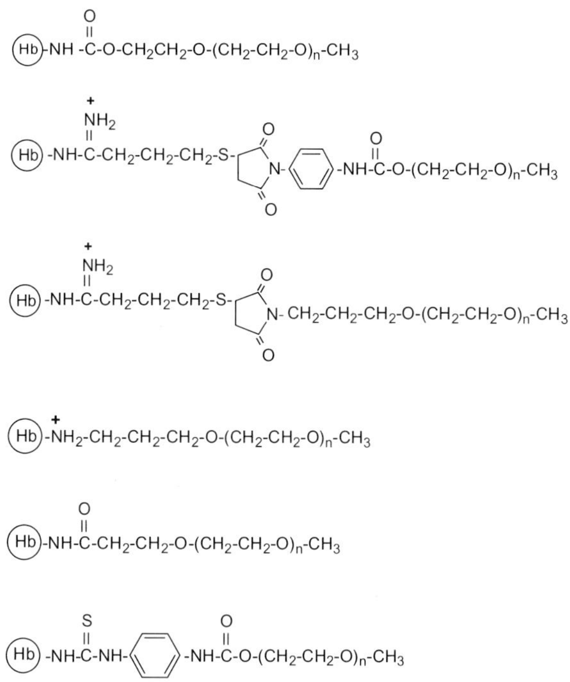

Fig. 3. Comparison of the chemical linkage between the protein core and the PEG chains of PEG shell in different PEGylation protocols used for PEGylation of Hb. The chemistry between the amino group of the protein core and the PEG chains is different in different chemistries used for PEGylation. The shortest linkage is in the Enzon PEGylated bovine Hb, just a urethane linkage as represented in "A." The distance between the PEG chain and the protein core is the maximum in the EAF P5K6 Hb of Einstein; maleimido phenyl urethane of PEG is used in this preparation. The structure of the extension arm and linker chemistry is shown in "B." This structure is distinct from that of Enzon bovine PEG Hb in that (δ-succinimido phenyl urethane) butirimidate replaces the urethane linkage between PEG and protein. Sangart's MP4 is also a hexaPEGylated Hb generated using extension arm chemistry. In making this product, maleimido propyl PEG 5K is used rather than maleimido phenyl urethane of PEG 5K. Accordingly the phenyl urethane linker chemistry (of functionalization of PEG) is replaced by propyl linkage. This structure is represented in "C." Figures D, E and F depict the amino propyl (reductive alkylation chemistry), propionamide (acylation chemistry) and thiourethane phenyl urethane (isothiocyanate chemistry) linkages between the protein amino groups and the PEG chains in some of the direct PEGylation platforms used for the PEGylation of Hb.

bovine Hb play in dictating the oxygen affinity and the *in vivo* vaso-activity of the respective PEG Hbs are not readily apparent.

The total mass of PEG conjugated to EAF PEGylated Hb is comparable to that in the Ajinomoto PEG Hb, which carries 10 copies of PEG 3.4K.[4] However the latter is a low oxygen affinity species.[4] Thus the relative roles of the oxygen affinity of the PEG Hb, the total mass of PEG (size of the PEG chains *x* number) and the pattern of PEGylation have to be delineated to fully understand the molecular aspects of PEGylation induced structure of Hb that attenuates the vasoconstrictive activity of acellular Hb.

EAF hexaPEGylation of Hb Attenuates the PEGylation Induced Weakening of the Quaternary (Inter Dimeric) Interactions of Hb

Enzon PEGylated bovine Hb is the first derivative of uncrosslinked Hb to be advanced as a blood substitute and a molecule that did not filter through the kidney. Either the PEGylation has stabilized the inter dimeric interactions or the increase in the molecular size of the PEGylated αβ dimer does not allow filtration through the kidney. Our studies established that direct PEGylation of Hb does indeed induce extensive weakening of interdimeric interactions into uncrosslinked Hb, accordingly the absence of the nephrotoxicity is apparently a consequence of the enhanced hydrodynamic volume of the PEGylated dimers.

In an extensive study we have established that reductive hexaPEGylation of Hb using PEG 5K propionaldehyde results in the formation of PEGylated Hb dimers. The influence of direct hexaPEGylation of uncrosslinked Hb on its quaternary structure has been compared with that of intramolecularly crosslinked Hb (αα-fumaryl Hb, intramolecular crosslinking prevents the dissociation into αβ dimers on PEGylation) using size exclusion chromatography. These studies determined that in the absence of tetramer stabilizing intramolecular crosslinking, all direct PEGylation platforms (reductive PEGylation, acylation chemistry based, urethane chemistry based PEGylation and thiocarbmoylation chemistry based PEGylation) weaken the inter dimeric interactions of hexaPEGylated Hb.[7–13] A similar comparison of EAF hexaPEGylation of uncrosslinked Hb with that of intramolecularly crosslinked Hb has clearly established that the engineering the extension arms attenuates the influence of the PEG shell on the interdimeric interactions, and EAF hexaPEGylated Hb is present under the physiological conditions exclusively as PEGylated tetramer. The extension arms apparently reduce the interactions of PEG chains with the protein surface (Fig. 4) and attenuate the perturbation of the protein hydration layer. Accordingly, we propose that the extension arms in EAF PEG-Hb act as the shock absorbers of the protein core

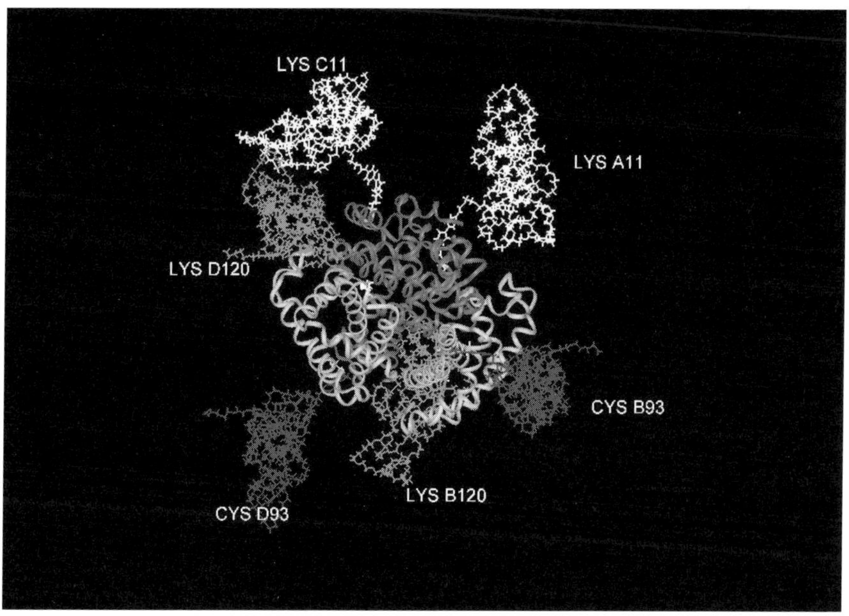

Fig. 4. Molecular model of EAF P5K6-Hb. The thiol group of Cys-93(β) and the ε-amino group of Lys-11(α) and Lys-120(β) have been used as the choice sites of PEGylation for the molecular model building. The modification on Cys-93(β) represents a direct PEGylation, and the reaction at the Lys residues is through extension arm chemistry. The α chains are shown in red, and the β-chains are shown in green. It may be seen that there is very little interaction between either the PEG chains or between the PEG chains and the protein surface.

to attenuate the perturbation of the structure/conformation of the protein core by the PEG-shell.[14]

Thus, it is concluded that EAF hexaPEGylation Hbs results in little or no perturbation of the quaternary structure of the protein and accordingly EAF hexaPEGylated Hb conserves its tetrameric structure.[6,7] On the contrary, direct hexaPEGylation induces a considerable weakening of the inter dimeric interactions within the quaternary the structure of Hb, and this results in the formation of significant amount of the PEGylated αβ-dimers under physiological conditions.

The molecular radius of nonhypertensive EAF hexaPEGylated Hb determined by dynamic light scattering is 6.0–6.2 nm, which is nearly double that of the molecular radius of uncrosslinked Hb (3 nm). The molecular radius of EAF hexaPEGylated αα-fumaryl Hb is also around 6 nm. This confirms the role of EAs sandwiched between the PEG chains and the protein amino groups in attenuating the influence of PEG-shell on the inter dimeric interactions of the tetramer. On the other hand, the molecular radius of decaPEG 5 K urethane of bovine Hb calculated from its COP (the colligative property of the solution of PEG Hb) is around 15 nm. The molecular radius of the same material by dynamic light scattering is only around 5.4 nm, i.e. it is even smaller than the EAF hexaPEGylated Hb even though the former molecule carries 10 copies of PEG 5K chains. Since this is a PEGylated bovine Hb with 10 copies of PEG 5K chains conjugated by direct PEGylation (urethane chemistry) and is smaller than EAF P5K6 Hb, these results show the advantage of Extension Arm Chemistry for the PEGylation of Hb on the molecular dimensions of the resulting PEGylated Hb.[14]

At the stage of EAF hexaPEGylation of Hb generated using maleimido phenyl urethane of PEG 5K, very slight but nonetheless noticeable weakening of the inter dimeric interactions is seen when the Cys-93(β) is PEGylated (Fig. 5). But if the EAF hexaPEGylation is targeted exclusively to the ε-amino groups of Lys residues of Hb, the dissociation constant of EAF hexaPEGylated Hb is indistinguishable from the value of the unmodified Hb (Fig. 4). Accordingly, we conclude that the weakening of the interdimeric interactions Hb

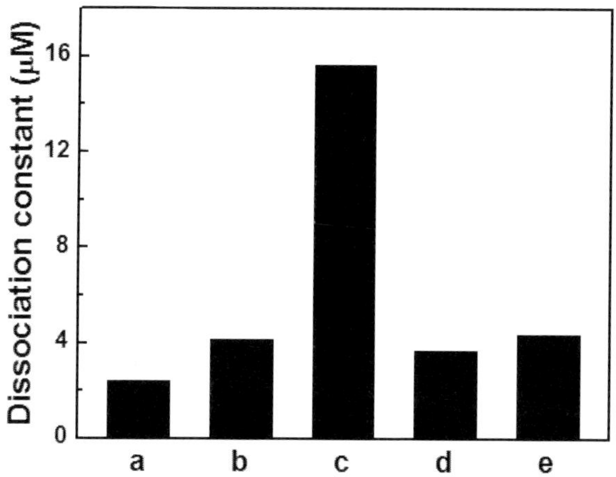

Fig. 5. Influence of PEGylation on Cys-93(β) in the EAF P5K6-Hb on the tetramer–dimer dissociation of EAF PEGylated Hb. Human Hb was subjected to EAF hexaPEGylation with and without reversible blocking of Cys-93(β). To generate control products without PEG chains, the same studies were carried out using N-ethyl maleimide instead of maleimido phenyl urethane of PEG 5K. (a) Control human Hb; (b) EAF PEG Hb generated with reversible blocking of Cys-93(β), in the final PEGylated Hb the thiol of Cys-93(β) is free; (c) EAF PEGylation of Hb without reversible blocking of Cys-93(β) of Hb, the final PEGylated Hb the thiol of Cys-93(β) is directly PEGylated; (d) NEM control for (b); and (e) NEM control for (c). The stability of the tetrameric structure of Hb is not influenced when the EAF PEGylation is targeted exclusively to Lys residues by reversibly blocking the Cys-93(β) during EAF PEGylation. The EAF NEM modification of Hb has no influence on the tetameric structure of Hb, with or without the modification of Cys-93(β). Thus EAF hexaPEGylation of Hb without weakening the inter dimeric interactions of Hb can be achieved by avoiding the modification of Cys-93(β) during EAF PEGylation.

noted by Mozzarelli[16] and his colleagues when EAF PEGylation is carried out in oxy conditions as compared to the hexa/hepta PEGylated Hb generated under deoxy conditions is related to the direct PEGylation of Hb at Cys-93(β) by maleimide PEG during EAF PEGylation of Hb under oxy conditions.[6]

An important question is to establish whether the neutralization/attenuation of the influence of PEG shell on the inter dimeric interactions of Hb discussed above is an absolute function of the presence of extension arms or if it is simply a function of the extent of

Table 1. Dissociation Constant of EAF-PEG Hb as s Function of Number of PEG Chains

Sample	Dissociation Constant (μM)
Hb	4.0
SP-P5K2-Hb	12.5
SP-P5K4-Hb	13.5
SP-P5K6-Hb	23.2
SP-P5K8-Hb	2019.0
SP-P5K10-Hb	2078.0

EAF PEGylation. The dissociation constant of EAF PEGylated Hb has been investigated as a function of number of PEG 5K chains conjugated to Hb through extension arm chemistry (Table 1). Increasing the extent of EAF PEGylation weakens the inter dimeric interactions. As the EAF PEGylation using maleimide PEG 5 K is increased beyond the stage of hexaPEGylation to octa, deca and dodeca stages, the weakening of the inter dimeric interactions increased as a function of increase in the number of PEG chains conjugated; the dodeca EAF PEGylated Hb appears to be essentially PEGylated dimers.

Influence of EAF PEGylation on the Oxygen Affinity of Hb

EAF hexaPEGylated Hb is a high oxygen affinity Hb species. This may be either a consequence of the direct PEGylation of Cys-93(β) during EAF PEGylation or a PEG shell induced effect on the oxy to deoxy conformational transition of the molecule. The oxygen affinity of the diPEGylated Hb with the PEG chains exclusively on thiol group of Cys-93(β) is also a high oxygen affinity species with an oxygen affinity comparable to that of the EAF hexaPEGylated Hb. Besides, the EAF hexaPEGylated Hb generated with a reversible protection of Cys-93(β) of Hb, where all the PEGylated exclusively to the ε-amino groups of Lys resides is also high an oxygen affinity species. Accordingly, the primary structural element for the PEGylation induced high oxygen affinity.

The EAF PEG-shell has been shown to stabilize the oxy conformation of Hb increasing the high oxygen affinity of the PEGylated Hb species. Accordingly, since the PEGylation of Cys-93(β) by itself induces a high oxygen affinity, the oxygen affinity of EAF PEGylated Hb represents the combination of the PEG shell induced stabilization of the oxy conformation and that induced by the maleimide modification Cys-93(β). Consistent with this (i) EAF hexaPEGylation of Hb[Cys-93(β)>Ala] also generates a PEGylated molecular species with high oxygen affinity, and (ii) the reversible protection of Cys-93(β) of Hb during EAF hexaPEGylation generates a high oxygen affinity molecular species.[8] The differences in the oxygen affinity of the two products are marginal, suggesting that two molecular events that induce an increase in the oxygen affinity of Hb are neither synergistic nor fully additive.

On the other hand, EAF hexaPEGylation of αα-fumaryl Hb with a reversible protection of Cys-93(β) during PEGylation generates a product with a P50 around 20 mmHg. But when Cys-93(β) is also PEGylated, the product generated from αα-fumaryl Hb has P50 ~12 mmHg. The αα-fumaryl Hb is a low oxygen affinity species generated by the intramolecular crosslinking of Hb under deoxy conditions with a P50 ~30 mmHg, the control value for unmodified Hb being ~14 mmHg. The intramolecular αα-fumaryl crosslink stabilizes the protein in the deoxy conformation. The high oxygen affinity inducing activity of maleimide PEG modification of Cys-93(β) is considerably stronger with αα-fumaryl Hb as compared to the EAF hexaPEGylation of αα-fumaryl Hb (the PEG 5 K chains are exclusively on the ε-amino groups of Lys residues). Accordingly reversible protection of Cys-93(β) of low oxygen affinity Hb during EAF PEGylation and having stability for its deoxy conformational state favored significantly more relative to that in human Hb even after PEGylation appears to be two critical structural designs that needs to be integrated in order to generating EAF PEGylated Hbs with lower oxygen affinity as compared to current versions of EAF hexa PEGylated uncrosslinked Hbs.[7]

PEGylated Low Oxygen Affinity Hbs are also Nonhypertensive

The selection of present design of EAF hexaPEGylated Hb as the test model is based on the assumptions that generation of the nonhypertensive PEG Hb is result of (i) engineering of plasma expander like properties of Hb by PEGylation, and (ii) EAF hexaPEGylation increases oxygen affinity of Hb so that the autoregulatory principles are not activated. Low oxygen affinity species of Hb are optimal for targeted oxygen delivery to hypoxic regions. EAF hexaPEGylated Hb is used at a significantly lower concentration (a 4 gm% solution) as compared to previously developed blood substitutes (about 12 gm% solution), lowering oxygen carrying capacity of EAF PEG-Hb nearly threefold, i.e. the remaining red blood cells become responsible for most of the oxygen carrying capacity. In the extreme hemodilution model, the acellular EAF PEG Hb represents only ~20% of total.

A number of hexaPEGylated species of Hb with oxygen affinity lower than that of EAF hexaPEGylated Hb (with P50s ranging from 12 to 30 mmHg) have been designed[12,18] and tested for their vasoactivity in the extreme hemodilution model. In these systems, lowering of oxygen affinity of hexaPEGylated Hb from 7 mmHg to 30 mm has no influence on the PEGylation induced neutralization of *in vivo* vasoactivity of Hb. Apparently, the increase in the anticipated oxygen unloading in the arterial side of the circulation by these synthetic materials is not adequate to activate the auto regulatory mechanism and cause vasoconstriction. It should be noted that in the extreme hemodilution model the oxygen carrying capacity is not overabundant, and it is very restricted. The total amount of Hb in the circulation is just about 5 g Hb/dL, and only about 20% (about 1 g of total Hb/dL) is present as free Hb in the plasma. The oxygen-carrying capacity is apparently significantly reduced and consistent with this, the lowering of the oxygen affinity of PEGylated Hbs does not increase tissue oxygenation beyond that we have seen with diPEGylated Hb, P5K2 Hb, a species of Hb that carries two copies of PEG chains, either 5 K or 10 K.[19]

In normal physiological conditions, the concentration of Hb is 10 to 12 gm of Hb/dL blood, oxygen-carrying capacity is overabundant and only about 20% of the overall oxygen-carrying capacity is used to achieve normal tissue oxygenation. As the total Hb content is reduced to about 7 gm/dL (transfusion trigger), blood loss has to be taken care of by blood volume expansion using plasma expanders. Once the Hb level is ~7 gm/dL, oxygen-carrying capacity of the system is compromised and blood loss needs to be treated with extrinsic oxygen-carrying capacity, blood transfusion, i.e. by increasing the oxygen-carrying capacity.

Optimization of Structure of PEG-Hb for Tissue Oxygenation in Extreme Hemodilution

PEG-Hbs used to correlate the oxygen affinity with the vasoconstrictive activity are hexaPEGylated Hbs and use different PEGylation chemistries. The Hb used for PEGylation is intramolecularly crosslinked Hb, i.e αα-fumaryl Hb. An increase in the P50 of PEGylated Hbs has been made possible by having the low oxygen affinity inducing αα-fumaryl intramolecular crosslinking of Hb and avoiding the PEGylation of Cys-93(β). An exception to this is the P5K2 αα-fumaryl Hb that exhibits a P50 value that is only slightly lower than that of the P2K2 Hb. Previously, we have seen that among the uncrosslinked PEGylated Hbs, the best level of tissue oxygenation is given by P5K2 Hb (~7 mmHg) and by P10K2 Hb, while tissue oxygenation achieved with MP4 is considerably lower. Interestingly, this material has the lowest amount of PEG and consequently induces highest blood pressure, functional capillary density and tissue oxygenation. Even though our thinking for the design of novel PEGylated Hb blood substitutes has been to generate an EAF hexaPEGylated Hb molecule with P50 ~15 mmHg, the observation that P5K2 Hb exhibits a tissue oxygenation better than EAF P5K6 Hb that has a P50 ~7 mmHg suggests that the higher level of PEGylation may have an attenuating impact on the ability of PEG Hb to deliver the oxygen.

It should be noted that though engineering αα-fumaryl intramolecular crosslinking into P5K2 Hb slightly increases P50 as compared to P5K2, it does not increase the tissue oxygenation. Accordingly, it appears that attenuating the flexibility of the Hb molecule to access oxy conformation through deoxy conformation stabilizing intramolecular crosslinking attenuates the tissue oxygenation by PEG Hbs in the extreme hemodilution model much in the same way as the increasing the PEG mass in PEGylated uncrosslinked Hb with comparable oxygen affinity.

The influence of site specific PEGylation of Cys-93(β) is considerably higher in the presence of αα-fumaryl intramolecular crosslinking of Hb than in the EAF hexaPEGylation of crosslinked Hb targeted specifically to the ε-amino groups of Hb. The increase in oxygen affinity of αα-fumaryl Hb by PEGylation of Cys-93(β) is also accompanied by higher changes in the co-operativity of the molecule, presumably a consequence of increased rigidity of the protein core of the molecule. In the case of uncrosslinked Hbs the influence of increasing the PEG mass in the PEG shell of EAF PEG Hb seen on the oxygen affinity is much more when the influence of PEGylation on the oxygen affinity is investigated in the presence of allosteric effectors as a function of EAF PEGylation. Accordingly, while attempting to optimize the structure of EAF PEG Hbs as blood substitutes for tissue oxygenation, we need to consider the potential impact of changes in the rigidity of the protein core as well as the packing density of the PEG shell of the EAF PEG Hb.

Behavior of EAF PEGylated Hbs under Non-Ideality Situations and Potential Consequence of this in Extreme Hemodilution Model

Blood (in particular plasma) is a non-ideal solution, and in developing the extreme hemodilution models, nearly 75% of the volume of circulatory blood is replaced by dextran 70 (6 gm% solution). Accordingly the non-ideality of the circulatory system impacted in the extreme hemodilution model relative to blood besides the change in hematocrit. The presence of dextran introduced in developing the extreme hemodilution models could also influence the properties of PEGylated Hbs. Both viscosity and the COP of PEGylated Hbs has been shown to increase significantly in the presence of dextran, and this increase is also exponential as a function of the concentration of the PEGylated Hbs. This phenomenon is also seen with albumin and EAF PEG albumin (Acharya and Intaglietta, unpublished results) besides dextran. We suggest that this is a consequence of the decrease in the water content in the solution due to the presence of dextran that exhibits a very high hydrodynamic volume. Accordingly, we need to exercise extreme care in evaluating the molecules designed to understand results obtained in different model systems used to study the microcirculatory properties and translating it to develop a more generalized mechanism for the unusual clinically beneficial microcirculatory properties of EAF PEGylated Hbs. It will also be important to evaluate the possible influence of the changes in non-ideality aspects of the expanded plasma layer on the microcirculatory and oxygen delivery properties of EAF PEG Hbs on transfusion of these solutions instead of blood.

Novel Aspects and Concepts Emerging from the Design and Development of EAF hexaPEGylated Hb as a Blood Substitute

The discussions make it clear that the design of EAF PEG Hb as oxygen therapeutics represents a complete departure from the conventional way of thinking and designing blood substitutes, namely intramolecular (or combined intra and inter molecular) crosslinking to overcome nephrotoxicity and keeping the oxygen affinity of the product comparable to that of red blood cells. EAF hexaPEGylated Hb has been designed to have a high oxygen affinity (to minimize the off-loading of oxygen by the acellular Hb on the arterial side) by having a maleimido phenyl urethane of PEG 5K, besides the EAF PEGylation at the ε-amino groups of Lys resides to make it act as plasma expanders and maintain excellent functional capillary density. Besides, these do not carry intramolecular crosslinking to avoid nephrotoxicty, but the unusually high hydrodynamic volume of the

EAF hexaPEGylated Hb makes them refractory to glomerular filtration, hence do not exhibit nephrotoxicity. The high oxygen affinity of the material makes it a poor molecule for general tissue oxygenation of the body, but an excellent molecule for targeted oxygenation of tissues that are oxygen starved. As these molecules will work in the presence red blood cells to enhance their efficacy to delivery oxygen tissues, these are referred to in recent years as "oxygen therapeutics" rather than as blood substitutes.

Maleimide PEG Modified Hb (Hemospan and MP 4)

HexaPEGylated human Hb from Sangart (San Diego, CA), known as Hemospan, is the most extensively studied PEGylated Hb and has moved up to Phase III clinical trials.[20–22] This product is developed based on the Einstein Extension Arm Chemistry for PEGylation that uses 2-iminothiolane to introduce the extension arms and maleimide PEG to target the modification of the new thiols on Hb.[5] However, Sangart has used maleimido propyl PEG-5K generated by Nippon Oil and Fats Co. Ltd., Tokyo, Japan (NOF) in their clinical trial. Some of the preclinical studies were with the maleimido phenyl urethane of PEG 5K used just as in the studies of Einstein[23] to generate EAF SP P5K6 Hb (SP = succinimido phenyl urethane). In Hemospan, the linker chemistry of the PEG reagent in Sangart's approach is aliphatic, while that in EAF SP P5K6 Hb is aromatic and combines both hydrophobic (phenyl moiety) and hydrophilic (urethane linkage) characteristics (Fig. 3). Thus in the Einstein EAF PEG Hb, there may be a potential contribution of the urethane in modulating the non-covalent interactions of PEG chains with protein surface; this should be absent in MP4. Small structural differences have been noticed in the structure of the protein when the phenyl urethane linker is replaced by ethyl linker.[24,25]

A 4 gm% solution of Hemospan is referred to as MP4 (4 g% solution of Maleimide-Polyethylene-Glycol modified hemoglobin; the 4 g% is based on Hb present in the sample, if PEG mass is considered this is a 6 g% solution of hexaPEGylated human Hb), an extensively studied PEG Hb molecule. MP4 is not vasoconstrictive. In a multi-center Phase II clinical study, the safety of the product was evaluated at two dosages, 250 and 500 mL of MP4. The study data indicated that MP4 was well tolerated on the patients enrolled. In another clinical study, patients were given 200, 400, 750 and 1000 mL of MP4, and there were no adverse effects reported in this study as well. The disappearance of MP4 from the plasma was dose dependent, ranging from 14 to 23 hr. Also it was noted that the plasma oxygen content was slightly enhanced even though the doses of MP4 are modest, thus administration of MP4 might offer the possibility of continuous oxygen delivery to the oxygen-starved tissues to maintain a tissue oxygenation close to the P50 of the PEG Hb in circulation.

Nonconservative EAF PEGylation may be a Preferable Platform to the Current Conservative Version towards the Design of EAF PEG Hb

The presence of extension arms between the amino groups of Hb and PEG chains, as discussed above, attenuates the influence of the PEG shell on the weakening of the inter-dimeric interactions of Hb. Accordingly, the EAF PEGylation is a preferred approach over the direct PEGylation platforms. The 2-iminothiolane mediated EAF PEGylation approach engineers the extension arm on the protein in a conservative mode. The use of succinimido caproyl maleimide engineers the extension arm on Hb in a nonconservative mode; the isopeptide linkage between the protein amino groups and the extension arm abolishes the positive charge of the amino group that is conserved in the 2-iminothiolane mediated engineering of the extension. The thermal transition pattern of the EAF hexa-PEGylated Hb generated by conservative and nonconservative mode has been compared using differential scanning microcalorimetry. The thermal stability of nonconservative EAF hexa-PEGylated Hb is higher than that of PEGylated Hb generated by conservative EAF PEGylation. During nonconservative EAF PEGyation, Cys-93(β) is reversibly protected from being PEGylated by the bifunctional maleimide reagent as mixed disulfide with dithiopyridine. Accordingly, the observed difference in the thermal stability of two versions of EAF hexa PEGylated Hb may be a consequence of Cys-93(β) being free in the EAF hexaPEGylated product. The maleimide modification of Cys-93(β) of Hb is known to weaken the inter dimeric interactions of the tetrameric structure of Hb. To assess the influence of the maleimide modification (PEGylation) of Cys-93(β) in the conservatively EAF PEGylated Hb, the thermal stability of the conservative EAF PEGylation of Hb was evaluated with and without reversible protection of Cys-93(β). The thermal stability of conservative EAF PEG Hb is not influenced significantly on keeping the Cys-93(β) free in the final product. The difference in the thermal stability is a correlate of the conservative vs. non-conservative mode of engineering the extension arms on Hb.[26]

Accordingly, it is concluded that the influence of the PEG shell on the structure of the protein core in the EAF PEG Hb not only reflects the influence of the presence of extension arm between the PEG shell and protein, but also the nature of the chemical linkage between the amino group and the extension arm. The influence of the nature of the chemical linkage can be approximated to that of the chemical modification, and in this case the modification of the amino groups by 2-iminotholation mediated thiolation and its modification by maleimide (without PEG-chains). A prototype of conservative EAF PEGylated Hb with the PEGylation exclusively targeted to the ε-amino groups of the protein has also been generated by replacing the maleimido phenyl urethane of PEG with N-ethyl maleimide. Thermal transition studies with this class of modified Hb established that the PEG shell introduced in the EAF PEGylation plays little role in the thermal stability of Hb, and stability of EAF PEGylated Hb correlates with the nature of the surface chemical modification. This would imply that the nature of electrostatic modification of the surface of Hb that accompanies the EAF PEGylation of Hb plays a dominant role in the overall thermal stability of EAF PEGylated Hb, at least in this case nonconservative EAF PEGylation affords a higher stability to the EAF PEGylated protein. Whether this is a general phenomenon that will be seen with every protein or is unique to Hb is not clear at this stage. Accordingly more detailed studies of nonconservative EAF hexaPEGylated Hb are needed to establish the preference of conservative or nonconservative EAF PEG Hb in terms of overall stability, supra perfusion and tissue oxygenation.

Polynitroxylation of Hb also Neutralizes the *In Vivo* Vasoconstrictive Activity

Conjugation of free radical scavenger nitroxy moieties to αα-fumaryl Hb that function as SOD mimetic small molecular weigh species, Hb (the most potent vasoconstrictive Hb derivative) neutralizes the *in vivo* vasoactivity of αα-fumaryl Hb in hemorrhagic shock models. The ability of the free radical scavenger groups suggests that antioxidant activity conjugated to Hb can also neutralize the vasoactivity of αα-fumaryl Hb. It is suggested that the oxidative stress induced by the presence of Hb derivatives in plasma is counteracted by the presence of the free radical scavengers conjugated to Hb. This is a surprising result and confirms that acellular Hb induced vasoconstriction *in vivo* is not a simple and direct correlate of the NO scavenging, since EAF PEGylation by itself has no noticeable *in vitro* effect on the interaction of NO with PEG-Hb. Vasocontriction by acellular Hb *in vivo* is a multifactorial event. The alternate explanation could be there is a commonality in the molecular aspect in the neutralization of *in vivo* vasoconstrictive activity induced by EAF PEGylation as by polynitroxylation of Hb. This also confirms the concept that the *in vitro* rate of autoxidation of EAF PEG Hb does not correlate with the *in vivo* effects of PEG Hbs.[27,28] Further studies are needed to dissect the molecular mechanisms that operate on EAF PEGylation and polynitroxylation of Hb to result in the neutralization of vasocontrictive activity of acellular Hbs. The Somatogen approach that synergizes the targeted efforts of simple recombinant approaches (to attenuate the interaction of Hb with NO *in vitro*) with the oligomerization of the recombinant Hb to attenuate the *in vivo* vasoconstructictive activity of acellular Hbs suggests the interplay of multiple factors in the *in vivo* neutralization of the vasoconstrictive activity.[29]

EAF Polynitroxylation EAF PEG Hb

As noted above the extension arm chemistry not only simplifies the conjugation of PEG chains to Hb since it attenuates the influence of the PEG shell on the structure/conformation of protein core as reflected by the attenuation of the PEG-shell mediated weakening of the interdimeric interactions of Hb. In view of these observed beneficial effects of EAF conjugation, we have undertaken the development of EAF polynitroxylation of EAF PEG-Hb. The approach designed and used by Synzyme (Irvine, CA) uses bromomoprylamido tempol as the nitroxylating agent. This reaction is not very efficient and is carried out at high temperatures and expected to damage Hb, in particular EAF PEG Hb. Accordingly, we designed and developed the EAF chemistry based Polynitroxylation chemistry that involves the thiol maleimide reaction (using maleimido tempol) as simpler and more efficacious and quantitative protocol that can be carried out either at room temperature or at 4°C (Fig. 6 and Table 2).

A nitroxylated Hb, with nitroxylation on Cys-93(β) and hexa-EAF hexa-nitroxylated EAF hexaPEGylated Hb was generated and the *in vitro* SOD mimetic activity of the tempol has been investigated

Table 2. Extension Arm Facilitated Conjugation of Tempol to EAF P5K6 Hb Comparison of EAF Protocol with Bromoacetamido Chemistry (direct) Protocol

Properties	EAF-Thiol Maleimide Chemistry	Bromoacetamido Chemistry
Reaction conditions	pH 7.4, 4°C, 4 to 6 hr	pH 7.4, 60°C, 10 hr
Charge of the adduct	conservative,	non-conservative,
Site selectivity	high, thiolated Amino,	low, Amino, thiol, imidazole,
Location	~1 nM away from the amino on the protein surface	directly on the amino groups on the protein surface
Extension arm	present	absent
Interaction with protein	none or limited	extensive
Mobility of tempol	mobile	rigid

Table 3. SOD Mimetic Activity of EAF Hb-Tempol

Sample	Tempol Removing Rate (NT/min)	(%/min)
EAF P5K6-Hb-EAF-T6	–0.278	–0.046
PEG6-Hb-T1	–0.036	–0.036

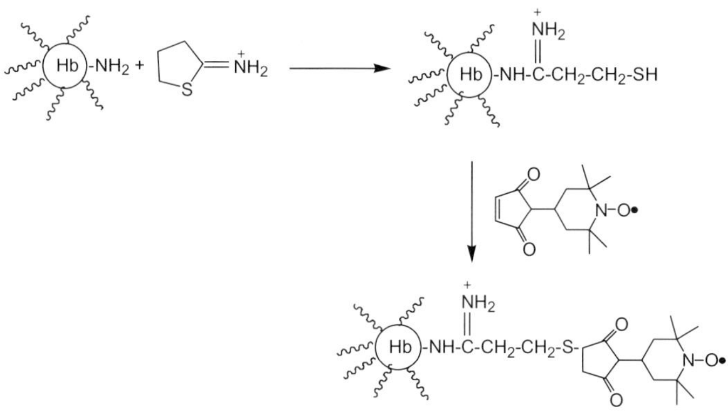

Fig. 6. Schematic representation of EAF nitroxylation of EAF P5K6 Hb. As with EAF PEGylation of Hb, this reaction is also a two-step process, but can be carried out in one step by incubating maleimido tempol with EAF P5K6 Hb in the presence of 2-imino thiolane.

(Fig. 6). This comparison of the specific SOD mimetic activity per tempol units conjugated site selectively to Cys-93(β) of Hb and of EAF Polynitroxylated EAF PEG Hb, suggests that the tempol at Cys-93(β) exhibits significantly higher SOD mimetic specific activity as compared to the tempol moieties conjugated to the ε-amino groups through extension arm chemistry (Table 3). The tempol units conjugated to ε-amino groups of albumin through extension arm chemistry were found to exhibit a SOD mimetic specific activity per mole that is nearly twice of that of tempol conjugated to albumin without the extension arm chemistry (Table 3). Accordingly, it follows that the very high specific activity of tempol on Cys-93(β) reflects the uniqueness of the microenvironment of this residue in Hb.

EAF PEGylation of Hb Induces Vasodilation Besides the Attenuation of Its *In Vivo* Vasoconstrictive Activity

PEGylation of Hb does not impact the *in vitro* NO scavenging of Hb, but the PEGylated Hbs are nonhypertensive. The exact molecular mechanisms for the attenuation of the *in vivo* vasoactivity of Hbs as a result of PEGylation is not clear, but the plasma expander-like properties to PEG Hb and the very large increase in the hydrodynamic volume of Hb (for the conjugation of a PEG mass of 30 K) by PEGylation has been advanced as a novel strategy to generate nonhypertensive Hbs as potential blood substitutes.[3] Besides the increase in the functional capillary density that improves perfusion, the administration of EAF PEG Hb also induces a vasodilation of veins, while arteries remain vasoneutral. The large hydrodynamic volume of EAF PEG Hb attenuates the extravasation of this Hb derivative, and the concomitant attenuation of the NO scavenging in the tissues may be contributing to the vasoneutrality of EAF P5K6 Hb. However the vasodilation of veins is interesting.

The molecular weight of albumin and Hb are nearly the same. Accordingly, it is assumed that PEGylation effects on microcirculation, i.e. functional capillary density and on vasodilation of arteries and/or veins, could be delineated by generating EAF hexaPEGylated albumin. The studies with EAF hexaPEGylated albumin have established that vasodilation is an extrinsic property endowed to Hb and albumin on EAF hexaPEGylation.

Molecular Aspects of the PEGylation-induced Vasodilatory Activity of EAF PEG Hb

How does PEGylation induce the vasodilatory activity to Hb? Two structural concepts have been advanced to explain the vasodilatory activity of EAF PEGylated Hb. The first hypothesis advances the concept that PEGylation of Hb induces an increase in the nitrite reductase activity of PEGylated Hbs.[30] An increase in the nitrite reductase activity has been seen with EAF hexaPEGylated Hb that supports this concept.[31] However, the increase in the vasoactivity of EAF PEG Hb was shown to be unique only to this PEGylated Hb and is a consequence of the chemical modification of Cys-93(β) by PEG maleimide.[26] In fact, increase in nitrite reductase activity can be induced to Hb by n-ethyl maleimide modification of Cys-93(β).[26] The EAF PEGylated Hb in which Cys-93(β) is not modified (by carrying out the EAF PEGylation of Hb under deoxy conditions or by reversibly protecting the Cys-93(β) during EAF PEGylation under oxy conditions) is still vasoinactive and is also a vasodilator, but without an increase in the nitrite reductase activity.[26] Besides, the vasodilatory activity of the EAF PEG albumin cannot be explained by the PEGylation-induced nitrite reductase activity. Thus, we conclude that the vasodilator activity of EAF PEGylated Hb is not a direct correlate of the PEGylation-induced increase in the *in vivo* nitrite reductase activity of Hb.

The PEGylation-induced increase in the mechano-transduction mediated endothelial NO production is the second mechanism that has been advanced as a possible molecular basis of the vasodilatory activity of the EAF hexaPEGylated Hb. Three forms of modified Hbs designed as potential blood substitutes, namely αα-fumaryl Hb, polymerized bovine Hb and EAF PEG Hb, have been subjected to comparative studies. All these modified Hbs caused reductions in perivascular NO levels, but that did not correlate with the microvascular responses. PEG-Hb (largest molecular volume) maintained blood flow, while the others caused vasoconstriction and reduced perfusion. All solutions increased mean arterial pressure due to vasoconstriction and blood volume expansion except for PEG-Hb, which increased blood pressure due to blood volume expansion and maintenance of cardiac output. In conclusion, perivascular NO reduction is similar for all Hb solutions because NO binding affinities are similar. However, effects on vascular resistance are related to the type of molecular modification, molecular volume, and oxygen affinity. In the case of PEG-Hb, besides the influence of COP on functional capillary density, the PEGylation-induced increase in the viscosity of EAF PEG Hb is anticipated to increase the mechano-transduction mediated endothelial NO production and hence vasodilation. In this concept the vasodilatory aspect of PEGylated Hb and the increased viscosity of the plasma layer can explain supra perfusion by a single mechanism.

Viscosity of EAF PEG Hb as a Correlate of Pattern of PEGylation

The enhanced rate, supra perfusion, seen with EAF PEG Hb and EAF PEG albumin has been considered as a correlate of the increase in the viscosity of the solution of Hb as a result of PEGylation. Our recent studies with EAF PEG albumin established that PEGylation-induced solution properties of PEG albumin are a function of the pattern of the PEGylation and not of the total mass of the PEG conjugated.[26] Accordingly we hypothesized that the supra perfusion of EAF PEGylated Hb and EAF PEGylated albumin can be optimized further by manipulation of the pattern of PEGylation to induce the best viscosity effects by engineering the molecular shape of the EAF PEGylated proteins. Consistent with this conceptual thinking, the hydrodynamic volume of EAF PEGylated Hb with a total PEG mass of 30 K conjugated is indeed a function of pattern of PEGylation (number and size of the PEG chain) as reflected by size-exclusion chromatography. The hydrodynamic volume of EAF (SP-P30K1)-αα-Hb is higher than that of EAF (SP-P5K6)-αα-Hb as reflected by size-exclusion chromatography.

EAF PEG Hb is a Low-Viscosity Plasma Expander, but still Induces Supra Perfusion Just Like High-Viscosity Plasma Expanders

The vasodilatory activity and supra perfusion seen with EAF PEG Hb and EAF PEG albumin are the properties of high-viscosity plasma expanders such as dextran 500 and alginate, both of which exhibit significantly higher viscosity than EAF PEG Hb as well as higher hydrodynamic volume. How then does a low-viscosity solution of EAF PEG Hb induce vasodilation and supra perfusion? The viscosity of EAF PEG Hb is only around 2.2 cP. Hetastarch and dextran 70 are also low-viscosity plasma expanders with a viscosity around 2.8 cP, and these do not induce vasodilation. On the other hand, high-viscosity plasma expanders like dextran and alginate induce vasodilation. Recent studies have shown that high-viscosity plasma expanders like Dextran 500 induce endothelial NO production.[34] Notably, 4 g% EAF hexa PEGylated albumin is a low-viscosity solution; it nonetheless induces endothelial NO production and has vasodilatory activity, whereas dextran 70 does not do either of these.[35] Low-viscosity plasma expanders like dextran 70 do not induce endothelial NO production or vasodilation. Thus the uniqueness of the structure of PEG albumin endows this molecule with the properties of high-viscosity plasma expanders. The uniqueness in the structure of EAF PEG albumin and EAF PEG Hb is that is a semi synthetic hybrid biopolymer consisting of a high packing density protein core and a low packing density PEG shell with an intermediate zone of extension arms. The differences in the packing density of the two regions may be expected to facilitate the rearrangement of the molecular shape of the low viscosity globular structure of PEG protein to a high-viscosity ellipsoidal structure under external influence, i.e. the ability of these hybrid biopolymers to undergo molecular shape transitions during circulation as function of the shear stress and such changes will be associated with changes in the viscosity of the molecules.[32]

Shear Thinning Effect of EAF PEG Hb

The shear thinning effect is a well-known phenomenon in blood rheology. The shear thinning effect refers to the phenomenon of the decrease in the viscosity of a solution as a function of increasing shear stress. The materials that exhibit this phenomenon are non-Newtonian and are referred to as pseudoplastic materials. Both EAF hexa PEGylated albumin and EAF hexa-PEGylated Hb (Fig. 7) are such materials. Red blood cells also exhibit this effect and allow the circulatory system to be able self-regulate flow to some degree. In clinical conditions the presence of either EAF PEG Hb or EAF PEG Albumin

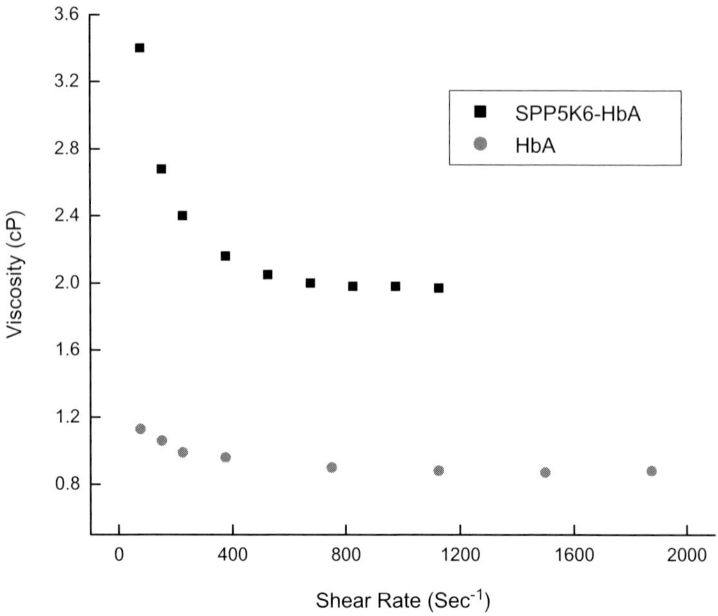

Fig. 7. Pseudoplasticity of EAF PEG Hb. The viscosity of a 4 gm% of the solution of the protein is shown as a function of shear stress. The EAF PEGylation of Hb has induced the shear thinning effect to Hb, a phenomenon representing an inverse correlate of viscosity of a material with the shear stress.

in the cell free plasma layer can modulate the density as an inverse correlate of microvessel wall shear stress and thus facilitate increasing endothelial NO production. The pseudoplasticity of the EAF PEG Hb and EAF PEG Albumin, low-viscosity plasma expanders, makes them efficient plasma expanders to achieve supra perfusion just as high-viscosity plasma expanders that do not exhibit shear thinning effects being low-viscosity materials.[33]

Even though both EAF PEG-Hb and EAF PEG Albumin exhibit pseudoplasticity, the vasodilation seen with EAF PEG Hb on the arterial side of the circulation is less pronounced as compared with that of EAF PEG albumin. We attribute this to the NO-scavenging effect of EAF PEG-Hb *in vivo* and absence of that in the EAF PEG albumin control. This conceptual recognition of the difference in the vasodilatory differences between EAF PEG Albumin and EAF PEG Hb triggers a new design strategy; if the NO-scavenging activity of Hb is fully reduced or neutralized, we may be to generate EAF PEGylated Hbs that exhibit supra perfusion at a level comparable to that of EAF PEG albumin.

Mutant Hbs with attenuated NO-scavenging activity could be used to generate EAF PEG Hbs, and when these are EAF PEGylated, novel molecules with supra perfusion properties better than current versions of EAF PEG Hb generated using wild type Hb could be developed. The increase in perfusion that can be achieved with such molecular species is expected to increase the efficacy of oxygen extraction, and by modulating the affinity of the EAF PEGylated Hbs by appropriately selecting the EAF PEGylation platform, will result in a new generation of oxygen-carrying plasma expanders bringing us closer to the reality of having a blood substitute.

Conclusions

EAF PEGylation of Hb distinguishes itself from the previous (direct) PEGylation platforms in engineering extension arms between the chains of the PEG shell and the protein core. This approach has been advanced as a simple and cost-effective PEGylation platform to induce plasma expander-like properties to Hb. The unique advantage of EAF PEGylation over direct PEGylation is that the extension arms result in a nearly complete attenuation of the direct PEGylation-induced weakening of inter dimeric interactions (quaternary structure) of the tetrameric Hb. This makes EAF PEG Hbs (and PEGylated Hbs in general) a novel class among PEGylated Hbs that are non-hypertensive and with a high potential for clinical application as blood substitutes. EAF PEG Hb and its prototype EAF PEG Albumin are also low-viscosity plasma expanders that mimic the physiological activity of high-viscosity plasma expanders to induce endothelial NO production. Hence these are also vasodilators, besides increasing the functional capillary density. These PEGylated molecules are also unusual molecular species in that they are pseudoplastic, i.e. they exhibit a shear thinning effect, and we speculate that presence of these nanoparticles with their ability to change shapes in response to shear stress in the plasma layer facilitate the modulation of the reduction in shear stress of RBC to induce supra perfusion. The chemistry of EAF PEGylation and the structure of Hb have been redesigned to generate these novel oxygen-carrying plasma expanders with desired oxygen affinity to customize the transport and delivery of oxygen to targeted tissues. Vasodilation of arteries seen with EAF hexaPEGylated Hb is lower than that with the EAF P5K6 Albumin, and we attribute this to the intrinsic NO-scavenging activity of acelluar Hbs and its absence in albumin. We anticipate the development of a new generation of EAF PEGylated Hb using recombinant Hbs with attenuated NO-scavenging activity. These oxygen-carrying plasma expanders are expected become a new class of oxygen-carrying plasma expanders (EAF PEG Hbs) with an ability to induce vasodilation comparable to that of their counterparts, EAF PEG Albumin. The new class of EAF PEG Hbs, the oxygen therapeutics, are expected to be used to treat blood loses wherein the oxygen-carrying capacity of RBC becomes limiting. In all other situations, EAF PEG Albumin, a non-oxygen carrying semisynthetic plasma expander, can treat blood losses very effectively.

Acknowledgments

Studies presented here are supported in part by USPHS Bioengineering Research Partnership grant R24-HL 064395 (MI), grants R01-HL 062354 (MI) and USAMRAA award W81XWH1120012 (AGT).

References

1. Jahr JS, Sadighi A, Doherty L, *et al.* (2011) Hemoglobin based oxygen carrier: history, limits, brief summary of the state of the art including clinical trials. In Mozzarelli A, Bettati S (eds), *Chemistry and Biochemistry of Oxygen Therapeutics*, John Wiley, pp. 301–326.
2. Tsai AG, Vandergriff KD, Intaglietta M, Winslow RM. (2003) Targeted oxygen delivery by low P_{50} hemoglobin, a new basis for oxygen therapeutics. *J Physiol Heart Circ Physiol* **285:** H1411– H1419.
3. Intaglietta M, Cabrales P, Tsai AG. (2006) Microvascular perspective of oxygen-carrying and noncarrying blood substitutes. *Annu Rev Biomed Eng* **8:** 289–321.

4. Yabuki A, Yamaji K, Ohki H, Iwashita Y. (1990) Characterization of a pyridoxylated hemoglobin-polyoxyethylene conjugate as a physiologic oxygen carrier. *Transfusion* **30:** 516–520.
5. Acharya SA, Manjula BN, Smith P. (2000) hemoglobin crosslinkers, US patent 6,017,943, January 25, 2000.
6. Acharya SA, Manjula BN. (2006) Surface decoration of hemoglobin with polyethylene glycol. In: Winslow RM (ed), *Blood Substitutes,* Elsevier, Inc., London, pp. 460–469.
7. Li D, Hu T, Manjula BN, Acharya SA. (2009) Extension arm facilitated pegylation of alphaalpha-hemoglobin with modifications targeted exclusively to amino groups: functional and structural advantages of free Cys-93(beta) in the PEG-Hb adduct. *Bioconjug Chem* **20:** 2062–2070.
8. Li D, Manjula BN, Ho NT, *et al.* (2007) Molecular aspects of the high oxygen affinity of non-hypertensive hexa pegylated hemoglobin, [(SP-PEG5K)(6)-Hb]. *Artif Cells Blood Substit Immobil Biotechnol* **35:** 19–29.
9. Hu T, Prabhakaran M, Acharya SA, Manjula BN. (2005) Influence of the chemistry of conjugation of poly(ethylene glycol) to Hb on the oxygen-binding and solution properties of the PEG-Hb conjugate. *Biochem J* **392:** 555–564.
10. Hu T, Manjula BN, Li D, *et al.* (2007) Influence of intramolecular cross-links on the molecular, structural and functional properties of PEGylated haemoglobin. *Biochem J* **402:**143–151.
11. Meng F, Manjula BN, Tsai AG, *et al.* (2009) Hexa-thiocarbamoyl phenyl PEG5K Hb: vasoactivity and structure: influence of rigidity of the conjugation linkage on the PEGylation induced plasma expander-like solution properties of PEG-Hb adducts. *Protein J* **28:**199–212.
12. Li D, Hu T, Manjula BN, Acharya SA. (2008) Non-conservative surface decoration of hemoglobin: influence of neutralization of positive charges at PEGylation sites. *Biochim Biophys Acta* **1784:** 1395–1401.
13. Li D, Hu T, Manjula BN, Acharya SA. (2006) Extension arm facilitated PEGylation of hemoglobin: correlation of the properties with the extent of PEGylation. *Protein J* **25:** 263–274.
14. Ananda K, Acharya SA. (2008) Role of extension arm in PEG-Hb conjugates on the stability of the tetramer: non-conservative EAF maleimide thio-PEG mediated PEGylation. *Artif Cells Blood Substit Immobil Biotechnol* **36:** 499–512.
15. Vandegriff KD, McCarthy M, Rohlfs RJ, Winslow RM. (1997) Colloid osmotic properties of modified hemoglobins: chemically cross-linked verses polyethylene glycol surface-conjugated. *Biophysical Chemistry* **69:** 23–30.
16. Caccia D, Ronda L, Frassi R. (2009) PEGylation promotes tetramer dissociation. *Bioconjug Chem* **20(7):** 1256–1366.
17. Manjula BN, Malavalli A, Smith PK, *et al.* (2000) Cys-93-betabeta-succinimidophenyl polyethylene glycol 2000 hemoglobin A. Intramolecular cross-bridging of hemoglobin outside the central cavity. *J Biol Chem* **275:** 5527–5534.

18. Li D, Hu T, Manjula BN, Acharya SA. (2009) Extension arm facilitated PEGylation of alphaalpha-hemoglobin with modifications targeted exclusively to amino groups: functional and structural advantages of free Cys-93(beta) in the PEG-Hb adduct. *Bioconjug Chem* **20:** 2062–2070.
19. Cabrales P, Kanika ND, Manjula BN, *et al.* (2004) Microvascular PO2 during extreme hemodilution with hemoglobin site specifically PEGylated at Cys-93(β) in the hamster window chamber. *Am J Physiol Heart Circ Physiol* **287(4):** H1609–H1617.
20. Vandegriff KD, Malavalli A, Wooldridge J, *et al.* (2003) MP4, a new nonvasoactive PEG-Hb conjugate. *Transfusion* **43:** 509–516.
21. Winslow R, Kramer G. (2003) Maleimide-polyethylene glycol-modified human hemoglobin. *Crit Care Med* **31:** 2815–2816.
22. Winslow RM. (2004) MP4, a new nonvasoactive polyethylene glycol-hemoglobin conjugate. *Artif Organs* **28:** 800–806.
23. Acharya AS, Intaglietta M, Tsai AG, *et al.* (2005) Enhanced molecular volume of conservatively PEGylated Hb: $(SP\text{-}PEG5K)_6$-HbA is non-hypertensive. *Art Cells Blood Subs Immob Biotech* **33:** 239–255.
24. Juszczak LJ, Manjula BN, Bonaventura C, *et al.* (2002) UV resonance Raman study of β–93 modified hemoglobin A. Chemical modification specific effects and added influence of attached poly (ethylene glycol) chains. *Biochemistry* **41:** 376–395.
25. Khan ID, Danskar U, Samoni AJ, *et al.* (2002) Cys-93(β) modified hemoglobin: kinetic and conformational consequences. *Biochemistry* **40:** 7581–7592.
26. Acharya SA, Intaglietta M, Tsai AG, *et al.* (2011) Engineering the molecular shape of PEG-Hb adducts for supra perfusion. In: Mozzarelli A, Bettati S (eds), *Chemistry and Biochemistry of Oxygen Therapeutics*, John Wiley, pp. 345–369.
27. Vandegriff KD, Malavalli A, Minn C, *et al.* (2006) Oxidation and haem loss kinetics of poly(ethylene glycol)-conjugated haemoglobin (MP4): dissociation between *in vitro* and *in vivo* oxidation rates. *Biochem J* **399:** 463–471.
28. Hu T, Li D, Manjula BN, Acharya SA. (2008) Autoxidation of site-specifically PEGylated hemoglobins: role of PEG-chains and sites of PEGylation in the autoxidation. *Biochemistry* **47:** 10981–10990.
29. Doyle MP, Apostol I, Kervin BA. (1999) Glutaraldehyde modification of recombinat human hemoglobin alters its hemodynamic properties. *J Bio Chem* **274:** 2583–2591.
30. Cabrales P, Friedman JM. (2007) PEGylated hemoglobins mechanisms to avoid vasoconstriction and maintain perfusion. *Transfusion Alternatives in Transfusion Medicine* **9:** 281–293.
31. Lui FE, Kluger R. (2009) Enhancing nitrite reductase activity of modified hemoglobin: bis-tetramers and their PEGylated derivatives. *Biochemistry* **48:** 11912–11919.

32. Ananda K, Manjula BN, Meng F, *et al.* (2011) Packing density of PEG in the PEG-shell of PEG albumins: PEGylation induced viscosity and COP are inverse correlate of packing density. *Art Cells Blood Substitutes and Biotech* **40:** 14–27.
33. Acharya AS, Acharya VN, Meng F, *et al.* (2011) Molecular aspects of *in vivo* "Active supra perfusion" by EAF P5K6 Alb and EAF P5K6 Hb: deformability of PEG-protein, its potential role in enhancing perfusion, *XIII International Symposium on Blood Substitutes and Oxygen Therapeutics*, Boston, MA, USA, 2011.
34. Tsai AG, Acero C, Nance PR, *et al.* (2005) Elevated plasma viscosity in extreme hemodilution increases perivascular nitric oxide concentration and microvascular perfusion. *Am J Physiol Heart Circ Physiol* **288:** H1730–H1739.
35. Sriram K, Tsai AG, Cabrales P, *et al.* (2012) PEG-Albumin supra plasma expansion is due to increased vessel wall shear stress induced by blood viscosity shear thinning. *Am J Physiol Heart Circ Physiol* **302:** H2489–H2497, doi:10.1152/ajpheart.01090.2011

Chapter 6

Biocompatibility of a Highly Concentrated Fluid of Hemoglobin-Vesicles as a Transfusion Alternative

Hiromi Sakai*,†,‡

1. Introduction

Blood transfusion systems have greatly benefited human health and welfare. Nevertheless, some problems remain: possibility of infection, blood type mismatching, immunological response, and a short shelf life that is insufficient for stockpiling for emergency situations. Realization of artificial O_2 carriers is anticipated to solve such problems. The most abundant protein in blood is Hb ([Hb] = 12–16 g/dL in healthy human blood), indicating that oxygen transport to tissues is the most important function of blood. To design an oxygen-carrying fluid to substitute the function of blood, the Hb concentration of the fluid should be high and comparable with that of blood Hb concentration. Chemically modified and cell-free Hb-based oxygen carriers (HBOCs), such as intramolecularly crosslinked, polymerized, and polymer-conjugated Hbs, were synthesized to prevent toxic effects of cell-free Hbs.[1] The hydrodynamic radius of such cell-free HBOCs is less than 20 nm. On the other hand, hemoglobin-vesicles (HbV) or

*Waseda Bioscience Research Institute in Singapore, Biopolis, Republic of Singapore.
†Organization for University Research Initiatives, Waseda University, Tokyo, Japan.
‡Department of Chemistry, Nara Medical University, Kashihara, Japan.

so-called liposome-encapsulated Hb, which encapsulate a concentrated Hb solution in phospholipid vesicles, are developed.[2,3] One particle of about 250–280 nm encapsulates nearly 30,000 Hb molecules. In the case of the chemically modified cell-free HBOCs, they are "**dissolved**" in aqueous solutions like plasma proteins. On the other hand, HbV as a cellular HBOC is "**dispersed**" in aqueous solutions like blood cells. The difference between dissolution and dispersion results in a considerable difference in physicochemical characteristics of the fluids containing a high concentration of Hb. The fluid properties should be adjusted within the biocompatible and physiological conditions for a massive blood exchange. In this chapter I first summarize some views of physicochemical differences between cell-free and cellular HBOCs. Second, biocompatibility of the dispersed particle is of course important. The blood compatibility, biodegradability, excretion, and immunological responses to the massive injection of such dispersed small particles are summarized.

2. Biocompatible Solution Properties of HbV Fluids

2.1. *Colloid Osmotic Pressure*

Albumin, dissolved in a blood plasma at ca. 5 g/dL, provides sufficient colloid osmotic pressure (COP, ca. 20 Torr) to play an important role in equilibrating COP between blood and interstitial fluid, thereby maintaining the overall blood volume. This COP is one requisite for a transfusion alternative to sustain blood circulation for transporting oxygen and metabolites. The extremely high concentration of the HbV suspension ([Hb]) 10 g/dL; [lipids] 6 g/dL, volume fraction, ca. 40 vol % imparts an O_2 carrying capacity that is comparable to that of blood. The HbV suspension does not possess a colloid osmotic pressure (COP), because one HbV particle (ca. 250 nm diameter) contains about 30,000 Hb molecules. In fact, HbV acts as a particle, not as a solute. Therefore, HbV must be suspended in or coinjected with an aqueous solution of a plasma substitutes. This requirement is identical to that for emulsified perfluorocarbon, which does not possess COP[4,5]; it contrasts to characteristics of other Hb-based O_2 carriers, intramolecular cross-linked Hbs,

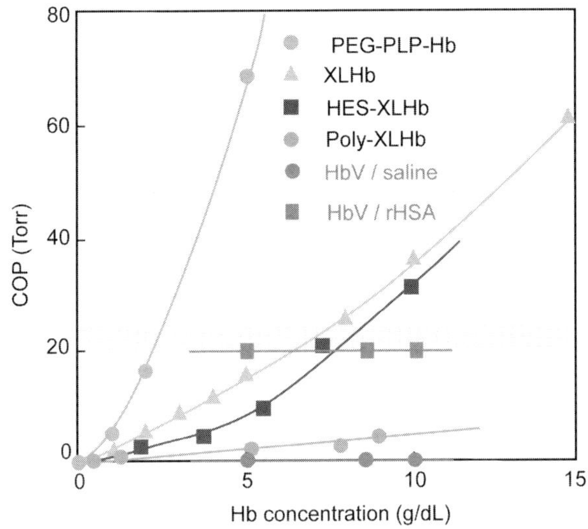

Fig. 1. Colloid osmotic pressures of cell-free and cellular HBOCs.[6] COP of HbV is determined by the suspending medium. For example, 5% albumin has 20 torr COP. It does not exceed the physiological condition at any Hb concentration.

polymerized Hbs, and polymer-conjugated Hbs, which all possess very high COP as protein solutions[6,7] (Fig. 1). These chemically modified Hb solutions can be categorized as "O_2-carrying plasma expanders"[8] because they have the oxygen-carrying capacity and colloid osmotic pressure. However, the problem with PEG-modified Hb solutions is that the COP is too high such that the Hb concentration of the resulting fluid is as low as 4–6 g/dL. On the other hand, HbV suspended in any plasma expander shows COP of the suspending medium, at any Hb concentration. When HbV is suspended in 5%-human serum albumin solution (HSA), COP is nearly 20 mmHg, which is in a physiological range of COP.

According to the guideline for safer blood transfusion, a transfusion trigger (the critical Hb level) is 6 g/dL to minimize unnecessary transfusion strictly or to avoid allogeneic transfusion as long as possible. But the problem of HBOCs with low Hb concentration is that injection of HBOCs cannot increase blood Hb level. In fact, according to the retrospective description of Nosé, pyridoxalated Hb polyoxyethylene conjugated (PHP) had the same problem, and it was not easily approved for clinical study during the negotiation of the FDA.[9] The Hb concentration of HbV is adjusted to 10 g/dL, which is higher than the concentration of transfusion trigger.

2.2. *Flocculate Formation and Viscosity Increase in the Presence of Plasma Expanders*

Animal tests of HbV suspended in plasma-derived HSA or recombinant HSA (rHSA) showed an O_2 transporting capacity that is comparable to that of blood.[10,11] We reported that HbV suspended in plasma-derived HSA or rHSA was almost Newtonian: no aggregation was detected microscopically.[12,13] In Japan, rHSA was very recently approved for clinical use, in May 2008,[14] but various plasma substitutes are used worldwide, such as hydroxylethyl starch (HES), dextran (DEX), and modified fluid gelatin (MFG). The selection among these plasma substitutes is best determined not only according to their safety and efficacy, but also according to their associated price, experience of clinicians, and customs of respective countries. Water-soluble polymers generally interact with particles such as polystyrene beads, liposomes, and RBCs to induce aggregation or flocculation.[15,16] As for the cell-free HBOCs dissolved in saline, they are Newtonian fluids, and do not interact with plasma expanders. In the case of cellular HBOC dispersed in saline, it is important to determine the compatibility of HbV with these plasma substitutes. With that background, we studied rheological properties of HbV suspended in these plasma substitute solutions using a complex rheometer and a microchannel array.[17] The rheological property of an HBOC is important because the infusion amount is expected to be considerably large, which might affect the blood viscosity and hemodynamics. The HbV suspended in rHSA was nearly Newtonian (Fig. 2). Its viscosity was similar to that of blood, and the mixtures with RBCs at various mixing ratios showed viscosities of 3–4 cP. Other polymers, HES, DEX, and MFG, induced flocculation of HbV, possibly by depletion interaction, and rendered the suspensions as non-Newtonian with the *shear-thinning* profile. These HbV suspensions showed high viscosity and a high storage modulus (G')

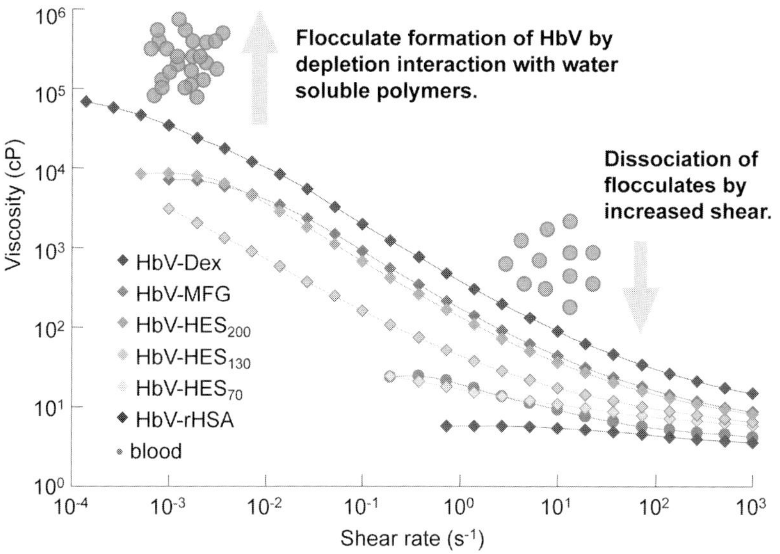

Fig. 2. Shear-thinning profiles of HbV suspended in a series of plasma expanders. [Hb] = 10 g/dL.[17]

because of the presence of flocculated HbV. On the other hand, HbV suspended in rHSA exhibited a very low G'. The viscosities of HbV suspended in DEX, MFG, and high-molecular-weight HES solutions responded quickly to rapid step changes of shear rates of 0.1–100 s^{-1} and a return to 0.1 s^{-1}, indicating that flocculation formation is both rapid and reversible. Microscopically, the flow pattern of the flocculated HbV perfused through microchannels (4.5 μm deep, 7 μm wide, 20 cm H_2O applied pressure) showed no plugging.

The mechanism of flocculate formation of liposome is controversial.[18] However, we believe PEG-modified liposomes are flocculated by depletion interaction. The flocculation level increased with hydrodynamic radius (R_h) or radius of gyration (R_g) of series of HES or DEX with different molecular weights at a constant polymer concentration (4 wt%). However, the flocculation level differed markedly among the polymers (Fig. 3). A crowding index (C_i) representing the crowding level of a polymer solution is defined as (excluded volume of one polymer) × (molar concentration) × Avogadro's number, using R_h

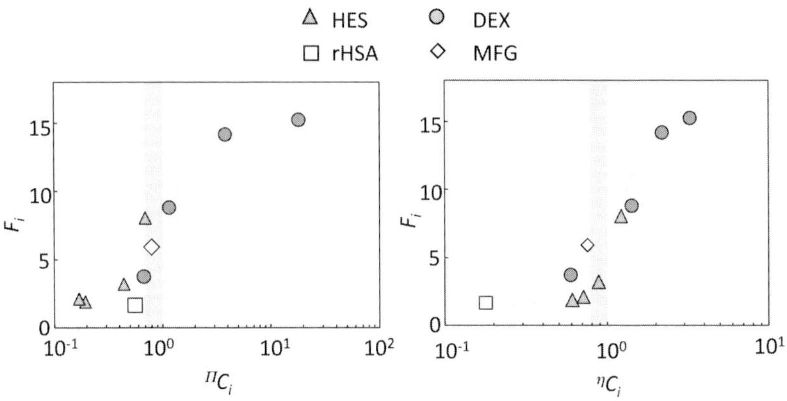

Fig. 3. Flocculation index (F_i) increases with the addition of dextran HES, MFG, or HSA, showing that the crowding index, C_i, of the polymer aqueous solution is the determining factor of flocculate formation.[18] We defined the flocculation index (F_i) as $F_i = (\eta_{10} - \eta_0)/(\eta_{1000} - \eta_0)$. In that equation, η_{10} and η_{1000} respectively represent the viscosity at the shear rates of 10 and 1000 s^{-1}. Crowding index ($^{\Pi}C_i$, and $^{\eta}C_i$) representing the crowding level of a polymer solution is defined using R_h and R_g, respectively, as [(excluded volume of one polymer) × (molar concentration) × Avogadro's number]. Adapted with permission from Sakai H, Sato A, Takeoka S, Tsuchida E. (2009) Mechanism of flocculate formation of highly concentrated phospholipid vesicles suspended in a series of water-soluble biopolymers. *Biomacromolecules* **10:** 2344–2350. Copyright (2009) American Chemical Society.

or R_g. All polymers' flocculation level increases when C_i approaches 1: when the theoretical total excluded volume approaches the entire solution volume, the excluded HbV particles are forced to flocculate.

2.3. *In Vivo Study of Co-Injection of HbV and a Series of Plasma Expanders*

It remained unknown whether such flocculate formation of HbV in blood might affect an animal's hemodynamics. Using a rat model, we tested infusion of a series of plasma expanders (MFG, HES_{670}, HES_{130}, HES_{70}, rHSA) to maintain the blood volume (level of blood exchange led to 60%) at repeated hemorrhages and the subsequent infusion of HbV (20 mL/kg, 36% of blood volume).[19] (In this experiment we did not use dextran because rats show anaphylactic reaction to dextran.[20]) All rats survived for 4 hr after the infusion of HbV; hemodynamic and respiratory functions were preserved, indicating that the flocculation does not induce capillary embolism. Blood exchange with rHSA and subsequent infusion of HbV showed more stable systemic parameters because of the longer retention of rHSA in blood than other plasma substitutes, indicating that rHSA is suitable for combination with HbV in this experimental model.

2.4. *Solution Properties Affects on Reactions of Hb and NO*

It has been regarded that lower blood viscosity after hemodilution is effective for tissue perfusion. However, microcirculatory observation shows that, in some cases, lower "plasma viscosity" decreases shear stress on the vascular wall, causing vasoconstriction and reducing the functional capillary density.[21] Therefore, an appropriate viscosity might exist which maintains the normal tissue perfusion level. The large molecular dimension of HbV can result in a transfusion fluid with high viscosity. A large molecular dimension is also effective to reduce vascular permeability and to minimize the reaction with NO and CO as vasorelaxation factors.[22–25]

Increased fluid viscosity of a solution of hemoglobin-based oxygen carriers (HBOCs) reduces vasoconstrictive effects because increased shear stress on the vascular wall enhances the production of vasorelaxation factors such as NO. Nevertheless, on a microcirculatory level, it remains unclear how viscosity affects the reaction of HBOCs and NO. To clarify the effect of viscosity on the NO-binding, different HBOCs were perfused through narrow gas-permeable tubes (25 μm inner diameter at 1 mm/s centerline velocity; hemoglobin concentration [Hb]=5 g/dL).[26] The reaction was examined microscopically based on the Hb visible-light absorption spectrum. When immersed in a NO atmosphere, the NO-binding of deoxygenated Hb solution (viscosity, 1.1 cP at 1000 s^{-1}) in the tube occurred about twice as rapidly as that of red blood cells (RBCs): 1.6 cP (Fig. 4). Binding was reduced by PEGylation (PEG-Hb, 7.7 cP), by addition of a high molecular weight hydroxyethyl starch (HES) (2.8 cP), and by encapsulation to

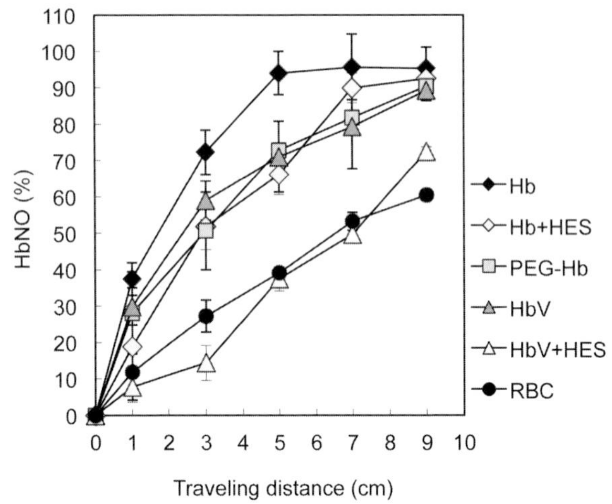

Fig. 4. Change of the level of NO-binding reactions of the Hb containing fluids, Hb solution, PEG-Hb, HbV, Hb+HES, HbV+HES, and RBC (black circles) with traveling distance.[26] (Permission obtained from Elsevier)

form Hb-vesicles (HbVs, 1.5 cP; particle size 279 nm). However, the reduction was not as great as that shown for RBCs. A mixture of HbVs and HES (6.2 cP) showed almost identical NO-binding to that of RBCs. Higher viscosity and particle size might reduce lateral diffusion when particles are flowing. The HbVs with HES showed the slowest NO-binding. Furthermore, Hb encapsulation and PEGylation, but not HES-addition, tended to retard CO-binding. Increased viscosity reportedly enhances production of endothelium NO. In addition, our results show that the increased viscosity also slows down the reaction with NO. Each effect might mitigate vasoconstriction.

3. Biocompatibility of HbV in Terms of Immunological Responses

3.1. *Complement Activation*

A so-called injection reaction, or pseudo-allergy, resulting from complement activation after injection of a small amount of liposome is well known, giving rise to anaphylatoxins, which trigger various hypersensitivity reactions.[27–29] Transient thrombocytopenia and pulmonary hypertension in relation to complement activation is an extremely important hematologic effect observed in rodent and porcine models after infusion of LEH (containing DPPG) developed by the US Naval Research Laboratory.[30] Neo red cells (Terumo Corp.) containing stearic acid showed pulmonary hypertension in beagle and porcine models,[31] but not in monkeys. In our group, exchange transfusion of prototype HbV (containing DPPG, no PEG modification) in anesthetized rats engendered transient thrombocytopenia and slight hypertension.[32] The transient reduction in platelet counts and increase of thromboxane B_2 caused by complement-bound liposomes was also associated with sequestration of platelets in the lung and liver.[30] In the present formation of HbV, we use a negatively charged lipid (DHSG) instead of DPPG. It does not induce thrombocytopenia or complement activation in animal experiments,[33,34] probably because it contains PEGylated lipids and a different type of negatively charged lipid (DHSG), instead of DPPG or a fatty acid. The *in vitro* human blood compatibility of HbV has been extensively studied.[33,35–37] The present PEG-modified HbV containing DHSG does not affect the extrinsic or intrinsic coagulation activities of human plasma, although HbV-containing DPPG and no PEG-modification tends to shorten the intrinsic coagulation time. The kallikrein–kinin cascade of plasma was activated slightly by the prototype DPPG-HbV, but not by the present PEG–DHSG–HbV. The exposure of human platelets to high concentrations of this HbV (up to 40%) *in vitro* does not cause platelet activation and does not affect adversely the formation and secretion of prothrombotic substances or proinflammatory substances that are triggered by platelet agonists.[38] These results imply that HbV, at concentrations of up to 40%, do not have aberrant interactions with either unstimulated or agonist-induced platelets. It can be concluded that the PEG–DHSG–HbV described here have higher blood compatibility.

3.2. *RES Trap, Degradation, and Excretion*

Biodistribution of HbV was examined using ^{99m}Tc-conjugated homocysteine or glutathione containing HbV[39] and HbV containing

^{125}I-labeled Hb.[40] These experiments show that HbV are finally captured by macrophages, mainly in the spleen and liver. Electron microscopic observation can detect the presence of Hb-encapsulating particles in the phagosomes of macrophages because of the high densities of protein and electrons (derived from Fe) in the particles such as RBCs. The HbV particles disappear in one week.[41] Immunohistochemical staining with antihuman Hb antibody and antimethoxy-PEG indicates that Hb and PEG of HbV disappear in two weeks.[41–43] It was shown recently that ^{125}I-labeled Hb and ^{3}H-labeled cholesterol in HbV have identical blood clearance, indicating that HbV retains its integrity in the bloodstream, and distributes to the reticuloendothelial system together. However, ^{125}I mainly appears in urine, and ^{3}H in feces, showing different metabolic routes in the macrophages.[44]

3.3. *Transient Immunosuppressive Effect*

Accumulation of considerable amounts of liposome in a RES can affect immunologic response. Actually, phagocytic index measured by carbon particle clearance in rats showed significant reduction of phagocytic index one day after injection of HbV. While, it increased considerably three days after injection,[41] indicating the increased defense function. On the other hand, HbV showed transient suppressive effect on the proliferation of rat splenic T cells. Takahashi *et al.* of Hokkaido Red Cross Blood Center[45] elucidated the mechanism underlying that phenomenon and its effect on both local and systemic immune response. HbV was injected intravenously at a volume of 20% of whole blood into rats. Then their spleens were removed, and T cell responses to concanavalin A (Con A) or keyhole limpet hemocyanin (KLH) were evaluated by measuring the amount of [^{3}H]thymidine incorporated into DNA. Results showed that T cell proliferation in response to Con A or KLH was inhibited from 6 hr to 3 days after the liposome injection. The phagocytosis of the large load of liposomes by rat CD11b/c+, class II immature monocytes temporarily renders them highly immunosuppressive, but most importantly, the systemic immune response was unaffected.

4. Conclusion

Liposomes are clinically utilized for cancer and antifungal therapies, and other usages as a vehicle of functional molecules are developed aggressively. HbV is one liposomal product, but the differences from such conventional liposomal products are that it is a highly concentrated fluid, and it inevitably requires a massive dose (like 20 mL/kg body weight) as it will be utilized as a substitute for a RBC concentrate. Therefore, injection of HbV would affect spontaneously on hemorheology, hemodynamics, immune system, phagocytosis, gas exchange reactions between tissue and blood, etc. It is also important to have stability as a capsule during storage and during blood circulation to shield a toxic effect of molecular Hb. It also requires instability to be decomposed by macrophages and complete excretion from a body. In this chapter we discuss such important biocompatibilities of HbV. We believe the above mentioned biocompatibility of HbV guarantees the safety of HbV and a potential for versatile clinical application.

Acknowledgments

The author greatly appreciates the late Emeritus Professor Eishun Tsuchida, Waseda University, for his support for the Project of Oxygen Infusion. Research of Hb-vesicles has been conducted by an academic consortium comprising many domestic and overseas research institutes. The author acknowledges the contribution of the collaborators. This research has been supported by Health and Sciences Grants from Ministry of Health, Labour and Welfare, Japan; and a Grant-in-Aid for Scientific Research from Japan Society for the Promotion of Sciences (JSPS).

Disclosure

Hiromi Sakai is an inventor holding some patents related to the production and utilization of Hb-vesicles.

References

1. Chang TMS. (1997) *Blood Substitutes: Principles, Methods, Products and Clinical Trials.* S. Karger AG, Basel.
2. Djordjevich L, Miller IF. (1977) Lipid encapsulated hemoglobin as a synthetic erythrocyte. *Fed Proc* **36:** 567.
3. Sakai H, Sou K, Tsuchida E. (2009) Hemoglobin-vesicles as an artificial oxygen carrier. *Methods Enzymol* **465:** 363–383.
4. Nolte D, Pickelmann S, Lang M, *et al.* (2000) Compatibility of different colloid plasma expanders with perflubron emulsion: an intravital microscopic study in the hamster. *Anesthesiology* **93:** 1261–1270.
5. Jouan-Hureaux V, Audonnet-Blaise S, Lacatusu D, *et al.* (2006) Effects of a new perfluorocarbon emulsion on human plasma and whole-blood viscosity in the presence of albumin, hydroxyethyl starch, or modified fluid gelatin: an *in vitro* rheologic approach. *Transfusion* **46:** 1892–1898.
6. Sakai H, Yuasa M, Onuma H, *et al.* (2000) Synthesis and physicochemical characterization of a series of hemoglobin-based oxygen carriers: objective comparison between cellular and acellular types. *Bioconjugate Chem* **11:** 56–64.
7. Vandegriff KD, McCarthy M, Rohlfs RJ, Winslow RM. (1997) Colloid osmotic properties of modified hemoglobins: chemically cross-linked versus polyethylene glycol surface-conjugated. *Biophys Chem* **69:** 23–30.
8. Cabrales P, Tsai AG, Intaglietta M. (2008) Isovolemic exchange transfusion with increasing concentrations of low oxygen affinity hemoglobin solution limits oxygen delivery due to vasoconstriction. *Am J Physiol Heart Circ Physiol* **295:** H2212–H2218.
9. Nosé Y. (1998) Oxygen-carrying macromolecules: therapeutic agents for the treatment of hypoxia. *Artif Organs* **22:** 618–622.
10. Sakai H, Masada Y, Horinouchi H, *et al.* (2004) Hemoglobin vesicles suspended in recombinant human serum albumin for resuscitation from hemorrhagic shock in anesthetized rats. *Crit Care Med* **32:** 539–545.
11. Terajima K, Tsueshita T, Sakamoto A, Ogawa R. (2006) Fluid resuscitation with hemoglobin vesicles in a rabbit model of acute hemorrhagic shock. *Shock* **25:** 184–189.
12. Sakai H, Takeoka S, Park SI, *et al.* (1997) Surface-modification of hemoglobin vesicles with poly(ethylene glycol) and effects on aggregation, viscosity, and blood flow during 90%-exchange transfusion in anesthetized rats. *Bioconjugate Chem* **8:** 23–30.
13. Sakai H, Tsai AG, Kerger H, *et al.* (1998) Subcutaneous microvascular responses to hemodilution with red cell substitutes consisting of polyethylene glycol-modified vesicles encapsulating hemoglobin. *J Biomed Mater Res* **40:** 66–78.
14. Kobayashi K. (2006) Summary of recombinant human serum albumin development. *Biologicals* **34:** 55–59.

15. Meyuhas D, Nir S, Lichtenberg D. (1996) Aggregation of phospholipid vesicles by water-soluble polymers. *Biophys J* **71:** 2602–2612.
16. Neu B, Meiselman HJ. (2002) Depletion-mediated red blood cell aggregation in polymer solutions. *Biophys J* **83:** 2482–2490.
17. Sakai H, Sato A, Takeoka S, Tsuchida E. (2007) Rheological property of hemoglobin vesicles (artificial oxygen carriers) suspended in a series of plasma substitute aqueous solutions. *Langmuir* **23:** 8121–8128.
18. Sakai H, Sato A, Takeoka S, Tsuchida E. (2009) Mechanism of flocculate formation of highly concentrated phospholipid vesicles suspended in a series of water-soluble biopolymers. *Biomacromolecules* **10:** 2344–2350.
19. Sakai H, Miyagawa N, Horinouchi H, *et al.* (2011) Intravenous infusion of Hb-vesicles (artificial oxygen carriers) after repetitive blood exchange with a series of plasma expanders (water-soluble biopolymers) in a rat model. *Polymer Adv Technol* **22:** 1216–1222.
20. Koller ME, Reed RK. (1992) Increased negativity of interstitial fluid pressure in rat trachea in dextran anaphylaxis. *J Appl Physiol* **72:** 53–57.
21. Tsai AG, Friesenecker B, McCarthy M, *et al.* (1998) Plasma viscosity regulates capillary perfusion during extreme hemodilution in hamster skinfold model. *Am J Physiol Heart Circ Physiol* **275:** H2170–H2180.
22. Goda N, Suzuki K, Naito M, *et al.* (1998) Distribution of heme oxygenase isoforms in rat liver. Topographic basis for carbon monoxide-mediated microvascular relaxation. *J Clin Invest* **101:** 604–612.
23. Sakai H, Hara H, Yuasa M, *et al.* (2000) Molecular dimensions of Hb-based O_2 carriers determine constriction of resistance arteries and hypertension. *Am J Physiol Heart Circ Physiol* **279:** H908–H915.
24. Sakai H, Sato A, Masuda K, *et al.* (2008) Encapsulation of concentrated hemoglobin solution in phospholipid vesicles retards the reaction with NO, but not CO, by intracellular diffusion barrier. *J Biol Chem* **283:** 1508–1517.
25. Nakai K, Ohta T, Sakuma I, *et al.* (1996) Inhibition of endothelium-dependent relaxation by hemoglobin in rabbit aortic strips: comparison between acellular hemoglobin derivatives and cellular hemoglobins. *J Cardiovasc Pharmacol* **28:** 115–123.
26. Sakai H, Okuda N, Takeoka S, Tsuchida E. (2011) Increased viscosity of hemoglobin-based oxygen carriers retards NO-binding when perfused through narrow gas-permeable tube. *Microvasc Res* **81:** 169–176.
27. Chonn A, Cullis PR, Devine DV. (1991) The role of surface charge in the activation of the classical and alternative pathways of complement by liposomes. *J Immunol* **146:** 4234–4241.
28. Loughrey HC, Bally MB, Reinish LW, Cullis PR. (1990) The binding of phosphatidylglycerol liposomes to rat platelets is mediated by complement. *Thromb Haemost* **64:** 172–176.
29. Szebeni J, Baranyi L, Savay S, *et al.* (2005) Complement activation-related cardiac anaphylaxis in pigs: role of C5a anaphylatoxin and adenosine in

liposome-induced abnormalities in ECG and heart function. *Am J Physiol Heart Circ Physiol* **290:** H1050–H1058.

30. Phillips WT, Klipper R, Fresne D, *et al.* (1997) Platelet reactivity with liposome-encapsulated hemoglobin in the rat. *Exp Hematol* **25:** 1347–1356.
31. Pape A, Kertscho H, Meier J, *et al.* (2008) Improved short-term survival with polyethylene glycol modified hemoglobin liposomes in critical normovolemic anemia. *Intensive Care Med* **34:** 1534–1543.
32. Izumi Y, Sakai H, Takeoka S, *et al.* (1997) Evaluation of the capabilities of a hemoglobin vesicle as an artificial oxygen carrier in a rat exchange transfusion model. *ASAIO J* **43:** 289–297.
33. Abe H, Azuma H, Yamaguchi M, *et al.* (2007). Effects of hemoglobin vesicles, a liposomal artificial oxygen carrier, on hematological responses, complement and anaphylactic reactions in rats. *Artif Cells Blood Substit Immobil Biotechnol* **35:** 157–172.
34. Sou K, Tsuchida E. (2008). Electrostatic interactions and complement activation on the surface of phospholipid vesicle containing acidic lipids: effect of the structure of acidic groups. *Biochim Biophys Acta* **1778:** 1035–1041.
35. Abe H, Fujihara M, Azuma H, *et al.* (2006) Interaction of hemoglobin vesicles, a cellular-type artificial oxygen carrier, with human plasma: effects on coagulation, kallikrein-kinin, and complement systems. *Artif Cells Blood Substit Immobil Biotechnol* **34:** 1–10.
36. Wakamoto S, Fujihara M, Abe H, *et al.* (2001) Effects of poly(ethylene glycol)-modified hemoglobin vesicles on agonist-induced platelet aggregation and RANTES release *in vitro*. *Artif Cells Blood Substit Immobil Biotechnol* **29:** 191–201.
37. Wakamoto S, Fujihara M, Abe H, *et al.* (2005) Effects of hemoglobin vesicles on resting and agonist-stimulated human platelets *in vitro*. *Artif Cells Blood Substit Immobil Biotechnol* **33:** 101–111.
38. Ito T, Fujihara M, Abe H, *et al.* (2001) Effects of poly(ethylene glycol)-modified hemoglobin vesicles on N-formyl-methionyl-leucyl-phenylalanine-induced responses of polymorphonuclear neutrophils *in vitro*. *Artif Cells Blood Substit Immobil Biotechnol* **29:** 427–437.
39. Sou K, Klipper R, Goins B, *et al.* (2005). Circulation kinetics and organ distribution of Hb vesicles developed as a red blood cell substitute. *J Pharmacol Exp Ther* **312:** 702–709.
40. Taguchi K, Maruyama T, Iwao Y, *et al.* (2009). Pharmacokinetics of single and repeated injection of hemoglobin-vesicles in hemorrhagic shock rat model. *J Control Release* **136:** 232–239.
41. Sakai H, Horinouchi H, Tomiyama K, *et al.* (2001) Hemoglobin-vesicles as oxygen carriers: influence on phagocytic activity and histopathological changes in reticuloendothelial system. *Am J Pathol* **159:** 1079–1088.

42. Sakai H, Masada Y, Horinouchi H, *et al.* (2004). Physiologic capacity of reticuloendothelial system for degradation of hemoglobin-vesicles (artificial oxygen carriers) after massive intravenous doses by daily repeated infusions for 14 days. *J Pharmacol Exp Ther* **311:** 874–884.
43. Sakai H, Seishi Y, Obata Y, *et al.* (2009). Fluid resuscitation with artificial oxygen carriers in hemorrhaged rats: profiles of hemoglobin-vesicle degradation and hematopoiesis for 14 days. *Shock* **31:** 192–200.
44. Taguchi K, Urata Y, Anraku M, *et al.* (2009) Pharmacokinetic study of enclosed hemoglobin and outer lipid component after the administration of hemoglobin-vesicles as an artificial oxygen carrier. *Drug Dispos Metabol* **37:** 1456–1463.
45. Takahashi D, Azuma H, Sakai H, *et al.* (2011) Phagocytosis of liposome particles by rat splenic immature monocytes makes them transiently and highly immunosuppressive in *ex vivo* culture condition. *J Pharmacol Exp Ther* **337:** 42–49.

Chapter 7

Molecular Basis of Haptoglobin and Hemoglobin Complex Formation and Protection against Oxidative Stress and Damage

Yiping Jia* and Abdu I. Alayash*

Abstract

Haptoglobin (Hp) is a plasma protein that clears cell free hemoglobin (Hb) from circulation and reduces the formation of harmful oxidative species by Hb. Three phenotypes of Hp can be found in human plasma, and these are Hp 1–1, Hp 2–1, and Hp 2–2; all have been shown to tightly but reversibly bind with liganded Hb. Rapid kinetic studies have examined the binding characteristics of Hp with Hb, as well as with isolated Hb subunits. The molecular mechanisms of Hp and Hb complex formation were elaborated in relation to the altered Hb ligand reactivities and redox properties in these studies. The Hp antioxidant activity was recently proposed to alleviate oxidative stress and cell/tissue damage caused by cell free Hb, and Hb-based oxygen carriers (HBOCs). In this chapter, the physicochemical characteristics of the Hp complex formation with free and modified Hbs and the potential utility of the enzymatic activities of the complex in protecting

*Laboratory of Biochemistry and Vascular Biology, Center for Biologics Evaluation and Research, Food and Drug Administration, Bethesda, MD 20892, USA.

against Hb oxidative toxicity and in maintaining tissue homeostasis are reviewed.

Introduction

In the past two decades, the development of many complicated technologies to manipulate and analyze matter at nanometer scales has led to rapid growth and wide applications in nanomedicine.[1–3] Biological molecules can be assembled into 1 to 100 nm dimensions by nanobiotechnological approaches.[4] Cross-linking reagents such as glutaraldehyde have been used to stabilize Hb tetramers in the form of Hb polymers in the development of Hb-based oxygen therapeutics. Catalase and superoxide dismutase may be included in the nanoscale complex to provide antioxidant activities in addition to oxygen-carrying capability. Hb can also undergo polyethylene glycol (PEG) surface modifications, or in some cases the Hb tetramers were cross-linked by dextran to form soluble nanoparticles with various dimensions.[5,6] This review is focused on complex formation of another macromolecule, Hp and Hb, with a special focus on the effect of the complex formation on the molecular binding characteristics as revealed by rapid kinetic analysis. The impact of complex formation on Hb redox reactions, e.g. radical stabilization and other enzymatic activities, is discussed in the context of potential nanomedicine-based therapeutic applications.

Nanobiotechnology Strategies to Develop Hemoglobin-Based Oxygen Therapeutics

Numerous nanobiotechnological approaches have been developed to generate nanoscale soluble Hb derivatives as potential blood substitutes.[7,8] Conventional chemical cross-linking reagents such as bis(3,5-dibromosalicyl) Fumarate were used to stabilize Hb in the tetrameric form in an effort to prevent Hb from dissociating into dimers that can induce adverse renal effects. Hb molecules can also be intramolecularly stabilized and inter-molecularly cross-linked to form a wide range of polymers as seen in the reactions with glutaldyhyde, *O*-raffinose, or in the zero-link reactions.[9] To attenuate the toxicity stemming from heme pocket redox activity, superoxide dismutase and catalase were included in the complex with Hb to provide protective reducing activities similar to the red blood cell environment.[10] PEGylation methods have been developed to stabilize Hb tetramers and increase Hb molecular size through surface modifications. Other reagents such as dextran and polyethylene glycol have also been tested in the development of chemical cross-linking methods for Hb stabilization and conjugation. Additionally, protein site-specific modifications can be achieved by site-directed mutagenesis for certain desired physicochemical property changes, and in some cases cysteine mutations can lead to Hb polymer formation.[11] The Hb-based oxygen carrier (HBOC) products based on diverse modification methods have reached different developmental stages, i.e. physicochemical characterizations, preclinical studies, clinical trials, and product license applications to the Food and Drug Administration (FDA). Despite the tremendous efforts in the field and the potential benefits of such a blood substitute product, the toxicity issues appear to be highly complicated and are yet to be fully understood.[12] The safety data of these products have not yet met the FDA regulatory review criteria to warrant approval for clinical use.[13,14]

A New Protective Strategy Based on Nano-Complex Formation of Hemoglobin and Haptoglobin

The plasma protein Hp has long been known to tightly bind to Hb and to take part in the clearance pathway of Hb from circulation.[15,16] The protective mechanism of Hp in reducing oxidative and other toxic effects of extracellular Hb has recently been published by our group using canine and guinea pig models.[17] In these studies, strong evidence was presented to suggest a remarkable protective role that Hp plays in addition to the clearance of Hb from circulation through macrophages. The high levels of endogenous Hp induced artificially in dogs by glucocorticoid stimulation prevented systemic high blood pressure and vascular injuries that are typically follow the infusion of cell free Hb. The co-infusion of exogenous Hp and purified Hb into guinea pigs dramatically reduced systemic hypertension as well as tissue and organ damage when compared to the infusion of purified Hb alone.[17] As a result, raising plasma Hp levels by either Hp administration or inducing the endogenous Hp was suggested as a potential effective therapeutic treatment for the adverse effects associated with excessive circulating cell free Hb in various medical conditions such as hemolytic anemias, sickle cell disease,[18] or malaria. Moreover, an alternative strategy for developing safe and effective HBOCs was put forth to include Hp complexed with Hb in lieu of extensive chemical modifications of Hb molecules.

Hemoglobin and Haptoglobin Complex Formation

The haptoglobins, a group of acute-phase plasma proteins, are alpha 2-glycoproteins that bind Hb with extremely high affinity that could be equivalent to that of antigen–antibody interactions.[19] There are three phenotypes of haptoglobins, Hp 1–1, Hp 2–1, and Hp 2–2, determined by the composition of the Hp 1 allele and Hp 2 allele. The Hp polymorphism can be demonstrated using non-denaturing gel electrophoresis, PCR-based assays or ELISA to analyze serum or plasma samples.[20] The molecular nature of Hp and Hb interactions was investigated previously in both equilibrium and kinetic studies.

The affinity and stoichiometry of Hb and Hp reactions were examined using techniques such as ultracentrifuge, electrophoresis, and static fluorescence quenching.[21] All types of Hp were shown to bind oxyHb[22,23] so tightly that the overall equilibrium dissociation constant (K_d) could only initially be estimated as lower than the magnitude of 10^{-6} M[24] due to Hb dimer formation, and was later reported as low as 10^{-12} M[25] or 10^{-15} M.[26] Hp and Hb stay bound after the complex is deoxygenated, but deoxyHb does not seem to react with Hp. In addition, the Hb dimer combines with Hp in the same way as the oxygenated Hb tetramer. The separated α and β chains of Hb can also associate with Hp but not as strongly as oxyHb, and β chains have lower binding affinity than α chains. The Hp combination with one type of isolated Hb chains can be greatly enhanced in the presence of the other type of Hb chains.[27] Although the affinities were determined to be very high for oxygenated Hb, Hb dimer, and α chains, the Hb and Hp binding appears to be reversible as confirmed in the dilution or displacement experiments.

Rapid Kinetic Studies of Hb and Hp Association

Hb and Hp complex formation was among the first protein–protein interactions that have been kinetically studied during the development of the techniques to measure rapid reaction rates. Hb and Hp reactions were followed by monitoring in a stopped-flow apparatus the fluorescence quenching of aromatic amino acids in Hp molecule upon binding with Hb.[27,28] The binding of the purified Hp 1–1 with HbA, as well as isolated Hb α and β chains, was examined in these studies. The kinetics of these reactions were measured by varying both Hp and Hb concentrations, or keeping one concentration constant and varying the other. The second-order rate constant of the Hb and Hp association was determined at the magnitude of 10^6 M^{-1} s^{-1} that could fluctuate under different experimental conditions. Moreover, the effects of other factors such as temperature, salt concentration, and associated ligands on Hb and Hp reaction were investigated. It was revealed that metHb, HbCO, and oxyHb all reacted with Hp identically in these kinetic measurements under the same conditions.[29]

Fluorescence quenching of Hp with isolated Hb α chains exhibits rapid second-order kinetics, whereas no significant fluorescence change could be measured when Hp was mixed with Hb β chains.[28] However, a rapid fluorescence change was observed when β chains were mixed with pre-incubated Hp and isolated Hb α chains. In addition, the reaction rate of Hp with isolated Hb α chains was similar to that of Hp and Hb β chains pre-incubated, then followed by the addition of α chains. These kinetic data are in general agreement with the equilibrium study results which showed isolated Hb α chains interact with Hp more readily than β chains, while strong binding of β chains can follow as a secondary reaction.[27]

The relations between Hb and Hp binding and Hb subunit dissociation were further explored by stopped-flow kinetic methods. A deoxygenated Hb solution was mixed with a CO-saturated Hp solution to detect any fluorescence change resulting from the combination of deoxyHb or HbCO with Hp. This was based on the observation that the dissociation constant of deoxy Hb tetramers to dimers is about one order of magnitude slower than that of liganded Hb. Since the CO reaction with Hb is much faster than the subunit dissociation of Hb, the observed lag phase of fluorescence quenching suggested that the Hb dimer formation is the prerequisite of Hb and Hp association. Numerical analysis provided satisfactory fitting of the kinetic data to support a mechanism with the combination rate constant of Hb dimers and Hp at about $5.4 \times 10^5\ M^{-1}\ s^{-1}$, the dissociation constant of Hb tetramers to dimers at about 1.3×10^{-6} M, and the tetramer–dimer dissociation rate constant at about 2 s^{-1}. These values are remarkably close to the results obtained from previous experimental measurements of these parameters.[24]

Hb tetramers appeared to be excluded as a form that can bind with Hp in these kinetic studies, although specific chemical identification of the reaction species could not be provided due to the nature of the techniques. Hp was subsequently used in several studies to determine whether Hb was present as dimers or tetramers in solution.[30] For instance, the kinetic measurements of Hb subunit dissociation of partially oxygenated Hb have been carried out using this approach.[31] It was reported in this study that the oxygenation of only the first subunit of Hb tetramer increased the Hb dissociation rate 50 fold relative to that of fully deoxygenated Hb. This result suggested that the oxygenation changed the Hb conformation, and possibly disrupted the intersubunit $\alpha 1\beta 1$ contact surface. The lack of interaction between deoxy Hb and Hp implies that deoxy Hb dimers do not bind with Hp or shift the equilibrium of deoxy Hb subunit dissociation between tetramers and dimers. In addition to the low level of Hb subunit dissociation in a fully deoxygenated solution, the Hb dimers in the liganded and unliganded states may differ in protein conformations that determine Hb and Hp binding affinity.

Ligand Binding Properties of Hp-Hb Complex

The oxygen equilibrium binding of Hb can be drastically altered after the Hb and Hp combination. The equilibrium studies of the Hb and Hp complex have shown large increase in oxygen affinity, the absence of ligand binding cooperativity indicated by non-sigmoidal oxygen equilibrium curves, as well as the demolishing of Bohr effects, i.e. the pH dependence of oxygen affinity.[23] Although the heme groups do not appear to participate in the interactions of Hb and Hp globin chains, there are apparently interruptions of electrostatic and hydrogen bonds of Hb that lead to overall conformational and functional alterations.

The oxygen equilibrium binding of the Hb and Hp complex was also measured as a function of oxygen concentration in temperature-jump experiments.[32] The binding process followed approximately second-order reaction kinetics that yielded an oxygen association rate constant of $5 \times 10^7\ M^{-1}\ s^{-1}$, which is close to the values of fast reacting Hb state, or the isolated Hb α and β subunits. However, the relaxation time leveled off at high oxygen concentrations (>0.15 mM) probably due to the constraints that the Hp imposed on Hb conformation.

The high oxygen affinities of other Hb forms relative to that of HbA are usually accompanied by increased ligand association rates. Not surprisingly, the CO association rate constant of the Hb and Hp complex was 10 times higher than that of HbA as measured in the stopped-flow rapid-mixing or flash photolysis studies. The Hp 2–2 species apparently showed twice as much effect as Hp 1–1 on the CO association rate of Hb. Consistent with the equilibrium study results, no pH dependence was observed in CO binding with the Hb and Hp complex.[33]

The rapid kinetic measurements of oxygen dissociation from the Hb and Hp complex revealed two distinct phases in stopped-flow experiments.[32] This reaction has deviated from the approximate single exponential process of Hb oxygen dissociation under the same conditions. The observed heterogeneity was probably due to the elimination of Hb ligand cooperativity by Hp binding, and the difference in intrinsic oxygen off rates of the Hb α and β chains.

Hp Structure and Polymorphism

The Hp protein structure and genetic variations have been the focus of many studies and reviews.[34,20] Hp molecules consist of two types of polypeptide chains, β chains (~45 kd), which are highly conservative, and α chains, which have two variations, i.e. α1 (~9 kd) and α2 (~16 kd). Each β chain associates with one α chain via a disulfide linkage to form an Hp monomer. Due to the variations of α chains, there are two types of Hp monomers, i.e. α1β (Hp 1) and α2β (Hp 2). Furthermore, the α chain can covalently link to another α chain via disulfide bonds. Because the α1 chain is monovalent and the α2 chain is bivalent in forming disulfide linkages, three Hp phenotypes, i.e. Hp 1–1, Hp 2–1, and Hp 2–2, can be formed. Hp 1–1 is the simplest Hp form that is composed of two α1β monomers, whereas Hp 2–1 and Hp 2–2 consists of multiple α2β units that give high molecular weight species of Hp.

The genes that determine Hp protein structures and polymorphism are located on chromosome 16q22.[35] The Hp 1 allele consists of five exons encoding amino acid sequences for both α1 and β subunits. The Hp 2 allele, which appears to have been derived from the Hp 1 allele, possesses duplications of the exons 3 and 4 of the Hp 1 allele originated from a rare gene crossover.[36] As a result, the α2 chain of a Hp protein molecule is not only larger than the α1 chain, but also bivalent to allow the formation of Hp polymers. In addition, there are two α1 chain variants differing at amino acid position 54, glutamic acid in $\alpha 1_S$ and lysine in $\alpha 1_F$, which can be resolved by gel electrophoresis. Other rare Hp variants with abnormal α or β chain, and a phenotype mixed with Hp 2–1 and Hp 2–2 have also been described in previous studies.

The Hp 2 allele can only be found in humans, and there is no evidence of its existence in other species. Although the Hp polymorphism is unique to humans, the occurrence frequencies may vary among populations worldwide. All major Hp phenotypes were found in every ethnic group examined. However, there are significant differences in the frequencies of the Hp 1_S, Hp 1_F, and Hp 2 alleles. This finding was employed in forensic medicine to identify subjects until the more accurate DNA fingerprinting methods were developed. The Hp polymorphism in human and the phenotype distribution among populations may have been resulted from genetic drift and natural selection. It is theoretically possible to link a particular disease to certain Hp type. For instance, the high frequency of Hp 2–2 was associated with selective pressures in genetics against infectious diseases in Southeast Asian populations.[37] Previous studies have reportedly demonstrated associations of certain heart conditions, liver disease, and infections with Hp subtypes.[38] However, any firm association of a disease and Hp variation has yet to be established due to the limitation of sample sizes and conflicting reports.

The main function of Hp proteins is to remove cell-free Hb from circulation and to reduce its toxicity to organs and tissues. Previous studies have demonstrated functional differences among Hp phenotypes that may have correlations with many biological activities and various diseases.[37] However, the molecular basis for these differences has not been fully understood in part because no protein crystal structure is available for any Hp variants. In addition, the previous studies of Hp molecular reactivity were largely focused on Hp 1–1 subtype, and thus more detailed analyses are needed to elucidate the functional variations among Hp phenotypes. These studies will aid in the effort to determine whether there are specific links between Hp subtypes and diseases.[38]

Hb Binding Contact Regions and Oxidative Protection of Amino Acids

Although the precise structural details have yet to be elucidated regarding the binding sites of Hp and Hb, the nature of the protein–protein interactions in this complex has been the subject of numerous studies. It has been determined that the Hp β chain binds tightly with Hb, whereas the Hp α chain does not appear to react with Hb or its subunits.[39] However, the Hp α chain is essential in maintaining both the structural integrity of Hp molecules and the high affinity of the Hp β chain with Hb.

The isolated Hb β subunit has been shown to have much lower affinity with Hp than the Hb α subunit. Nevertheless, the synthetic peptide overlapping studies identified Hp binding contact regions in both Hb α and β subunits. The Hp binding sites were located to the region α121–127 in the Hb α subunit, and the regions β11–25 and β131–146 in the β subunit. Based on the locations of these regions in Hb tetramer molecule, a major conformational alteration was required to expose α1β1 interface upon binding with Hp.[40]

Quantitative mass spectrometry studies revealed that extensive oxidative damage occurred to the Hb peptide regions near αTyr42, βTyr145, and βCys93 in the Hb reaction with hydrogen peroxide.[41] The surface amino acid βCys93 appeared to be the major end point of the free radical migration pathway emanating from the heme pocket. The association of native Hb with Hp regardless of the phenotypes prevented the oxidative modifications of these amino acids, supporting the antioxidant properties of Hp. In contrast, the Hp protection was not detected in a modified Hb in which the two α subunits were cross-linked and thus stabilized as non-dissociable Hb tetramers with low Hp affinity. It would therefore be of great value to further examine the Hp interaction with chemically modified Hbs, developed as blood substitutes in relation to antioxidant properties of Hp and other potential therapeutic applications.

Hp Antioxidant Properties that may Reduce Free Radical Damages

In addition to its strong binding and clearance of Hb from circulation, Hp may play a role in inhibiting Hb radical formation and subsequent reactions that can lead to cell and tissue damage. It was reported in early studies that Hp can inhibit lipid peroxidation stimulated by Hb, but not the presence of heat-denatured Hb or hemin. This antioxidant function was not observed with albumin, transferrin, or desferrioxamine. Therefore, it was proposed that Hp is an important component of the extracellular antioxidant system *in vivo* against oxygen radical formation and propagation caused by redox reactions of cell-free Hb that has escaped the erythrocyte protective enzymes including superoxide dismutase, catalase, GSH or glutathione peroxidase.[42]

Consistent with the Hp antioxidant function, the Hp-Hb complex has recently been shown to have significant peroxidase activity.[43] In addition, a previous study indicated that Hp 1–1 is more effective in the inhibition of Hb oxidation reactions than Hp 2–2. The difference may be intensified *in vivo* due to their size variations, and thus the capabilities to diffuse into extravascular spaces.[44] Because the Hb pseudoperoxidase redox cycle that results in the consumption of H_2O_2 does not appear to be affected in the Hb-Hp complex, Hp may function as antioxidant by stabilizing the ferryl oxidation state, or by preventing the radicals from dissipating to surrounding proteins or other susceptible biological molecules.

Our recent studies provided evidence to support the hypothesis that Hp can indeed stabilize the highly reactive ferryl Hb formed as intermediate in the Hb oxidative reaction.[45] Figure 1 shows the ferryl

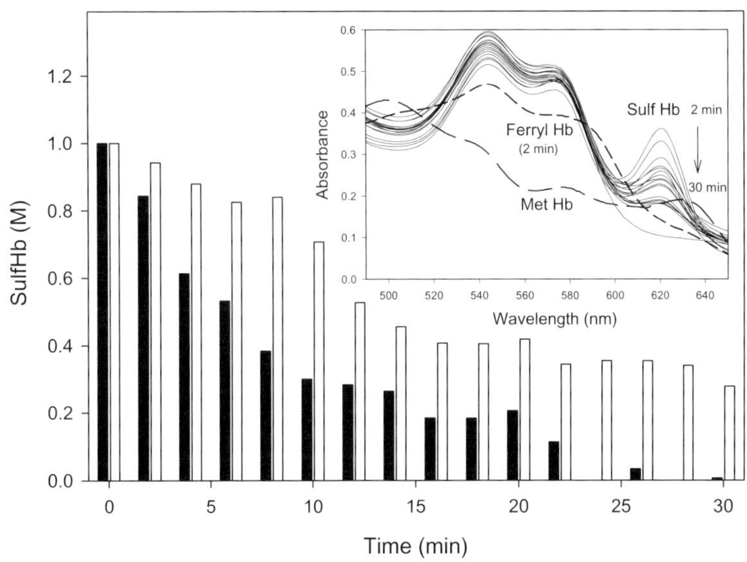

Fig. 1. Ferryl Hb formation and stabilization measured as sulfHb concentrations in the reaction of metHb and H_2O_2. MetHb (50 μM) was mixed with 1:5 molar ratio of H_2O_2 in 50 mM potassium phosphate buffer, pH 7.4, to initiate the oxidative reaction in a cuvette monitored on an Agilent 8453 diode array spectrophotometer. The ferryl Hb level was detected by adding 6 μL of 2 mM sodium sulfide (Na_2S) to form sulfHb at specific reaction time. The sulfHb concentrations were calculated using the extinction coefficient of 20.8 mM^{-1} cm^{-1} at 620 nm.[46] The black and white bars in the graph represent the sulfHb concentrations at indicated time intervals of metHb oxidative reaction in the absence and presence of Hp (in slight excess), respectively. The inset showed the initial metHb spectrum, the representative spectrum of ferryl Hb intermediate at 2 min, and the sulfHb spectra from 2 min to 30 min.

Hb levels detected as sulfHb in the reaction of metHb and H_2O_2 that can be quantified in a spectrophotometer using an extinction coefficient for sulfHb of 20.8 mM^{-1} cm^{-1}.[46] The intermediate ferryl Hb species reached maximum level shortly after the reaction was initiated, and started to decay over time. In the absence of Hp, the amount of ferryl Hb was reduced to baseline level at 30 min of reaction time. In contrast, the presence of Hp in slightly excess vs. Hb significantly slowed down the ferryl Hb decay under the same conditions. Since Hp does not alter the reactivity of Hb with H_2O_2, i.e. the pseudoperoxidase activity, [47] the stabilization of ferryl heme intermediate in the Hb-Hp complex may therefore explain the well-known antioxidant role of Hp, and may provide mechanistic insights on the protection against Hb-related oxidative damage. As Hb is cleared by the strong binding Hp, the Hb-Hp complex may also function as a shield absorbing Hb radicals and prevent oxidative damage to surrounding environment.

Potential Therapeutic Applications of Hp Antioxidant Activities

The development of therapeutic applications based on this newly discovered Hp antioxidant and protective properties against cell-free Hb toxicities appears to hold promise for the treatment of clinical conditions such as severe hemolysis or the use of HBOCs. Over the last several decades, tremendous effort has been devoted to the research and development of a wide range of HBOC product candidates including unmodified purified native Hb, internally stabilized Hb tetramers, and intermolecularly cross-linked or chemically modified Hb polymers. Several such HBOC candidates have demonstrated considerable therapeutic benefits in both animal studies and clinical trials, and thus reached advanced stages of drug development, i.e. Phase III clinical trial and Biological License Application (BLA). However, the safety criteria for licensed biological products have not been met by any of these products up to date in the United States largely due to high mortality rate and severe adverse events.[13,14]

The toxicities of HBOCs observed in animal models and clinical settings can be largely attributed to the uncontrolled intrinsic reactivities of cell-free Hb. For instance, the pressor effect of infused Hb is believed to be due to the rapid molecular reaction of the heme group with NO and which leads to the depletion of this vasodilatory signaling molecule. Subsequently, chemical cross-linking and polymerization methods were developed not only to increase the circulation time of the stabilized Hb but also to reduce the extravasation of free Hb and ultimate NO scavenging. Additionally, the autoxidation and redox reactions of the heme group are accompanied by the generation of reactive oxygen species (ROS) and free radicals that can cause extensive tissue and organ damage and could potentially contribute to toxicities and serious adverse events seen animals and in humans. The various modification approaches developed so far have not been able to attenuate the intrinsic toxicities of free Hb for the safe *in vivo* use of these products. Hp association with modified Hbs is one of the potential protective mechanisms that can be explored in the development of safe HBOCs.[48]

Haptoglobin Interactions with Modified Hemoglobins

The combination of Hp and native Hb has been described as a part of the detoxification and clearance pathway of cell-free Hb from circulation. To explore the possible protective benefits Hp may provide to cell-free Hb-based blood substitutes, a recent study has evaluated first the binding capabilities of Hp with several chemically modified Hbs using surface plasmon resonance (SPR) techniques.[49] Chemical modifications in Hb derivatives usually involved intramolecular cross-linking through either αα-subunits or ββ-subunits to stabilize the tetramer, and additional surface decoration or polymerization to increase the molecular size. It has been demonstrated that Hp 2–2 binds to all Hbs with 4-fold lower affinity than Hp 1–1 in this study. More interestingly, while native Hb displayed the strongest binding as predicted, the αα-cross-linked Hbs showed very weak, or even no binding when polymerized with Hp. In contrast, the ββ-cross-linked Hbs demonstrated much reduced Hp affinities compared to native Hb, but stronger than that of αα-cross-linked Hbs. These results indicate that the dissociation of Hb into αβ dimers may not be a prerequisite for Hp binding as previously reported, whereas the integrity of Hb α subunit structure is important for Hp interaction.

We have recently measured the Hp association with modified Hbs in solution based on the quenching of the Hp fluorescence.[45,50] Figure 2 shows the stopped-flow analysis data that are consistent with the observations described above. The kinetic trace of the rapid combination reaction of Hp and native Hb was illustrated in Panel A of Fig. 2. The fluorescence quenching followed an exponential process and that can be completed within 5 sec under the conditions employed. Conversely, no fluorescence change could be detected at any time scale after the mixing of Hp solution with αα-cross-linked Hb (αα-DBBF), indicating no significant binding has occurred (Panel B). The ββ-cross-linked Hb (ββ-DBBF) on the other hand showed a slower but clear binding reaction with Hp as confirmed by the comparison of second order rate constants (Panel C and D). The presence of the slow second kinetic phase in Panel C is consistent with higher polymer formation observed previously between Hp and ββ-cross-linked Hb.[50] More extensive kinetic analysis is necessary to elucidate the characteristics of Hp interaction with modified Hbs.

Effects of Hp Binding on Modified Hb Ligand Reactivities

Since the Hp binding with ββ-DBBF was clearly demonstrated, it is interesting to evaluate its effect on ligand reactions in comparison with native Hb. We carried out the oxygen equilibrium binding studies of native and modified Hbs with and without Hp (Fig. 3). The typical oxygen equilibrium curve (OEC) of native Hb is of sigmoidal shape providing a p50 value (the oxygen partial pressure at which Hb is half saturated) of 13.5 mmHg and a Hill coefficient for oxygen cooperative binding at 2.6 under our experimental conditions.[51] The presence of Hp clearly induced changes in both the OEC position and

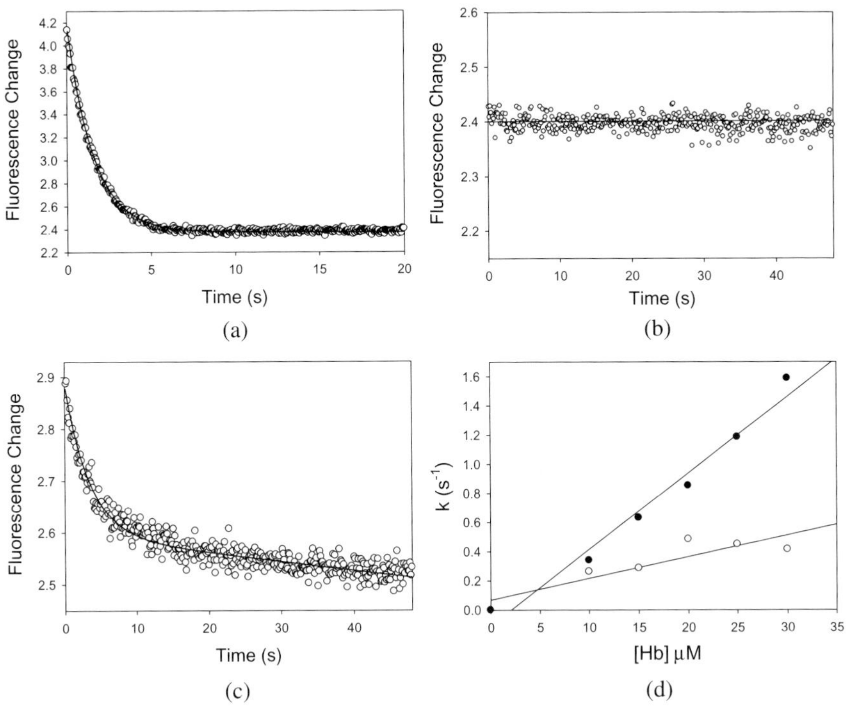

Fig. 2. Stopped-flow fluorescence measurement of the Hb and Hp binding. The reaction of Hb and Hp was monitored in an Applied Photophysics stopped-flow instrument (Leatherhead, UK). Hp solutions (0.1 mg/mL) were mixed with Hb solutions of increasing concentrations up to 30 μM in 50 mM sodium phosphate buffer, pH 7.4. The fluorescent change of the reaction was measured with an excitation wavelength at 285 nm, and a cutoff filter >360 nm for emission as function of time. The time courses of Hp binding with 15 μM of Hbs were fitted to exponential equations to obtain the pseudo-first-order rate constants for each reaction: (a) native Hb; (b) αα subunits cross-linked Hb by bis(3,5-dibromosalicyl) fumarate (αα-DBBF); and (c) ββ subunits cross-linked Hb by bis(3,5-dibromosalicyl) fumarate (ββ-DBBF). D. The second order rate constants of Hp binding with native Hb (close circle) and ββ-DBBF (open circle) were obtained from the slope of the linear relationship of the apparent association rate constants as a function of Hb concentration.

shape of native Hb with the curve left shifted and less sigmoidal. The p50 value and Hill coefficient of the Hp-Hb complex were reduced to 4.8 mmHg and 1.1, respectively. In contrast, the chemical modifications significantly reduced the oxygen affinity of αα-DBBF

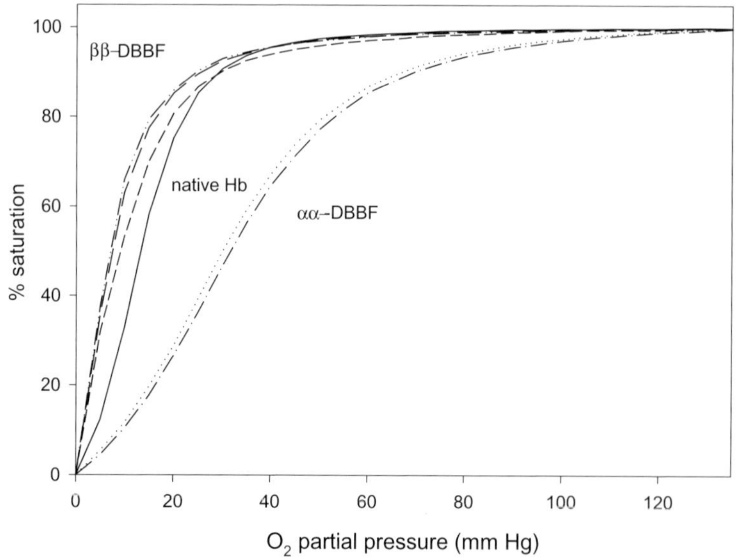

Fig. 3. Oxygen equilibrium curves (OECs) of native Hb, αα-DBBF, and ββ-DBBF in the presence and absence of Hp. The OECs were measured in the Hemox Analyzer (TCS Scientific, New Hope, PA) equipped with a Clark oxygen electrode (Model 5331 oxygen probe, Yellow Springs Instrument, Yellow Springs, OH) in 0.01 M phosphate buffer, pH 7.4, containing 0.1 M NaCl, anti-foaming agent, and Hayashi enzymatic system at 37°C. The Hb concentration of the reaction was typically between 60–75 μM (heme) and the Hp concentration is slightly in excess to form Hb-Hp complex. The OECs were generated by calculating the Hb oxygen percent saturation (%) vs. increasing oxygen partial pressures (mmHg): solid line – native Hb; short dash – native Hb + Hp; dash dot – αα-DBBF; dotted line – αα-DBBF + Hp; long dash – ββ-DBBF; and dash dot dot – ββ-DBBF + Hp.

with the right-shifted OEC (p50 = 31 mmHg) as expected, while the addition of Hp did not alter oxygen equilibrium binding properties of αα-DBBF, probably due to lack of binding. Interestingly, while ββ-cross-linking shifted the OEC to the left with a p50 value at about 7 mmHg, the Hp binding with ββ-DBBF did not change the oxygen binding characteristics under equilibrium conditions. Our preliminary results suggested that the chemical nature of Hp interactions with native Hb and ββ-DBBF may be different as reflected in their functional disparities. More detailed studies are necessary to characterize the equilibrium and kinetic ligand reactivities, as well as redox properties of modified Hbs complexed with Hp.

Concluding Remarks and Future Directions

Manufactured oxygen carriers that can safely substitute for blood transfusion have broad potential therapeutic applications. However, previous HBOC candidates have not been adequately demonstrated to be sufficiently safe for human use in extensive biochemical analyses, animal studies, and clinical trials over a long time period. It is imperative to bring about new protective strategies in designing detoxified HBOC products. One such approach that has been proposed is based on the natural protective mechanism of Hp association with Hb. The Hp interaction with Hb and its effects on Hb ligand binding and redox reactions have been well studied and could serve as the basis for further design of safe and effective blood substitutes. Various Hb molecular modifications are typically required for the development of HBOCs, however, much less knowledge has been acquired about the Hp and modified Hbs interaction and associated reactivity changes. Furthermore, the neutralization mechanism of highly reactive Hb radicals provided by Hp has yet to be clearly elaborated. The lack of fundamental understanding of underlying molecular mechanisms is a proven hindrance for explaining the toxicities discovered in animal studies and clinical trials at later stages of drug development.

References

1. Drexler KE, Peterson C, Pergamit G. (1991) *Unbounding the Future: The Nanotechnology Revolution.* William Morrow, New York.
2. Chang TMS. (2007) *Artificial Cells: Biotechnology, Nanotechnology, Blood Substitutes, Regenerative Medicine, Bioencapsulation, Cell/Stem Cell Therapy.* World Science Publisher/Imperial College Press, Singapore/London.
3. Gupta AS. (2011) Nanomedicine approaches in vascular disease: a review. *Nanomedicine* **7:** 763–779.
4. Chang TM. (2009) Nanobiotechnology for hemoglobin-based blood substitutes. *Crit. Care Clin.* **25:** 373–382.
5. Alayash AI. (2004) Oxygen therapeutics: can we tame haemoglobin? *Nature Rev* **3:** 152–159.

6. Chang TM. (2010) Blood replacement with nanobiotechnologically engineered hemoglobin and hemoglobin nanocapsules. *Wiley Interdiscip Reviews* **2**:418–430.
7. Riess JG. (2001) Oxygen carriers ("blood substitutes") — raison d'etre, chemistry, and some physiology. *Chem Rev* **101:** 2797–2920.
8. Buehler PW, Alayash AI. (2008) All hemoglobin-based oxygen carriers are not created equally. *Biochim Biophys Acta* **1784:** 1378–1381.
9. Jia Y, Alayash AI. (2009) Effects of cross-linking and zero-link polymerization on oxygen transport and redox chemistry of bovine hemoglobin. *Biochim Biophys Acta* **1794:** 1234–1242.
10. D'Agnillo F, Chang TM. (1998) Polyhemoglobin-superoxide dismutase-catalase as a blood substitute with antioxidant properties. *Nature Biotechnol* **16:** 667–671.
11. Vasseur-Godbillon C, Sahu SC, Domingues E, *et al.* (2006) Recombinant hemoglobin betaG83C-F41Y. *FEBS J* **273:** 230–241.
12. Buehler PW, D'Agnillo F, Schaer DJ. (2010) Hemoglobin-based oxygen carriers: from mechanisms of toxicity and clearance to rational drug design. *Trends Mol Med* **16:** 447–457.
13. Silverman TA, Weiskopf RB. (2009) Hemoglobin-based oxygen carriers: current status and future directions. *Anesthesiology* **111:** 946–963.
14. Natanson C, Kern SJ, Lurie P, *et al.* (2008) Cell-free hemoglobin-based blood substitutes and risk of myocardial infarction and death: a meta-analysis. *JAMA* **299:** 2304–2312.
15. Lim SK. (2001) Consequences of haemolysis without haptoglobin. *Redox Rep* **6:** 375–378.
16. Kristiansen M, Graversen JH, Jacobsen C, *et al.* (2001) Identification of the haemoglobin scavenger receptor. *Nature* **409:** 198–201.
17. Boretti FS, Buehler PW, D'Agnillo F, *et al.* (2009) Sequestration of extracellular hemoglobin within a haptoglobin complex decreases its hypertensive and oxidative effects in dogs and guinea pigs. *J Clin Invest* **119:** 2271–2280.
18. Ostrowski RS, Travis JC, Talley ES. (1987) The association of Hp 1 and sickle cell disease. *Hum Hered* **37:** 193–195.
19. Javid J. (1978) Human haptoglobins. *Curr Top Hematol* **1:** 151–192.
20. Levy AP, Asleh R, Blum S, *et al.* (2010) Haptoglobin: basic and clinical aspects. *Antioxid Redox Signal* **12:** 293–304.
21. Chiancone E, Alfsen A, Ioppolo C, *et al.* (1968) Studies on the reaction of haptoglobin with haemoglobin and haemoglobin chains. I. Stoichiometry and affinity. *J Mol Biol* **34:** 347–356.
22. Jayle MF, Moretti J, Mouray H. (1962) Metabolism of haptoglobin and its role in the metabolism of hemoglobin. *Nouv Rev Fr d'Hematol* **2:** 473–482.
23. Nagel RL, Whittenberg JB, Ranney HM. (1965) Oxygen equilibria of the hemoglobin-haptoglobin complex. *Biochim Biophy Acta* **100:** 286–289.

24. Nagel RL, Gibson QH. (1971) The binding of hemoglobin to haptoglobin and its relation to subunit dissociation of hemoglobin. *J Biol Chem* **246:** 69–73.
25. Polticelli F, Bocedi A, Minervini G, Ascenzi P. (2008) Human haptoglobin structure and function — a molecular modelling study. *FEBS J* **275:** 5648–5656.
26. Hwang PK, Greer J. (1980) Interaction between hemoglobin subunits in the hemoglobin . haptoglobin complex. *J Biol Chem* **255:** 3038–3041.
27. Nagel RL, Gibson QH. (1967) Kinetics and mechanism of complex formation between hemoglobin and haptoglobin. *J Biol Chem* **242:** 3428–3434.
28. Alfson A, Chiancone E, Wyman J, Antonini E. (1970) Studies on the reaction of haptoglobin with hemoglobin and hemoglobin chains. II. Kinetics of complex formation. *Biochim Biophys Acta* **200:** 76–80.
29. Nagel RL, Rothman MC, Bradley TB Jr, Ranney HM. (1965) Comparative haptoglobin binding properties of oxyhemoglobin and deoxyhemoglobin. *J Biol Chem* **240:** 4543–4545.
30. Michalski LA, McDonald MJ. (1988) An investigation of human oxyhemoglobin beta tetramer dissociation using haptoglobin binding. *Biochem Biophys Rese Commun* **156:** 438–444.
31. Romeo PH, Thillet J, Tsapis A, Rosa J. (1982) Kinetics of subunit dissociation of partially oxygenated hemoglobin determined by haptoglobin binding. *Biophys Res Commun* **105:** 1354–1360.
32. Chiancone E, Antonini E, Brunori M, *et al.* (1973) Kinetics of the reaction between oxygen and haemoglobin bound to haptoglobin. *Biochem J* **133:** 205–207.
33. Nagel RL, Gibson QH. (1966) Kinetics of the reaction of carbon monoxide with the hemoglobin-haptoglobin complex. *J Mol Biol* **22:** 249–255.
34. Sadrzadeh SM, Bozorgmehr J. (2004) Haptoglobin phenotypes in health and disorders. *Am J Clin Pathol* **121 Suppl:** S97–S104.
35. Teisberg P, Gjone E. (1974) Probable linkage of LCAT locus in man to the alpha haptoglobin locus on chromosome 16. *Nature* **249:** 550–551.
36. Yang F, Brune JL, Baldwin WD, *et al.* (1983) Identification and characterization of human haptoglobin cDNA. *Proc Natl Acad Sci USA* **80:** 5875–5879.
37. Langlois MR, Delanghe JR. (1996) Biological and clinical significance of haptoglobin polymorphism in humans. *Clin Chem* **42:** 1589–1600.
38. Carter K, Worwood M. (2007) Haptoglobin: a review of the major allele frequencies worldwide and their association with diseases. *Int J Lab Hematol* **29:** 92–110.
39. Valette I, Waks M, Wejman JC, *et al.* (1981) Haptoglobin heavy and light chains. Structural properties, reassembly, and formation of minicomplex with hemoglobin. *J Biol Chem* **256:** 672–679.
40. McCormick DJ, Atassi MZ. (1990) Hemoglobin binding with haptoglobin: delineation of the haptoglobin binding site on the alpha-chain of human hemoglobin. *J Protein Chem* **9:** 735–742.

41. Pimenova T, Pereira CP, Gehrig P, *et al.* (2010) Quantitative mass spectrometry defines an oxidative hotspot in hemoglobin that is specifically protected by haptoglobin. *J Proteome Res* **9:** 4061–4070.
42. Gutteridge JM. (1987) The antioxidant activity of haptoglobin towards haemoglobin-stimulated lipid peroxidation. *Biochim Biophys Acta* **917:** 219–223.
43. Cerda S, Oh SK. (1990) Methods to quantitate human haptoglobin by complexation with hemoglobin. *J Immunol Methods* **134:** 51–59.
44. Melamed-Frank M, Lache O, Enav BI, *et al.* (2001) Structure-function analysis of the antioxidant properties of haptoglobin. *Blood* **98:** 3693–3698.
45. Jia Y, Wood F, Buehler PW, *et al.* (2013) Haptoglobin preferentially binds β but not α subunits cross-linked hemoglobin tetramers with minimal effects on ligand and redox reactions. PLoS One **8:** e59841.
46. Carrico RJ, Peisach J, Alben JO. (1978) The preparation and some physical properties of sulfhemoglobin. *J Biol Chem* **253:** 2386–2391.
47. Buehler PW, Abraham B, Vallelian F, *et al.* (2009) Haptoglobin preserves the CD163 hemoglobin scavenger pathway by shielding hemoglobin from peroxidative modification. *Blood* **113:** 2578–2586.
48. Alayash AI. (2011) Haptoglobin: old protein with new functions. *Clin Chim Act* **412:** 493–498.
49. Buehler PW, Vallelian F, Mikolajczyk MG, *et al.* (2008) Structural stabilization in tetrameric or polymeric hemoglobin determines its interaction with endogenous antioxidant scavenger pathways. *Antioxid Redox Signal* **10:** 1449–1462.
50. Benesch RE, Ikeda S, Benesch R. (1976) Reaction of haptoglobin with hemoglobin covalently cross-linked between the alpha beta dimers. *J Biol Chem* **251:** 465–470.
51. Jia Y, Ramasamy S, Wood F, *et al.* (2004) Cross-linking with O-raffinose lowers oxygen affinity and stabilizes haemoglobin in a non-cooperative T-state conformation. *Biochem J* **384:** 367–375.

Chapter 8

Polynitroxylated Hemoglobin as a Multifunctional Therapeutic for Critical Care and Transfusion Medicine

Li Ma* and Carleton Jen Chang Hsia†

Abstract

Polynitroxylation (PN) technology is a chemical modification process, which covalently links multiple nitroxides to various macromolecules, including hemoglobin-based oxygen carriers (HBOCs). Like free nitroxides, PN products possess superoxide dismutase mimetic activity. In addition, when redox coupled with heme centers, polynitroxylated hemoglobins have catalase and peroxidase mimetic activities. PN of HBOCs allows the therapeutic benefits of nitroxides to be target delivered intravascularly. More significantly, it corrects the intrinsic toxicity of HBOCs and adds therapeutic neurovascular protective, anti-oxidant, and anti-inflammatory activities to these products. Results from multiple pre-clinical efficacy studies suggest that PN improves oxygen delivery by HBOCs through the correction of inadequate blood flow via prevention of both nitric oxide depletion and reperfusion/inflammation injuries after ischemia. PN of conjugated hemoglobins

* Department of Physics, Georgia Southern University, Statesboro, GA 30460.
† SynZyme Technologies LLC, Irvine, CA 92618.

with hyper-colloid effects, such as polyethylene glycol or dextran conjugated hemoglobins, converts these HBOCs into small volume multifunctional hemodynamic therapeutics. Polynitroxylated pegylated hemoglobin (PNPH), in particular, has the potential to be a high-therapeutic-index next-generation HBOC for treatment of civilian unmet medical needs and military combat casualty care. Efficacy of PNPH has been demonstrated in models of traumatic brain injury with hemorrhagic shock, ischemic stroke, and sickle cell disease. This chapter traces the research leading to the development of PNPH.

Introduction

Transfusion of homologous donated blood has long been associated with potential disease transmission and more recently with increased mortality due to red cell aging.[1] The outbreak of the AIDS epidemic stimulated a major drive to develop hemoglobin-based oxygen carriers (HBOCs) in the early 1980s, and this development was specifically encouraged by the United States (US) military for combat casualty care. A major R&D effort on HBOCs at the Lettermen Army Institute of Research was initiated in the late 1980s. The drive to develop possible HBOCs followed different strategies. Some of the strategies were: diaspirin cross-linked human hemoglobin (DCLHb, aka, HemAssist® or ααHb) by Baxter Healthcare,[2] polymerized bovine hemoglobin by BioPure Corporation,[3] pegylated bovine hemoglobin by Enzon,[4] polymerized human hemoglobin by Northfield Laboratories[5] and Hemosol Inc.,[6] pegylated human hemoglobin by Sangart Inc.,[7] and recombinant hemoglobin by Somatogen Inc.[8] These HBOCs were all aimed at creating an oxygen carrier to reduce the need of homologous blood transfusion under the guidance of the US FDA (Points to Consider on the Safety Evaluation of Hemoglobin-Based Oxygen Carriers, dated August 27, 1990). Baxter's DCLHb clinical trial showed increased mortality in the treatment group leading to the conclusion that DCLHb was not an effective resuscitative fluid.[9] Even though other current generation HBOC developers attempted to minimize the intrinsic cell free hemoglobin vasoconstrictive activity from nitric oxide scavenging, these attempts were largely unsuccessful, which ultimately manifested as increased mortality in the advanced clinical trials of all current generation HBOCs.[10] This increased mortality led to the 2008 Food and Drug Administration/National Institutes of Health (FDA/NIH) cosponsored workshop on HBOCs calling for the development of higher-therapeutic-index next-generation HBOCs (FDA/NIH Workshop: Hemoglobin based oxygen carriers: current status and future directions. Bethesda, Maryland, Wednesday, April 29–30, 2008) and the FDA putting a hold on clinical development of current generation HBOCs. This in turn led to a high attrition rate among HBOC developers and to further clinical development of HBOCs being done outside of the US. Time will tell whether these HBOCs being developed outside the US will be approved for use in the US.

An evolutionarily selected cell free invertebrate polymeric oxygen carrier of marine in origin is being developed by Hemarina SA.[11] Co-polymerization of human hemoglobin with superoxide dismutase (SOD), catalase (CAT) and carbonic anhydrase is also being attempted.[12] Again the jury is out regrading whether or not the evolutionarily developed oxygen carrier or chemically co-polymerized hemoglobin with three major red blood cell enzymes will have the safety and efficacy to be the next-generation HBOCs called for by FDA.

Thus, there still remains a need for a safe and effective next-generation HBOC for treatment of hemorrhagic shock, trauma, and other types of ischemia/reperfusion injury. The ideal high-therapeutic-index next-generation HBOC would not require refrigeration and cross-matching, could be manufactured in large quantities, would have a long shelf life, and would reduce the risk of iatrogenic infection. However, the history of poor clinical efficacy and increased risk of death and myocardial infarction[10] has resulted in the negative views of HBOCs held by the medical, regulatory, and investment communities. Many published reviews by experts in the HBOC field have discussed extensively the past, current and future of HBOC development.[13–19] Going forward, the intrinsic toxicity of hemoglobin must be controlled in order to harvest its unique oxygen-carrier efficacies. The acute toxicity of HBOCs is thought to be due primarily to vasoconstriction leading to inadequate blood flow and the resulting organ ischemia. Although multiple mechanisms may contribute to excessive vasoconstriction induced by HBOCs, the most important factor is almost certainly their nitric oxide (NO) scavenging property.[20] It has been discussed that an imbalance between NO and superoxide ($O_2^{\bullet-}$) causes inadequate blood flow and oxidative stress in the reperfusion of ischemic tissue. In support of this hypothesis, it is well known that cell-free hemoglobin binds NO. Additionally, the redox reaction of hemoglobin generates hydrogen peroxide and causes over-production of $O_2^{\bullet-}$, which reacts with NO to produce peroxynitrite. These factors cause NO depletion and oxidative stress. As a result of NO depletion at the endothelium, vasoconstriction occurs, causing inadequate blood flow, which neutralizes the net oxygen-carrying capacity of the transfused HBOCs. An intact red blood cell contains hemoglobin and an elaborate enzymatic system, including (1) SOD, to detoxify the free radical $O_2^{\bullet-}$ generated by normal binding and release of oxygen by hemoglobin by dismutating the $O_2^{\bullet-}$ to hydrogen peroxide (H_2O_2), and (2) CAT, which converts the H_2O_2 to O_2 and water. Purified cell free hemoglobin, as used in current generation HBOCs, does not have these antioxidant enzyme activities.

This chapter presents polynitroxylated hemoglobin (PNH), developed by SynZyme Technologies, as a class of high-therapeutic-index next-generation HBOCs that may have the potential to move forward from pre-clinical development to IND and to bedside as a drug in the US. PNH is a hemoglobin molecule covalently decorated by multiple nitroxides. Polynitroxylation (PN) maintains the oxygen-carrying and delivery capabilities of the hemoglobin while adding multiple therapeutic activities such as vascular and multiple-organ protective activities including restoration of inadequate blood flow without oxidative stress or injuries. The nitroxide and heme iron act in concert to perform antioxidant activities that mimic the behavior of the two physiologically important antioxidant enzymes that are found in red blood cells, SOD and CAT. Thus PN technology restores the missing red cell enzymatic functionality to the HBOC, thereby reducing $O_2^{\bullet-}$ concentration, H_2O_2 concentration, and NO depletion, which may overcome the barrier to the development of current generation HBOCs in the US.

From this class of PNHs, a polynitroxylated pegylated hemoglobin (PNPH) has been selected for development as a next generation therapeutic HBOC by SynZyme Technologies.[13] PNPH has recently been described as a nanopartical where a polyethylene glycol hydration shell surrounds a core consisting of a single hemoglobin molecule, which has been covalently labeled with multiple nitroxides. The multifunctional hemodynamic therapeutic efficacy of PNPH has been demonstrated in several models. In a mouse model of traumatic brain injury (TBI) complicated by hemorrhagic shock (HS), PNPH reduces the volume requirements for attaining hemodynamic stability, maintains cerebral oxygen delivery and provides neuroprotection.[21] Neuroprotection was also demonstrated in preliminary results with PNPH in a rat model of ischemic stroke.[22] Furthermore, PNPH was found to correct global NO deficiency in transgenic sickle cell disease mice via removal of over production of superoxide from chronic red cell hemolysis.[23] The pathophysiology of the stroke model appears to involve correction of inadequate cerebral blood flow by PNPH with minimized reperfusion injury to the penumbra region of the core of ischemic brain. This mechanism is supported by the published results for polynitroxylated albumin (PNA).[24] The pre-clinical efficacies of PNPH in these three major disease models demonstrate the high therapeutic index needed for development as a next-generation HBOC. The sections below review the background research leading to the development and selection of PNPH as a next-generation therapeutic HBOC.

Nitroxide and PN Technology

PN covalently links multiple nitroxides to various macromolecules including a number of HBOCs. The nitroxides on a macromolecule have distinct therapeutic advantages over low molecular weight nitroxides, such as Tempo (where R=H) and Tempol (where R=OH) shown in Fig. 1(a).[25] Nitroxides have been shown to act as mimics of SOD[26,27] and CAT,[28] inhibit peroxidation of lipids and lipoproteins,[29,30] and inhibit peroxynitrite-mediated nitration.[31] They are effective agents in the protection of cells against reactive oxygen species mediated damage

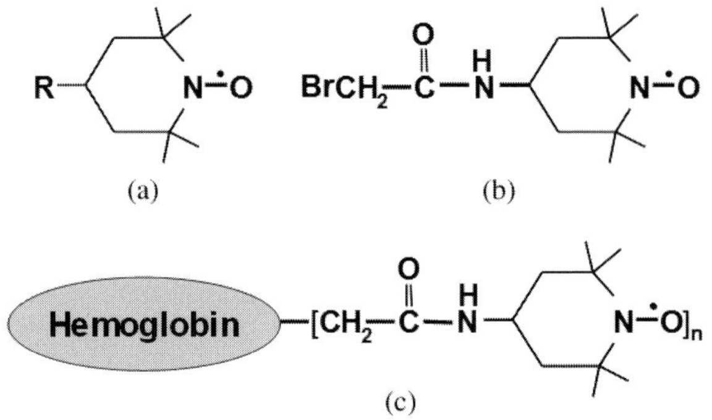

Fig. 1. Structures of BrAcTPO, PNH, and low molecular weight nitroxides. (a) Low molecular weight nitroxides: TEMPO (R=H), Tempol (R=OH), (b) 4-(2-bromoacetamido)-2,2,6,6-tetramethyl-1-piperidinyloxy (BrAcTPO), (c) Polynitroxylated hemoglobin, where hemoglobin is ααHb, ββHb, recombinant cross-linked, O-raffinose cross-linked, glutaraldehyde polymerized, Dextran conjugated, or polyethylene glycol conjugated hemoglobin.

including inflammation,[32,33] and inhibition of injury caused by ionizing radiation[34,35] and by ultraviolet radiation.[36] Recently, their antihypertensive effects,[37] their vasodilatation effects,[38] their neuroprotective effects in animal models of TBI,[39] and their protection of spinal cord mitochondria from oxidative stress[40] have been reported. The therapeutic efficacy and mechanism of physiological effects of nitroxides has been comprehensively reviewed.[41] Unfortunately, the problem with low molecular weight nitroxides is that they show poor pharmacokinetics because they are isotropically distributed *in vivo* and rapidly bio-inactivated to their reduced hydroxylamine form with half-life on the order of minutes.

Thus, PN is a novel technology for improving the therapeutic index of nitroxides by linking multiple nitroxides to macromolecules for delivery to the vasculature. Linking multiple nitroxides to macromolecules such as albumin to yield PNA has been shown to prolong the half-life of the nitroxide *in vivo*.[42,43] PNA also inhibits post-ischemic leukocyte-endothelial cell adhesion and leukocyte migration in a mesenteric model of ischemia/reperfusion in the rat[44] and inhibits vaso-occulsion in transgenic mouse models of sickle cell disease through the mechanisms of its inhibition of adhesion of the sickle red blood cells, white blood cells and platelets to endothelium.[45,46] Additionally, PNA was shown to profoundly reduce infarct volume following transient middle cerebral artery occlusion in the rat by mechanisms that may include both improvement of cerebral perfusion and inhibition of the post-ischemic reperfusion and inflammatory injuries.[24,47] In other recent studies of PNA, PNA treatment was shown to reduce early hyperemia and improve cerebral blood flow and neurologic outcome in a rat model of pediatric asphyxial cardiac arrest,[48,49] reduce hyperoxia induced oxidative injuries from breathing 100% oxygen in a rat model,[49] and improve hemodynamics and survival and reduce neuronal death in a TBI model.[50]

Through research agreements, PN technology was applied to the products of most major HBOC developers, such as Baxter's HemAssist®, Somatogen's Optro®, Hemosol's Hemolink®, and Northfield's PolyHeme® to form PN$\alpha\alpha$Hb, PNrHb1.1, PNHL, and PNP, respectively. PN was also applied to SynZyme's Dextran conjugated hemoglobin, SynZyme's polyethylene glycol conjugated human hemoglobin, and Prolong's polyethylene glycol conjugated bovine hemoglobin to form PNDxHb, hPNPH, and PNPH, respectively. In general PN does not alter the oxygen-carrier function of the HBOCs. In the case of PNPH, its oxygen-binding capability, P_{50}, viscosity, and hyper-colloid properties are essentially the same as polyethylene glycol conjugated hemoglobin (PegHb). Table 1 lists physical and biochemical properties of several PN modified HBOCs. PNHs can be stored at room temperature in their carboxyl forms with P_{50} about 10 mmHg or below. PN technology converts a pro-oxidant hemoglobin to an antioxidant hemoglobin regardless of the type of other modifications present in the parent HBOC.

In general, PN is accomplished via a reaction of 4-(2-bromoacetamido)-2,2,6,6-tetramethyl-1-piperidinyloxy (BrAcTPO) (Fig. 1(b)) with macromolecules in which Br is a leaving group forming PN products with the molecular structure shown in Fig. 1(c). In the case of PNPH, the macromolecule or HBOC is PegHb. To confirm that the reaction occurs, electron paramagnetic resonance

Table 1. **Characteristics of Polynitroxylated Products**

Characteristic	Polynitroxylated Products				
	PNααHb	PNββHb	PNDxHb	PNPH	*hPNPH
Hb Concentration (%)	10	6	6	4	4
Nitroxides per Heme	4	3.5	2.5–3	~3	3–3.5
P_{50} (mm Hg)	~6	~12	~8	~10	~10
Average Molecular Weight (kDa)	~67	~67	~210	120	~110
% Hb Tetramer	>98	>98	<2	<2	<2
Viscosity	~1	~1	~3	3–4	3–4
Colloid Osmotic Pressure (mm Hg)	~40	~25	~45	~45	~45

*hPNPH is human polynitroxylated PEGHb with 2 × 20kDa polyethylene glycols.

(EPR) spectroscopy is used. EPR is a sensitive method to monitor the reaction, control the quality of final product and analyze pharmacokinetic samples. Nitroxides are a class of stable free radicals with one unpaired electron per molecule. The unpaired electron energy level will split into two levels in a magnetic field, spin up and spin down. The energy difference between these two levels falls into the microwave range. Normally, the unpaired electron of the nitroxide stays at the lower spin energy state. When the nitroxide absorbs microwave energy, the spin of the unpaired electron will flip to the higher energy state. The intensity and the shape of the EPR spectrum will give quantitative and qualitative information of the PN products. Figure 2(a) shows the EPR spectrum of nitroxides before reaction with PegHb, and Fig. 2(b) is the EPR spectrum of PNPH after the reaction. A difference in the shape and intensity of the two spectra is expected and reflects the reduced mobility of the nitroxides when covalently labeled to PegHb. The covalent bonds between the nitroxide and the amino groups of macromolecules are quite stable *in vivo*. Figures 2(c) and 2(d) show the EPR spectra of PNA before and after circulation for 15 days in dogs. Similar *in vivo* stability of the EPR spectra of PNH[51] was shown in guinea pigs 36 hr post transfusion.

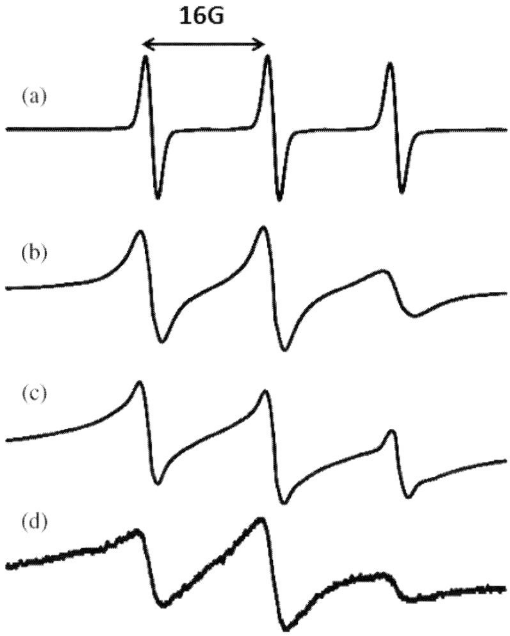

Fig. 2. Electron Paramagnetic Resonance (EPR) spectra of: (a) BrAcTPO in dimethyl sulfoxide, (b) Polynitroxylated polyethylene glycol conjugated bovine hemoglobin (PNPH), (c) Polynitroxylated human albumin (PNA), (d) Dog plasma sample 15 days after PNA administration at 10 mL/kg dose. EPR conditions: microwave power, 2 mW; modulation amplitude, 0.5 G; modulation frequency, 100 kHz; receiver gain, 4.0×10^4; temp, 300 K.

Polynitroxylated αα Cross-Linked Hemoglobin (PNααHb)

Baxter's DCLHb (ααHb) was designed to have high hemoglobin concentration, high P_{50}, and low viscosity as ααHb tetramers. The ααHb showed promising results in animal models but failed in two large clinical trials due to increased mortality in the treatment group.[9] Pressor activity induced by ααHb administration was believed to be responsible for the lack of efficacy in the clinical trials. Using PN technology to attach approximately 16 nitroxides to each ααHb tetramer, a PNααHb was prepared and shown to reduce the vasopressor effect and oxidative stress. Selected physical and chemical characteristics are listed in Table 1. PN significantly attenuated the pressor effect of ααHb and showed anti-inflammatory effects in both *in vitro* and *in vivo* studies.

PN Attenuates Vasoconstriction

To test the pressor effect of PNααHb, Sprague Dawley rats were chronically cannulated for hemodynamic monitoring. Test preparations were: PNααHb (10 g/dL, $n = 5$), and ααHb (10 g/dL, n = 5). Test articles were infused via the tail vein at a dose of 600 mg/kg (6 mL/kg) top load over 10 min. The rats were allowed to freely move in the cages. Mean arterial pressures (MAP) were recorded starting 15 min before infusion to collect baseline data ($n = 4$ in each group) and continued up to 90 min after the start of infusion. The result showed 5 ± 2 mmHg or 23 ± 5 mmHg increases in MAP on average during 10–90 min post administration of PNααHb or ααHb, respectively. PNααHb significantly attenuated vasoconstriction compared to ααHb. The PNααHb had a purity of greater than 98% hemoglobin tetramer. Complete elimination of the pressor effect in a preparation of PNDxHb free of 64 kDa tetramer is discussed later.

In a separate study, PN of Somatogen's recombinant hemoglobin (rHb1.1) was performed and the resulting PNrHb1.1 had 16 covalently labeled nitroxides per rHb1.1 with more than 97% purity of hemoglobin tetramer. The systemic hemodynamic responses to rHb1.1 ($n = 8$) and PNrHb1.1 ($n = 4$) were then compared in awake rats at 350 mg/kg top-load dose. The results indicated the following (i) there were no observable signs of discomfort in the PNrHb1.1 treated group; (ii) the change in mean arterial pressure elicited by PNrHb1.1 is less than that observed with rHb1.1 (Fig. 3(a)); (iii) the change in heart rate induced by PNrHb1.1 was nearly identical to that induced by rHb1.1; (iv) the cardiac output improved but did not reach a statistically significant difference; and (v) total peripheral resistance elicited by PNrHb1.1 was markedly less than that elicited by rHb1.1 (Fig. 3(b)).

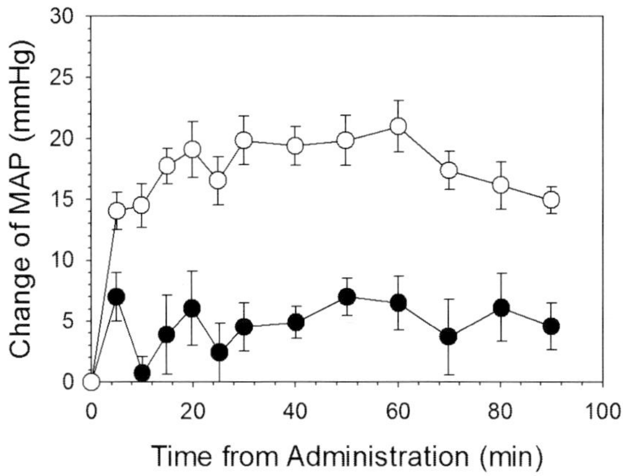

(a)

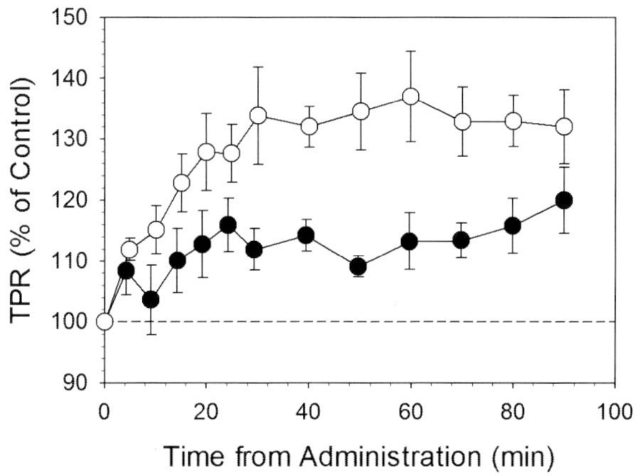

(b)

Fig. 3. Change in MAP and peripheral resistance. (a) Change in mean arterial pressure (MAP) measured in mmHg, (b) Change in total peripheral resistance (TPR) expressed as percent of control. Change is measured after administration of 350 mg/kg top load in rats of rHb1.1 ($n = 8$, ○) and PNrHb1.1 ($n = 4$, ●). Data are shown as mean ± s.e.

NO Binding Unaffected by PN

In order to answer the question of whether or not PN changes NO binding to the hemoglobin molecule, NO binding reactions of different hemoglobin formulations were measured using stop-flow and flash

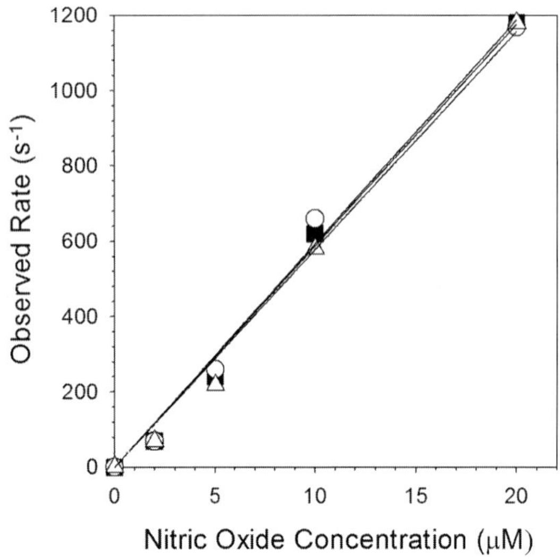

Fig. 4. The observed rate of nitric oxide binding by Hemoglobin A_0, (■) ααHb (○), and PNααHb (Δ). Rate is measured by flash photolysis in 0.1 M phosphate buffer, pH 7, 20°C.

photolysis techniques.[52] Sample stability was confirmed by optical absorption spectra measured before and after the flash photolysis measurements. No differences in the reaction rates were observed among hemoglobin A_0, ααHb, and PNααHb (see Fig. 4). We conclude that PN does not change the NO binding rate *in vitro*. Because PN of HBOCs eliminates the vasoconstriction that normally is associated with their transfusion, these results support the hypothesis that NO binding may not be the sole cause of vasoconstriction during exposure of blood vessels to HBOCs *in vivo*.

Selected In Vivo and In Vitro Studies of PNααHb

We have observed that PNααHb inhibited H_2O_2 and $O_2^{\bullet-}$ -mediated neutrophil adherence to human endothelial cells.[53] The study compared the effects of ααHb and PNααHb on $O_2^{\bullet-}$ and H_2O_2 induced neutrophil-endothelial adhesion *in vitro*. At 30 μM hemoglobin, ααHb significantly increased $O_2^{\bullet-}$ induced adhesion to above 45%, while PNααHb inhibited the adhesion to below 25%. The inhibition showed a dose response with maximal inhibition at 10 μM of hemoglobin concentration. In a hamster model of microcirculatory dysfunction, PNααHb significantly reduced the formation of leukocyte/platelet aggregates in the blood stream during cigarette smoke exposure using the dorsal skin fold chamber for intravital microscopic observation.[54] In contrast to ααHb, PNααHb significantly inhibited pro-adhesive processes secondary to cigarette smoke. Furthermore, we have shown that PNααHb performed best as a resuscitation fluid when tested against ααHb, 7.5% albumin, 300% Ringer's lactate (RL) and 0.9% normal saline in a rat model of severe hemorrhage. PNααHb was found to be less hypertensive and improved systemic hemodynamics, regional blood circulation and base deficit.[55] PNααHb prolonged animal survival time following resuscitation until secession of heart rhythm in this study.

Polynitroxylated Dextran Conjugated Human Hemoglobin

PNDxHb Preparation

Stroma-free hemoglobin is cross-linked prior to PN and conjugation. The cross-linking is performed using bis(3,5-dibromosalicyl) fumarate (DBBF) under carbon monoxide according to a published method[56] to yield ββHb or under nitrogen atmosphere to yield ααHb. Both cross-linked hemoglobins can be conjugated by Dextran-20. This procedure was modified from a previously published method.[57] Dextran conjugation gives a greater than 98% yield, which is free of 64 kDa hemoglobin as demonstrated by size-exclusion chromatography.[57] The PN step is performed by allowing DxββHb or DxααHb to react with 4-(2-bromoacetamido)-2,2,6,6-tetramethyl-1-piperidinyloxy to form PNDxHb. EPR result showed an average of 10–12 nitroxides per hemoglobin tetramer. Same as shown for PNααHb, PNDxHb showed SOD and CAT activities *in vitro* including a culture assay of oxidative neutrophil adhesion to human endothelial cells. In addition, other studies showed PNDxHb had antioxidant activities against oxidant challenges and also had neuron protective activities.

Reduction of Heme Degradation during H_2O_2 Challenge

It is accepted that uncontrolled autoxidation of hemoglobin leads to change in heme geometry and subsequent degradation. This degradation can be detected by fluorescence spectroscopy.[58] Similar fluorescent spectra showing heme degradation can also be detected with H_2O_2 challenge of hemoglobin. Two fluorescent compounds have been defined: Compound I — the emission spectrum collected from 400–560 nm at an excitation wavelength of 321 nm, and Compound II — the emission spectrum collected from 490–570 nm at an excitation wavelength of 460 nm. PNDxHb and DxHb were tested for the formation of fluorescent heme degradation compounds during H_2O_2 challenge. PNDxHb had 50% and 70% attenuation of the formation rates of heme degradation compound I and II, respectively, compared to control DxHb without PN.

The Inhibition of Iron-Induced Toxicity in Cortical Neurons

It is well documented that stroma-free hemoglobin is neurotoxic in cell culture systems.[59,60] This cytotoxicity reflects the pro-oxidant activity of the heme iron. However, when primary cortical neuron cell cultures were exposed to ααHb, DxHb, and PNDxHb at 10 μM for 24 hr, neuron survival was $60 \pm 5.7\%$, $71 \pm 2.4\%$ and $91 \pm 2.8\%$ of untreated control values, respectively. These survival rates represent an improvement over the neuron survival in a stroma-free hemoglobin treated control sample, which was about 50% in this experiment. The neuron survival in a culture without any hemoglobin treatment was considered as 100% survival. These results demonstrate that PN decreases the neurotoxicity of HBOCs.

PNDxHb Supports Life in a Mouse Model of 90% Blood Exchange Transfusion

It had been reported that DxHb could support the survival and full recovery in room air of dogs that have been subjected to exchange transfusion of DxHb for blood that reduced hematocrit (HCT) to 2%.[57] PNDxHb was tested in a mouse model of 90% blood exchange. Mice weighing 28–30 g were anesthetized with pentobarbital at 30 mg/kg. Blood volume was estimated at 6% of body weight. Twelve mice were divided into two groups, a PNDxHb treatment group and a control human albumin (5%) group. For the treatment group, blood was withdrawn at a volume of 1% of body weight via the tail vein (about 0.2 mL per 20 g). Then, an equal volume of PNDxHb was infused. The next withdrawal and infusion was carried out 2 min later. After two PNDxHb exchanges, one albumin exchange was given to compensate for plasma loss. The set of three exchanges is considered as one cycle. This cycle was continued until the desired HCT was reached and the total withdrawn blood had been replaced by 2/3 volume of PNDxHb and 1/3 volume of albumin. The HCT was measured after each infusion until it was less than 6% of starting HCT equaling a 90% blood exchange. For the control group, blood withdrawal protocol was the same as above, followed by only albumin infusion at 1% of body weight. The cycles were repeated until HCT was less than 6% (animal death will occur immediately if HCT drops much below 6%). All the mice treated with albumin alone died within 2 hr after exchange to 6% HCT. All PNDxHb treated mice were alive after 48 hr without any blood cell infusion or any other treatment. Two mice out of six died after 72 hr in the treatment group. The blood chemistry and HCT of the surviving mice returned to normal 7 days after the exchange transfusion with PNDxHb.

Inhibition of Brain Infarct in a Mouse Hemorrhagic Shock Model

We have developed a mouse model of cerebral infarct induced by hemorrhagic shock. Even though the incidences of this type of brain damage are not high, it is a good experimental model to test the neuron toxicity of cell free hemoglobin especially with large volume use of a HBOC under hemorrhagic conditions.

C57BL mice weighing 22–27 g were used. Under isoflurane anesthesia, the right ascending carotid artery was cut in a controllable way to attain 60% hemorrhage in 10 min and was ligated afterwards. The hypotensive and hypovolemic animals were exposed to hemorrhagic shock without fluid resuscitation for 60 min. PNDxHb or control ααHb at the shed blood volume were used for resuscitation at the end of the 60 min. Thirty percent of the animals died during the 60 min. The survivors were assigned randomly to PNDxHb or ααHb groups. Twenty-four hours after resuscitation, the mice were decapitated and each brain was cut into 6 slices for TTC staining. Infarct volume of the brain was evaluated based on the following scoring: 0, no infarct; 1, small infarct in caudoputamen (CP); 2, small infarct in CP and cortex; 3, medium infarct in both CP and cortex; 4, more than a half brain infarct. Due to the mortality that occurred during the resuscitation, the two groups have uneven animal numbers, $n = 25$ for PNDxHb and $n = 20$ for ααHb. *T* test was used and significance was accepted at $p < 0.05$. Mice had an average infarct score of 0.88 ± 0.23 SE and 1.95 ± 0.23 SD when treated with PNDxHb and ααHb, respectively ($p = 0.007$ by *T* test).

PNDxHb Completely Eliminated Pressor Effect

Using the same experimental protocol described above (see PN Attenuates Vasoconstriction), the PN effect on hemoglobin-induced vasoconstriction was tested with PNDxHb which had less than 2% hemoglobin tetramer ($n = 4$). The result showed MAP was unchanged (the average change was 0 ± 3 mmHg) with a 600 mg/kg dose (10 mL/kg) of PNDxHb. Data was the average MAP during minutes 10–90 post initiation of the exchange transfusion of PNDxHb.

Efficacy of PNDxHb in a Rat Model of Hemorrhagic Shock

PNDxHb was tested against six control agents including oxygen carriers and non-oxygen carriers in a rat model of hemorrhagic shock.

The seven agents tested were: Saline ($n = 8$), 300% RL ($n = 8$), human albumin ($n = 11$), ααHb ($n = 10$), DxHb + tempol ($n = 10$), whole blood ($n = 7$), and PNDxHb ($n = 10$), and the methodology has been described previously.[61] Basically, shock was induced via controlled hemorrhage according to the following protocol. Animals were bled at a rate of 1 mL/min until a mean arterial blood pressure (MAP) of 35–40 mmHg was achieved. As the animals' compensatory mechanisms came into play, further blood was withdrawn as needed to maintain a MAP of 35–40 mmHg for 30 min. The total hemorrhage volume was typically 40–45% of estimated blood volume. After 30 min of hypotension, the rats received full-volume resuscitation; a volume equal to the volume of shed blood. The following parameters were monitored: heart rate, MAP, cardiac output, total peripheral resistance, stroke volume, base-deficit, survival time and organ blood flow. Animals were monitored until they succumbed, and survival time was recorded. The results indicate that PNDxHb prolongs survival time better than all other fluids tested during the first 10 hr after resuscitation. The group treated with PNDxHb was the only group to have a 100% survival rate at 6 hr post administration. Ten-hour survival rates are 0%, 0%, 0%, 10%, 10%, 28%, and 70% in saline, 300% RL, human albumin, ααHb, DxHb+tempol, whole blood, and PNDxHb, respectively. All surviving animals were terminated at 10 hr after resuscitation. The results showed that only the PNDxHb-treated animals recovered all cardiovascular parameters to baseline values within 60 min of resuscitation. The PNDxHb-treated animals exhibited no significant change in base deficit values at the end of resuscitation, while the other treatment groups did show a significant change. Base deficit in normal rats before hemorrhage was 3.3 ± 0.52 mMol/L. Hemorrhage significantly increased base deficit to 13.34 ± 0.68 mMol/L on average in all groups. Following resuscitation with saline, 300% RL, human albumin, ααHb, DxHb+tempol, whole blood, and PNDxHb base deficit changed to 15.5 ± 0.1, 15.1 ± 0.2, 5.7 ± 0.8, 7.1 ± 0.9, 5.8 ± 1.0, 8.4 ± 0.5 and 4.7 ± 0.5 ($p > 0.01$) mMol/L, respectively, at 90 min post resuscitation. It was clear that PNDxHb performed better even than whole blood as a resuscitative agent in this rat model. It was also demonstrated that bound nitroxide performed better than an equal amount of free nitroxide with hemoglobin.

Polynitroxylated Pegylated Bovine Hemoglobin for Traumatic Brain Injury with Hemorrhagic Shock

There are 1.4 million people who sustain a TBI each year in the US. Secondary insults such as HS and hypotension are common, especially in severe TBI, and have a devastating impact on outcome.[62] Vulnerability of brain-injured patients to hypotension and hypoxemia is well recognized and identifies a key need for new approaches. Traditional fluids such as RL have shortcomings, i.e. exacerbation of brain edema, limited oxygen delivery, and the need for large volumes.[63] Colloid resuscitation (Hextend, recommended for field use in combat casualty care) has the advantage of requiring a lower volume but is not specifically neuroprotective. A novel therapeutic agent is needed to improve outcome and ameliorate the secondary damage in the pre-hospital setting. Recent studies suggest that TBI plus HS represents a special opportunity for resuscitation with neuroprotective HBOC solutions, and that benefit on intracranial pressure (ICP) or neuropathology can be achieved.[21,64] PNPH is being developed as a small volume, multifunctional resuscitation fluid for pre-hospital use in the setting of TBI complicated by hypotension. PNPH is a neuroprotective, hyper-colloid, oxygen-carrying therapeutic which is produced by applying PN to polyethylene glycol conjugated hemoglobin. There are ~10 polyethylene glycol (Peg) chains and ~12 nitroxides per bovine hemoglobin tetramer. While PN technology corrects intrinsic toxicity of current HBOCs and adds neuroprotective and antioxidant properties, PN technology does not alter the hyper-colloid properties provided by the Peg chains. The hyper-colloid effects of Peg chains results from the fact that each Peg moiety attracts (by hydrogen bonding) ~100 water molecules, thus each PNPH binds ~1000 water molecules. This produces a potent colloid effect that mediates the remarkable small volume resuscitation effects of PNPH.

Confirmation of Antioxidant Activities of PNPH

SOD and CAT mimetic activities of nitroxides have been discussed. H_2O_2 has been identified as the most potent oxidant during the hemoglobin redox reaction. To demonstrate PN added antioxidant activities, PNPH was compared with PegHb in assays of (a) CAT mimetic activity, (b) oxo-ferryl (Fe^{4+}) formation, and (c) free iron release. An unusual peroxidase activity of PNPH has also been described.[65]

CAT mimetic activity: Hemoglobin redox reactions generate H_2O_2, which can cause direct cellular damage to DNA and protein. H_2O_2 crosses cell membranes rapidly and reacts with iron and copper ions to form much more damaging reactive oxygen species, hydroxyl radicals. It can also cause heme degradation. Inside the red cell, CAT converts H_2O_2 to O_2 and H_2O. Using the rate of oxygen evolution as a measure of CAT mimetic activity as previously described,[66] the CAT mimetic activity of PNPH and PegHb was determined after adding various concentrations of H_2O_2. Data showed PNPH has a higher rate of oxygen release than that of PegHb at each H_2O_2 concentration.

PNPH reduces oxo-ferrylformation: With cell free hemoglobin, the formation of Fe^{4+} through the reaction of ferric (Fe^{3+}) heme with H_2O_2 has been identified as an intermediate in the pathway leading to cellular, protein and heme group degradation. Fe^{4+} can arise during ischemia and reperfusion.[67] PN can reduce the formation of Fe^{4+} after adding H_2O_2. Data showed that, over time, Fe^{4+} formation, measured by absorption at 545 nm, is much smaller in PNPH than in PegHb.

PNPH reduces free iron release: D'Agnillo and Chang[68] reported that the complex formed by co-polymerization of hemoglobin with endogenous red cell SOD and CAT released less free iron in the presence of H_2O_2 than hemoglobin alone. Using the same assay method, free iron release was measured. The absorption measured at 562 nm was used for free iron release against standard calibration curve. The data showed that PNPH released less free iron than PegHb at different H_2O_2 concentrations.

Neuroprotective Effect of PNPH in Primary Cortical Neuron Culture Models

It is well known from the classic studies of Regan and Panter that cell free hemoglobin is neurotoxic in neuronal cell culture models.[59,60] After TBI, a hemoglobin used in resuscitation could extravasate into brain tissue, raising concerns about the use of hemoglobin in TBI resuscitation. However, data shows that PNPH is not only safe but also neuroprotective.[21] *In vitro* studies were carried out with primary cortical neurons in three models: (a) a cytotoxicity study by incubating neurons with stroma free hemoglobin, PegHb, or PNPH for 24 hr; (b) study of PNPH protection against excitotoxicity in a glutamate/glycine-induced neuronal death model; and (c) study of PNPH protection against TBI-like stretch injury. The results of these studies were remarkable. First, unlike either stroma-free hemoglobin or PegHb, which were neurotoxic, there was a lack of neurotoxicity of PNPH across a wide range of concentrations as measured by LDH release. Surprisingly, neuroprotection by PNPH in both glutamate/glycine induced neurotoxicity and neuronal stretch injury was demonstrated. The results from stretch injury model are also positive by immunochemistry assay. These data suggest that PNPH is a unique hemoglobin-based product with neuroprotective rather than neurotoxic properties. This represents a paradigm shift in the potential utility of HBOCs in TBI resuscitation and other CNS insults.

Summary

After years of attempting to use the PN technology to tame the toxicity of current generation HBOCs to yield safer oxygen carriers, an unexpected outcome was realized: that already highly substituted PegHb could still be polynitroxylated. This finding resulted in the development of PNPH as a multifunctional hemodynamic therapeutic agent. The accumulation of efficacies in various challenging preclinical animal models supports this development of PNPH as a next-generation HBOC with multifunctional therapeutic activities. The key attributes of PNPH are: (a) PNPH improves oxygen delivery by controlling the oxidative cascades and NO depletion of cell free hemoglobin; (b) PNPH is a neuro-protectant against reactive oxygen species mediated injury, excitotoxicity, and TBI-like injury; and (c) PNPH maintains the advantages of pegylated hemoglobin as a hyper-colloid resuscitation fluid. PNPH may well have therapeutic potential to be a high-therapeutic-index next-generation HBOC for use in a number of un-met medical indications and transfusion medicine and also in critical care medicine including combat casualty care with or without hemorrhagic stress.

Acknowledgments

We thank Dr. J. Olson for the NO binding study, Drs. A Gulati and P.W. Buehler for PNDxHb studies in rat hemorrhagic shock model, and Dr. P.M. Kochanek for PNPH neuron protection and TBI studies. Most study results presented this chapter were supported by NIH and DOD (grant numbers R44 NS36119, R44 NS036119, N00014-98-C-0140, DAMD17-01-1-0001, DAMD17-02-1-0636, and W81XWH-06-1-0247).

References

1. Pettilä V, Westbrook AJ, Nichol AD, *et al.* (2011) Age of red blood cells and mortality in the critically ill. *Crit Care* **15(2):** R116. Epub 2011 Apr 15.
2. Nelson D, Azari M, Brown R, *et al.* (1992) Preparation and characterization of diaspirin cross-linked hemoglobin solutions for preclinical studies. *Biomater Artif Cells Immobilization Biotechnol* **20(2–4):** 423–437.
3. Vlahakes GJ, Lee R, Jacobs EE Jr, *et al.* (1990) Hemodynamic effects and oxygen transport properties of a new blood substitute in a model of massive blood replacement. *J Thorac Cardiovasc Surg* **100(3):** 379–388.
4. Gilbert C, Nho K, Johnson M, *et al.* (1994) Hemoglobinuria in rats: a sensitive test of renal filtering and absorption of PEG-hemoglobin, a red blood cell substitute. *Artif Cells Blood Substit Immobil Biotechnol* **22(3):** 535–541.
5. Gould SA, Sehgal LR, Sehgal HL, Moss GS. (1992) Artificial blood: current status of hemoglobin solutions. *Crit Care Clin* **8(2):** 293–309.
6. Hsia JC, Song DL, Er SS, *et al.* (1992) Pharmacokinetic studies in the rat on a o-raffinose polymerized human hemoglobin. *Biomater Artif Cells Immobilization Biotechnol* **20(2–4):** 587–595.

7. Vandegriff KD, Malavalli A, Wooldridge J, *et al.* (2003) MP4, a new nonvasoactive PEG-Hb conjugate. *Transfusion* **43(4):** 509–516.
8. Hoffman SJ, Looker DL, Roehrich JM, *et al.* (1990) Expression of fully functional tetrameric human hemoglobin in *Escherichia coli. Proc Natl Acad Sci USA* **87(21):** 8521–8525.
9. Sloan EP, Koenigsberg M, Gens D, *et al.* (1999) Diaspirin cross-linked hemoglobin (DCLHb) in the treatment of severe traumatic hemorrhagic shock: a randomized controlled efficacy trial. *JAMA* **282(19):** 1857–1864.
10. Natanson C, Kern SJ, Lurie P, *et al.* (2008) Cell-free hemoglobin-based blood substitutes and risk of myocardial infarction and death: a meta-analysis. *JAMA* **299:** 2304–2312.
11. Rousselot M, Delpy E, Drieu La Rochelle C, *et al.* (2006) Arenicola marina extracellular hemoglobin: a new promising blood substitute. *Biotechnol J* **1(3):** 333–345.
12. Bian Y, Rong Z, Chang TM. (2011) Polyhemoglobin-superoxide dismutase-catalase-carbonic anhydrase: a novel biotechnology-based blood substitute that transports both oxygen and carbon dioxide and also acts as an antioxidant. *Artif Cells Blood Substit Immobil Biotechnol* **39(3):** 127–136.
13. Hsia CJ, Ma L. (2012) A hemoglobin-based multifunctional therapeutic: polynitroxylated pegylated hemoglobin. *Artif Organs* **36(2):** 215–220.
14. Elmer J, Alam HB, Wilcox SR. (2012) Hemoglobin-based oxygen carriers for hemorrhagic shock. *Resuscitation* **83:** 285–292.
15. Chang TM. (2010) Blood replacement with nanobiotechnologically engineered hemoglobin and hemoglobin nanocapsules. *Wiley Interdiscip Rev Nanomed Nanobiotechnol* **2(4):** 418–430.
16. Greenburg AG, Light WR, Dubé GP. (2010) Reconstructing hemoglobin-based oxygen carriers. *Transfusion* **50(12):** 2764–2767.
17. Chen JY, Scerbo M, Kramer G. (2009) A review of blood substitutes: examining the history, clinical trial results, and ethics of hemoglobin-based oxygen carriers. *Clinics (Sao Paulo)* **64(8):** 803–813.
18. Winslow RM. (2008) Cell-free oxygen carriers: scientific foundations, clinical development, and new directions. *Biochim Biophys Acta* **1784(10):** 1382–1386.
19. Buehler PW, Alayash AI. (2008) All hemoglobin-based oxygen carriers are not created equally. *Biochim Biophys Acta* **1784(10):** 1378–1381.
20. Alayash AI. (1999) Hemoglobin-based blood substitutes: oxygen carriers, pressor agents, or oxidants? *Nat Biotechnol* **17:** 545–549.
21. Shellington DK, Du L, Wu X, *et al.* (2011) Polynitroxylated pegylated hemoglobin: a novel neuroprotective hemoglobin for acute volume-limited fluid resuscitation after combined traumatic brain injury and hemorrhagic hypotension in mice. *Crit Care Med* **39(3):** 494–505.
22. Koehler RC, Cao S, Zhanf J, Kwansa H. (2009) Effect of cell-free hemoglobin on the cerebral circulation. *XIIth ISBS,* Parma Italy, August 25–27.

23. Hsia CJC, Ma L. (2009) PNPH — A therapeutic for inadequate blood flow and superoxide, nitric oxide dependent vascular dysfunctions. *XIIth ISBS,* Parma Italy, August 25–27.
24. Beaulieu C, Busch E, Röther J, *et al.* (1998) Polynitroxyl albumin reduces infarct size in transient focal cerebral ischemia in the rat: potential mechanisms studied by magnetic resonance imaging. *J Cereb Blood Flow Metab* **18(9):** 1022–1031.
25. Rozantsev EG. (1970) *Free Nitroxyl Radicals.* Plenum Press, New York.
26. Samuni A, Krishna CM, Mitchell JB, *et al.* (1990) Superoxide reaction with nitroxides. *Free Radic Res Commun* **9(3–6):** 241–249.
27. Krishna MC, Grahame DA, Samuni A, *et al.* (1992) Oxoammonium cation intermediate in the nitroxide-catalyzed dismutation of superoxide. *Proc Natl Acad Sci USA* **89(12):** 5537–5341.
28. Krishna MC, Samuni A, Taira J, *et al.* (1996) Stimulation by nitroxides of catalase-like activity of hemeproteins. Kinetics and mechanism. *J Biol Chem* **271(42):** 26018–26025.
29. Damiani E, Paganga G, Greci L, Rice-Evans C. (1994) Inhibition of copper-mediated low density lipoprotein peroxidation by quinoline and indolinone nitroxide radicals. *Biochem Pharmacol* **48(6):** 1155–1161.
30. Nilsson UA, Olsson LI, Carlin G, Bylund-Fellenius AC. (1989) Inhibition of lipid peroxidation by spin labels. Relationships between structure and function. *J Biol Chem* **264(19):** 11131–11135.
31. Carroll RT, Galatsis P, Borosky S, *et al.* (2000) 4-Hydroxy-2,2,6,6-tetramethylpiperidine-1-oxyl (Tempol) inhibits peroxynitrite-mediated phenol nitration. *Chem Res Toxicol* **13(4):** 294–300.
32. Tsuhako MH, Augusto O, Linares E, *et al.* (2010) Tempol ameliorates murine viral encephalomyelitis by preserving the blood-brain barrier, reducing viral load, and lessening inflammation. *Free Radic Biol Med* **48(5):** 704–712.
33. Cuzzocrea S, Pisano B, Dugo L, *et al.* (2004) Tempol reduces the activation of nuclear factor-kappaB in acute inflammation. *Free Radic Res* **38(8):** 813–819.
34. Hahn SM, Tochner Z, Krishna CM, *et al.* (1992) Tempol, a stable free radical, is a novel murine radiation protector. *Cancer Res* **52(7):** 1750–1753.
35. Goffman T, Cuscela D, Glass J, *et al.* (1992) Topical application of nitroxide protects radiation-induced alopecia in guinea pigs. *Int J Radiat Oncol Biol Phys* **22(4):** 803–806.
36. Bernstein EF, Kong SK, Brown DB, *et al.* (2001) The nitroxide tempol affords protection against ultraviolet radiation in a transgenic murine fibroblast culture model of cutaneous photoaging. *Exp Dermatol* **10(1):** 55–61.
37. Wilcox CS, Pearlman A. (2008) Chemistry and antihypertensive effects of tempol and other nitroxides. *Pharmacol Rev* **60(4):** 418–469.
38. Simonsen U, Christensen FH, Buus NH. (2009) The effect of tempol on endothelium-dependent vasodilatation and blood pressure. *Pharmacol Ther* **122(2):** 109–124.

39. Deng-Bryant Y, Singh IN, Carrico KM, Hall ED. (2008) Neuroprotective effects of tempol, a catalytic scavenger of peroxynitrite-derived free radicals, in a mouse traumatic brain injury model. *J Cereb Blood Flow Metab* **28(6):** 1114–1126.
40. Xiong Y, Singh IN, Hall ED. (2009) Tempol protection of spinal cord mitochondria from peroxynitrite-induced oxidative damage. *Free Radic Res* **43(6):** 604–612.
41. Wilcox CS. (2010) Effects of tempol and redox-cycling nitroxides in models of oxidative stress. *Pharmacol Ther* **126(2):** 119–145.
42. Kuppusamy P, Wang P, Shankar RA, *et al.* (1998) *In vivo* topical EPR spectroscopy and imaging of nitroxide free radicals and polynitroxyl-albumin. *Magn Reson Med* **40(6):** 806–811.
43. Kuppusamy P, Wang P, Zweier JL, *et al.* (1996) Electron paramagnetic resonance imaging of rat heart with nitroxide and polynitroxyl-albumin. *Biochemistry* **35(22):** 7051–7057.
44. Russell J, Okayama N, Alexander JS, *et al.* (1998) Pretreatment with polynitroxyl albumin (PNA) inhibits ischemia-reperfusion induced leukocyte-endothelial cell adhesion. *Free Radic Biol Med* **25(2):** 153–159.
45. Mahaseth H, Vercellotti GM, Welch TE, *et al.* (2005) Polynitroxyl albumin inhibits inflammation and vasoocclusion in transgenic sickle mice. *J Lab Clin Med* **145(4):** 204–211.
46. Kaul DK, Liu XD, Zhang X, *et al.* (2006) Inhibition of sickle red cell adhesion and vasoocclusion in the microcirculation by antioxidants. *Am J Physiol Heart Circ Physiol* **291(1):** H167–H175.
47. Sugawara T, Yu F, Ma L, *et al.* (2001) Delayed treatment with polynitroxyl albumin reduces infarct size after stroke in rats. *Neuroreport* **12(16):** 3609–3612.
48. Manole MD, Kochanek PM, Foley LM, *et al.* (2012) Polynitroxyl albumin and albumin therapy after pediatric asphyxial cardiac arrest: effects on cerebral blood flow and neurologic outcome. *J Cereb Blood Flow Metab* **32(3):** 560–569.
49. Walson KH, Tang M, Glumac A, *et al.* (2011) Normoxic versus hyperoxic resuscitation in pediatric asphyxial cardiac arrest: effects on oxidative stress. *Crit Care Med* **39(2):** 335–343.
50. Exo JL, Shellington DK, Bayir H, *et al.* (2009) Resuscitation of traumatic brain injury and hemorrhagic shock with polynitroxylated albumin, hextend, hypertonic saline, and lactated Ringer's: effects on acute hemodynamics, survival, and neuronal death in mice. *J Neurotrauma* **26(12):** 2403–2408.
51. Buehler PW, Haney CR, Gulati A, *et al.* (2004) Polynitroxyl hemoglobin: a pharmacokinetic study of covalently bound nitroxides to hemoglobin platforms. *Free Radic Biol Med* **37(1):** 124–135.
52. Rohlfs RJ, Mathews AJ, Carver TE, *et al.* (1990) The effects of amino acid substitution at position E7 (residue 64) on the kinetics of ligand binding to sperm whale myoglobin. *J Biol Chem* **265(6):** 3168–3176.

53. Okayama N, Park JH, Coe L, *et al.* (1999) Polynitroxyl alphaalpha-hemoglobin (PNH) inhibits peroxide and superoxide-mediated neutrophil adherence to human endothelial cells. *Free Radic Res* **31(1):** 53–58.
54. Saetzler RK, Arfors KE, Tuma RF, *et al.* (1999) Polynitroxylated hemoglobin-based oxygen carrier: inhibition of free radical-induced microcirculatory dysfunction. *Free Radic Biol Med* **27(1–2):** 1–6.
55. Buehler PW, Mehendale S, Wang H, *et al.* (2000) Resuscitative effects of polynitroxylated alphaalpha-cross-linked hemoglobin following severe hemorrhage in the rat. *Free Radic Biol Med* **29(8):** 764–774.
56. Shibayama N, Imai K, Hirata H, *et al.* (1991) Oxygen equilibrium properties of highly purified human adult hemoglobin cross-linked between 82 beta 1 and 82 beta 2 lysyl residues by bis(3,5-dibromosalicyl) fumarate. *Biochemistry* **30(33):** 8158–8165.
57. Tam SC, Blumenstein J, Wong JT. (1978) Blood replacement in dogs by dextran-hemoglobin. *Can J Biochem* **56:** 981–984.
58. Nagababu E, Rifkind JM. (1998) Formation of fluorescent heme degradation products during the oxidation of hemoglobin by hydrogen peroxide. *Biochem Biophys Res Commun* **247(3):** 592–596.
59. Regan RF, Panter SS. (1993) Neurotoxicity of hemoglobin in cortical cell culture. *Neurosci Lett* **153(2):** 219–222.
60. Regan RF, Panter SS. (1996) Hemoglobin potentiates excitotoxic injury in cortical cell culture. *J Neurotrauma* **13(4):** 223–231.
61. Gulati A, Sen AP. (1998) Dose-dependent effect of diaspirin cross-linked hemoglobin on regional blood circulation of severely hemorrhaged rats. *Shock* **9(1):** 65–73.
62. Chesnut RM, Marshall LF, Klauber MR, *et al.* (1993) The role of secondary brain injury in determining outcome from severe head injury. *J Trauma* **34:** 216–222.
63. Tan PG, Cincotta M, Clavisi O, *et al.* (2011) Review article: prehospital fluid management in traumatic brain injury. *Emerg Med Australas* **23(6):** 665–676.
64. Rosenthal G, Morabito D, Cohen M, *et al.* (2008) Use of hemoglobin-based oxygen-carrying solution-201 to improve resuscitation parameters and prevent secondary brain injury in a swine model of traumatic brain injury and hemorrhage: laboratory investigation. *J Neurosurg* **108:** 575–587.
65. Stoyanovsky DA, Kapralov A, Huang Z, *et al.* (2010) Unusual peroxidase activity of polynitroxylated pegylated hemoglobin: elimination of H_2O_2 coupled with intramolecular oxidation of nitroxides. *Biochem Biophys Res Commun* **399(2):** 139–143.
66. Krishna MC, Samuni A. (1994) Nitroxides as antioxidants. *Methods Enzymol* **234:** 580–589.

67. Svistunenko DA. (2005) Reaction of haem containing proteins and enzymes with hydroperoxides: the radical view. *Biochim Biophys Acta* **1707(1):** 127–155.
68. D'Agnillo F, Chang TM. (1998) Polyhemoglobin-superoxide dismutase-catalase as a blood substitute with antioxidant properties. *Nat Biotechnol* **16:** 667–671.

Chapter 9

Surface-Mediated Drug Delivery

Benjamin M. Wohl,*,† Betina Fejerskov,*
Siow-Feng Chong* and Alexander N. Zelikin*,†

1. Introduction

Accelerated development of biomaterials science and biomedical opportunities associated with therapeutic implants has led to recognition of interfacial phenomena as being decisive in the success of implantation and therapeutic intervention as such. When placed in contact with body fluids and tissues, implants are met with stringent mechanisms of recognition exerted by the body, and unprepared surfaces often elicit inflammatory responses and further severe side effects. In the preceding two decades, biomaterials science has catered for a resolution of these complications achieved through modification of surfaces and rendering them low fouling, non-cell adhesive, benign, etc. A further biomedical opportunity is associated with surface-mediated drug delivery (SMDD),[1] a strategy whereby the interface exerts a therapeutic benefit through a release and/or presentation of biologically active cargo.

This book chapter aims to provide a brief presentation of opportunities offered by SMDD and reveal the paradigm of biointerfaces

*Department of Chemistry and †iNano Interdisciplinary Nanoscience Centre, Aarhus University, Aarhus, Denmark.
† zelikin@chem.au.dk

eliciting therapeutic response. The techniques relevant to the context of SMDD are versatile and often well established as stand-alone disciplines (e.g. bioconjugation, protein immobilization, etc.); for an in-depth discussion of these, readers are referred to existing reviews on the respective subjects. Also, to keep presentation within limits in terms of scope, this discussion does not consider controlled drug release from degrading matrices: this biomedical opportunity is similar to SMDD (especially in the case of surface eroding materials) yet differs in that for the latter it is possible to unequivocally distinguish between the bulk material and the therapeutically active surface coating. Finally, to limit the presentation in size, to present SMDD we discuss only a few selected examples and do not aim to provide an exhaustive overview of existing literature.

The drug types discussed below are small molecules and oligopeptides, proteins, and gene cargo. For each of these, presentation is structured to present existing methods of cargo immobilization, followed by successful approaches towards a controlled drug release (or presentation), and concluded with few examples of particular biomedical success. A significant part of this discussion deals with a *sequential polymer deposition* technique (also known as layer-by-layer technique, LbL), an approach to surface functionalization achieved through adsorption of polymers onto an underlying surface with pre-immobilized complementary polymer (Fig. 1). The underlying phenomenon is simple and is readily understood on an example of charged polymers (polyelectrolytes): polycations adsorb onto negatively charged surfaces and through an oversaturation lead to a new, positively charged surface, the latter lending itself for an adsorption of polyanions, etc. The same can be achieved using polymers which interact through hydrogen bonding, form covalent linkages, etc., and the field is well reviewed and updated from the standpoint of materials science.[1–5] As for other applications, the advantages offered by this technique for SMDD include the widest imaginable array for candidate polymers, including biomacromolecules (polysaccharides, proteins, DNA and RNA, etc.); control over the thicknesses of polymer coatings through the number of adsorption cycles; multiple polymers can be co-adsorbed within one layer or

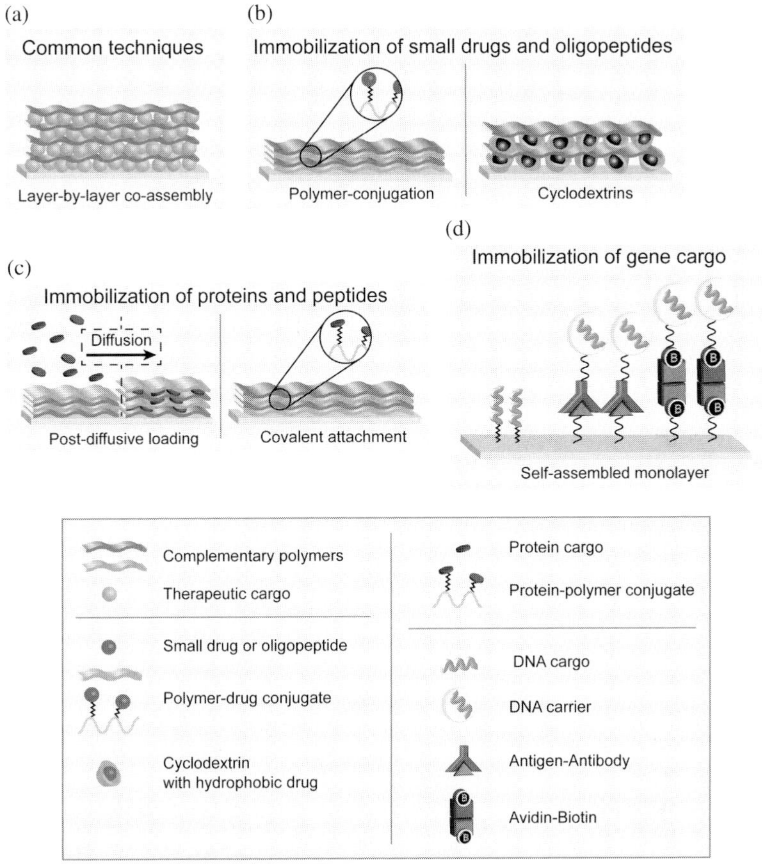

Fig. 1. Immobilization of therapeutic cargo within/on surface coatings has been achieved through a variety of techniques. In this schematic the different approaches are exemplified on the basis of the well-studied layer-by-layer assembly, except for the immobilization of gene cargo. A general approach used in all type of therapeutic cargo (i.e. small drugs, proteins, peptides or gene cargo) to form a drug-loaded thin film is co-assembly (a). The immobilization of small drugs is generally achieved through conjugation to a carrier polymer or association with supramolecular species (b). An approach that has received less attention for the incorporation of peptides and proteins: post-diffusion of therapeutic into a preformed thin film. Additionally, the retention of proteins and peptides in polymer thin films has also been shown to increase through conjugation to polymer carriers (c). Together with a highly successful LbL, immobilization of gene cargo to surfaces has been achieved through self-assembled monolayers (SAM); either through direct covalent conjugation or specific interactions (e.g. antigen-antibody, avidin-biotin interactions) (d).

within subsequent layers within the film architecture; adsorption at a desired cycle number positions the cargo at a desired depth for a potential time controlled release, etc. All these advantages are available for coatings assembled on surfaces with virtually no restrictions on the substrate geometry, which make it powerful to construct therapeutically active coatings on implants as well as any other given biointerface.[1] We emphasize that while accomplishments of LbL are sound and early achievements on the biomedical arena are also encouraging, the overall potential of SMDD is only beginning to unravel. We therefore hope that the discussion below will stimulate further developments in the field.

2. Small Drugs and Oligopeptides

Surface-mediated delivery of low molecular weight drugs holds biomedical relevance in the context of drug eluting stents whereby a controlled release of e.g. anti-inflammatory and anti-proliferative drugs contributes to a favorable and/or decreases the unfavorable tissue response.[6] As with most routes of administration, a fast distribution of small molecules throughout the body results in a decreased localization of the drug at the desired site of action and may also result in side effects. In this light, a controlled release of drugs achieved from the surface presents itself as a unique opportunity to communicate directly to the cells of interest and aims to control the cellular processes directly at a biointerface. However, while low molecular weight drugs constitute a majority of therapeutics on the market, these drugs are arguably the most challenging for SMDD, largely due to a high diffusivity of these molecules through the surface coatings and poor opportunities in controlled drug retention and release. The following sections provide a brief overview of the different methods used for small molecule immobilization and release followed by successful examples of SMDD. For the sake of brevity, this section also discusses examples of oligopeptide cargo with retention characteristics and immobilization techniques similar to that of low molecular weight drugs.

2.1. *Methods of Immobilization*

Inasmuch as small molecules typically exhibit ill-controlled immobilization, retention and release from surface coatings, successful examples in SMDD of small drugs often rely on the advances of associated techniques of macromolecular design, e.g. polymer therapeutics. This biomedical paradigm investigates conjugation of drugs to carrier polymers with the aim to achieve a favorable change in the pharmacokinetics of the drug.[7] The tools of polymer therapeutics produce polymer–drug conjugates amenable for implementation in SMDD, specifically through the assembly of surface coatings via sequential polymer deposition technique. From a different perspective, learning in part from the developments in composite hydrogels,[8] immobilization of small drugs on the surface can also exploit drug reservoirs, which, in turn, are assembled on the interface in a desired combination with polymers.

These points are well illustrated on the example of small hydrophobic candidate drugs, which often suffer from poor aqueous solubility. Immobilization of these therapeutics is well established using supramolecular carrier species such as cyclodextrins (CDs).[9,10] These molecules have a hydrophobic core to isolate the drug within and a hydrophilic exterior amenable for chemical modification towards incorporation of the CDs into the surface coatings. The hydrophobic pockets enhance inclusion of hydrophobic drugs for SMDD and make it possible to achieve a controlled sustained release of hydrophobic drugs. The second approach to achieve this relies on a covalent binding of the drug to a charged carrier polymer via a hydrolysable bond in a so-called pro-drug approach.[11,12] This technique thereby renders the drug more soluble for a controlled immobilization onto a surface using solution based coating methods and achieves a controlled release through a choice of the hydrolysable linkage between the drug and the polymer.

When dealing with small charged molecules, the electrostatics can be used directly for incorporation of the cargo into a surface coating polymer films. The positively charged drugs have often been complexed with polyanionic species either prior or during LbL deposition.[9,10,13–18]

It is also possible to achieve immobilization of charged cargo molecules through adsorption from their solution into the preformed surface coating,[19] however the latter method presents fewer tools of control over the kinetics of drug release. Examples of small positively charged molecules include many antibiotics. The positively charged aminoglycoside antibiotic gentamycin can be incorporated into films of e.g. cross-linked poly(methacrylic acid) (PMAA).[15] Films containing the negatively charged Prussian Blue (PB) nanoparticles have also been used to fabricate LbL assemblies with gentamycin.[14] A poly(β-amino ester) series of polycations, rendering the LbL film hydrolytically degradable, have been widely used in SMDD of small drugs.[9,10,16–18] When incorporating small, positively charged antibiotics, a counter polyanion is required, successful examples include biological polymers, e.g. dextran sulfate, chondroitin sulfate, and alginate.[16] Correspondingly, when considering small, negatively charged molecules, complexing with polycationic species can be performed before film assembly,[13,20] or they may be bound to the free positively charged sites following assembly.

Recently, polymer multilayer coatings for controlled local co-delivery of antibiotics and anti-inflammatory agents were assembled with the aim of a simultaneous targeting of infection and inflammation. Vancomycin was incorporated into a polymer coating film comprised of hydrolytically degradable cationic poly(β-amino esters) and polyanions such as dextran sulfate as mentioned above directly through its net positive charge. An anti-inflammatory drug diclofenac is a hydrophobic small molecule, and its retention required an anionic polymer carrier, poly(carboxymethyl-β-cyclodextrin) (polyCD).[10]

2.2. *Controlled Drug Release*

Delivery of biologically active compounds mediated from a surface can involve their passive diffusion from the film, hydrolysis of the polymer multilayer causing surface erosion,[9,10,16–18] affinity-based systems,[21] or environmentally stimulated release such as pH-triggered[11,15] or an electrical potential[14] (see Fig. 2(a)–(c) for schematics of these concepts). With multilayered polymer coatings, the release rate of a drug can be controlled from several seconds to several weeks through changes in the number of layers and the nature of the biodegradable building blocks

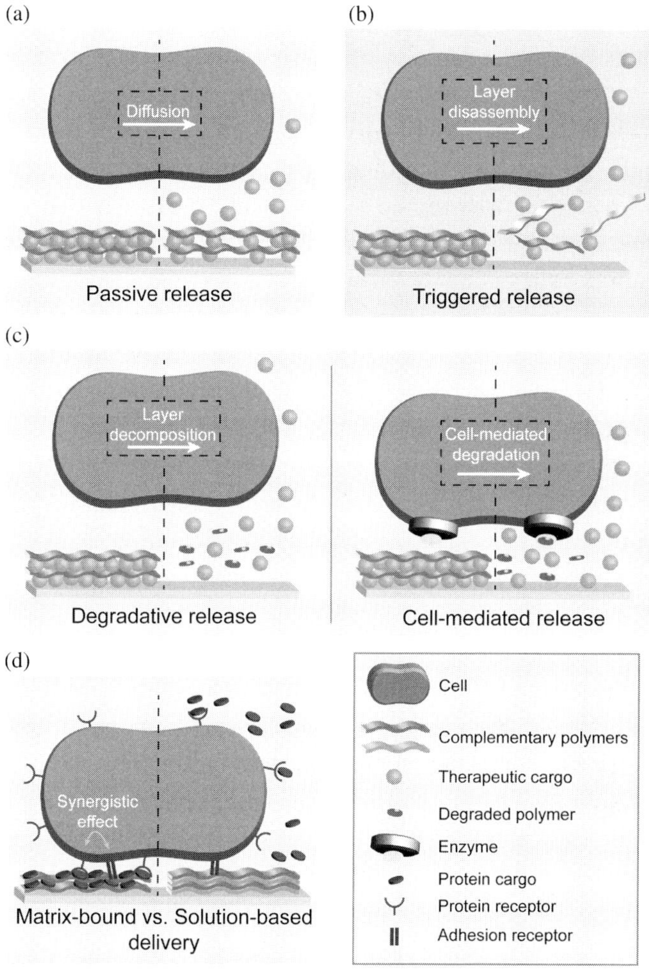

Fig. 2. Release of therapeutic cargo from thin film surface coatings has been achieved via a range of mechanisms, which are illustrated schematically herein on the example of LbL polymer thin films. These mechanisms include (a) a passive release through diffusion, (b) release upon film disassembly triggered via environmental or external stimuli, and (c) degradative release in films constituting a (bio-)degradable film component. A special case of degradative release is cell-mediated release where the thin film is actively degraded by cellular enzymes (c). (d) shows that films presenting proteins in an insoluble matrix-bound form are able to retain their signal transduction ability and that close proximity of adhesion and protein receptors facilitates a synergistic effect, increasing the effectiveness of the protein.

used during multilayer assembly. Below, we discuss some of the successful mechanisms employed to control the kinetics of drug release.

For the case of environmental stimuli, antimicrobial substances can be released from cross-linked PMAA hydrogel thin films in response to changes in pH and salt concentration in the surrounding medium. It is also possible to further tune antimicrobial release by varying the type of cross-linker. Using adipic acid dihydrazide as a cross-linker supplies centers for hydrogen bonding, and compared to the films cross-linked by ethylene diamine also increases the retention of loaded antibacterial.[15] For the case of an electrical potential, applying different electric potential profiles could control both the dose and the kinetics of drug release. Besides application of a constant potential, the drug-release behavior under influence of short electric potential allows the release of gentamycin to be turned on and off on demand for a pulsatile drug release profile.[14]

Degradable polyelectrolyte multilayer films built by alternating the selected drug with polyelectrolytes containing hydrolytically degradable ester bonds releases the incorporated cargo as the layers degrade and the drug escapes. These systems afford sustained release of small molecule therapeutics with a linear profile and kinetics of drug release is under a control of e.g. the hydrophobicity of poly(β-amino esters).[9,22] Both loading and release of vancomycin mediated by degradable multilayer assemblies were possible to be controlled by the amount of polymer layers or the method of assembly (polymer spraying vs. dipping the substrate into a polymer solution) and typically films assembled through polymer spraying exhibited a markedly higher drug loading than the dipped films. The bolus release and large drug loads in the sprayed films assembled are useful for immediate eradication of bacteria. The longer and more linear release of the films assembled via substrate dipping is appropriate for preventing infection or re-colonization of a wound.[10,16]

2.3. *Highlighted Examples*

By and large, most of the small molecule drugs investigated so far in the context of SMDD aim to combat infectious and inflammatory conditions. For this reason, most examples of practical utility of SMDD and small molecule therapeutics relate to *in vitro* antibacterial research. For example, vancomycin release from cross-linked PMAA hydrogel films strongly inhibited the bacterial colonization at surfaces.[15] With the attempt of releasing antibiotics by applying an electrical potential, it became apparent that the gentamycin released from films of chitosan and Prussian Blue particles retained a comparable level of bactericidal activity compared to unmodified, commercial gentamycin.[14] In another example, a piece of coated bandage of dextran sulfate/alginate and vancomycin was applied to an agar plate coated with *S. aureus* and showed a clear inhibition zone as the commercial available vancomycin loaded disc.[16] Concurrent therapeutic efficacy of bandages containing both vancomycin and nonsteroidal anti-inflammatory drugs against infectious and inflammatory conditions has also been demonstrated.[10] The latter coatings were successfully deposited on a number of model substrates such as intraocular lenses, bandages and sutures, highlighting the versatility of LbL and a potential of SMDD for diverse biomedical applications.

One example of a successful *in vivo* attempt was conducted on a rabbit bone model targeting implant-related osteomyelitis caused by an infection of *S. aureus.*[17] For this, gentamycin was incorporated into polyelectrolyte multilayers of a poly(β-amino ester) and poly(acrylic acid) (PAA) for a direct implant exchange operation. The release profile was obtained from both diffusion of gentamycin through the outer layers, causing a burst release, and erosion of the film causing a more sustained release over weeks of time. This profile is proposed to achieve an immediate sterilization of the implant site followed by a prevention of survival bacteria to re-colonize the site. Titanium implants coated with the polyedectrolyte multilayer films were introduced into New Zealand White rabbits bone and, after exchange, these antimicrobial-coated devices provided a significantly lower degree of infection.

3. Proteins and Peptides

Proteins and peptides are an essential part of every living organism and are also recognized as potent, highly selective therapeutics.

Enzymatic catalysis, structural integrity of tissues, signaling, modulation of cellular differentiation, proliferation, and adhesion are among the areas where protein and peptide therapeutics deliver practical biomedical benefits, to name a few. Advances in genomics and proteomics identify an ever-increasing number of potential targets for intervention using this class of drugs. In parallel, technological developments deliver viable production techniques and commercial availability of protein and peptide drugs for wider biomedical applications. A significant remaining challenge relates to a controlled and targeted delivery of proteins and peptides, therapeutics highly susceptible to degradation.

A particular area of utility of protein and peptide cargo in the context of SMDD relates to therapeutic implants. Close control over cell line differentiation and vascularization has significant importance for an efficient *in vivo* acceptance of implants, tissue engineering and cell therapy.[23–25] It is desirable to present growth factors in a localized and sustained manner while keeping systemic concentrations low to avoid adverse effects like malformed and non-functional blood vessels.[26] These challenges are met using the tools of SMDD, as described below.

Protein immobilization on substrates and surfaces is a well-established biotechnological discipline which delivers practical benefits to applications such as sensing, patterning, drug screening, etc. For biomedicine, immobilization of therapeutic proteins and affinity-based drug delivery has also received considerable research attention,[25,27–29] the latter recently reviewed by Wang *et al.*[21] The advent of sequential polymer deposition (LbL) delivered a novel paradigm for SMDD of protein cargo with points of advantage including all-aqueous assembly condition, coating of complex geometries, and control over protein loading and release. In the discussion below we present the methods of protein immobilization within therapeutic surface coatings, followed by an overview of delivery mechanisms to the cells of interest and a few select examples of biomedical applications of these techniques. While the overall focus of the presentation is set on LbL coatings, non-LbL examples will be discussed where relevant.

3.1. *Methods of Immobilization*

The three dominating approaches to the loading of protein cargo into polymer thin films are: post-diffusion of the protein into the pre-assembled film, co-assembly of the protein with the constituents of the films during polymer deposition, and covalent attachment to the film or substrate (see Figs. 1(a) and 1(c)).

Post-diffusion

Loading via post-diffusion is arguably the simplest technique for protein immobilization within surface adhered polymer thin films. The benefits of this approach include a choice of protein concentration, solution properties for immobilization, possibility to concentrate and accumulate the cargo from dilute solutions into surface coating, etc. However, utility of this approach to protein immobilization in SMDD is significantly limited by a pronounced burst release of protein from the film upon incubation in physiologically relevant media and an uncontrollable loss of cargo.[26,30,31] This general observation has notable exceptions. Crouzier *et al.* achieved a loading of recombinant human bone morphogenetic protein 2 (rhBMP-2) into a poly(L-lysine)/hyaluronan (PLL/HA) LbL film via protein post-diffusion conducted using an acidic protein solution with a low salt concentrations, i.e. conditions where the protein exhibits high solubility.[31] Subsequent change to neutral pH drastically lowered the solubility of the protein, releasing only about 10% of the cargo and effectively immobilizing the payload. This approach was also implemented for the loading of TAT peptide in which case post-diffusion afforded markedly higher loading capacities then co-assembly of the peptide within exponentially grown LbL films.[30]

Co-assembly

Significant advantages of the co-assembly of a protein with polymers constituting the bulk of the film include a control over immobilized payload through repetitive adsorption and, most importantly, protein immobilization at a specific depth of the film. This latter possibility was well demonstrated by Chluba *et al.* with the hormone α-melanocortin.[32] Short-term peptide activity was shown to depend on the integration depth, while long-term activity remained constant. However, to achieve protein incorporation via co-assembly, the protein needs to be able to participate in LbL assembly in the first place; e.g. possess sufficient charge to interact with polyelectrolytes and sustain the integrity of the resulting film. To achieve this, in the above case α-melanocortin was incorporated in the form of its conjugate to a carrier polymer, PLL.[32] Similarly, conjugation of a cationic nonaarginine protein transduction domain (R_9) to the protein was shown to allow a deposition of R_9 functionalized protein into the LbL film.[33] R_9 also facilitated a cellular uptake of the protein upon release from the surface coating.[33] These films have later been applied to microneedles to achieve highly localized and minimally invasive protein delivery to the skin.[34]

Covalent attachment

Covalent attachment of the protein to a pre-formed surface coating offers an advantage of combining the functional properties of the two components and an assembly of a multi-functional coating with sustained, lasting functionality. Wang *et al.* demonstrated a conjugation of BMP-2 and vascular endothelial growth factor (VEGF) onto a number of polysaccharide films previously grafted on titanium substrates.[35–37] While no account of an active fraction of the growth factor was provided, polysaccharide films were able to prevent bacterial adhesion, whereas BMP-2 and VEGF facilitated increased osteoblast functions, both effects observed *in vitro*. An inherent drawback of covalent immobilization relates to a protein loss of function. Lorenz *et al.* recently achieved a high efficiency in covalent immobilization of BMP-2 in a pre-formed polymer film, yet biological activity was retained only on a 4–8% level as compared to the pristine protein.[29]

3.2. *Controlled Drug Release*

Matrix-bound delivery

SMDD of protein cargo is unique compared to the delivery of small drugs and nucleic acids as this type of cargo does not always require internalization. In matrix-bound delivery, a term coined by the group of Picart,[38] the therapeutic protein is immobilized on or within the surface, allowing receptors on the cell membrane to bind them, but preventing internalization (Fig. 2(d)). Whereas soluble ligands are internalized thus reducing the duration of signal transduction, prevention of internalization has been suggested to prolong the therapeutic effect of the protein.[39,40] Several research groups showed that proteins are able to retain their biological activity once attached to a surface coating and induce a localized therapeutic effect in contacting cells.[25,33,36,40,41] Furthermore, an increase in effectiveness of matrix-bound versus soluble proteins has been demonstrated *in vitro*.[41,42]

Surface-based delivery

A different mode of SMDD of protein cargo relies on a degradation of the polymer film, hydrolytic or enzymatic (Fig. 2(c)), and a delivery of the payload into the immediate vicinity of the cultured cells. In the case of enzymatic degradation, this effect can be achieved by the proteases released into the microenvironment by the cultured cells. Jessel *et al.*[43] demonstrated a degradation of the multilayered polypeptide substrate by the cultured cells. Protein A was immobilized within the film under polypeptide capping layers and shown not to diffuse vertically, indicating that protein release was dependent on cell-mediated degradation of the substrate.[43] Su *et al.* demonstrated transcutaneous protein delivery achieved using a multilayered coating assembled using a model protein and a hydrolytically degradable poly(β-aminoester).[44] Surface mediated release affords a high and sustained local concentration of the protein,[30,33] which leads to a high efficiency of protein delivery.[26]

3.3. *Highlighted Examples*

In vitro, Crouzier *et al.* developed an efficient non-covalent immobilization of BMP-2 in multilayered polymer surface coating and achieved an efficient dose-dependent differentiation of progenitor cells into osteoblasts through BMP-2 signal transduction.[31] Bioactivity was shown to persist during 12 days of continued cell culturing with three consecutive cell populations, albeit to a reducing degree.[31] Prolonging bioactivity could be achieved through an increase in the amount of loaded protein, the latter attained using a higher loading concentration or through the use of blend multilayers where heparin was co-deposited with hyaluronan.[31,15] A further finding of this group relates to a variation of the film stiffness as a tool of control over cell adhesion, spreading and migration. Stiff films showed similar cell behavior (i.e. good cell adhesion and spreading) as standard cell-culture substrates regardless of the presence of BMP-2 in solution or within the surface coating. In contrast, on soft films, greatly increased cell adhesion and spreading was observed on films with matrix-bound BMP-2, whereas soluble BMP-2 did not deliver significant changes, thus indicating a synergistic effect between adhesion and BMP-2 receptors (Fig. 2(d)). The presence of matrix-bound BMP-2 also strongly increased cell migration. These effects on cell behavior were attributed to BMP-2 induced signal transduction: on the one hand, no significant changes in film properties were observed with matrix-bound BMP-2, and on the other hand these effects were only exhibited by cell lines capable of BMP-2 signal transduction.[38]

Hammond *et al.* recently demonstrated a highly efficient immobilization of BMP-2 into a multilayered surface coating assembled using poly(β-aminoester) and poly(acrylic acid) and deposited onto a 3D β-tricalcium phosphate/polycaprolactone scaffold. *In vitro*, a sustained release of BMP-2 over a period of two weeks was achieved through a slow degradation of the poly(β-aminoester). Exposing pre-osteoblasts to the release media resulted in efficient differentiation of the cells into osteoblasts, and the effect was more pronounced as compared to administration of an equal concentration of BMP-2 in a solution form. *In vivo* implantation into rat quadriceps muscle showed efficient recruitment of host progenitor cells and their differentiation into maturing bone, wherein bone formation and maturation continued for at least nine weeks and was only observed using implants incorporating BMP-2.[46] In a subsequent publication it was also revealed that co-formulation of VEGF into these multilayered coatings led to a faster bone maturation as well as a bone formation throughout the scaffold, indicating that VEGF intensified cell recruitment into the interior of the scaffold.[47]

For non-polymer based coatings, VEGF immobilized through co-precipitation with calcium phosphate onto biphasic calcium phosphate ceramics led to an increased vascularization of the matrix. In contrast to surface-adsorbed VEGF, co-precipitated proteins promoted formation of functional blood vessels as well as formation and maturation of bone tissue throughout the scaffold.[26] Majid *et al.* conducted an *in vivo* study on spinal fusion in rabbits using collagen sponges with BMP-2 incorporated through co-precipitation.[48] Even though new bone mass was comparable across different protein-loading mechanisms (i.e. superficial adsorption vs. co-precipitation), when formed from co-precipitated BMP-2 it was more localized to its intended site. Furthermore, the initial high doses of protein released from superficially adsorbed BMP-2 lead to much higher recruitment of osteoclasts and thus bone resorption.[48]

4. Surface-mediated Gene Delivery

Surface-mediated delivery of gene cargo, also termed reverse transfection[49] or solid-phase transfection,[50] enables a localized and sustained transfection of adhering or surrounding cells and has already found utility in both, fundamental science and therapeutic applications.[49,51] In basic science, manipulation of gene expression allows understanding of specific cellular mechanisms and pathways as well as modification of cell behavior.[52] For therapeutic applications, the promise of gene delivery rests in a potential to treat genetic disorders as well as aid in tissue regeneration.[53] For SMDD of gene cargo, particular successes are associated with such surface modification techniques as self-assembled monolayers (SAM),[54] deposited nanocomposites[55] and

LbL polymer deposition.[56–59] These tools are now widely employed in biotechnology in e.g. microarrays techniques[49] and also have biomedical utility in e.g. drug eluting stents.[60,61] The following sections provide a brief overview of the methods and techniques used for DNA immobilization, mechanism of DNA release and the current development of surface-mediated gene delivery.

4.1. *Methods of Immobilization*

Several methods have been employed for DNA immobilization including DNA entrapment in gelatin,[49] co-deposition through biomineralization,[55] and tethering through specific recognition.[62] Surface modification with the use of SAM[54] provides a versatile and convenient platform to immobilize genes of interest through covalent bonding,[62] antigen–antibody interactions,[61] and the avidin–biotin system[63,64] (Fig. 1(d)). However, particular successes of SMDD of gene cargo are associated with the advent of sequential polymer deposition technique. LbL assembly has shown significant advantages over many other techniques such as all aqueous deposition, high stability of the resulting films, precise control over DNA loading, and control over DNA release kinetics, to name a few.[3] DNA is a negatively charged polymer and has been widely used in LbL films paired with diverse polycations and deposited through electrostatic interaction[56–59] (Fig. 1(a)). Two distinctly different approaches to the incorporation of DNA through LbL deposition exist, namely deposition of "naked" DNA[3] or pre-formulated with a polycation.[65] While the former remains much better explored and widely used, the latter allows incorporation of DNA associated with a reagent assisting internalization of gene cargo. For SMDD, LbL technique allows incorporation of different types of genes within the layers of the film. For example, Jessel *et al.*[59] and Zhang *et al.*[66] demonstrated that multiple plasmids could be immobilized within the multilayer architectures. Gene expression was well correlated with the embedding level of DNA within the layers, i.e. depths of deposition within the surface coating. These findings demonstrate an unprecedented opportunity for a programmed, time-delayed transfection using multiple vectors.

SMDD of siRNA is similar to reverse transfection in that immobilization of these two types of nucleic acids follows the same rules and trends. Specifically, successful adsorption of siRNA achieved in multiple subsequent deposition steps afforded a greater payload, i.e. an increased quantity of RNA therapeutic, which subsequently yielded a greater therapeutic response.[67] Taken together, LbL technique appears to be unparalleled as a tool of SMDD of gene cargo and other nucleic acid based drugs and is well suited to address diverse biotechnological and biomedical challenges.

4.2. *Controlled Drug Release*

Controlled release of gene cargo determines the feasibility of surface-mediated gene delivery and potentially enables transfection to be undertaken at specific and local sites. For LbL assembly, the mechanism of surface-mediated gene release relies on a degradation of the polymer thin film where the polycation degrades under reductive condition,[58] hydrolysis[66,68] or upon degradation by cellular enzymes[57] (Fig. 2(c)). For example, Blacklock *et al.*[58] used the exofacial reducing environment of plasma membrane to disassemble LbL films composed of DNA and reducible polycations, which results in transfection of surrounding tissue. Zhang *et al.*[66] investigated LbL films composed of hydrolytically degradable polycations and compared duration of gene release from coating assembled with poly(β-amino ester) and side-chain functionalized polyamines. Hu *et al.*[69] reported on a profile of gene release from biodegradable chitosan/DNA multilayer film and showed that gene release could be tuned by varying the number of layers.

For SAM, kinetics of gene release ranged from 24 hr to eight days based on the interactions between substrate and DNA complexes. For example, hydrophobic surfaces exhibit lower release than hydrophilic substrates.[70] It was also reported that release rates of complexes are independent of the surface densities of the substrate, however it could be affected by the components within the solution which may displace the complexes and bind through specific interactions to the underlying SAMs.[71]

4.3. *Highlighted Examples*

Of the three classes of drugs discussed in this chapter, SMDD of gene cargo has arguably progressed the most and is now well suited to solve practical biomedical challenges. For both DNA[56] and siRNA,[67] surface-mediated delivery was shown to be more effective than solution-based administration, further highlighting the power of SMDD. A few select, recent examples discussed below aim to attract interest from a broad readership to surface-mediated gene transfer; for an up-to-date account on the subject, we refer readers to current reviews.

In vitro, Richard *et al.*[72] demonstrated the utility of SMDD of gene cargo to control sequential expression of multiple genes. A multi-layered polyelectrolyte surface coating based on hyaluronic acid and chitosan was assembled to contain plasmids encoding for Neuropilin-1 (NRP-1) and Plexin-A1 (Plx-A1), two proteins which interact to form a co-receptor for semaphorin 3A (sema3A). The resulting surface coating was successful in delivery of embedded genes to Cos-1 cells resulting in a sustained expression of the gene products and a sema3A-induced collapse phenotype of Cos-1 cells upon co-expression of NRP-1 and Plx-A1.

In vivo, Saurer *et al.*[68] reported on a localized tissue transfection in the carotid arteries of rats using LbL-coated catheter balloons. The authors fabricated an ultrathin hydrolytically degradable surface coating which promoted surface-mediated release of DNA on catheter balloons and demonstrated that the plasmid DNA delivered via this approach remained functional upon insertion and inflation of balloons. Blacklock *et al.*[58] have also described a successful implementation of SMDD and gene delivery *in vivo*. A subcutaneously implanted substrate released a plasmid encoding for a secretable alkaline phosphatase (SEAP) from a reductively degradable surface coating. Successful transfection and detectable levels of SEAP in blood circulation were observed in rats after five days of implantation. More importantly, *in vivo* transfection in the vicinity of the implant suggested the possibility to spatially regulate delivery of genes.[58]

5. Outlook

With the above presentation, we aimed to demonstrate that surface-mediated drug delivery has fast progressed from its advent to a mature platform, from being a peculiar research tool into a sound biomedical paradigm with documented successes *in vitro* and *in vivo*. For each class of the drugs considered in this chapter, namely small molecules, peptides and proteins, and nucleic-based therapeutics, there now exist flexible, adaptable tools for SMDD. We strongly believe that SMDD is well fit to address current challenges of biomedical research. To achieve this, a stronger alliance between materials scientists and biology/clinical research units is highly warranted. We hope that this presentation aids in the establishment of these ties and motivates further research into the field of surface-mediated drug delivery.

References

1. Zelikin AN. (2010) Drug releasing polymer thin films: new era of surface-mediated drug delivery. *ACS Nano* **4(5):** 2494–2509.
2. Boudou T, *et al.* (2010) Multiple functionalities of polyelectrolyte multilayer films: new biomedical applications. *Adv Mater* **22(4):** 441–467.
3. Jewell CM, Lynn DM. (2008) Multilayered polyelectrolyte assemblies as platforms for the delivery of DNA and other nucleic acid-based therapeutics. *Adv Drug Deliv Rev* **60(9):** 979–999.
4. Pavlukhina S, Sukhishvili S. (2011) Polymer assemblies for controlled delivery of bioactive molecules from surfaces. *Adv Drug Deliv Rev* **63(9):** 822–836.
5. Wohl BM, Engbersen JFJ. (2012) Responsive layer-by-layer materials for drug delivery. *J Control Release* **158(1):** 2–14.
6. Acharya G, Park K. (2006) Mechanisms of controlled drug release from drug-eluting stents. *Adv Drug Deliv Rev* **58(3):** 387–401.
7. Duncan R. (2003) The dawning era of polymer therapeutics. *Nat Rev Drug Discov* **2(5):** 347–360.
8. Hoare TR, Kohane DS. (2008) Hydrogels in drug delivery: progress and challenges. *Polymer* **49(8):** 1993–2007.
9. Smith RC, *et al.* (2009) Layer-by-layer platform technology for small-molecule delivery. *Angew Chem Int Ed* **48(47):** 8974–8977.
10. Shukla A, Fuller RC, Hammond PT. (2011) Design of multi-drug release coatings targeting infection and inflammation. *J Control Release* **155(2):** 159–166.

11. Kim BS, *et al.* (2009) Hydrogen-bonded multilayer of pH-responsive polymeric micelles with tannic acid for surface drug delivery. *Chem Commun* **2009(28):** 4194–4196.
12. Thierry B, *et al.* (2005) Delivery platform for hydrophobic drugs: prodrug approach combined with self-assembled multilayers. *J Am Chem Soc* **127(6):** 1626–1627.
13. Chen H, *et al.* (2007) To construct "ion traps" for enhancing the permselectivity and permeability of polyelectrolyte multilayer films. *Macromolecules* **40(3):** 653–660.
14. Schmidt DJ, Moskowitz JS, Hammond PT. (2010) Electrically triggered release of a small molecule drug from a polyelectrolyte multilayer coating. *Chem Mater* **22(23):** 6416–6425.
15. Pavlukhina S, *et al.* (2010) Polymer multilayers with pH-triggered release of antibacterial agents. *Biomacromolecules* **11(12):** 3448–3456.
16. Shukla A, *et al.* (2010) Tunable vancomycin releasing surfaces for biomedical applications. *Small* **6(21):** 2392–2404.
17. Moskowitz JS, *et al.* (2010) The effectiveness of the controlled release of gentamicin from polyelectrolyte multilayers in the treatment of *Staphylococcus aureus* infection in a rabbit bone model. *Biomaterials* **31(23):** 6019–6030.
18. Chuang HF, Smith RC, Hammond PT. (2008) Polyelectrolyte multilayers for tunable release of antibiotics. *Biomacromolecules* **9(6):** 1660–1668.
19. Vodouhê C, *et al.* (2006) Control of drug accessibility on functional polyelectrolyte multilayer films. *Biomaterials* **27:** 4149–4156.
20. Thierry B, *et al.* (2003) Bioactive coatings of endovascular stents based on polyelectrolyte multilayers. *Biomacromolecules* **4(6):** 1564–1571.
21. Wang NX, von Recum HA. (2011) Affinity-based drug delivery. *Macromol Biosci* **11(3):** 321–332.
22. Smith RC, *et al.* (2009) Hydrophobic effects in the critical destabilization and release dynamics of degradable multilayer films. *Chem Mater* **21(6):** 1108–1115.
23. Van Den Beucken JJJP, *et al.* (2007) *In vitro* and *in vivo* effects of deoxyribonucleic acid-based coatings functionalized with vascular endothelial growth factor. *Tissue Eng* **13:** 711–720.
24. Lee K, Silva EA, Mooney DJ. (2011) Growth factor delivery-based tissue engineering: general approaches and a review of recent developments. *J R Soc Interface* **8:** 153–170.
25. Liu Y, *et al.* (2011) A synthetic substrate to support early mesodermal differentiation of human embryonic stem cells. *Biomaterials* **32:** 8058–8066.
26. Wernike E, *et al.* (2010) Vegf incorporated into calcium phosphate ceramics promotes vascularisation and bone formation *in vivo*. *Eur Cell Mater* **19:** 30–40.

27. Tao C, *et al.* (2009) Surface modification of diamond-like carbon films with protein via polydopamine inspired coatings. *Appl Surf Sci* **256:** 294–297.
28. Wu G, *et al.* (2010) Biomimetic coating of organic polymers with a protein-functionalized layer of calcium phosphate: the surface properties of the carrier influence neither the coating characteristics nor the incorporation mechanism or release kinetics of the protein. *Tissue Eng - Part C Methods* **16:** 1255–1265.
29. Lorenz C, *et al.* (2011) Coating of titanium implant materials with thin polymeric films for binding the signaling protein BMP2. *Macromol Biosci* **11:** 234–244.
30. Wang X, Ji J. (2009) Postdiffusion of oligo-peptide within exponential growth multilayer films for localized peptide delivery. *Langmuir* **25:** 11664–11671.
31. Crouzier T, *et al.* (2009) Layer-by-layer films as a biomimetic reservoir for rhBMP-2 delivery: controlled differentiation of myoblasts to osteoblasts. *Small* **5:** 598–608.
32. Chluba J, *et al.* (2001) Peptide hormone covalently bound to polyelectrolytes and embedded into multilayer architectures conserving full biological activity. *Biomacromolecules* **2:** 800–805.
33. Jewell CM, *et al.* (2007) Multilayered films fabricated from an oligoarginine-conjugated protein promote efficient surface-mediated protein transduction. *Biomacromolecules* **8:** 857–863.
34. Saurer EM, *et al.* (2010) Layer-by-layer assembly of DNA- and protein-containing films on microneedles for drug delivery to the skin. *Biomacromolecules* **11:** 3136–3143.
35. Shi Z, *et al.* (2009) Titanium with surface-grafted dextran and immobilized bone morphogenetic protein-2 for inhibition of bacterial adhesion and enhancement of osteoblast functions. *Tissue Eng - Part A* **15:** 417–426.
36. Hu X, *et al.* (2010) An *in vitro* assessment of titanium functionalized with polysaccharides conjugated with vascular endothelial growth factor for enhanced osseointegration and inhibition of bacterial adhesion. *Biomaterials* **31:** 8854–8863.
37. Shi Z, *et al.* (2009) Surface functionalization of titanium with carboxymethyl chitosan and immobilized bone morphogenetic protein-2 for enhanced osseointegration. *Biomacromolecules* **10:** 1603–1611.
38. Crouzier T, *et al.* (2011) Presentation of BMP-2 from a soft biopolymeric film unveils its activity on cell adhesion and migration. *Adv Mater* **23:** H111–H118.
39. Lidke DS, *et al.* (2004) Quantum dot ligands provide new insights into erbB/HER receptor-mediated signal transduction. *Nat Biotech* **22:** 198–203.
40. Fan VH, *et al.* (2007) Tethered epidermal growth factor provides a survival advantage to mesenchymal stem cells. *Stem Cells* **25:** 1241–1251.
41. Kuhl PR, Griffith-Cima LG. (1996) Tethered epidermal growth factor as a paradigm for growth factor-induced stimulation from the solid phase. *Nat Med* **2:** 1022–1027.

42. Chen TT, *et al.* (2010) Anchorage of VEGF to the extracellular matrix conveys differential signaling responses to endothelial cells. *J Cell Biol* **188:** 595–609.
43. Jessel N, *et al.* (2003) Bioactive coatings based on a polyelectrolyte multilayer architecture functionalized by embedded proteins. *Adv Mater* **15:** 692–695.
44. Su X, *et al.* (2009) Layer-by-layer-assembled multilayer films for transcutaneous drug and vaccine delivery. *ACS Nano* **3(11):** 3719–3729.
45. Crouzier T, *et al.* (2010) Polysaccharide-blend multilayers containing hyaluronan and heparin as a delivery system for rhBMP-2. *Small* **6:** 651–662.
46. Macdonald ML, *et al.* (2011) Tissue integration of growth factor-eluting layer-by-layer polyelectrolyte multilayer coated implants. *Biomaterials* **32(5):** 1446–1453.
47. Shah NJ, *et al.* (2011) Tunable dual growth factor delivery from polyelectrolyte multilayer films. *Biomaterials* **32:** 6183–6193.
48. Majid K, *et al.* (2011) Biomimetic calcium phosphate coatings as bone morphogenetic protein delivery systems in spinal fusion. *Spine J* **11:** 560–567.
49. Ziauddin J, Sabatini DM. (2001) Microarrays of cells expressing defined cDNAs. *Nature* **411(6833):** 107–110.
50. Bielinska AU, *et al.* (2000) Application of membrane-based dendrimer/DNA complexes for solid phase transfection *in vitro* and *in vivo*. *Biomaterials* **21(9):** 877–887.
51. Salvay DM, Zelivyanskaya M, Shea LD. (2010) Gene delivery by surface immobilization of plasmid to tissue-engineering scaffolds. *Gene Ther* **17(9):** 1134–1141.
52. Stark GR, Gudkov AV. (1999) Forward genetics in mammalian cells: functional approaches to gene discovery. *Hum Mol Genet* **8(10):** 1925–1938.
53. Anderson WF. (1998) Human gene therapy. *Nature* **392(6679 Suppl.):** 25–30.
54. Wang CHK, Pun SH. (2011) Substrate-mediated nucleic acid delivery from self-assembled monolayers. *Trends Biotech* **29(3):** 119–126.
55. Shen H, Tan J, Saltzman WM. (2004) Surface-mediated gene transfer from nanocomposites of controlled texture. *Nat Mater* **3(8):** 569–574.
56. Zhang X, *et al.* (2008) Transfection ability and intracellular DNA pathway of nanostructured gene-delivery systems. *Nano Lett* **8(8):** 2432–2436.
57. Dimitrova M, *et al.* (2007) Adenoviral gene delivery from multilayered polyelectrolyte architectures. *Adv Funct Mater* **17(2):** 233–245.
58. Blacklock J, *et al.* (2009) Gene delivery *in vitro* and *in vivo* from bioreducible multilayered polyelectrolyte films of plasmid DNA. *Biomaterials* **30(5):** 939–950.
59. Jessel N, *et al.* (2006) Multiple and time-scheduled *in situ* DNA delivery mediated by β-cyclodextrin embedded in a polyelectrolyte multilayer. *Proc Natl Acad Sci* **103(23):** 8618–8621.
60. Fishbein I, *et al.* (2006) Bisphosphonate-mediated gene vector delivery from the metal surfaces of stents. *Proc Natl Acad Sci USA* **103(1):** 159–164.

61. Klugherz BD, *et al.* (2000) Gene delivery from a DNA controlled-release stent in porcine coronary arteries. *Nat Biotech* **18(11):** 1181–1184.
62. Peckys DB, *et al.* (2008) End-specific strategies of attachment of long double stranded DNA onto gold-coated nanofiber arrays. *Nanotechnology* **19(43):** 435301.
63. Segura T, Shea LD. (2002) Surface-tethered DNA complexes for enhanced gene delivery. *Bioconjug Chem* **13(3):** 621–629.
64. Segura T, Volk MJ, Shea LD. (2003) Substrate-mediated DNA delivery: role of the cationic polymer structure and extent of modification. *J Control Release* **93(1):** 69–84.
65. Meyer F, *et al.* (2006) Polyplex-embedding in polyelectrolyte multilayers for gene delivery. *Biochim Biophys Acta-Biomembranes* **1758(3):** 419–422.
66. Zhang J, *et al.* (2007) Multilayered films fabricated from plasmid DNA and a side-chain functionalized poly(β-amino ester): surface-type erosion and sequential release of multiple plasmid constructs from surfaces. *Langmuir* **23(22):** 11139–11146.
67. Dimitrova M, *et al.* (2008) Sustained delivery of siRNAs targeting viral infection by cell-degradable multilayered polyelectrolyte films. *Proc Natl Acad Sci* **105(42):** 16320–16325.
68. Saurer EM, *et al.* (2011) Delivery of plasmid DNA to vascular tissue *in vivo* using catheter balloons coated with polyelectrolyte multilayers. *Biomaterials* **32(2):** 610–618.
69. Hu Y, *et al.* (2009) Surface mediated *in situ* differentiation of mesenchymal stem cells on gene-functionalized titanium films fabricated by layer-by-layer technique. *Biomaterials* **30(21):** 3626–3635.
70. Yamauchi F, Kato K, Iwata H. (2004) Micropatterned, self-assembled monolayers for fabrication of transfected cell microarrays. *Biochim Biophys Acta – General Subjects* **1672(3):** 138–147.
71. Pannier AK, Anderson BC, Shea LD. (2005) Substrate-mediated delivery from self-assembled monolayers: effect of surface ionization, hydrophilicity, and patterning. *Acta Biomateri* **1(5):** 511–522.
72. Richard D, *et al.* (2010) Polyelectrolyte multilayer-mediated gene delivery for semaphorin signaling pathway control. *Small* **6(21):** 2405–2411.

Chapter 10

Use of Tunable Pores for Accurate Characterization of Micro- and Nanoparticle Systems in Nanomedicine

James Eldridge,[*,†] Aaron H. Colby,[‡] Geoff R. Willmott,[*,†] Sam Yu[*,§] and Mark W. Grinstaff[‡]

Abstract

The use of micro- and nanoparticles for diagnostic and therapeutic applications has considerable potential to improve human health, quality of life, and patient outcomes. In particular, such particles show promise for the treatment of tumors, delivery of gene therapies or vaccinations, as contrast agents in imaging, and in diagnostic assays. As with any technology used in medical applications, synthesis of and treatment with these particles is subject to stringent quality control and safety standards. Implicit in this is the need to accurately characterize the physical properties of particle populations in solution; such measurements present an especially pertinent research topic. Numerous techniques are used to this end including: dynamic light scattering,

[*] The MacDiarmid Institute for Advanced Materials and Nanotechnology, Victoria University of Wellington, PO Box 600, Wellington 6140, New Zealand.
[†] Callaghan Innovation, PO Box 31310, Lower Hutt 5040, New Zealand.
[‡] Departments of Biomedical Engineering and Chemistry, Boston University, 590 Commonwealth Avenue, Massachusetts 02215, United States.
[§] Izon Science, PO Box 20189, Bishopdale, Christchurch 8543, New Zealand.

nanoparticle tracking analysis, differential centrifugal sedimentation, flow cytometry, suspended microchannel resonators, and electron microscopy. However, these methods can be limited by requirements for monodispersity, sample preparation, and lack of versatility. Relatively recently, tunable pores (TPs) have emerged as an effective tool for particle size, charge and concentration measurements using resistive pulse sensing (RPS). TPs are single conical holes in an elastomeric membrane, where the pore geometry can be reversibly "tuned" *in situ* by stretching and relaxation of the membrane. This technology has specific technical strengths for measurement, such as real-time monitoring of changes in a particle population, close control of electrokinetic and pressure-driven transport, and non-ensemble particle-by-particle sensing. The latter is particularly important for nanomedicine, where outlying particles can have a major effect on the safety and efficacy of the particle population. In this chapter, we review resistive pulse sensing with tunable pores in the context of other characterization techniques and the types of particles used in nanomedicine. Specifically, we examine the use of this method to detect and quantify synthetic inorganic nanoparticles, such as those used for on-bead diagnostics and liposomes used in gene delivery. We also review the characterization of biological particles of interest such as cells, platelets, bacteria, viruses, phages, large biomolecules, and ultimately single molecules for proteomics and genomics.

1. Introduction

Nanomedicine is a broad term describing the use of nanotechnology, technology which is based on materials and devices <100 nm in dimension, to improve the standard of patient care.[1] Nanomedicine concepts are actively investigated and evaluated through *in vitro*, *ex vivo* and *in vivo* experiments with the following goals: (1) to detect and diagnose disease states; (2) to improve the efficacy of existing drugs by altering the mode of delivery; and, (3) in some fields, to begin the transition toward individually, as opposed to epidemiologically, guided patient care. Excitingly, a few nanotechnologies have advanced to the clinic and are being used today. In this chapter, we review some of the methods researchers use to characterize nanoparticles in order to better understand and design these technologies, paying special attention to the use of resistive pulse sensing (RPS) with tunable pores (TPs).

Of the many nanotechnologies, nanoparticles are frequently studied for use in imaging and diagnostics. Contrast agents such as gold nanoparticles, quantum dots, carbon nanotubes, magnetic nanoparticles, nanoshells, ceramic nanoparticles, and targeted radiolabeled nanoparticles are used, for example, to track tumor size, progression, and metastases, observe intracellular interactions, and detect imminent heart attacks *in vivo*.[2,3] Often it is the increased spatial and temporal resolution, as compared to traditional contrast agents, that has been the driving force for development of these technologies. Synthetic micro- and nanoscale beads are also being used increasingly in point-of-care diagnostics.[4] Beads are often used in assays, interacting with target analytes to, for example, enable identification via labels or aggregation. Magnetic beads are also useful for separation technologies and general sample handling.[5]

In addition to diagnostic and imaging applications, nanomedicines are also being used to improve the efficacy of systemically administered therapeutics by concentrating these to specific sites in the body. In particular, nanoformulations such as polymer nanoparticles or lipid-based micelles are used to encapsulate drugs for the purpose of modifying and improving their delivery to provide increased overall efficacy (Fig. 1). There exist numerous animal studies and select human examples in which new methods of delivering drugs have shown increased efficacy or equivalent efficacy with reduced side effects compared to the traditional drug alone. Some examples of this class approved by the FDA and in clinical use include: *Doxil* — PEGylated liposomal doxorubicin used in breast and ovarian cancers as well as multiple myeloma and Kaposi sarcoma with a circulation time 100× longer than that of free doxorubicin; *Abraxane* — albumin-bound paclitaxel nanoparticles used in breast cancer with higher response rates and tolerable doses than free paclitaxel in Cremophor E/L; and, *Oncaspar* — PEG-L-asparaginase used for leukaemia with higher response rates and lower hypersensitivity reactions.[1,6] Many

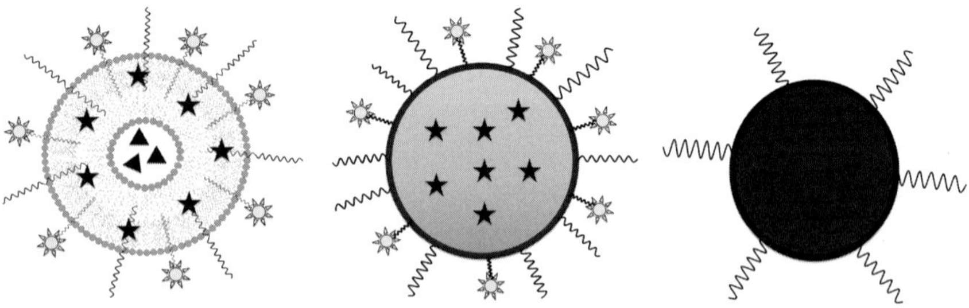

Fig. 1. Schematic representation of nanoformulations for drug delivery and imaging: (a) liposome with hydrophilic drug (triangles) encapsulated within inner aqueous core, hydrophobic drug (stars) encapsulated in lipid bilayer, fluorophore or antibody (sun), and polyethylene glycol (PEG) (wavy arm) coating; (b) nanoparticle with hydrophobic drug encapsulated, fluorophore or antibody attached at the surface, and PEG coating; (c) iron oxide quantum dot with PEG coating. Figures not to scale.

approved formulations incorporate polyethylene glycol (PEG) to improve their pharmacokinetics by increasing circulation time and reducing clearance of the particle-drug system by the body's reticuloendothelial system (RES), which consists of cells (macrophage and monocytes) and tissues (lymph nodes, liver, spleen).

In general, these nanoformulations confer several advantages over traditional *in vivo* imaging modalities or drug-alone therapies. They have the ability to target specific sites within the body, alter biodistribution, and minimize adverse side effects — useful in both diagnostics and imaging. Nanoformulations also afford improved drug solubility, stability, and pharmacokinetics, increased drug half-life, and controlled and sustained drug delivery at therapeutic concentrations. In designing nanomedicines to achieve these performance specifications, engineers control various characteristics including particle material, size, surface charge/coating, concentration, and, in the case of drug delivery, drug loading and release. These factors play important roles in determining how each nanotechnology will interact with the body and its immune and RES systems. As stated by Davis *et al.*: "The ultimate fate of nanoparticles within the body can be determined by the interactions of nanoparticles with their local environment, which depends on a combination of size and surface properties."[6]

Particle size affects filtration by the body's RES. Though nanotechnologies are defined in the size range <100 nm, the upper limit is somewhat arbitrary; in practice, sizes up to and over 1000 nm are frequently studied, though even this upper limit is sub-cellular in dimension. In their review of particle uptake and biodistribution, Alexis *et al.* note that particles smaller than 10 nm are cleared from the body via the kidneys while particles 10–150 nm are sequestered in the bone marrow; particles larger than 200 nm can undergo splenic clearance; particles 50–500 nm in diameter can experience clearance via the liver while particles larger than 500 nm are removed by the macrophage and monocytes of the RES.[7,8] Cellular uptake of nanoparticles is also size-dependent with smaller sizes yielding higher uptake. Rejman *et al.* found that 50 nm fluorescent latex particles were internalized immediately by B16 tumor cells while 100 nm and 200–500 nm particle uptake were 3–4 times and 8–10 times less, respectively; 1000 nm beads were not internalized at all.[9]

Surface charge also affects the *in vivo* fate of nanoformulations. It is generally accepted that positively charged particles are more quickly opsonized and removed from the circulation than negatively charged particles, which tend to repel from the negatively charged surface of vasculature and cells of the RES resulting in longer circulation times.[6,7] Surface modification with non-ionic surfactants also alters biodistribution. For example, Peracchia *et al.* showed that 24 hr after injection, 40% of nanoparticles coated with PEG, a hydrophilic, non-ionic surfactant, were sequestered in the liver while 90% of uncoated, negatively charged particles were found in the liver only 3 min after injection.[10] However, due to the complexities of a biological system and the vast array of different types of particles investigated, further studies will be required to gain a complete understanding of how surface charge effects particle biodistribution, uptake, and clearance of nanoparticles.

Due to the important and interconnected role that particle size and surface charge play in determining nonspecific binding, macrophage removal, particle filtration, cellular uptake, and ultimate biodistribution, it is critical for researchers to have precise and accurate means of measuring these characteristics. Researchers need to know, for example, how size and surface properties will alter nanoparticle access to novel locations inaccessible to larger microparticles previously studied. The importance of accurate particle concentration measurements is also often underestimated, even though concentration is critical for determining dosages and diagnostic sensitivity. In addition to aiding researchers in understanding the biological interactions of nanoparticles, improved particle characterization will be required to ensure quality control over manufactured particles for both safety and efficacy concerns as well as for business considerations. In the case of manufacturing, the most useful techniques will be those that can provide accurate, inexpensive, fast, and reliable measurements.

In this review we will briefly discuss the fundamental concepts behind each of the major modern nanoparticle characterization techniques and consider their advantages and limitations with specific regard to nanomedicine. In particular, we discuss a developing technique for nanoparticle characterization, resistive pulse sensing (RPS) using tunable nanopores (TPs). The underlying principles of this technique are described, its use in biotechnology and nanomedicine is reviewed, and future research directions opening up in the field of nanomedicine are discussed.

2A. Common Nanoparticle Characterization Techniques

2A(i) *Dynamic Light Scattering (DLS) and Related Techniques*

A range of dynamic light scattering (DLS)-based techniques are presently finding widespread use for nanoparticle size and charge measurements. The capabilities and limitations of these systems are broadly summarized and compared with other particle characterization methods in Table 1 and Fig. 2. In DLS, electromagnetic radiation is scattered by small moving particles in a regime where particle dimensions are much smaller than the wavelength of the radiation.[11,12] Brownian motion of the particles can be measured by monitoring the intensity fluctuation of the scattered light. The hydrodynamic radius

Table 1. Summary of Typical Parameters for Nanoparticle Characterization Techniques. Values are Indicative, based on Information in the References Provided. Here RPS refers to resistive pulse sensing using tunable pores.

	Used to Measure						Sample Volume		
Technique	Size	Charge	conc	Measured Quantity	Size Range nm	Ensemble or Individual	μL	Indicative Particle Concentration mL^{-1}	References
DLS	√			Brownian motion	<1–6000	Ensemble	2–3000	$<10^{10}–10^{13}$	12, 13, 25
LDE/PALS		√		Electrophoretic mobility	<1–6000	Ensemble	2–3000	10^{8}	15
NTA	√		√	Brownian motion	10–8000	Individual	~300	$>10^{8}–10^{9}$	13
DCS	√			Sedimentation velocity	2–80	Ensemble	~100	$10^{11}–10^{13}$	19, 25
FCM	√		√	Light scattering	>500	Individual	> 50	$>10^{6}$	14
EM	√			Electron scattering	>0.05	Individual	negligible	$\sim10^{8}$ (n.i)	23, 24
SMR	√	√	√	Cantilever resonance	~50–10000	Individual	< 20	$<10^{8}$	20, 21, 22
RPS	√	√	√	Occluded ionic current	>~50	Individual	~40	$10^{5}–10^{12}$	26, 27, 28

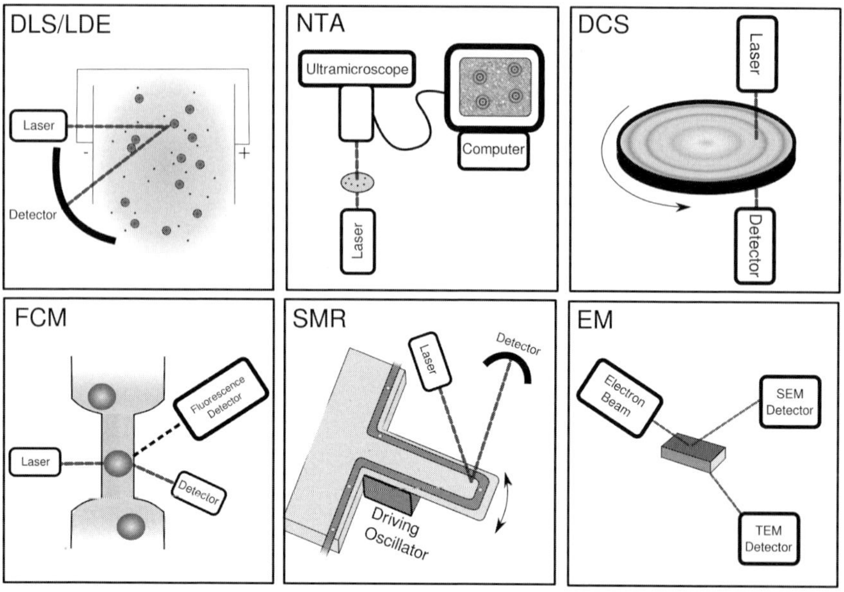

Fig. 2. Summary schematic diagrams outlining the general working method for the nanoparticle characterization methods discussed in the text. Acronyms refer to dynamic light scattering (DLS), laser Doppler electrophoresis (LDE), nanoparticle tracking analysis (NTA), differential centrifugal sedimentation (DCS), flow cytometry (FCM), suspended microchannel resonator (SMR), and electron microscopy (EM).

of the particles is calculated from the autocorrelation of this signal, coupled with the Einstein–Stokes relation, which describes the motion of spherical particles through a fluid at low Reynolds number.

Polydispersity, aggregation and the presence of large contaminants are key limitations for DLS techniques.[11,13] In the Rayleigh approximation, the intensity of scattered light is proportional to r^6, where r is the particle radius, so that (for example) the intensity of scattered light produced by one 400 nm particle is matched by the light from 64 similar 200 nm particles. Data relating to an underlying distribution of small particles can be swamped or skewed by light scattered from a comparatively low concentration of large particles. Some recently-developed DLS techniques partially compensate for large particles by detecting backscatter, because larger particles preferentially scatter radiation forwards. Alternatively, DLS can be coupled to microfluidic flow field fractionation in order to separate particles based on size prior to light-scattering, reducing size bias effects.[14] This technique has yet to achieve widespread adoption however, due to additional complexity and cost.

The quality of DLS data is enhanced when the nanoparticle shape and refractive index are known, and the concentration of particles (e.g. $< 1\%$ by volume) and electrolyte (e.g. < 0.1 w/v%) are relatively low.[11,12] Shape and refractive index are used in calculations, while low concentrations prevent multiple scattering events and screen electronic interactions between particles. DLS is an example of an "ensemble" characterization method: the data describe a large population of particles. Ensemble methods provide the statistical benefits of examining a large population, providing an "intensity-weighted" average in the case of DLS. Alternative particle-by-particle measurement techniques can provide more precise information regarding the nature of individual particles in solution, and are not so susceptible to skewing of the data by outliers.

Similar techniques to DLS (e.g. commercially available as laser Doppler electrophoresis (LDE) and phase analysis light scattering (PALS)) are used to determine the surface charge of nanoparticles, usually quantified as the zeta potential ($\zeta_{particle}$) in the electrical double layer at the particle surface. Surface charge is important for determining the stability and agglomeration potential of nanoparticles and emulsions, and is a key indicator of surface functionality under varying environmental conditions.[15] For this measurement, the particle-containing solution is placed in an oscillating electric field (**E**) of sufficient strength so that electrokinetic behavior dominates Brownian motion. This electrokinetic motion causes a Doppler shift in the scattered light frequency which allows calculation of the particle velocity (**v**). The particle electrophoretic mobility is:

$$\mu = \frac{|\mathbf{v}|}{|\mathbf{E}|} = \frac{\varepsilon \zeta_{particle}}{\eta}, \tag{1}$$

where ε is the absolute electrolyte permittivity and η is viscosity. This expression allows calculation of $\zeta_{particle}$ under Smoluchowski's approximation, where the extent of the electrical double layer is much smaller than the particle radius. High resolution charge measurements can be obtained by detecting the phase shift between the incident and scattered radiation, rather than the direct shift in frequency. To compensate for electro-osmotic effects at the sample walls, the frequency of the oscillating field can be varied. Shorter oscillatory periods do not provide sufficient time for electro-osmotic flow to stabilize, whilst the long oscillatory periods allow direct measurement of electro-osmotic flow at the container wall. As with light-scattering sizing techniques, large particles can dominate and reduce the accuracy of light-scattering charge measurements.

DLS is widely used as a generic size and charge measurement tool for nano- and microparticles, including those used in nanomedicine. However, the size effects described above can lead to inaccuracies when DLS is used as a quick, quality-control tool. For example, mixtures of drug delivery capsules made using bulk synthesis, a popular technique due to simplicity and scalability, are commonly polydisperse.[11] DLS is also used to monitor particle dynamics in processes such as aggregation, where polydispersity is again an important factor that can affect the accuracy and consistency of the results.

2A(ii) *Nanoparticle Tracking Analysis (NTA)*

Nanoparticle tracking analysis (NTA) is a widely-applicable technique for size and concentration measurements. NTA is tolerant of low level impurities (unlike DLS) and is particularly suitable for particles dispersed in non-aqueous solvents such as toluene and heptane. In NTA, individual nanoparticles are identified and tracked in the optical field of an ultramicroscope. The diffracted light patterns produced by the nanoparticles are projected onto a CCD camera, and particle motion is tracked algorithmically by video processing software.[16,17] As with DLS, characterization of the Brownian motion enables calculation of the particle hydrodynamic radius. NTA requires that the particle refractive index differs sufficiently from that of the solution, and that the ultramicroscope is isolated from mechanical vibration.[17]

NTA can accurately characterize polydisperse specimens, and is able to resolve populations with less than 40 nm difference in their radii.[16] Concentration measurements can be carried out by simply counting the total number of particles producing diffraction patterns in a known fluid volume. NTA is limited by some selection bias towards larger particles, which block smaller particles from the view of the CCD.[13] The technique is also dependent on user-defined adjustments to the algorithmic detection parameters, which can alter the minimum particle size detected and thus affect the particle sampling process. The number of individual particles that can be measured in a practical time frame is limited, and typically ~1000 sampled particles are required to effectively describe an assumed log-normal distributed population.[13]

2A(iii) *Differential Centrifugal Sedimentation (DCS)*

Gravitational sedimentation has long been a staple technique for separating particles based on size. However, its use is typically restricted to particles larger than 1 μm, as the settling times for smaller particles are prohibitively long. With centrifuge technology however, it is possible to separate particles in the 2 nm–80 μm size range, with resolution down to 2% of particle diameter.[18,19] In a differential centrifugal sedimentation (DCS) experiment, suspended particles are introduced to the center of a transparent rotating disc with a radial fluid density gradient. Rotational motion causes the (assumed spherical) particles to migrate at a well-defined speed v, as defined by sedimentation under centripetal acceleration a, from Stokes' law:

$$\nu = \frac{2a'^2(\rho_P - \rho_F)\alpha}{9\eta}, \tag{2}$$

where ρ_P and ρ_F are particle and fluid densities respectively, a' is the particle radius and η is the fluid viscosity.[19] An accurate particle size distribution can be determined from the velocity of the particle ensemble. Velocities are calculated (using Raleigh or Mie scattering theory) from absorbance measurements obtained from a monochromatic light source and detector, positioned perpendicular to the plane of rotation towards the edge of the disc. Typically, short wavelength (400–500 nm) radiation is used for detection of <100 nm particles. DCS is an ensemble method, which can efficiently produce data that is reproducible and resistant to size bias, although it requires the use of standard calibration particles, as well as knowledge of the solution viscosity and density. DCS can reliably characterize particles with neutral buoyancy and lower densities, which are otherwise difficult to measure.[19]

2A(iv) *Flow Cytometry (FCM)*

Flow cytometry (FCM), which is used for detecting and sorting microscopic particles (> 500 nm), has been commercially available since the late 1960s. The FCM technique uses light emitted or scattered by microstructures suspended in a hydrodynamically focused stream of fluid to determine a range of properties. These can include the size and refractive index of individual particles, and the internal structure of cells, depending on the location of the collected light.[14] Cells or particles can be simultaneously categorized on the basis of fluorescent or radioactive expression markers or tags. Although FCM relies on the detection of individual particles, results are algorithmically sorted to enable categorization of thousands of cells or microparticles per second. This high throughput ensures relatively low uncertainly in population characteristics. For best quality results, FCM requires high purity samples and low particle concentration to minimize coincident events. FCM is often included as part of integrated systems for cell sorting, optoacoustic and microfluidic applications.[14]

2A(v) *Suspended Microchannel Resonator (SMR)*

Suspended microchannel resonators (SMRs) measure the mass of individual particles. Solutions containing particles are passed through a microchannel embedded within a resonating cantilever in a vacuum. Resonant cantilever oscillations are driven by an electrical signal and optically monitored by a laser focused on the reflecting cantilever surface. When a particle passes through the channel, a shift in the resonant frequency is observed. The system's resonant frequency is:

$$f = \frac{1}{2\pi}\sqrt{\frac{k}{m^{*} + \beta\Delta m}}, \tag{3}$$

where k is the spring constant, m^* is the effective cantilever mass, β is a geometric factor and Δm is the change in mass.[20] SMR is reportedly able to detect mass changes with a precision of 300 attograms, over six orders of magnitude more precise than high-end quartz-crystal microbalances or surface plasmon resonance based systems.[21] Mass measurements can be used to calculate (assumed spherical) particle diameter, to monitor size changes in real time, and to study particles bound to the microchannel walls. SMRs have been used to characterize a wide range of particle types, including polystyrene (1.5 μm) and gold (100 nm) particles, *Escherichia coli* (0.8 × 2 μm), *Bacillus subtilis* (1 × 2.5 μm), human red blood cells (~6 × 2 μm) and to monitor antibody binding at concentrations as low as 1 nM.[21] By applying an oscillating electric field to the resonator, electrophoretic mobility and thus zeta potential can be measured, and concentration can be determined by counting particles, given knowledge of the volume flow rate in the microchannel.[22] The technique is limited by the need for low particle concentrations to prevent multiple particles concurrently occupying the cantilever resonator, and the associated low sampling rate.

2A(vi) *Electron Microscopy (EM)*

For imaging of any submicron structure, use of electron microscopy (EM) is standard practice and ubiquitous. To create an EM image, a beam of electrons is focused on to a substrate, and the pattern of the deflected electrons is detected and analyzed to reconstruct the substrate structure. Scanning electron microscopy (SEM) collects low energy (<10 keV) electrons backscattered from the substrate, and is suitable for imaging objects from ~1 nm–2 μm.[23] Transmission electron microscopy (TEM) utilizes electron energies up to 100 keV to penetrate a thin substrate, affording imaging resolution down to 0.05 nm.[24] For nanoparticle characterization, the greatest drawback of these techniques is sample preparation. Imaging is carried out in a near-vacuum, with particles deposited on a dry solid surface. Coatings are applied to non-conducting specimens, and specialized preparation techniques are required for TEM in particular. Consequently, the number of particles that can be observed is restricted by time and cost, thereby limiting the statistical significance of quantitative measurements. Particles may undergo size changes or surface modification during sample preparation, and there can be errors associated with observation of a 2D projection of 3D particles. Recently-developed techniques, such as Wet-STEM, show promise for imaging particles in solution, but can also induce undesirable artifacts. Overall, EM is best suited for accurate observation of small collections of particles. Aside from size measurements, EM provides clear advantages such as visualization of contaminants or aggregation, and will remain important for nanoparticle characterization.

2B. Resistive Pulse Sensing (RPS)

2B(i) *Introduction and History*

Recent progress with resistive pulse sensing (RPS) has enabled efficient, accurate and high-throughput measurement of the size, concentration and charge of individual suspended nanoparticles. This development presents important new opportunities for nanomedicine, extending from single molecule detection to rapid characterization of micron-sized particles, depending on the size of the pore used. Here we briefly introduce RPS research, while drawing the reader's attention to reviews elsewhere, and noting that the measurement processes are detailed in Section 2C.[29–33]

An RPS experiment (Fig. 3(a)) uses an individual micro- or nanopore which tunnels between the surfaces of a thin membrane. The membrane is immersed in aqueous electrolyte, wetting the pore, and when a voltage is applied to electrodes on either side of the membrane, a current is carried through the pore by mobile ions in the electrolyte. When particles are introduced to the system and migrate through the pore, their presence restricts the flow of ions and thus is

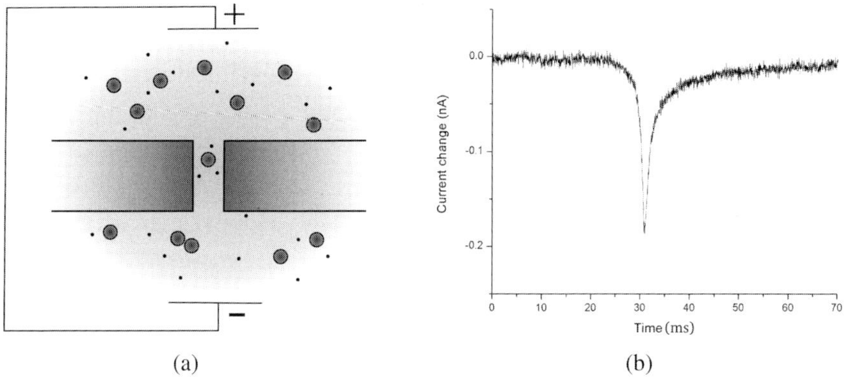

Fig. 3. (a) Schematic sectional view of a typical RPS experimental arrangement. (b) A typical resistive pulse, caused by a 200 nm polystyrene particle in 0.1 M KCl electrolyte, passing through a conical tunable pore (see Section 2C).

detectable as a transient decrease in the recorded current (Fig. 3(b)), known as a "resistive pulse" corresponding to a "blockade" or "translocation" event. Events are usually characterized by three metrics: current magnitude (ΔI), duration, and the frequency at which they occur. These measurements provide information about the particles in solution.

The first RPS devices were early Coulter counters, a type of device developed in the 1950s for counting and sizing cell populations which are still commercially available.[34] In a typical Coulter counter, pressure-driven flow is used to force cells through an aperture bored in a sapphire membrane and subsequently embedded in a glass tube. The resolution of these devices is dictated by the aperture size, and they are typically used for reliable measurements of particles larger than 500 nm, including microorganisms, blood cells, spermatozoa, platelets, algae and yeasts.[29] In the 1970s, DeBlois and co-workers used RPS to study polystyrene nanospheres down to 60 nm, and subsequently to detect T4 bacteriophages, using track-etched polycarbonate membranes (Section 2B(iii)).[35–37] The next major step forward in RPS research did not occur until the mid-1990s, by which time interest in biological and synthetic sub-micron particles had grown, and techniques suitable for nanoscale particle and pore fabrication had matured.

2B(ii) *Biological Pores*

In 1996, Kasianowicz *et al.* used RPS with an α-hemolysin (α-HL) nanopore embedded in a lipid bilayer to detect and differentiate polynucleotide sequences.[38] α-HL is a water-soluble, heptameric protein pore with minimum internal diameter of ~2 nm, secreted by *Staphylococcus aureus,* which spontaneously assembles within lipid bilayer membranes. Once assembled, α-HL is robust (relative to other biological pores) across ranges of temperature and pH. This study triggered a new generation of RPS work, using both biological and synthetic nanopores to study molecular-scale processes such as protein folding, ion channel mechanics, and DNA dehybridization.[33,39] However, the most attractive potential application is fast DNA sequencing, where RPS (or a similar technique) is used to read off bases in sequence as a strand of DNA passes through a nanopore.[30,40] Numerous biological nanopores can be used in place of α-HL, with diverse structures and functionalities that enable novel investigations of binding kinetics and ligand gating.[39] In some ion channels, the conformational change induced by a molecular binding event can cause restriction or expansion of the nanopore and consequently a shift in the current trace, leading to detection of an analyte with high specificity.[39] Biological nanopores operate at single molecule scales, and are therefore much more selective than many of the synthetic nanopores described below, and unsuitable for studying particles larger than a few nanometers.

2B(iii) *Solid-State Pores*

The use of biological nanopores brings some general limitations, such as their functional complexity and instability to extremes of temperature or pH, particularly when compared with nanopores manufactured in solid-state membranes. Solid-state nanopores have been synthesized using a range of techniques, and good control over pore size and geometry has been developed over many investigations. To demonstrate the wide variety of synthetic pore types and applications, we note that graphene nanopores have been used to detect the passage of ssDNA across a membrane a few atoms wide,[41] while in this chapter we are mostly concerned with conical polymeric nanopores used for rapid micro- and nanoparticle characterization.

The smallest synthetic nanopores have been fabricated in silicon-based membranes using beams of ions and electrons.[30] For example, Li *et al.* reported production of nanopores with diameters between 1–60 nm in silicon nitride membranes using "ion-beam sculpting."[42] Pores were made within larger (~100 nm) bowl-shaped cavities by removing the surface above the cavity apex with a beam of 3 keV Ar^+ ions. The pore diameter could be controlled by varying the duration of the ion-beam sculpting, and 5 nm diameter nanopores produced using this method were used to detect 500 base pair dsDNA using RPS. Ion beam sculpting has since been used to synthesize nanopores in silica, Al, Cr, poly(methyl methacrylate) (PMMA) and polyimide membranes.[33] Storm *et al.*, who also made nanopores with openings as small as 2 nm diameter in silicon wafers, used a 300 keV electron beam from a TEM to complete pore fabrication.[43] An initial pyramidal pore, less than 200 nm in size, was made by controlled wet etching of a 500 nm exposed square of silicon in a SiO_2 mask layer, defined using electron-beam lithography.

Relatively small pores can be formed in glass and silica without the use of such costly, time-consuming equipment. It has long been known that small channels can be produced by careful mechanical separation of the two halves of a heated glass nanopipette. This technique can create two nanopipettes with diameters as low as 40 nm, as used in patch-clamp studies of cellular ion channels.[44] Nanopores have also been constructed in glass membranes by implanting electrochemically sharpened Au or Pt electrodes into the glass, then thinning the membrane by polishing. This technique can produce nanopores with diameters down to 10 nm.[45] Silica nanotubes that can detect DNA molecules have also been created by controlled oxidation of chemical vapors deposited on silicon nanowires.[33]

The recent explosion of interest in single-layer carbon has also spilled over into synthetic pore fabrication methods. For example,

Ito *et al.* carried out extensive RPS experiments using 132 nm multi-walled carbon nanotubes (MWNTs) embedded in membranes of Si/Si_3N_4 and PDMS.[46] Graphene pores, which are currently the subject of much research interest, could have wide-ranging applications due to the minimal thickness of the membrane.[41] They are typically 5–10 nm in diameter, pass through several layers of graphene, and are sculpted using the electron beam of a TEM. Graphene pores show particular promise for rapid DNA sequencing applications — simulations have shown that RPS resolution should be sufficient for detecting and differentiating between single nucleotides. As well as difficulties associated with handling such thin membranes, graphene pore technology may be limited by slight permeability of the thin membrane to the conducting electrolyte, and direct molecular interactions.

Nanopores have also been made in many different types of polymer membrane, and indeed tunable pores (detailed below) are fabricated in thermoplastic polyurethane (TPU). Elsewhere, the track-etching fabrication method is commonly used, whereby energetic heavy ions are fired into polymer membranes as thick as 100 μm, creating damage tracks.[47] Chemical etching of the membrane preferentially erodes material along the damage track, creating pores with geometry that can be controlled by parameters such as temperature, etchant and etching duration. Track-etched polycarbonate pores were used by DeBlois *et al.*, and the technique has also been used to fabricate nanopores in poly(ethylene terephthalate) (PET), polyimide and mica.[33,36] Asymmetric etching produces conical nanopores which can have lower electrical resistance than equivalent cylindrical pores, can exhibit non-Ohmic behavior, and tend to generate event "peaks" rather than square pulses characteristic of cylindrical pores.[33] Photolithography and micromolding of polymers such as polydimethylsiloxane (PDMS) can be used for relatively rapid, cheap, reproducible pore production. Saleh and Sohn used this method for RPS detection of λ-phage DNA molecules using a 200 nm nanopore.[48] As a final example, laser heating can be used to make pores in both plastic and glass membranes.[49,50]

2C. Tunable Pores

2C(i) *Introduction to Tunable Pores*

Fabrication of tunable pores (TPs) was first reported by Sowerby *et al.*, who described how a chemically-etched tungsten needle can be used to mechanically puncture an elastomeric membrane with sufficient control to produce a conical nanopore.[51] As the needle is pushed through the membrane, a voltage is applied between the needle and a conducting electrolyte in contact with the opposite membrane surface. The resistance between these elements is monitored and used to control the final pore size (Fig. 4). Due to its simplicity, this technique lends itself to an efficient fabrication process, which has been developed commercially by Izon Science (Christchurch, New Zealand) in production of TPs for the *qNano* RPS system.[52] TPs have been made with constriction diameters small enough to resolve dsDNA, and are currently manufactured for detection of particles from ~50 nm up to several microns.[51]

The distinctive feature of TPs is that their geometry can be tuned on the nanoscale by stretching and relaxing the elastomeric membrane on macroscopic scales. The usual membrane material is thermoplastic polyurethane (TPU), a stable, inexpensive elastomer which can be extended to high strains without causing large plastic deformations or breakage.[53] Stretch is applied by pulling the arms of a cross-shaped specimen with a nanopore located at the intersection of the arms (Fig. 5).[28] When the arms are extended symmetrically in two directions (X and Y in Fig. 5), the membrane strain around the pore is near-isotropic and radial.[54,55] A key practical advantage of membrane stretching is modulation of pore diameter *in situ* to control the "event" to noise ratio, and hence optimize the accuracy, of an RPS measurement.[56] When the size of the nanopore is similar to that of particles being detected, "tuning" can screen measurements of larger particles or maximize the resistive pulse for smaller particles.[56] Stretching has also been used to gate dsDNA, and applications involving particle trapping, nanoscale confinement and fast actuation can be envisaged.[51,57] The elastomeric membranes do exhibit some

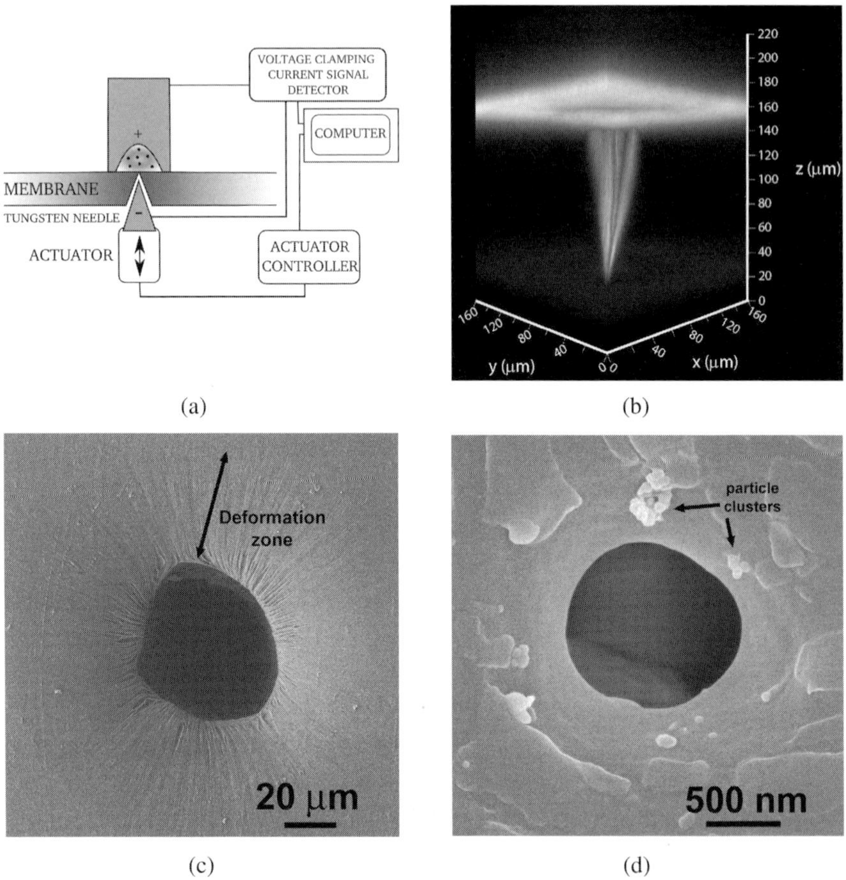

Fig. 4. Tunable pores in thermoplastic polyurethane. (a) Schematic control system for mechanical membrane penetration with a conical tungsten tip. (b) Confocal microscopy image of a tunable pore profile with ~12% strain applied macroscopically to the specimen arms. (c) SEM image of the larger opening of a conical pore at ~20% strain, and (d) Smaller pore opening under the same conditions. (b)–(d) are reproduced with permission from Ref. 28.

plastic deformation, so most RPS experiments with TPs use standard particle sets for calibration. However, near-reproducible actuation can be obtained by stretch-cycling the membrane prior to use, invoking the *Mullins effect*.[53]

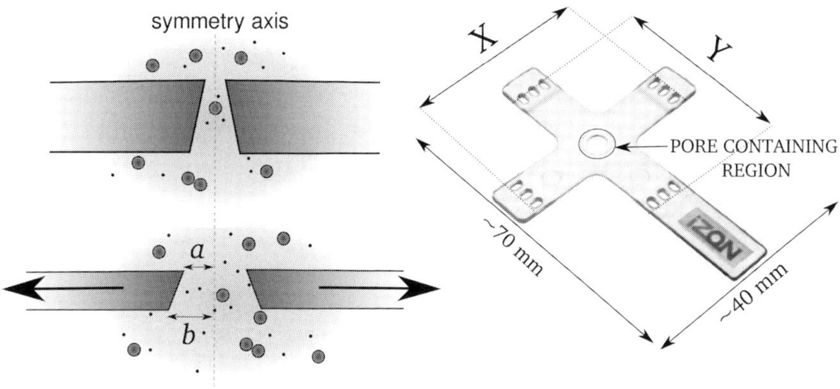

(a) (b)

Fig. 5. (a) Schematic membrane sections showing how the membrane thins and the pore size increases upon application of radial strain. (b) A cross-shaped TP specimen. Actuation is carried out by teeth placed in the holes on the specimen arms. Image courtesy of Izon Science.[52]

TPs have been imaged using atomic force microscopy, SEM, optical, confocal and fluorescence microscopy.[28,31,53,54,56] Confocal microscopy has been used to characterize the sectional profile of TPs at varying membrane stretch (Fig. 4).[28,31,56] Profiles are near-conical, but with slight curvature similar to that of a wine glass.[28] SEM images have been used to characterize pore openings, also at varying stretch, revealing radial striations which are formed adjacent to larger pore openings during pore manufacture (Fig. 3(c)).[28,54,58,59] With no stretch applied, a typical pore extends through a membrane of thickness 200 μm, joining a near-circular larger ("cis") opening of diameter ~20 μm to the submicron smaller ("trans") opening. In an electric field, conical pores create a localized region of concentrated field lines, and therefore heightened sensitivity, near the trans pore opening. The asymmetry of the conical profile also causes asymmetry in event profiles (Fig. 3).[57]

RPS measurements using TPs can be carried out at physiologically relevant pH, ionic strength and temperature, without any need for chemical labeling. Experiments to date have typically used a standard aqueous solution of 0.1 M KCl, with the minimum salt content used equivalent to 0.02 M KCl. The typical particle concentration to obtain a fast measurement rate while avoiding overlapping events is around 10^8–10^{10} mL^{-1}. Solutions are regularly filtered to remove contaminants and ultrasonicated to limit particle aggregation. The *qNano* system has fluid cells with Ag/AgCl electrodes which hold approximately 40 μL of particle-containing solution.

2C(ii) *Tunable Pores and RPS: Concentration, Size and Charge Measurements*

Standard data analysis techniques have been developed for measurement of particle size, concentration and charge using TPs in RPS experiments. Here, we briefly recount these methods, while noting that more detail regarding assumptions and analytic methods have been detailed elsewhere.[26–28,31,59] The analytical starting point is the current flowing when no particles are near the pore. A key assumption, that the electrolyte resistivity ρ is constant throughout the system, is typically justified because TP experiments take place in the high-salt limit, where the Debye length at the pore walls (~1 nm) is much smaller than the pore dimensions. This assumption is also consistent with the characteristic Ohmic response of TPs.[55] For conical geometry, the pore resistance R can then be written in closed analytic form,

$$R = \frac{\rho[l + 0.8(a+b)]}{\pi ab}, \tag{4}$$

where a and b are respectively the smaller and larger pore opening radii and l is the membrane thickness. This resistance is calculated for current flowing parallel to the central pore axis, integrated over cylindrical increments, and including the access resistance between the pore openings and a semi-infinite half space.[31,58] This type of analytic approach has been used, for example, to infer that TP opening sizes vary by as much as an order of magnitude during stretching, but geometric considerations also provide the major uncertainties for this approach.[54]

The resistance change ΔR caused by a spherical, insulating particle of diameter d inside a cylindrical nanopore of diameter D can be generally described by:

$$\frac{\Delta R}{R} = f\frac{V_{particle}}{V_{pore}} S\left(\frac{d}{D}\right), \tag{5}$$

where $V_{particle}$ and V_{pore} are the particle and pore volumes, f is a structure factor (equal to 3/2 for spherical particles) and S is a correction factor due to the non-linearity between pore and particle volumes.[31] This equation, along with the empirically-derived expression:

$$S\left(\frac{d}{D}\right) = \left[1 - 0.8\left(\frac{d}{D}\right)^3\right]^{-1}, \tag{6}$$

gives values broadly consistent with several methods of analytically determining ΔR.[36]

When using conical pores, ΔR can be analytically determined by integrating resistance along the pore, in a similar manner to the derivation of Eq. (4), but with an insulating sphere present.[60] This approach has been extended for TPs by using a more comprehensive semi-analytic model, in which sphere position can be varied, with the key outcome being that ΔR is largely proportional to $V_{particle}$ (as in Eq. (5)).[58,59] The non-linear correction only becomes significant when the particle size is very close to the pore size. Therefore, experimental measurements of ΔR for an unknown particle set can be compared with those for calibration particles in order to accurately, rapidly measure particle size.[59] Equation (5) also demonstrates that peak magnitude can be controlled by altering the pore size, allowing tuning of the signal resolution when using an elastomeric nanopore.

Sizing of nanoparticles using TPs was demonstrated by Vogel *et al.*, who compared standard sets of 100, 200, 400 and 780 nm carboxylated polystyrene nanospheres.[59] Complementary DLS and TEM size measurements were compared, and a population of adenovirus particles was sized. For the standard particles, the accuracy of size measurements was better than 6%. Peak blockade heights were extracted algorithmically from the current trace, and the mean peak magnitude (ΔI) normalized by the baseline current (I) provided comparison between calibration particle sets at any given stretch. The calibration step could be removed with precise knowledge of the pore geometry, but in practice it is more efficient to use the standard particles.

Polydispersity can be important for nanomedicine applications, whether caused by aggregation or the size distribution of individual particles. Measurements of a bimodal mixture of 84 nm and 150 nm polystyrene spheres have been reported, while Roberts *et al.* studied a trimodal solution containing 100 nm, 220 nm and 400 nm particles.[53,56] They were able to distinguish between the populations at constant stretch settings. In each case, researchers also demonstrated changes in the resistive pulse distribution due to changing pore size by varying the macroscopic stretch applied to the TP membrane, and used stretch to gate 80 nm particles from a mixture of 50 nm and 80 nm polystyrene nanospheres.

Transport of nanoparticles through a pore can be described by the Nernst–Planck equation, which can be written as a summation of (from left to right) diffusive, electrophoretic, electroosmotic and pressure effects, where all components are parallel to the pore's symmetry axis:

$$J = -D_{diff}\nabla C + \frac{\varepsilon C}{\eta}(\zeta_{particle}E - \zeta_{pore}E) + \frac{r^2}{8\eta d}CP. \quad (7)$$

Here J is the particle flux, C is the particle concentration, P is the net pressure (including the pressure head), D_{diff} is the diffusion coefficient and ζ_{pore} is the zeta potential of the pore wall. The pressure term in Eq. (7) assumes that the pore is a long cylinder. In typical TP experiments, diffusion is not significant because nanoparticles are relatively large compared to the pore opening.[28] The two electrokinetic mechanisms, summarized in Fig. 6, depend on the surface charges of the particle and the pore, as quantified by the zeta potential, and are typically considered in the "high-salt" (small Debye length) limit for TP experiments. Pressure-driven flow can be influenced by fluid menisci and gravitational heads in fluid cells containing a nanopore membrane. The *qNano* system for TPs can include a

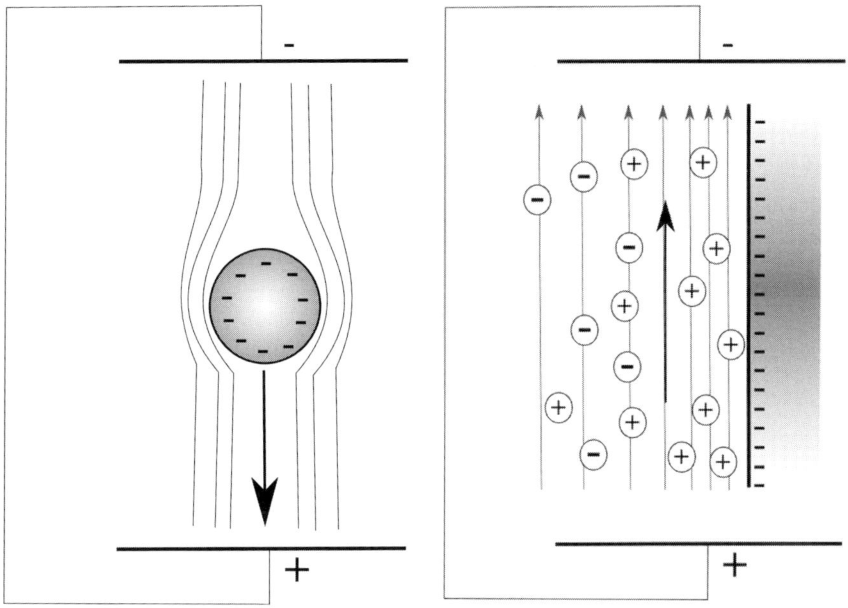

Fig. 6. (left) Electrophoresis: migration of charged particles against the fluid background due to non-zero surface charge. (right) Electro-osmosis: charge in the electrical double layer adjacent to the pore wall moves in an applied electric field, causing the bulk electrolyte to flow.

manometer system, to control the pressure across the membrane, enabling charge and concentration measurements.[26–28,61]

In some RPS regimes, pressure-driven flow dominates transport to the extent that other mechanisms can be neglected. Analytically, the volumetric flow rate Q can be related to the pressure across the membrane P by combining Sampson's result for the pressure drop between a semi-infinite space and a circular opening with application of the Hagen–Poiseuille equation within the conical pore, such that:

$$P = \frac{8\eta lQ}{3\pi a^3 b} + \frac{3\eta Q}{2a^3} + \frac{3\eta Q}{2b^3}, \tag{8}$$

where a and b are the radii of the smaller and larger pore openings respectively, and l is the pore length.[26] Equation (8) predicts that the rate of events should be proportional to the particle concentration and pressure, as long as the pore geometry is unchanged. Experiments using TPs have verified that the event rate is a linear function of both concentration and pressure, even when particles are of different sizes and types, enabling measurement of particle concentration to ~3% precision. The concentration of an unknown particle set (C_2) is most easily determined by comparison with a standard particle set of known concentration (C_1).[26,28,53,61] A plot of particle count against applied pressure yields a line with a gradient proportional to concentration. Comparing the gradients for an unknown particle solution (g_2) with the corresponding gradient of a known standard (g_1):

$$C_2 = \frac{g_2}{g_2} C_1. \tag{9}$$

Concentration can also be calculated directly from Eq. (8) when the pressure and pore geometry are known to high precision.

When a high-precision manometer is used, the pressure across the membrane can be used to isolate the electrokinetic (electrophoretic and electro-osmotic) components of Eq. (7). At a certain applied pressure, particle fluxes due to pressure and electrokinetic flow components exactly balance so that $J = 0$. If electro-osmotic effects are not significant, or are characterized (e.g. by measuring ζ_{pore}), the electrophoretic mobility of nanoparticles can then be isolated, allowing calculation of $\zeta_{particle}$ to characterize surface charge. Charge measurement with a TP has been demonstrated in practice, with the zero flux point identified by continuously varying the applied pressure over time and locating the pressure at which growth of cumulative event count was minimized.[27] In this study, $\zeta_{particle}$ values between −14 and −39 mV were determined for a range of polystyrene nanoparticle types, with experimental uncertainty of ~25% or less arising from geometric considerations and determination of electro-osmotic effects. Results were in good agreement with values obtained from titration calculations and PALS.

For nanoparticle zeta potential measurements, RPS has the key benefit of eliminating the size bias and contamination effects known to be present in light scattering techniques. RPS measurements can be carried out in under 10 min in relatively high concentration (0.1 M) electrolyte.[27] Ito *et al.* previously compared RPS measurements of nanoparticle size and electrophoretic mobility with DLS and PALS respectively.[62] These researchers used 28–90 nm carboxylated polystyrene nanoparticles with multi-wall carbon nanotubes, and reported greater accuracy with RPS than with light scattering methods. The authors noted the complexity of RPS charge measurements due to presence of both electrophoretic and electro-osmotic effects. The work described above using TPs has effectively met this challenge by isolating the electrophoretic component using precise application of pressure. In an earlier TP study, the influence of solution pH on the translocation rate of electrophoretically-driven silica particles was demonstrated.[58] Both carboxylate and amine surface modification was used, and the observed transport trends were broadly consistent with values of $\zeta_{particle}$ measured independently using light scattering.

Characteristics of individual resistive pulses provide additional metrics for generic development of future nanoparticle characterization methods. The duration of each translocation could reveal charge information on a particle-by-particle basis. Peak asymmetry allows identification of the translocation direction, enabling study of transport mechanisms near the zero flux point.[58,63] More detailed analysis of events may also prove useful for determining nanoparticle shapes.[64]

3A. Tunable Pores in Nanomedicine

The ability to accurately and rapidly characterize nanoparticles using RPS is clearly of interest for particle-based nanomedicine applications such as diagnostics and drug delivery. Standard measurement techniques are mostly developed using polystyrene sphere particle sets, but TPs have also been used to measure a plethora of different particle types, ranging from 2700 base pair dsDNA to microscale viruses and bacteria.[26,51] Those most relevant to nanomedicine and biotechnology are recounted here.

Direct measurement and characterization of biological particles is clearly relevant for future medical advances. Following Sowerby *et al.*'s work with DNA, TPs were used to detect lambda phage particles, with results suggesting that the "head" and "tail" of the phage could be distinguished in the resistive pulse.[53] Subsequently, Vogel *et al.* studied a population of adenovirus particles, which are near-spherical, 70–90 nm in diameter, and often used as gene vectors.[59] The method described in the previous section was used to obtain a particle size distribution (Fig. 7(a)) by comparison with calibration particles. This study reported that pore tuning could increase the resolution of size measurements by as much as a 60%.

TP concentration measurements have been demonstrated using both ~1 μm baculovirus in buffered electrolyte and ~600 nm *Prochlorococcus* cyanobacteria in seawater.[26] The reported accuracy was equal to or greater than cytometry and microscopy techniques when using the variable applied pressure technique (described above) over the range 0–1 kPa (Fig. 7(b)). The particle size distribution was measured at the same time as a concentration measurement, allowing simultaneous confirmation of data reliability and demonstrating the efficiency of TP-based RPS experiments. The benefits of small sample size (compared with cytometry) and quick measurement (compared with microscopy) were also apparent in this study. More recently, Garza-Licudine *et al.* have used TPs to detect synthesized unilamellar 200 nm liposomes, a type of capsule commonly used for drug delivery.[65] This study demonstrated that non-rigid particles used in nanomedicine can be characterized using TPs.

Characterization of synthetic nanoparticle surfaces is also essential to nanomedicine. Parameters and interactions that are important to measure include the detection of binding in an assay (i.e. dynamics), and quality control of particles manufactured with functionalized surfaces. Roberts *et al.* have shown that size measurements using a TP have sufficient resolution to differentiate between bare carboxyl-stabilized 220 nm organosilica spheres, and the same particles bio-functionalized with either streptavidin or l-DNA.[56] They also noted a 68% increase in event duration between bare particles and those functionalized with 48 kbp λ-DNA. These effects were attributed to the increased hydrodynamic radius of the particles, as the coiled λ-DNA has a radius of gyration of ~1 μm.

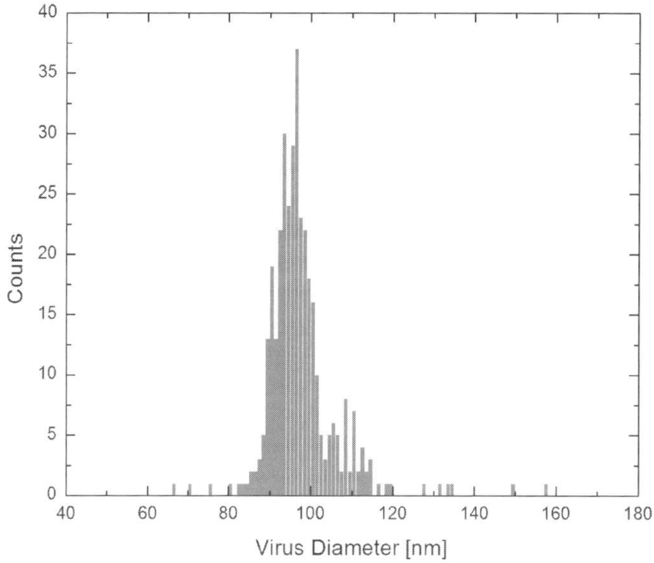

(a)

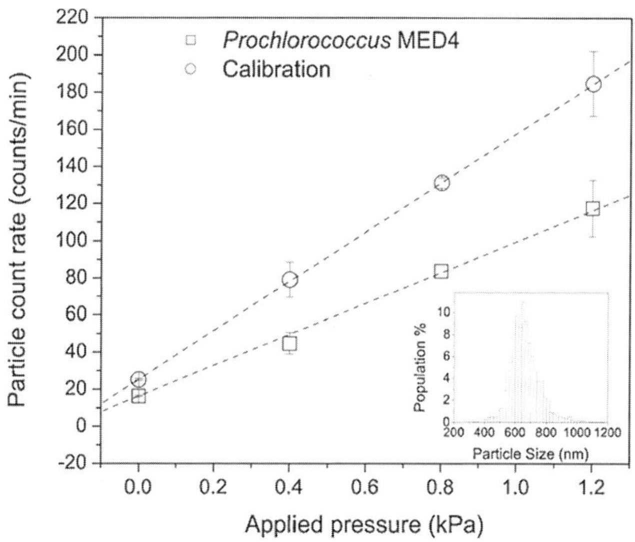

(b)

Fig. 7. (a) Adenovirus population sized using RPS within a tunable pore, reproduced with permission from Ref. 59. (b) Concentration measurements of a bacterial population of *Prochlorococcus*, with a size histogram inset. Comparison between calibration concentration particles and bacterial population, reproduced with permission from Ref. 26.

Aggregation, a key mechanism for assays and particle stability, was recently studied in TP experiments using 1 μm superparamagnetic nanospheres.[63] The study was able to follow magnet-induced aggregation over the course of ~1 min, and used simulations to distinguish between resistive pulses caused by individual particles and dimers. Competition between magnetic, pressure-driven and electrokinetic transport was explored, and evidence suggested that pressure-driven flow at the pore constriction forced separation of aggregates formed outside of the pore. Magnetic beads are widely used in sub-cellular diagnostics and medicine, because their motion in magnetic fields is independent from other fluidic and biological transport.[66] Their versatility and efficiency has resulted in several assay, separation and biological handling applications, and they are particularly suitable for studying aggregation because they are known to aggregate in an externally applied field.

Low *et al.* used TPs to investigate a bio-assay aggregation based on the interaction between a mixed base 22-mer peptide nucleic acid (PNA) and 50 nm citrate-capped gold nanoparticles.[67] The presence of the PNA, which binds to nanoparticle surfaces and reduces their surface charge, caused the nanoparticles to aggregate in a concentration-dependent manner. Aggregation was monitored using RPS, DLS and UV-vis spectrophotometry. RPS reportedly gave the best resolution of these techniques, with TPs being sensitive to size changes at PNA concentrations as low as 5 nM. The goal of this work is to produce an assay to detect a target DNA molecule that will prevent aggregation when bound to PNA on the surface of the gold nanoparticles. In another recent aggregation study using TPs, templated Au-Ni rods (approximate diameter 1 μm and length 2 or 4 μm) were used in two assays.[64] The first was a model biotin–avidin system, while the second used an aptameric capture probe to detect a platelet derived growth factor (PDGF) protein. PDGF is a growth factor overexpressed in cancer patients, particularly those with lung and pleural tumors. The shape of dimer aggregates was controlled by localizing capture probes on the Ni surfaces of the rods using a his-modified peptide. “End-to-end” and “side-on” orientations were distinguishable based on the resistive pulse size and duration. In the aptamer assay, PDGF was detected at fM levels within approximately 10 min, and specificity was verified using a control. This study was also notable for an experimental comparison between resistive pulses caused by individual rod-shaped particles and spheres of similar size.

3B. Potential Applications of RPS

Nanomedicine offers many potential areas for application of resistive pulse sensing techniques, including those that require efficient measurement of particle size, concentration, and surface charge. In the field of gene delivery, tunable nanopores and RPS are used by Grinstaff *et al.* to measure dipalmitoylphosphatidylcholine (DPPC) lipids that can be used to facilitate DNA transfection in cells. Following extrusion (a process used to create droplets of uniform size) through 200 nm and 800 nm pores, an ensemble population measurement using DLS measured a single modal peak with diameter of 456.4 ±25.2 nm. Measurements performed on the same sample using *qNano* nanopore technology showed a primary population at 350–400 nm and a distinct secondary population at 500–550 nm (Fig. 8). The DLS data did not resolve these sub populations.

Tunable nanopore and RPS techniques may also prove useful in studying the real-time interactions of nanoparticles and biomarkers. By measuring the increase in particle size as antibodies or molecular conjugates are added to a suspension of nanoparticles, researchers can determine the binding kinetics and changes in surface charge induced by this binding. Some preliminary examples of such experiments were reviewed in Section 3A, but the potential for dynamic, quantitative monitoring and measurement of specific biomarkers is largely untapped at present. RPS is also being used to examine physiological fluids; in particular, red blood cells in serum and the exosomes they secrete that may be involved in diagnosis and treatment of blood clotting and disease states.[52]

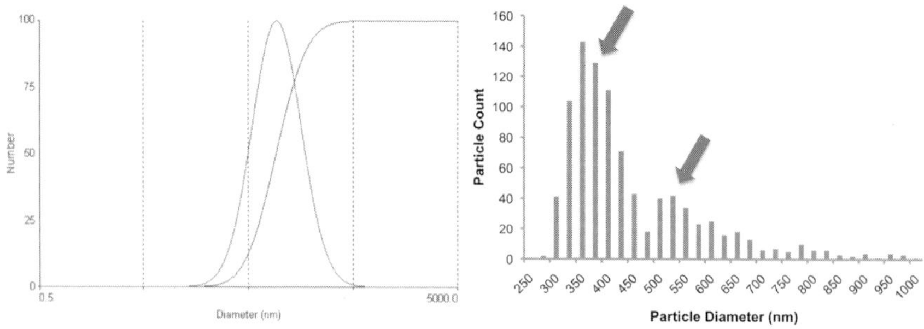

Fig. 8. DPPC liposomes at 1 μg/mL (w/w) in 100 mM PBS measured on DLS (left) and qNano (right) instruments. Arrows indicate sub-populations discernable with qNano.

Measurements of virus and bacteria in solution, as presented in Section 3A, are also of interest and may aid in vaccine development. By monitoring virus cell concentration and correlating this information with infectivity assays, the total "infective" percentage of a virus population may be determined and studied in further detail. Discrimination between infective and non-infective viruses will allow vaccine manufacturers to optimize synthesis and refine doses so that patients receive the minimum dose of infective virus necessary. More generally, virus and bacteria studies could contribute in areas such as public health, biosecurity and water quality assessment.[52]

While tunable pores and RPS have numerous applications to the field of nanomedicine, future applications will reach beyond the scope of this field. For example, researchers and engineers working in electronics, energy, optoelectronics, catalyst and ceramic industries may find use for the precise particle characterization afforded by RPS. These fields can take advantage of the colloid properties of particles, which affect the glossiness of films, the rheology and modulus of bulk materials, the surface area and packing density of particles, and the color of pigment solutions.[52] The diversity of applications to which tunable pores and RPS may be applied will lead to exciting new research and discoveries in the future.

4. Conclusion

While the field of nanomedicine continues to develop and advance new therapies, imaging modalities, and diagnostics, its success will, to a great degree, be coupled with the ability to understand, design and characterize structures on the nano scale. New drug delivery methods and imaging devices will become available faster as researchers and engineers develop a greater ability to characterize the size, surface charge, and concentration, among other properties, of nanoformulations. Standard techniques for particle characterization, such as DLS, PALS, SEM, TEM, and SMR provide adequate measurements under favorable conditions, but are limited by requirements for monodispersity, sample preparation, and solution characteristics, as well as by low throughput rate. Additionally, these techniques generally lack the versatility to measure particle size, charge and concentration concurrently. Tunable pores and resistive pulse sensing, however, are an advancing new technology that allows efficient measurement of all three metrics for particles ~50 nm to >1000 nm in diameter in biologically relevant suspensions, while also conferring the ability to resolve sub populations of particles. Additional information regarding particle concentration and dynamics, the real time measurement of changes in particle size and surface charge, will allow researchers to better understand the effects these nanoformulations have *in vivo*. The future development of TPs, RPS systems or other techniques that allow efficient *in situ* characterization of particles smaller than 50 nm on a particle by particle basis will help advance the field of nanomedicine even further.

Acknowledgments

This work has been supported by New Zealand's Ministry for Science and Innovation (NERF contract C08X0806 JE & GRW) and in part by NSF DMR-1006601 (MWG) and Boston University's Nanomedicine Program and Cross-Disciplinary Training in Nanotechnology for Cancer Program, NIH R25 CA153955 (AHC).

References

1. Lammers T, Kiessling F, Hennink WE, Storm G. (2011) Drug targeting to tumors: principles, pitfalls and (pre-) clinical progress. *J Control Release* **161**: 17–187.
2. Freitas RA. (2005) Current status of nanomedicine and medical nanorobotics. *J Comput Theor Nanosci* **2:** 1–25.
3. Huang HH, Barua S, Sharma G, *et al.* (2011) Inorganic nanoparticles for cancer imaging and therapy. *J Control Release* **155:** 322–357.
4. White KA, Rosi NL. (2008) Gold nanoparticle-based assays for the detection of biologically relevant molecules. *Nanomedicine* **3:** 543–553.
5. Gijs MAM. (2004) Magnetic bead handling on-chip: new opportunities for analytical applications. *Microfluid Nanofluid* **1:** 22–40.
6. Davis ME, Chen Z, Shin DM. (2008) Nanoparticle therapeutics: an emerging treatment modality for cancer. *Nat Rev Drug Discov* **7:** 771–782.
7. Alexis F, Pridgen E, Molnar LK, Farokhzad O. (2008) Factors affecting the clearance and biodistribution of polymeric nanoparticles. *Mol Pharm* **5:** 505–515.
8. Owens III DE, Peppas NA. (2006) Opsonization, biodistribution, and pharmacokinetics of polymeric nanoparticles. *Int J Pharm* **307:** 93–102.
9. Rejman J, Oberle V, Zuhorn IS, Hoekstra D. (2004) Size-dependent internalization of particles via the pathways of clathrin- and caveolae-mediated endocytosis. *Biochem J* **377:** 159–169.
10. Peracchia MT, Gref R, Minamitake Y, *et al.* (1997) PEG-coated nanospheres from amphiphilic diblock and multiblock copolymers: investigation of their drug encapsulation and release characteristics. *J Control Release* **46:** 223–231.
11. Brar SK, Verma M. (2011) Measurement of nanoparticles by light-scattering techniques. *Trends in Anal Chem* **30:** 4–17.
12. Malvern. (2011) "Zetasizer datasheet." www.malvern.com/LabEng/products/zetasizer/datasheet.htm (accessed Dec 2011).
13. Boyd RD, Pichaimuthu SK, Cuenet A. (2011) New approach to inter-technique comparisons for nanoparticle size measurements; using atomic force microscopy, nanoparticle tracking analysis and dynamic light scattering. *Coll Surf A: Physicochem Eng Aspects* **387:** 35–42.
14. Cho TJ, Hackley VA. (2010) Fractionation and characterization of gold nanoparticles in aqueous solution: asymmetric-flow field flow fractionation with MALS, DLS, and UV-Vis detection. *Anal Bioanal Chem* **398:** 2003–2018.
15. Theilbeer F, Donaldson K, Bradley M. (2011) Zeta potential mediated reaction monitoring on nano and microparticles. *Bioconjug Chem* **22:** 144–150.
16. Carr B. (2009) Real-time nanoparticle detection and analysis on a particle-by-particle basis. *SPIE* doi: 10.117/2.1200907.1429.
17. Filipe V, Hawe A, Jiskoot W. (2010) Fluorescent molecular rotors as dyes to characterize polysorbate-containing IgG formulations. *Pharm Res* **27:** 796–810.

18. Fitzpatrick ST. (1998) Structure and method for centrifugal sedimentation particle size analysis of particles of lower density than their suspended medium. United Stated Patent US 5,786,898.
19. Vegad H. (2010) An introduction to particle size characterization by DCS: Do you know the real size of your nano particles. *Analytik Ltd* http://www.analytik.co.uk/images/Whitepapers/Whitepaper-2–140110-CPS_Introduction_to_DCS.pdf (accessed Dec 2011).
20. Burg TP, Godin M, Knudsen SM, *et al.* (2007) Weighing of biomolecules, single cells, and single nanoparticles in fluid. *Nature* **446:** 1066–1069.
21. Godin M, Bryan AK, Burg TP, *et al.* (2007) Measuring the mass, density and size of particles and cells using a suspended microchannel resonator. *Appl Phys Lett* **91:** 123121.
22. Dextras P, Burg TP, Manalis SR. (2009) Integrated measurement of the mass and surface charge of discrete microparticles using a suspended microchannel resonator. *Anal Chem* **81:** 4517–4523.
23. Bogner A, Jouneau PH, Thollet G, *et al.* (2007) A history of scanning electron microscopy developments: towards "wet-STEM" imaging. *Micron* **38:** 390–401.
24. Erni R, Rossell MD, Kisielowski C, Dahman U. (2009) Atomic resolution imaging with a sub-50 pm electron probe. *Phys Rev Lett* **102:** 096101.
25. Fallabella JB, Cho TJ, Ripple DC, *et al.* (2010) Characterization of gold nanoparticles modified with single-stranded DNA using analytical ultracentrifugation and dynamic light scattering. *Langmuir* **26:** 12740–12747.
26. Roberts GS, Yu S, Chan LCL, *et al.* (2011) Tunable pores for measuring concentrations of synthetic and biological nanoparticle dispersions. *Biosens Bioelectron* **31:** 17–25.
27. Vogel R, Anderson W, Eldridge J, *et al.* (2012) A variable pressure method for charactererizing nanoparticle surface charge using pore sensors. *Anal Chem* **84:** 3125–3132.
28. Willmott GR, Vogel R, Yu SSC, *et al.* (2010) Use of tunable nanopore blockade rates to investigate colloidal dispersions. *J Phys Condens Matt* **22:** 454116.
29. Bayley H, Martin CR. (2000) Resistive-pulse sensing from microbes to molecules. *Chem Rev* **100:** 2575–2594.
30. Dekker C. (2007) Solid-state nanopores. *Nat Nanotechnol* **2:** 209–215.
31. Kozak D, Anderson W, Vogel R, Trau M. (2011) Advances in resistive pulse sensors: devices bridging the void between molecular and microscopic detection. *Nano Today* **6:** 531–545.
32. Henriquez RR, Ito T, Crooks RM. (2004) The resurgence of Coulter counting for analyzing nanoscale objects. *Analyst* **129:** 478–482.
33. Howorka S, Siwy Z. (2009) Engineered voltage-responsive nanopores. *Chem Soc Rev* **38:** 2360–2384.

34. Coulter WH. (1953) Means for counting particles suspended in a fluid. United States Patent US 2,656,508.
35. DeBlois R, Bean CP. (1970) Counting and sizing of submicron particles by the resistive pulse technique. *Rev Sci Instrum* **41:** 909–916.
36. DeBlois R, Bean CP, Wesley RKA. (1977) Electrokinetic measurements with submicron particles and pores by the resistive pulse technique. *J Coll Int Sci* **61:** 323–335.
37. DeBlois R, Uzgiris EE, Cluxton DH, Mazzone HM. (1978) Comparative measurements of size and polydispersity of several insect viruses. *Anal Biochem* **90:** 273–288.
38. Kasianowicz JJ, Brandin E, Branton D, Deamer DW. (1996) Characterization of individual polynucleotide molecules using a membrane channel. *Proc Natl Acad Sci USA* **93:** 13770–13773.
39. Majd S, Yusko EC, Billeh YN, *et al.* (2010) Applications of biological pores in nanomedicine, sensing and nanoelectronics. *Curr Opin Biotechnol* **21:** 439–476.
40. Branton D, Deamer DW, Marziali A, *et al.* (2008) The potential and challenges of nanopore sequencing. *Nat Biotechnol* **26:** 1146–1153.
41. Siwy ZS, Davenport M. (2010) News & views: making nanopores from nanotubes. *Nat Nanotechnol* **5:** 697–698.
42. Li J, Stein D, McMullan C, *et al.* (2001) Ion-beam sculpting at nanometre length scales. *Nature* **412:** 166–169.
43. Storm AJ, Chen JH, Sing XS, *et al.* (2003) Fabrication of solid-state nanopores with single nanometer precision. *Nat Mater* **2:** 537–540.
44. Lan W, Holden DA, Liu J, White HS. (2011) Pressure-driven nanoparticle transport across glass membranes containing a conical-shaped nanopore. *J Phys Chem* **115:** 18445–18452.
45. Zhang B, Galusha J, Shiozawa PG, *et al.* (2007) Bench-top method for fabricating glass-sealed nanodisk electrodes, glass nanopore electrodes, and glass nanopore membranes of controlled size. *Anal Chem* **79:** 4778–4787.
46. Ito T, Sun L, Crooks RM. (2003) Simultaneous determination of the size and surface charge of individual nanoparticles using a carbon nanotube-based coulter counter. *Anal Chem* **75:** 2399–2406.
47. Sexton LT, Horne LP, Martin CR. (2007) Developing synthetic conical nanopores for biosensing applications. *Mol Bio Syst* **3:** 667–685.
48. Saleh OA, Sohn LL. (2002) An artificial nanopore for molecular sensing. *Nano Lett* **3:** 37–38.
49. An R, Uram JD, Yusko EC, *et al.* (2008) Ultrafast laser fabrication of submicrometer pores in borosilicate glass. *Opt Lett* **33:** 1153–1155.
50. Wu S, Park SR, Ling XS. (2006) Lithography-free formation of nanopores in plastic membranes using laser heating. *Nano Lett* **6:** 2571–2576.
51. Sowerby SJ, Broom MF, Petersen GB. (2007) Dynamically resizable nanometre-scale apertures for molecular sensing. *Sens Actuators B* **123:** 325–330.

52. Izon. (2011) www.izon.com (accessed Dec 2011).
53. Willmott GR, Broom MF, Jansen ML, *et al.* (2011) In: Hayden O, Nielsch K (eds), *Molecular- and Nano-Tubes.* Springer, New York, pp. 209–262.
54. Willmott GR, Moore PW. (2008) Reversible mechanical actuation of elastomeric nanopores. *Nanotechnology* **19:** 475504.
55. Willmott G, Chaturvedi R, Cummins SJW, Groenewegen LG. (2011). (Under review).
56. Roberts GS, Kozak D, Anderson W, *et al.* (2010) Tunable nano/micropores for particle detection and discrimination: scanning ion occlusion spectroscopy. *Small* **6:** 2653–2658.
57. Jansen ML, Willmott GR, Hoek I, Arnold WM. (2011) Fast piezoelectric actuation of an elastomer micropore. *Measurement* DOI: 10.1016/*j.measurement.* 2013.05.023.
58. Willmott G, Parry B. (2011) Resistive pulse asymmetry for nanospheres passing through tunable submicron pores. *J Appl Phys* **109:** 094307.
59. Vogel R, Willmott G, Kozak D, *et al.* (2011) Quantitative sizing of nano/microparticles with a tunable elastomeric pore sensor. *Anal Chem* **83:** 3499–3506.
60. Heins EA, Siwy ZS, Baker LA, Martin CR. (2005) Detecting single porphyrin molecules in a conically shaped synthetic nanopore. *Nano Lett* **5:** 1824–1829.
61. Willmott G, Yu S, Vogel R. (2010) Pressure dependence of particle transport through resizable nanopores. *Proc ICONN* 128–131.
62. Ito T, Sun L, Bevan MA, Crooks RM. (2004) Comparison of nanoparticle size and electrophoretic mobility measurements using a carbon nanotube-based coulter counter, dynamic light scattering, transmission electron microscopy, and phase analysis light scattering. *Langmuir* **20:** 6940–6945.
63. Willmott GR, Platt M, Lee GU. (2011) Resistive pulse sensing of magnetic beads and supraparticle structures using tunable pores. *Biomicrofluidics* **6:** 014103.
64. Platt M, Willmott GR, Lee GU. (2012) Resistive pulse sensing of analyte-induced multicomponent rod aggregation using tunable pores. *Small* **8:** 2436–2444.
65. Garza-Licudine E, Deo D, Yu S, *et al.* (2010) Portable nanoparticle quantization using a resizable nanopore instrument — the IZON qNano™. *32nd Proc Annual International Conference of the IEE EMBS* **2010:** 5736–5739.
66. Gijs MAM. (2004) Magnetic bead handling on-chip: new opportunities for analytical applications. *Microfluid Nanofluid* **1:** 22–40.
67. Low M, Yu S, Han MY, Su X. (2011) Investigative study of nucleic acid-gold nanoparticle interactions using laser-based techniques, electron microscopy, and resistive pulse sensing with a nanopore. *Aust J Chem* **64:** 1229–1234.

Chapter 11

Microfluidic Probes to Process Surfaces, Cells, and Tissues

Mohammad A. Qasaimeh,[*,†] Sebastien G. Ricoult[†,‡] and David Juncker[*,†,‡]

Introduction

Over the last two decades, microfluidics has made significant contributions in cell biology research and tissue engineering by providing novel approaches and methods. Microfluidics is defined as the manipulation of fluidics at the micro-scale,[1] where the physics of fluid behavior is dominated by different phenomena than at the macro-scale, yet predictable and controllable.[2]

In this chapter, we briefly discuss the use of microfluidic devices in biological research, and highlight their use in processing surfaces, cells, and tissues. We explain the advantages that microfluidic devices can bring to biological experiments in comparison to the conventional methods such as Petri dishes, flasks, and microtiter plates (Fig. 1(a)). We also describe the limitations and challenges that are associated with microfluidic devices, mainly because of their closed-channel configuration. A microfluidic device can be described as a network of closed channels and chambers where at least one of the dimensions is

* Biomedical Engineering Department, †Genome Quebec Innovation Centre, ‡Department of Neurology and Neurosurgery, McGill University, Montréal, QC H3A OG1, Canada.

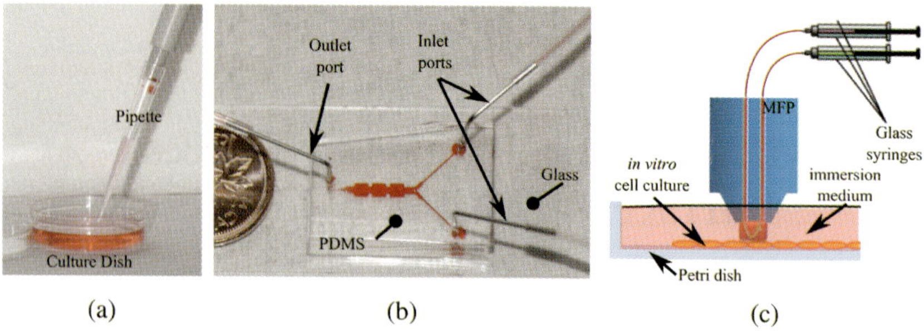

Fig. 1. Traditional and microfluidic methods for *in vitro* cell culture. (a) A conventional technique for culturing and stimulating a population of adherent cells. (b) A microfluidic device that encloses a network of closed channels and chambers with a set of input/output fluidic connections. The device is made out of a Polydimethylsiloxane (PDMS) layer that is bonded to a glass substrate. The channels are filled with red food dye to facilitate visualization. (c) Schematics of the channel-free microfluidic probe (MFP) operating on top of a substrate (i.e. Petri dish) and immersed in liquid. (Reproduced with permission of the Royal Society of Chemistry.[3])

less than 1 mm, and a set of fluidic access points to the network, which represent the fluid flow inlets and outlets (Fig. 1(b)).

The focus of this chapter is the microfluidic probe (MFP) and its biological applications. The MFP works in an open configuration and can be moved to areas of interest, thus can overcome various limitations that are associated with the use of microfluidic devices with closed channels (Fig. 1(c)). The MFPs are envisioned to be the next generation of microfluidic devices, and can be coupled with the traditional culture protocols and other technologies. Through the following sections, we emphasize on the use of MFPs in processing surfaces, cells, and tissues.

1. Microfluidic Devices for Biological Research

Microfluidic systems have many advantages that overcome the conventional tools used in biology laboratories. Notably, microfluidic systems offer miniaturization of the experimental footprint, speeding up of experiments, parallelization and thus high-throughput readout, reduced sample consumption, improved efficiency and robustness of experimentation by providing predictable, repeatable, and controllable fluidic behaviors and conditions.[4,5] As a result, microfluidic systems have become increasingly prevalent and have revolutionized numerous areas in biomedical research. To name a few, microfluidic devices have been used for the study of cell signaling[6] and fusion,[7] the study of cellular proteomics[8,9] and genomics,[10] the study of cellular mechanics[11] and dynamics,[12,13] and the study of cellular chemotaxis and electrotaxis.[14] Other areas where microfluidic devices are being developed and applied are the fields of tissue engineering,[15,16] surface patterning,[17–19] point of care diagnostics,[20,21] and drug discovery.[22,23] The efforts to develop robust microfluidic systems for these research areas and many others are currently performed by a vast number of groups around the world and commercialization of microfluidics has gained magnitude in the global market.[24–26]

Despite the powerful features that microfluidics bring, they entail different challenges and limitations, which can be attributed mainly to the materials employed to fabricate these chips and the closed channel configuration. Microfluidic chips are generally composed of a network of channels and chambers where their dimensions range from tens of micrometers to a few millimeters.[1,2] Based on the desired application and functionality, the dimensions and complexity of the microfluidic chips vary. Microfluidic chips are typically composed of a layer of polydimethylsiloxane (PDMS), which is a silicon-based organic polymer,[27] and a glass substrate (Fig. 1(b)). Patterning of PDMS can be achieved using the different photo- and soft-lithography techniques[28,29] or by using other micromachining methods.[30]

The miniaturized size of microfluidic networks is associated with high flow resistance that prevents rapid reagent exchange, consequentially the introduction of biological samples (such as cells) within the network is challenging. Furthermore, some regions within the network cannot easily or selectively be accessed as in open-surface setups.[31,32] Indeed, these issues become even more critical when aiming to enclose large biological samples inside the chip, which limits the use of these devices with samples such as embryos or tissue slices. Beyond the issues of closed microfluidic systems, the switch from performing experiments using the traditional cell culture protocols (i.e. Petri dishes) to applying them to the micro-scale systems is not straightforward. Cells in the microscale devices experience different physical environments than those present at the macro-scale.[33,34] Thus, modified protocols for long term cell culture in channels often need to be developed and require customization based on the cell type. Another issue is the PDMS material used to produce these devices; PDMS has many great attributes that make it the material of choice for fabricating microfluidic devices such as transparency, gas permeability, deformability and easy and inexpensive molding with sub-micrometer resolution.[27] However, several drawbacks are associated with the use of this polymer such as its tendency to absorb small molecules,[35] which could potentially change the concentration of reagents and alter experimental outcome. In addition, researchers showed that the un-crosslinked low molecular weight PDMS leak into the medium and interact with cultured cells.[36]

Recently, we developed novel microfluidic systems that can be integrated with the conventional cell culture protocols (Fig. 1(c)) to take advantage of the strengths of microfluidic systems while avoiding their drawbacks that are associated with the closed-channel configurations. These systems combine properties of microfluidics and of non-contact scanning probes.[31]

2. The Microfluidic Probe

The MFP[31] avoids several limitations that exist with closed-channel microfluidic systems which were discussed above. Typically, the MFP is placed above a substrate to form a narrow gap while immersed within the surrounding liquid (Fig. 1(c)). The gap between the MFP and the substrate is typically in the range of 1 μm to 100 μm. The MFP consists of a flat tip with a pair of apertures, fluidic ports, which are located within a few tens of micrometers of one another. One of the apertures is used to inject a specific solution, while the other is used for aspiration (Fig. 2(a)).

When the aspiration flow rate (Q_A) is considerably higher than the injection flow rate (Q_I), the injected solution is confined and focused by the concentric flow field and consequently deflected into a microjet that is aspirated entirely by the aspiration aperture. Thus, the microjet injected into the gap is confined by the hydrodynamic

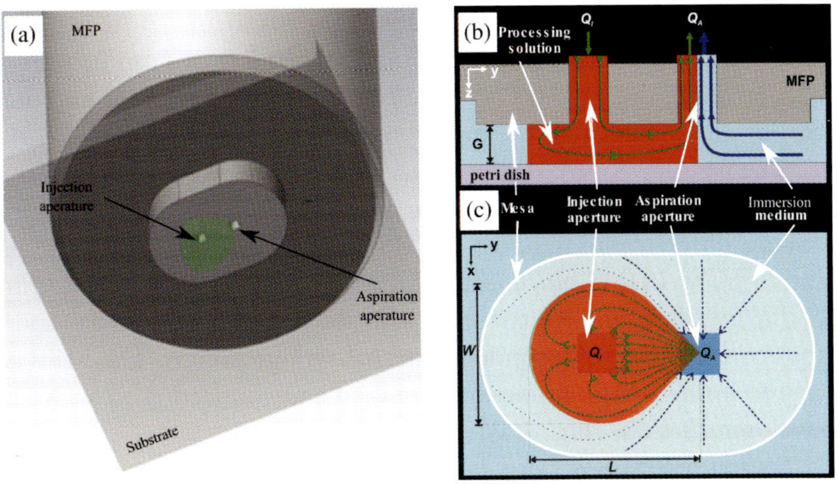

Fig. 2. Schematics of the microfluidic probe. (a) 3D representation of the MFP concept and setup. The injected solution is shown in green. The immersion medium is not shown. (b) Cross-sectional view and (c) bottom view of the MFP showing the push-pull working concept and the processing solution confinement (shown in red). The confinement of the processing solution by the immersion liquid is called hydrodynamic flow confinement. G is the gap between the MFP surface and the bottom substrate. Q_I and Q_A are the injection and aspiration flow rates, respectively. L and W are the length and the width of the tear shape of the confined injected liquid, respectively. (Reproduced with permission of The Royal Society of Chemistry.[3])

forces of the surrounding liquid in a push–pull configuration, called the hydrodynamic confinement, and therefore replaces the solid walls of closed microchannels (Fig. 2(b)). To retain the hydrodynamic confinement, the ratio Q_A/Q_I should remain higher than a certain threshold, otherwise the injected solution leaks into the immersion medium. Using the MFP, the solution microjet can be directly flushed across a sample of interest, which provides a precise tool for localized delivery of reagents. Thus, the MFP prevents the enclosure of samples in closed microfluidic networks; instead, samples can be cultivated using conventional methods (i.e. Petri dishes) and thereby eliminates the need of developing new culture protocols. At the time of experimentation, the MFP approaches the sample from above and delivers reagents to specified areas (Fig. 2(a)). In other words, the MFP defines a mobile channel-less microfluidic device. The MFP therefore can be used to process large surfaces by scanning across them. The tear shape of the hydrodynamic confinement and its size (L & W) can be tuned by controlling the gap (G) between the MFP and the sample and the ratio of the aspiration to the injection flow rates (Q_A/Q_I) (Fig. 2(c)).

The first developed MFP was made out of a microstructured Si tip which is composed of microfluidic apertures for injecting and aspiratitvng different fluids, and a PDMS block that is bonded to the Si chip.[31,32] The PDMS block in the MFP assembly serves as the world-to-chip interface, and is structured with receiving ports to insert capillaries connected to independent external syringe pumps. The MFP is secured inside a MFP holder for handling and connecting to an *XYZ* positioning system (Fig. 3(a)).

Several MFPs have been fabricated in different shapes and out of various materials. For example, we designed and fabricated a MFP out of a glass chip that is structured with fluidic apertures, and bonded to a PDMS block to serve as the world-to-chip interface.[37] Another MFP was completely made out of PDMS (Fig. 3(b)) for imaging purposes.[38] Glass and PDMS are transparent and allow for sample illumination through the MFP itself. Moreover, fabricating the MFP out of PDMS using multilayer soft-lithography techniques is inexpensive and less complicated than those made out of Si and glass, and made it easier to fabricate an array of apertures. Connecting different apertures to different reagents would permit for multiplexing and sequential/simultaneous application of different reagents on the same sample.

As various shapes and materials are used for different MFPs, different holders have been produced[32,37,38] to provide low dead volume interface assembly and a simple world-to-chip interface.

An automated setup[32] was developed to precisely control the injection and aspiration flow rates, the parallelism of the MFP surface with the bottom substrate, and the gap between the MFP and the bottom substrate. The setup comprises an inverted microscope that enables live imaging, a probe holder connected to a precise computer-controlled *XYZ*-stage (Probe Manipulation System), high-precision syringe pumps, and an environmental chamber to control the temperature, humidity and CO_2 levels (Fig. 4). The syringes, which are connected to the computer-controlled syringe pumps, are connected to the MFP via capillaries.

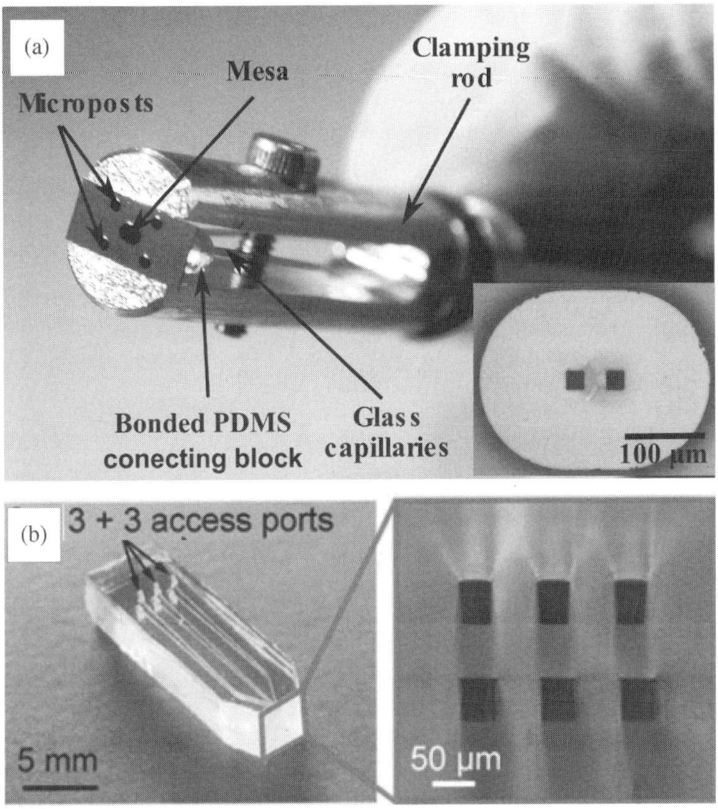

Fig. 3. Different designs and materials for making MFPs. (a) A photograph of the first developed MFP shown while clamped in the holder to connect with the micropositioning system. An enlarged view of the MFP tip and apertures is shown in the inset. This MFP was made out of a Si chip that was bonded to a PDMS block for input/output fluidic connections. The Si chip is etched on both sides to have microfluidic channels on one side, while having a tip shape (called mesa) on the other side. The mesa has two apertures and is surrounded by four microposts which facilitates the parallel alignment of the mesa to the bottom substrate. (b) A photograph of the PDMS-MFP. This MFP was fabricated by laminating three PDMS layers. The two outer layers are patterned using soft-lithography techniques, while the middle layer is a thin flat polymer sheet to separate the apertures from the two outer PDMS layers. This particular MFP has six apertures that could be connected to independent reagents sources for multiplexing. An enlarged view of the tip is shown on the right. (Reproduced with permission from The Royal Society of Chemistry.[3])

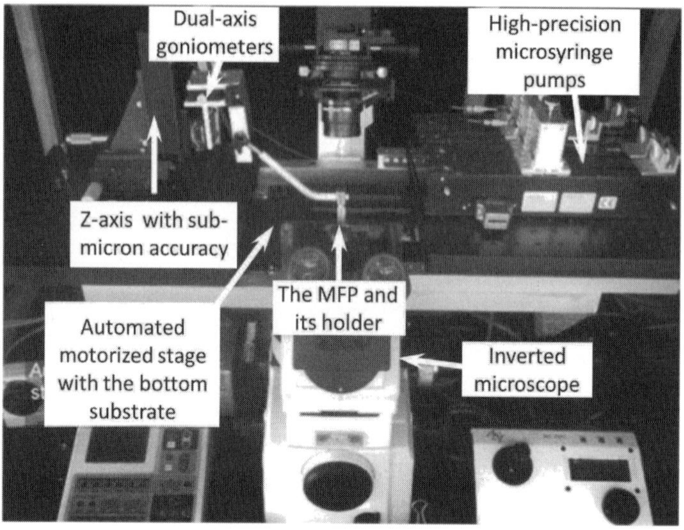

Fig. 4. The developed MFP station. The various components of the station are built around an inverted microscope to permit live imaging during experiments. The Environmental chamber is not shown.

3. Microfluidic Probes to Pattern Surfaces

Patterning surfaces, generally speaking, have wide applications in immunoassays, biosensors, cell biology, drug discovery, and tissue engineering.[39] There are vast methods and technologies to process and pattern surfaces. Notably, robotic inkjet printing and photolithography techniques[40] have been intensively used for processing surfaces. Even though inkjet printing has been the method of choice for microarray patterning, this approach entails some difficulties and challenges. For example, printing is performed in a dry environment, and consequently evaporation and drying of reagents is a major challenge,[40] which leads to patterns obtained with limited control. In addition, complex protein patterns are an issue with inkjet printing as it is mostly restricted to patterning spots on surfaces to form microarrays. On the other hand, photolithographic technologies are highly developed for patterning surfaces and have been frequently used for patterning proteins and cells.[41] Complex patterns can be generated using these methods[42]; nonetheless, several biological samples are not compatible with the typical conditions associated with photolithography, which limits the use of these methods with some sample. In addition, the high associated costs with its protocols and equipment renders it more challenging for biologists to adopt to and to employ.

Soft lithography, however, has allowed for more flexible and low-cost methods for tailoring and patterning surfaces. Mainly, microcontact printing[43] and patterning using microfluidic channels[43,44] are the most commonly used soft lithography methods for patterning surfaces. These techniques contributed drastically to the field; however, they are inadequate in terms of pattern-on-demand flexibility. After the device is designed and fabricated, it is limited to a single kind of pattern. If another — or a modified — pattern is desired, a new device needs to be designed and fabricated. In addition, issues of world-to-chip interface, complex flow systems, complications in obtaining combinatorial patterns on the same substrate, the nature of direct contact between the substrate and the device and the difficulties of processing large areas are limiting factors of the current soft lithography methods for patterning surfaces.[45] More convenient and on-demand patterning methods are required.

As the concept of the MFP is to remain functional while being immersed in a surrounding liquid, patterning surfaces with biological samples via the MFP overcomes several drawbacks associated with the previously mentioned techniques. Notably, the problems of reagent evaporation, drying, and protein denaturation are eliminated given that the biological samples are processed under physiological conditions. Likewise, because the MFP works in a non-contact mode and is thus mobile; large areas can be processed by scanning the MFP across them. Moreover, due to the small working gap between the MFP and the surface, enhanced mass transfer and molecular adsorption to the surface occur, which results in rapid patterning, low sample consumption, and high spot uniformity.[31] The single spot shape can be altered "on demand" by controlling the hydrodynamic confinement (Figs. 2(b) and 2(c)). Furthermore, complex and advanced pattern shapes can be achieved by controlling the MFP's scanning velocity and direction with respect to the jet flow direction, and the MFP's residency time on spots (the time the probe remains static).

As shown in Fig. 5(a), an array of two different proteins was patterned on a glass slide using a single MFP, with a density as high as 15,000 spots/cm^2. The MFP was programmed to scan across the

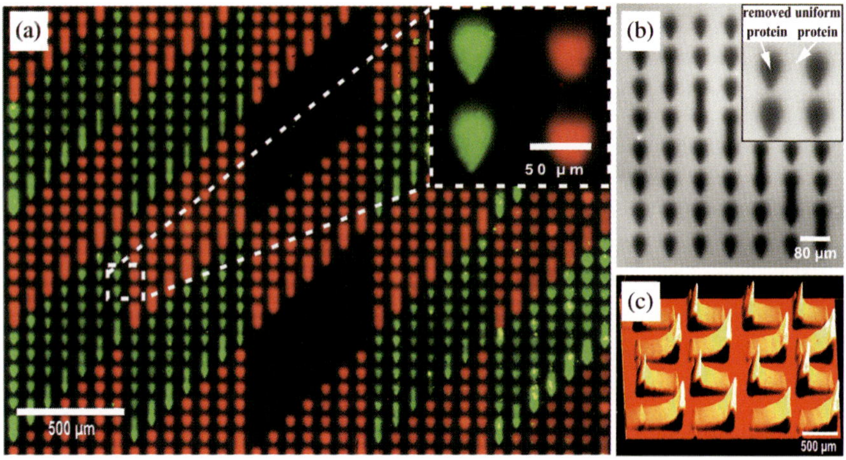

Fig. 5. MFP for advanced patterning of surfaces. (a) Fluorescence micrograph of an array of two different proteins patterned using the MFP. The first deposited protein was a TRITC-labeled goat IgG (red). The second protein, a rabbit IgG, was stained with a FITC-labeled anti-rabbit IgG (green) following the patterning process. (b) Fluorescence micrograph of an array of TRITC-labeled goat IgG achieved by subtractive patterning. The white color represents the homogenously adsorbed protein on the glass slide from solution, and the black spots represent the local removal of the IgG using the MFP by delivering a stream of solvent solution. (c) An intensity profile of the patterned array of TRITC-labeled goat IgGs gradient islands on a glass slide. The intensity clearly illustrates the achieved surface-density protein gradients obtained by ramping the MFP's scanning speed. (Reproduced with permission from The Royal Society of Chemistry.[3])

substrate. Instead of obtaining isolated spots by stopping the flow, they were achieved by the simple stop-and-go movement of the MFP. The residency time was 0.3 s for this specific case. The array of the first protein was performed initially, and subsequently the injection aperture was connected to a second syringe containing another protein solution that was deposited using the same procedure on the same substrate to obtain an array of two different proteins on a common substrate. To demonstrate the ability to control the single spot shape, we connected the last two spots of each column of each of the two proteins (making a protein segment), simply by scanning the MFP back-and-forth. Moreover, a unique feature of patterning using the MFP is subtractive patterning, whereby proteins were uniformly deposited on a glass slide and then selectively dissolved using the

MFP to achieve the negative of the previously mentioned patterns (Fig. 5(b)). Once more, the segment patterns were obtained by the back-and-forth scanning motion of the MFP.

More advanced patterns can be generated with the MFP. For example, by programming the MFP to continuously change the scanning speed from one point to another, we were able to control the amount of deposited protein on a glass substrate. By gradually decreasing the scanning speed, we successfully patterned protein gradients (Fig. 5(c)). Several distinct gradient islands were achieved through the rapid scan of the MFP while moving from one island to the next, which prevented the protein absorption within these regions of separation.

4. Microfluidic Probes to Alter Cellular Microenvironments

As we discussed in the previous section, there are many alternative possibilities that the MFP technology can bring to the field of surface patterning. In this section, we are describing how the MFP technology can be used to manipulate and label cells *in vitro* selectively, and how it can control the cellular microenvironment of the cell culture.

Conventionally, pipetting is one of the major steps used in experiments involving *in vitro* cell cultures. Drugs are typically pipetted into cell containers and eventually a uniform stimulation of the entire population of cells occurs. This state does not mimic the real *in vivo* processes, where the microenvironment is different from point to point and cell to cell. The local microenvironment is tightly controlled and regulated by the extracellular matrix molecules, soluble factors, and neighboring cells.[46] Moreover, equipment and user errors are main caveats in the procedure of pipetting. Microfluidics, however, is controllable and repeatable,[5] and allows for local and partial stimulation of cells,[47,48] which opens a new avenue to study cells under spatiotemporally-controlled stimulations. These techniques would greatly facilitate the design and achievement of novel and precise experimentation to better understand *in vivo* biology. Nonetheless, microfluidic devices suffer the closed-channel configuration which leads to several drawbacks as we described earlier in this chapter.

The MFP technology, conversely, is well suited to selectively controll cellular microenvironments at the subpopulation and subcellular levels. For example, selected cells in a culture can be detached, added, or moved to a new location, without affecting neighboring cells. As shown in Fig. 6, the MFP was placed atop of a cell culture in a dish filled with cell medium. The MFP was then moved above a selected cell, and used to locally flush the cell with a solution of Trypsin (Fig. 6(c)). A few minutes later, the selected cell was detached from the culture and dragged into the aspiration aperture (Fig. 6(d)). The aspiration flow could be switched to injection, and the cell could then be retrieved and deposited in a new location on the same or a different dish. This way, the MFP could be used as a tool to move cells around, and could be envisioned as a microfluidic tweezers. Indeed, adding or removing selected cells changes the local microenvironment of a cell population, and could

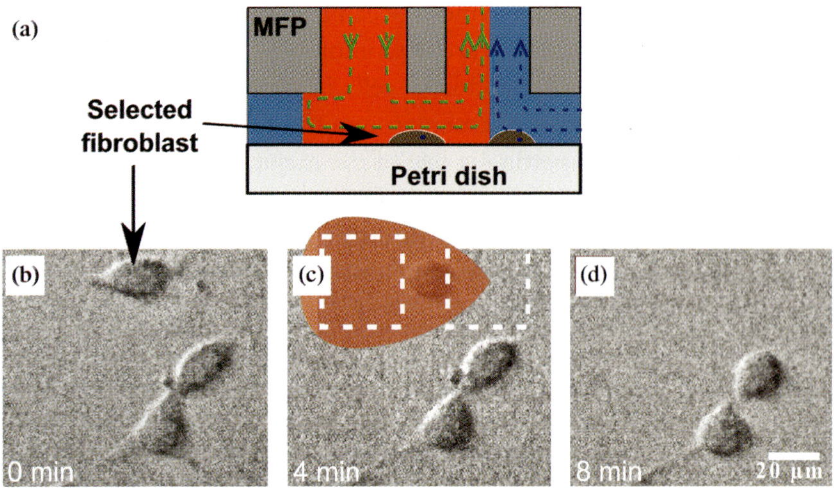

Fig. 6. The MFP is used to detach and collect selected living adherent cells. (a) A cross-sectional schematic of the MFP while placed atop a selected cell in the cell culture. (b) A micrograph showing the selected and two neighboring fibroblast cells. (c) Schematics are added to the cell micrograph to outline the position of the MFP apertures and the localized Trypsin microjet (red). (d) The selected cell was detached from the cell culture while neighboring cells remained unaffected. (Reproduced with permission from The Royal Society of Chemistry.[3])

empower many interesting studies in cell biology, such as research on cell–cell interactions.[49]

Moreover, the MFP can be used to selectively label subpopulation of cells in a dish (Fig. 7(a)). The process is similar to writing on a monolayer of cultured cells without any contact, while leaving other surrounding cells unlabeled. Similarly, living cells can be targeted and selected as single entities or subpopulations of cells to be labeled, Fig. 7(b), or even deactivated, Fig. 7(c). Using the same procedure, subpopulation of cells could be deactivated and then removed from the dish altogether, for the study of cellular migration, cellular division, or tissue reorganization.[50,51]

The MFP is well suited for the study of cellular dynamics, where selected cells — or a part of a cell — can be locally perfused with chemicals for several hours while being imaged to observe cellular dynamics. With the MFP, individual cells in the dish can be selected and the chemical perfusion can be precisely delivered to a chosen part of the cell. This flexibility and specificity is nonexistent in the closed-channel microfluidic system approach, where only part of the cell population is accessible.[47,48] More importantly, the open microfluidic configuration of the MFP brings the possibility of coupling it with other technologies and protocols at the time of the experiment. These features of the MFP make it an attractive tool for research in neuroscience, for example, where neurons are highly branched cells that connect to each other and form networks. For example, an axon from a fixed neuronal culture was targeted and selectively perfused with a drug that is taken up by the myelin sheaths surrounding the axon (Fig. 7(c)). This experiment will be extended to perfuse axonal bundles that were grown in culture for several months[52] to study of myelination and discover drugs that promote axonal insulation. Another example is the use of the MFP for the study of synaptic plasticity in a network of hippocampal neurons. A few axons were perfused with a drug to investigate the dynamics of synaptic receptors on the cell membrane (Fig. 7(d)). This experiment can be coupled with microelectrode array technology[53] for the local electrical stimulation/measurement of the neuronal activity.

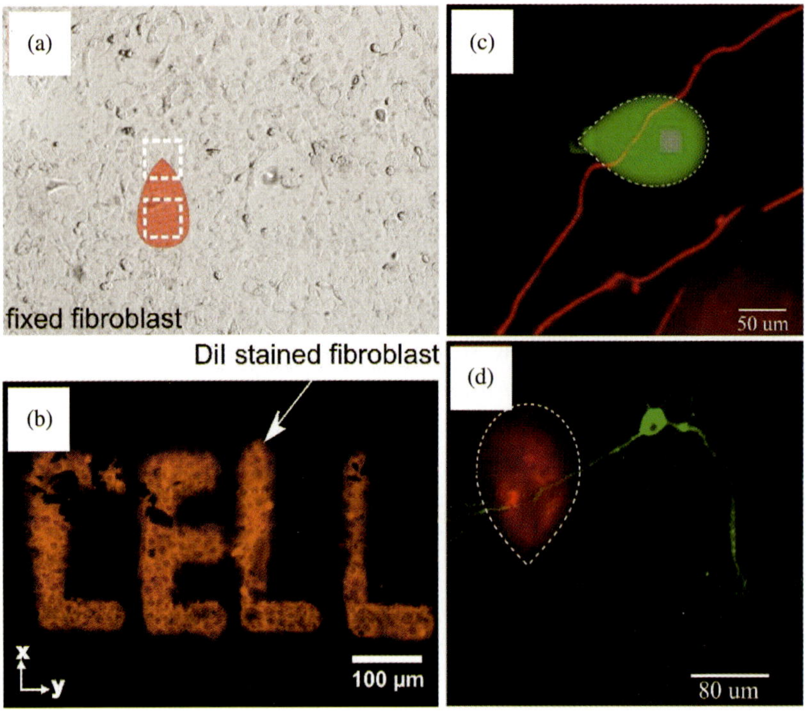

Fig. 7. MEPs to selectively label cells and to study cellular dynamics. (a)–(b) A fixed monolayer of fibroblasts was processed selectively with a MFP so that only selected cells were labeled with a fluorescent dye (DiI) to have the word "CELL" geometery, while others are left untouched. (a) An optical image of the cell monolayer with schematics showing the locations of the apertures and the microjet boundaries. (b) The final fluorescent image highlighting the labeled and unlabeled cells. (c) Local perfusion of a selected axon in a fixed culture of dorsal root ganglia neurons in a dish. Neurons were stained with FluoroMyelin (red) to label myelin. The MFP was perfusing a solution of Fluorescein (green) to demonstrate the local stimulation. (d) Stimulation of a selected dendrite of live neuronal network (rat hippocampal neurons) with tumor necrosis factor-alpha (TNF) for the studies of synaptic plasticity. Neurons are transfected with pHluorin-GluR1 (green) which labels GluR1 receptors. TNF is mixed with dextran (red) to visualize the local stimulation. (Reproduced with permission of The Royal Society of Chemistry.[3])

5. Microfluidic Probes for Local Processing of Tissue Slices

Laboratory research based on 2D *in vitro* cultures of dissociated cells is of significant value because of its contribution in shaping our current understanding of *in vivo* biology. Similarly, there is an elevated interest in studying tissue slices that are obtained through biopsies.[54]

Tissue slices, a 3D complex network of various cell types that possess the natural tissue-matrix configuration, represent the actual physiological settings of the organ far beyond those present in *in vitro* cultures of dissociated cells.[55] Thus, for a variety of studies, i.e. those applied in the pharmaceutical industry, there is high demand for experimentation on actual tissue slices.[56,57] For example, preclinical studies of new drug candidates on human tissue slices would be more relevant to the real *in vivo* systems than those achieved by studying the 2D cell-culture or animal models. Using tissue slices, information about metabolism and toxicity of the candidate drug can be obtained in an environment that better mimics the organ response *in vivo*.[58] Conventional methods for culturing and handling tissue slices are complicated by such factors as the lack of ability to control the tissue's microenvironment, the labor-intensive methods and the low throughput read out. Conversely, microfluidic technology can be used for studying tissue slices[59] and offers a diminished reagent consumption, multiplexing capabilities and an increased result throughput.[60] Microfluidics facilitates the spatiotemporal controllability of the tissue microenvironment and also allows for long-term tissue culture with continuous laminar perfusion.

Several microfluidic devices were developed to culture tissue slices and are used in a variety of experiments, such as for electrophysiological studies,[61] biomarker discoveries,[62] and toxicology studies.[63] Nonetheless, the use of microfluidic technology for tissue slice applications suffers from closed channels configurations. It is even more challenging to controllably place a large tissue slice inside a closed chamber than it is to seed dissociated cells in this setup. This limitation is the main reason for which the application of microfluidic devices for tissue slices has remained sparse in comparison to

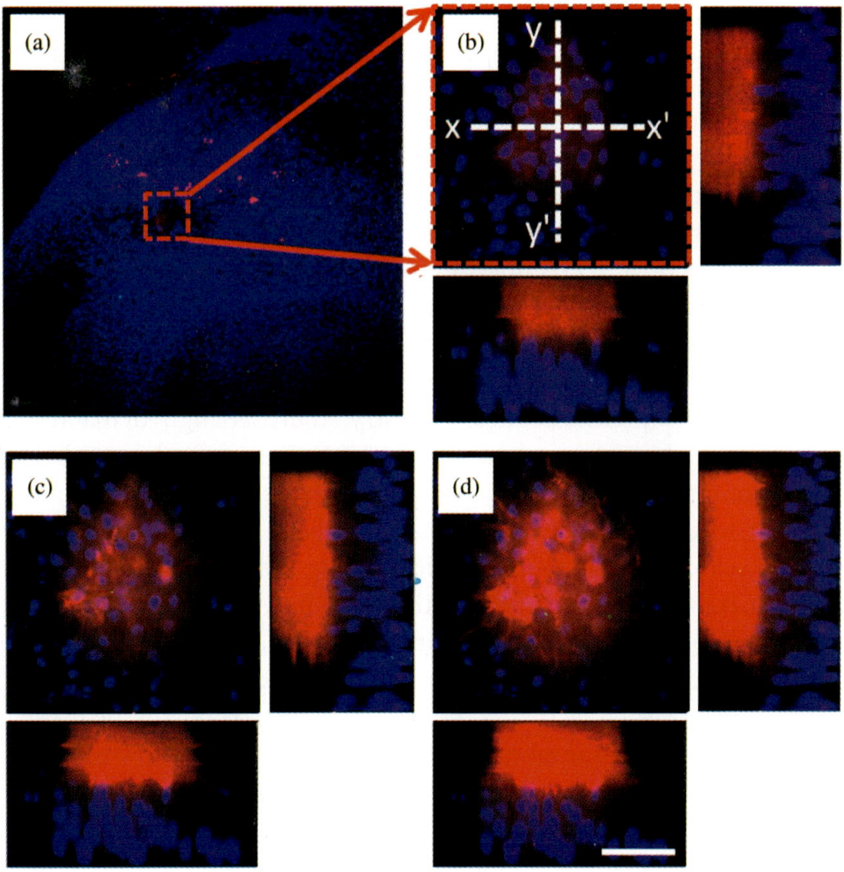

Fig. 8. A fixed organotypic slice from an L15 transgenic mouse was perfused locally with dextran for 19 min using a MFP. (a) An overview of a fixed and stained mouse organotypic hippocampal slice. The blue (DAPI) color represents the cell nuceli. The slice is placed in a chamber that has a coverslip on the bottom to permit high reslou-tion microscopy. (b)–(d) Time-lapse confocal images of the dextran (red) distribution at a depth of 32 μm inside the 70 μm thick slice. The main micrographs show the XY dextran distribution at 2 min (b), 10 min (c), and 19 min (d), following the perfusion. The insets are showing the YZ and XZ cross-sectional views of the slice and the dextran penetration through the slice. The scale bar is 40 μm.

widespread cell culture applications. A number of open microfluidic chambers have been developed for culturing and perfusing tissue slices,[63–65] however, closed channels or more complicated features and designs are needed when experiments require controlled local stimulation[66] and multiplexing.[60]

On the contrary, MFP technology works in open systems and is well suited for local stimulation, thus eliminating the need to cultivate tissue slices inside closed chambers. Instead, tissue slices are cultured in open perfusion systems where the MFP is applied. Having the open configuration, the MFP local stimulation can be coupled with other technologies at the time of experimentation, such as live microscopy imagining and electrophysiology experiments. This flexibility of the MFP technology renders it an ideal platform for tissue-based biological studies.

In order to demonstrate the functionality of MFPs with tissue slices, we used the MFP to locally perfuse an organotypic hippocampal brain slice with a fluorescent dye.[38] Organotypic brain slices mimic the brain's *in vivo* morphology, neuronal connectivity and synaptic transmission, and demonstrate preserved glial-neuronal complexes.[67] Organotypic brain slices have therefore been used to study learning and memory, neuronal regeneration following injuries, and they also have been used as disease models such as for epilepsy.[68] We placed the fixed organotypic hippocampal brain slice on a perfusion chamber that was designed to fit in an inverted confocal microscope for imaging. The MFP was then introduced from the top to locally perfuse fluorescent dextran (red). Using live confocal imaging, we studied the distribution of the dextran perfusion on the slice and its penetration in terms of depth within the slice (Fig. 8). Results showed that the dextran penetrated to a depth of 32 μm inside the 70 μm thick slice after 19 min of perfusion. Obviously, this experiment highlights the potential of using the MFP for local perfusion on live tissue slices, and could even be empowered by integrating other technologies such as two-photon fluorescence imaging.

6. Conclusion and Outlook

Throughout the different sections of this chapter, we discussed several benefits that microfluidic devices can bring in comparison to traditional methods used in biology laboratories and the biotechnology industry. We highlighted the associated limitations and drawbacks of the close-channel microfluidic devices and explained how the MFP technology can overcome them. The MFP technology is an open microfluidic system, which takes advantage of the controllable and predicable flow of fluids at the microscale, but also integrates the features of scanning probes by providing mobility in a non-contact nature. The principle of the hydrodynamic confinement was achieved by having the MFP form a narrow gap with the bottom substrate and maintaining an aspiration to injection flow ratio higher than a certain threshold value. Thus, the aspiration flow created a concentric flow that was strong enough to deflect and confine the whole injected solution without any physical walls. The hydrodynamic confinement can be controlled and localized precisely by controlling the flow rates and the gap between the MFP and the bottom substrate. In the future, internal pumps and other components such as proximity sensors could be integrated within the MFP itself to minimize the setup footprint and enhance MFP precision. We further introduced the different MFPs that were developed, which vary based on their designs and materials. We expect to see other new designs and materials defining the MFPs in the future, where depending on the experimental conditions, some designs and materials would be advantageous over others. For example, we recently developed a novel MFP with four apertures to generate concentration gradients within a quadrupolar fluid flow.[69]

The MFP technology's applicability is not limited to biology, but could also find applications in surface chemistry, polymer electronics and microfabrication. However, we focused this chapter on the MFP's biological application to remain focused on the general theme of the book. We showed how the MFP can be used to pattern proteins on surfaces with multiplexing ability. The patterning shape can be configured on demand, and complex shapes can be generated by controlling the scanning speed and direction. Problems of evaporation and denaturation were eliminated using the MFP. Furthermore, we showed how the MFP can be used to manipulate individual adherent cells, label single cells or subpopulations of cells and how it can be used for localized chemical stimulation at the subcellular level. There is no need to cultivate the biological samples inside channels or to develop new culturing protocols when using the MFP; the MFP works with the conventional protocols. We hypothesize that MFP technology will be of great use in studying sensitive cells such as primary neurons or stems cells.

We explained how the MFP technology suits large biological samples and can be used with tissue slices or embryos, for example. Using the MFP, tissue slices were perfused locally and in a selective manner. This way, local stimulation of tissue slices can be coupled with other techniques such as electrophysiology electrodes.

Acknowledgments

We are greatly thankful to Nageswara R Ghattamaneni, David Stellwagen, R. Anne McKinney, and Dalinda Liazoghli from McGill University and the Montreal Neurological Institute for supplying neuronal cultures and contributing with the use of the MFP in these cultures. M.A.Q. acknowledges support from Alexander Graham Bell Canada NSERC Scholarship and the BME Excellence Award, S.G.R. is funded through the NSERC-CREATE Integrated Sensors Systems Program and the NSERC-CREATE Program in Neuroengineering, while D.J. acknowledges support from a Canada Research Chair.

References

1. Whitesides GM. (2006) The origins and the future of microfluidics. *Nature* **442(7101):** 368–373.
2. Squires TM, Quake SR. (2005) Microfluidics: fluid physics at the nanoliter scale. *Rev Mod Phys* 77(**3**)**:** 977–1026.
3. Qasaimeh MA, Ricoult SG, Juncker D. (2013) Microfluidic probes for use in life sciences and medicine. *Lab Chip* **13**(1): 40–50.

4. Hansen C, Quake SR. (2003) Microfluidics in structural biology: smaller, faster ... better. *Curr Opin Struct Biol* **13(5):** 538–544.
5. Cooksey GA, Elliott JT, Plant AL. (2011) Reproducibility and robustness of a real-time microfluidic cell toxicity assay. *Anal Chem* **83(10):** 3890–3896.
6. Cheong R, Wang CJ, Levchenko A. (2009) Using a microfluidic device for high-content analysis of cell signaling. *Sci Signal* **2(75):** pl2.
7. Skelley AM, Kirak O, Suh H, *et al.* (2009) Microfluidic control of cell pairing and fusion. *Nat Meth* **6(2):** 147–152.
8. Faley SL, Copland M, Reboud J, Cooper JM. (2011) Cell chip array for microfluidic proteomics enabling rapid *in situ* assessment of intracellular protein phosphorylation. *Biomicrofluidics* **5(2):** 024106–024107.
9. Salehi-Reyhani A, *et al.* (2011) A first step towards practical single cell proteomics: a microfluidic antibody capture chip with TIRF detection. *Lab Chip* **11(7):** 1256–1261.
10. Kotz KT, *et al.* (2010) Clinical microfluidics for neutrophil genomics and proteomics. *Nat Med* **16(9):** 1042–1047.
11. Hou H, *et al.* (2011) Microfluidics for applications in cell mechanics and mechanobiology. *Cell Mol Bioeng* **4(4):** 591–602.
12. Albrecht DR, *et al.* (2010) Microfluidics-integrated time-lapse imaging for analysis of cellular dynamics. *Integr Biol* **2(5–6):** 278–287.
13. Bennett MR, Hasty J. (2009) Microfluidic devices for measuring gene network dynamics in single cells. *Nat Rev Genet* **10(9):** 628–638.
14. Li J, Lin F. (2011) Microfluidic devices for studying chemotaxis and electrotaxis. *Trends Cell Biol* **21(8):** 489–497.
15. Choi NW, *et al.* (2007) Microfluidic scaffolds for tissue engineering. *Nat Mater* **6(11):** 908–915.
16. Gillette BM, *et al.* (2008) *In situ* collagen assembly for integrating microfabricated three-dimensional cell-seeded matrices. *Nat Mater* **7(8):** 636–640.
17. Delamarche E, Bernard A, Schmid H, *et al.* (1997) Patterned delivery of immunoglobulins to surfaces using microfluidic Networks. *Science* **276(5313):** 779–781.
18. von Philipsborn AC, *et al.* (2006) Microcontact printing of axon guidance molecules for generation of graded patterns. *Nat Protocols* **1(3):** 1322–1328.
19. Goluch ED, Shaw AW, Sligar SG, Liu C. (2008) Microfluidic patterning of nanodisc lipid bilayers and multiplexed analysis of protein interaction. *Lab Chip* **8(10):** 1723–1728.
20. Fan R, *et al.* (2008) Integrated barcode chips for rapid, multiplexed analysis of proteins in microliter quantities of blood. *Nat Biotech* **26(12):** 1373–1378.
21. Chin CD, *et al.* (2011) Microfluidics-based diagnostics of infectious diseases in the developing world. *Nat Med* **17(8):** 1015–1019.
22. Dittrich PS, Manz A. (2006) Lab-on-a-chip: microfluidics in drug discovery. *Nat Rev Drug Discov* **5(3):** 210–218.

23. Kang L, Chung BG, Langer R, Khademhosseini A. (2008) Microfluidics for drug discovery and development: from target selection to product lifecycle management. *Drug Discov Today* **13(1–2):** 1–13.
24. Mark D, Haeberle S, Roth G, *et al.* (2010) Microfluidic lab-on-a-chip platforms: requirements, characteristics and applications. *Chem Soc Rev* **39(3):** 1153–1182.
25. Panikowska K, Tiwari A, Alcock J. (2011) Towards service-orientation — the state of service thoughts in the microfluidic domain. *Int J Adv Manuf Technol* **56(1):** 135–142.
26. Mukhopadhyay R. (2009) Microfluidics: on the slope of enlightenment. *Anal Chem* **81(11):** 4169–4173.
27. McDonald JC, *et al.* (2000) Fabrication of microfluidic systems in poly(dimethylsiloxane). *Electrophoresis* **21(1):** 27–40.
28. Qin D, Xia Y, Whitesides GM. (2010) Soft lithography for micro- and nanoscale patterning. *Nat Protocols* **5(3):** 491–502.
29. Kim P, *et al.* (2008) Soft lithography for microfluidics: a review. *Biochip J* **2(1):** 1–11.
30. Waldbaur A, Rapp H, Lange K, Rapp BE. (2011) Let there be chip-towards rapid prototyping of microfluidic devices: one-step manufacturing processes. *Anal Methods* **3(12):** 2681–2716.
31. Juncker D, Schmid H, Delamarche E. (2005) Multipurpose microfluidic probe. *Nat Mater* **4(8):** 622–628.
32. Perrault CM, *et al.* (2010) Integrated microfluidic probe station. *Rev Sci Instrum* **81(11):** 115107–115108.
33. Young EWK, Beebe DJ. (2010) Fundamentals of microfluidic cell culture in controlled microenvironments. *Chem Soc Rev* **39(3):** 1036–1048.
34. Walker GM, Zeringue HC, Beebe DJ. (2004) Microenvironment design considerations for cellular scale studies. *Lab Chip* **4(2):** 91–97.
35. Toepke MW, Beebe DJ. (2006) PDMS absorption of small molecules and consequences in microfluidic applications. *Lab Chip* **6(12):** 1484–1486.
36. Regehr KJ, *et al.* (2009) Biological implications of polydimethylsiloxane-based microfluidic cell culture. *Lab Chip* **9(15):** 2132–2139.
37. Qasaimeh MA. *et al.* (2008) The generation of stationary chemical gradients around stagnant points using a microfluidic probe. *The Twelfth International Conference on Miniaturized Systems for Chemistry and Life Sciences* (Chemical and Biological Microsystems Society), pp. 841–843.
38. Queval A, *et al.* (2010) Chamber and microfluidic probe for microperfusion of organotypic brain slices. *Lab Chip* **10(3):** 326–334.
39. Khademhosseini A, *et al.* (2003) Direct patterning of protein- and cell-resistant polymeric monolayers and microstructures. *Adv Mater* **15(23):** 1995–2000.
40. Delaney JT, Smith PJ, Schubert US. (2009) Inkjet printing of proteins. *Soft Matter* **5(24):** 4866–4877.

41. Hui EE, Bhatia SN. (2007) Microscale control of cell contact and spacing via three-component surface patterning. *Langmuir* **23(8):** 4103–4107.
42. Wang S, *et al.* (2009) Gradient lithography of engineered proteins to fabricate 2D and 3D cell culture microenvironments. *Biomed Microdevices* **11(5):** 1127–1134.
43. Kane RS, Takayama S, Ostuni E, *et al.* (1999) Patterning proteins and cells using soft lithography. *Biomaterials* **20(23–24):** 2363–2376.
44. Pla-Roca M, Juncker D. (2011) PDMS Microfluidic capillary systems for patterning proteins on surfaces and performing miniaturized immunoassays. In: Khademhosseini A, Suh K-Y, Zourob M. (eds), *Biological Microarrays,* Methods in Molecular Biology (Humana Press), Vol. 671, pp. 177–194.
45. Delamarche E, Juncker D, Schmid H. (2005) Microfluidics for processing surfaces and miniaturizing biological assays. *Adv Mater* **17(24):** 2911–2933.
46. Folch A, Toner M. (2000) Microengineering of cellular interactions. *Annu Rev Biomed Eng* **2(1):** 227–256.
47. Takayama S, *et al.* (2003) Selective chemical treatment of cellular microdomains using multiple laminar streams. *Chem Biol* **10(2):** 123–130.
48. Takayama S, *et al.* (1999) Patterning cells and their environments using multiple laminar fluid flows in capillary networks. *Proc Natl Acad Sci* **96(10):** 5545–5548.
49. Yin Z, Noren D, Wang CJ, *et al.* (2008) Analysis of pairwise cell interactions using an integrated dielectrophoretic-microfluidic system. *Mol Syst Biol* **4.**
50. Yarrow J, Perlman Z, Westwood N, Mitchison T. (2004) A high-throughput cell migration assay using scratch wound healing, a comparison of image-based readout methods. *BMC Biotechnol* **4(1):** 21.
51. van der Meer AD, Vermeul K, Poot AA, *et al.* (2010) A microfluidic wound-healing assay for quantifying endothelial cell migration. *Am J Physiol Heart Circ Physiol* **298(2):** H719–H725.
52. Liazoghli D, Roth AD, Thostrup P, Colman DR. (2011) Substrate micropatterning as a new *in vitro* cell culture system to study myelination. *ACS Chemical Neuroscience.*
53. Mernier G, De Keersmaecker K, Bartic C, Borghs G. (2007) On-chip controlled release of neurotransmitter molecules. *Microelectron Eng* **84(5–8):** 1714–1718.
54. McCarthy F, *et al.* (2009) An improved method for constructing tissue microarrays from prostate needle biopsy specimens. *J Clin Pathol* **62(8):** 694–698.
55. Kim J, Stein R, O'Hare M. (2004) Three-dimensional in vitro tissue culture models of breast cancer — a review. *Breast Cancer Res Treat* **85(3):** 281–291.
56. Pampaloni F, Reynaud EG, Stelzer EHK. (2007) The third dimension bridges the gap between cell culture and live tissue. *Nat Rev Mol Cell Biol* **8(10):** 839–845.
57. Davila JC, Rodriguez RJ, Melchert RB, Acosta D. (1998) Predictive value of *in vitro* model systems in toxicology. *Ann Rev Pharmacol Toxicol* **38(1):** 63–96.

58. van Midwoud PM, Merema MT, Verpoorte E, Groothuis GMM. (2011) Microfluidics enables small-scale tissue-based drug metabolism studies with scarce human tissue. *Journal of the Association for Laboratory Automation* **16(6):** 468–476.
59. van Midwoud PM, Verpoorte E, Groothuis GMM. (2011) Microfluidic devices for *in vitro* studies on liver drug metabolism and toxicity. *Integr Biol* **3(5):** 509–521.
60. Kim MS, *et al.* (2010) Breast cancer diagnosis using a microfluidic multiplexed immunohistochemistry platform. *PLoS ONE* **5(5):** e10441.
61. Blake AJ, Pearce TM, Rao NS, *et al.* (2007) Multilayer PDMS microfluidic chamber for controlling brain slice microenvironment. *Lab Chip* **7(7):** 842–849.
62. Song B, *et al.* (2009) Time-resolved lanthanide luminescence for lab-on-a-chip detection of biomarkers on cancerous tissues. *Analyst* **134(10):** 1991–1993.
63. van Midwoud PM, Groothuis GMM, Merema MT, Verpoorte E. (2010) Microfluidic biochip for the perifusion of precision-cut rat liver slices for metabolism and toxicology studies. *Biotechnol Bioeng* **105(1):** 184–194.
64. Rambani K, Vukasinovic J, Glezer A, Potter SM. (2009) Culturing thick brain slices: an interstitial 3D microperfusion system for enhanced viability. *J Neurosci Methods* **180(2):** 243–254.
65. Choi Y, McClain M, LaPlaca M, *et al.* (2007) Three dimensional MEMS microfluidic perfusion system for thick brain slice cultures. *Biomed Microdevices* **9(1):** 7–13.
66. Tang YT, Kim J, Lopez-Valdes HE, *et al.* (2011) Development and characterization of a microfluidic chamber incorporating fluid ports with active suction for localized chemical stimulation of brain slices. *Lab Chip* **11(13):** 2247–2254.
67. Gähwiler BH, Capogna M, Debanne D, *et al.* (1997) Organotypic slice cultures: a technique has come of age. *Trends Neurosci* **20(10):** 471–477.
68. Sundstrom L, Pringle A, Morrison B, Bradley M. (2005) Organotypic cultures as tools for functional screening in the CNS. *Drug Discov Today* **10(14):** 993–1000.
69. Qasaimeh MA, Gervais T, Juncker D. (2011) Microfluidic quadrupole and floating concentration gradient. *Nat Commun* 2: 464.

Chapter 12

Subcompartmentalized Systems Towards Therapeutic Cell Mimics

Leticia Hosta-Rigau[†] and Brigitte Städler[†]

1. Introduction

The paradigm shift from simply keeping patients alive towards improving their quality of life has become more and more central due to the increasing age and number of the average population, especially in the context of chronic age-related or chronic genetic diseases. Therefore, novel therapy concepts are in high demand to address this challenge. Therapeutic cell mimicry is an approach which aims to substitute for missing or lost cellular function, often in the context of enzymatic replacement therapy. The aim is to provide a nature-inspired, long-term solution for patients with not detrimental, but chronic diseases to enable or facilitate the pursuit of their daily life while relieving costs from the health care system.

In general, there are two conceptually different approaches: (1) encapsulation of living cells and (2) the bottom-up design of carriers from custom-made building blocks into a functional assembly. The former approach is interesting in its own right but beyond the scope of this chapter, and the reader is referred to recent reviews.[1–3] The latter idea was realized for the first time in the 1960s by encapsulating an enzyme into a microscopic dialysis bag and measuring the subsequent conversion of a

[†]iNANO The Interdisciplinary Nanoscience Center, Aarhus University, Aarhus, Denmark.

substrate.[3,4] This system was the first example of an encapsulated catalysis, i.e. although very basic, the first cell mimic. Follow up work included the encapsulation of adsorbents to remove toxins or drugs from the blood[5–7]; implanted enzyme-containing artificial cells for enzyme therapy in acatalasemic mice[8]; or artificial cells encapsulating the urease enzyme and ammonia adsorbent to lower systemic urea levels,[3] phenylanine ammonia lyase for phenylketonuria treatment,[9] which is the most common inborn error of metabolism, or xanthine oxidase to lower systemic hypoxantine levels in a patient with Lesch-Nyhan disease,[10] which is an extremely rare inborn metabolic disorder. This initial work has motivated considerable efforts by encapsulating enzymes or other biomolecules towards cell mimicry using single compartment systems for therapeutic applications.[11–15]

The aim of this book chapter is to outline the most recent developments and progress in therapeutic cell mimicry focusing on bottom-up assembled multi-compartment systems. The general idea and advantages of subcompartmentalized systems is outlined, followed by a review of the considered approaches. Finally, we will discuss one of the most promising platforms to date based on liposomal subunits encapsulated within a polymer carrier capsule. By doing so, we hope to demonstrate the pace at which such concepts develop and the promises they hold, while showing their limitations and future challenges which need to be addressed.

2. Subcompartmentalized Systems

Biological cells have optimized the possibility to conduct cascade reactions and/or parallel reactions in a confined space with perfect precision, timing, and conversion rate. The lipid bilayer entraps their organelles and together with membrane proteins controls the transport of biomolecules and other substances across the membrane. Artificial systems do not have to mimic the entire complexity of a biological cell, but a specific aspect, mostly depending on the application in mind. The subcompartmentalized structure is essential when the performance of triggered, spatially separated, encapsulated (enzymatic cascade) reactions are the aim. Despite the high potential of subcompartmentalized systems, only a small number of assemblies containing subunits confined in a larger carrier have been reported to date. Single composition systems (Scheme 1, top) include vesosomes (liposomes with in a carrier

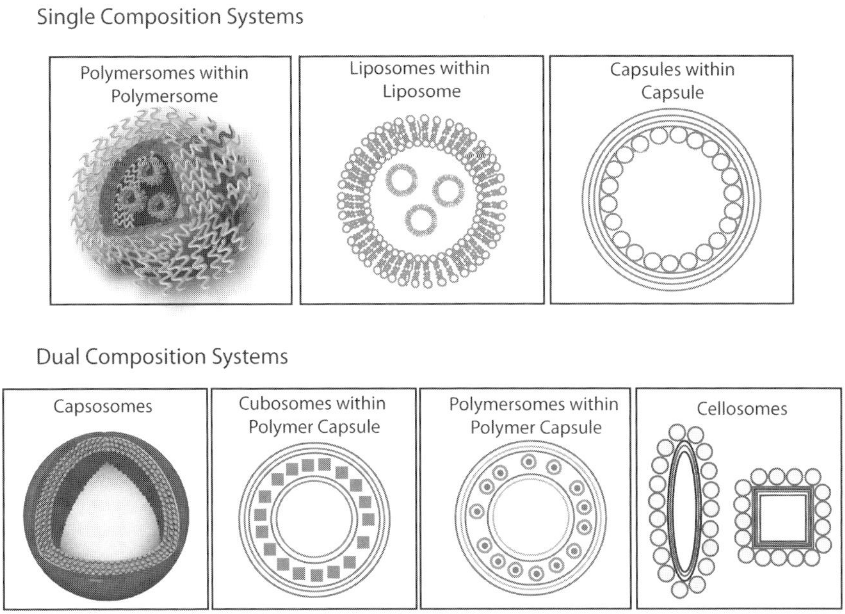

Scheme 1 Subcompartmentalized single (top) or dual composition systems towards therapeutic cell mimicry. Top, from left to right: polymersomes within polymersome; liposomes within liposome; and capsules within capsule. Bottom, from left to right: capsosomes, adapted from reference,[41] reproduced with permission; polymersomes within polymer capsule; and cellosomes.

liposome),[16] polymersomes within a polymersome,[17] or polymer capsules within a polymer capsule.[18,19] Vesosomes have been used to improve the performance of lipid-based carriers for drug delivery applications.[16] Polymersomes within polymersomes have been employed to encapsulate fluorescent probes and biomolecules, and were also equipped with a membrane protein which acted as specific gate between the carrier polymersome exterior and interior.[17,20] Polymer capsules within polymer capsules have been used to encapsulate fluorescently labeled dextran as a model drug for visualization.[18] This system was also considered to trap a model peptide in capsule subunits which were degradable using different chemical triggers, i.e. oxidative or reductive agents.[19] Although each of these three systems has reported beneficial properties, none of them considered the trapping of active biomolecules towards encapsulated catalysis.

Dual composition systems (Scheme 1, bottom) include polymer hydrogel capsules containing cubosomes[21] or polymersomes[22]; cellosomes, which consist of polyelectrolyte-coated yeast cells subcompartments linked to each other by electrostatic interactions[23]; and our recently introduced capsosomes,[24] which consist of thousands of liposomes embedded into a polymeric carrier capsule, and are the probably most established and thoroughly studied assembly of a functional bio-inspired systems reported to date.

3. Capsosomes: Lipsomal Subunits Entrapped within Polymer Carrier Capsules

3.1. *Assembly*

3.1.1. *Basic Aspects*

A successful subcompartmentalized cell mimic requires the trapping of a controlled number of subunits in a larger carrier to mimic the hierarchical structure of biological cells. For capsosomes, liposomes are employed as subunits in polymer carrier capsules. Due to the inherent difference of the two building blocks, their combination preserves the advantages of both systems while eliminating some of their drawbacks. While the liposomes not only equip the carrier's interior with subcompartments, they are also able to encapsulate both hydrophobic and hydrophilic cargo, and to provide a protective barrier for the encapsulated fragile or small payload. However, liposomes have some limitations such as low *in vivo* stability, lack of control over their degradability,[25,26] and they are largely impermeable to their surroundings.[27] On the other hand, the polymer carrier capsules prepared by the layer-by-layer (LbL) technique, which is based on the deposition of interacting polymers onto sacrificial core templates,[28] supplies the structural integrity, and the semipermeable nature of the polymer membrane allows the communication of the carrier's interior with the external milieu, thus making it possible that (small) molecules can freely diffuse across the polymer membrane while larger ones are excluded, which is a very important factor to allow for continuous parallel/cascade encapsulated reactions.

Capsosomes are assembled *via* the LbL technique, starting by the deposition of a polymer precursor layer onto silica template particles, followed by the adsorption of the first layer of liposomes (Scheme 2). The deposition of polymer separation layers is required prior to the

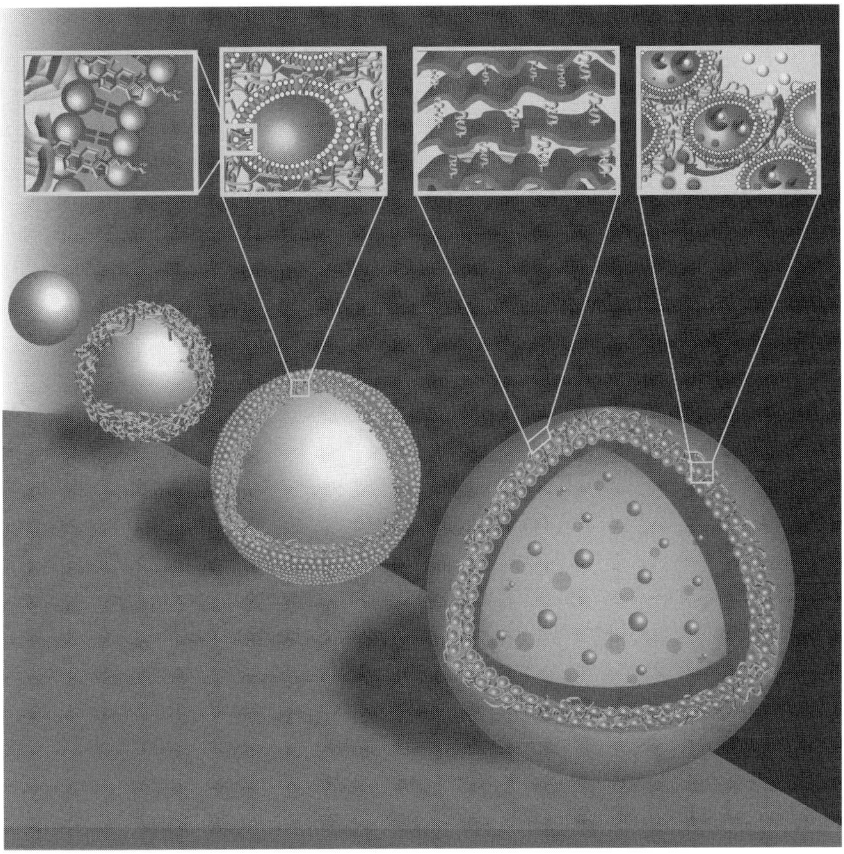

Scheme 2 Schematic illustration of the assembly of capsosomes. A colloidal template is coated with a precursor polymer layer followed by the deposition of layers of liposomes. After the adsorption of the capping layer and the alternate deposition of PVP and PMA_{SH} to assemble the carrier capsule, capsosomes are obtained upon removal of the core. Inset, from left to right: detail of the incorporation of cholesterol moiety of the precursor polymer layer into the liposomal membrane, liposomal subcompartments upon the cholesterol-containing precursor layer, detail of the disulfide cross-linked PMA layers, and enzymatic conversion of the substrate nitrocefin into its hydrolyzed product by the enzyme β-lactamase entrapped within the liposomal subcompartments.

adsorption of the next layer of liposomes, and these two steps are repeated until the required number of liposomes is deposited. After the adsorption of the last liposome layer, a polymer capping layer is deposited that acts as "connector" between the liposomes and the subsequent assembly of the polymer carrier capsule. This carrier capsule is assembled using several layers of interacting polymers. Upon dissolution of the silica core, capsosomes are obtained.

Depending on the capsosomes' application, a (bio)degradable or a non-degradable carrier capsule will be required. When capsosomes are expected to repetitively perform a reaction for an extended period of time acting as an artificial cell or organelle, a non-degradable carrier capsule might be preferable. However, if the aim is the deconstruction of the assembly once the capsosomes' task has been accomplished, the choice of the carrier capsule will be a polymer pair that can be dismantled by the body into its building blocks. The first generation of capsosomes consisted of the incorporation of 50 nm-sized unsaturated zwitterionic liposomes into a non-biodegradable polymer capsule assembled from poly(styrene sulfonate) (PSS) and poly(allylamine hydrochloride) (PAH).[29] Although successful, the carrier capsules for the advanced capsosomes are based on poly(*N*-vinyl pyrrolidone) (PVP) and thiol-modified poly(methacrylic acid) (PMA_{SH}). This polymer pair allows for the assembly of non-aggregated, colloidally stable, single-component PMA capsules upon cross-linking of the thiols of the PMA_{SH} and the PVP release at physiological conditions. This type of capsule is one of the few well-studied carrier system of this type and has proven to be nontoxic,[30] (bio)-degradable,[30] and was successfully used to encapsulated genes,[31,32] peptides,[33–35] or drug-loaded oil droplets.[36] In the latter two cases, eradication of colon cancer cells *in vitro*[37] and stimulation of T-cells for vaccination,[35,38] has been reported.

The thiols in the multilayered polymer film of the carrier capsule have to be cross-linked in order to obtain structurally stable capsosomes at physiological conditions.[19,39] Several methodologies for doing this have been considered, each important in its own right, and often depending on the nature of the encapsulating cargo, i.e. the resistance towards oxidation. Originally, these PMA-based multilayers were cross-linked using the oxidation agent Choramine T,[40] and this has proven to be a valid approach for capsosomes carrying hydrophobic cargo in the lipid membrane of the subunits.[41,42] However, this method can have detrimental consequences to encapsulated therapeutics and biomolecules that are susceptible to oxidation and can cause a loss of their functional performance. Therefore, benign techniques were developed either based on a thiol-disulfide exchange reaction by using PMA with thiol groups activated with pyridyl sulfide (PMA-PDA) as a cross-linking reagent,[24,43,44] or by the use of small cross-linking molecules such as 2,2'-dithiodipyridine (DTDP),[45] the homobifunctional thiol-reactive linker 1,8-bismaleimidodiethyleneglycol ($BM(PEG)_2$), or a reagent which creates noncleavable thioether bonds when reacting with thiol groups.[19,46,47]

3.1.2. *Assembly of the Liposomal Subunits*

The co-assembly of liposomes and polymers beyond electrostatic driven interactions was facilitated by a novel, non-covalent anchoring concept based on cholesterol-[24] or oleic acid-modified polymers,[41] with the aim to stably anchor the liposomes to the polymeric membrane. Further, the incorporation of cholesterol or oleic acid into the lipid membrane is a spontaneous process, largely independent of the lipid composition or the solution conditions. Cholesterol-modified poly(*L*-Lysine) (PLL_c), poly(methacrylic acid)-*co*-(cholesteryl methacrylate) (PMA_c), poly(*N*-vinyl pyrrolidone)-*block*-(cholesteryl acrylate) (PVP_c) and oleic acid-modified poly(methacrylic acid)-*co*-(oleyl methacrylate) (PMA_{oa}) have been tailor-made and were successfully used as precursor, separation and capping layers. When comparing oleic-acid and cholesterol-containing PMA, while both types of polymers are equally suited as precursor layers, PMA_c was found to be superior as capping layer.[41] Furthermore, the use of PMA_c, PMA_{oa} or PVP_c allows for the subsequent assembly of the polymer carrier capsule via hydrogen bonding of PVP and PMA_{SH}.

For any carrier vehicle acting as an artificial cell or organelle, in order to substitute for deficient or missing cellular function, the reactions need to occur continuously in a cascade/parallel manner, and thus require the encapsulation of a specific number of subcompartments containing cargo, e.g. distinctive enzymes or different/complementary drugs for co-administration. Further, depending on the targeted application, different types of liposomes might be required, charged or zwitterionic, as well as saturated or unsaturated ones. For example, for the encapsulation of enzymatic cargo, saturated liposomes are preferably used due to their higher phase-transition temperature (T_m), or, when encapsulating charged cargo, the encapsulation should be favored by using zwitterionic or oppositely charged liposomes. Thus, control over the number and type of liposomal subcompartments is essential for the development of capsosomes as a biomedical platform towards cell mimicry.

To this end, we identified the upper limit of liposome layers that could be entrapped in a capsosome for liposomes with four different lipid compositions by optimizing the polymer separation layers employed between the liposome layers.[48] Our results showed that when encapsulating negatively charged liposomes, the highest number of liposome multilayers, for both unsaturated and saturated lipids, was obtained when a combination of PMA_c and PLL was used as separation layer. When encapsulating zwitterionic liposomes, PMA_c as separation layer yielded the highest amount of entrapped liposomes. In all cases, non-aggregated, structurally intact capsosomes with up to 16,000 subunits were obtained (Fig. 1(a)).

Apart from gaining control over the number of entrapped liposomes, their position within the carrier capsules is of essential importance, similar to biological cells, where the location of the organelles contributes to their efficient and successful function. We have reported the assembly of capsosomes containing "membrane-associated" liposomes (sticking to the polymer carrier capsule) and "free-floating" liposomes (filling the cavity of the capsule), as well as the combination thereof.[49] Although the encapsulation efficiency is not supposed to be affected by the position of the liposomes, the capsosomes' functionality can take advantage, e.g. when fast mixing is required or when equal access of the reagents to the subcompartments is necessary, having the subcompartments "free-floating" within the capsule's wall could be very beneficial.

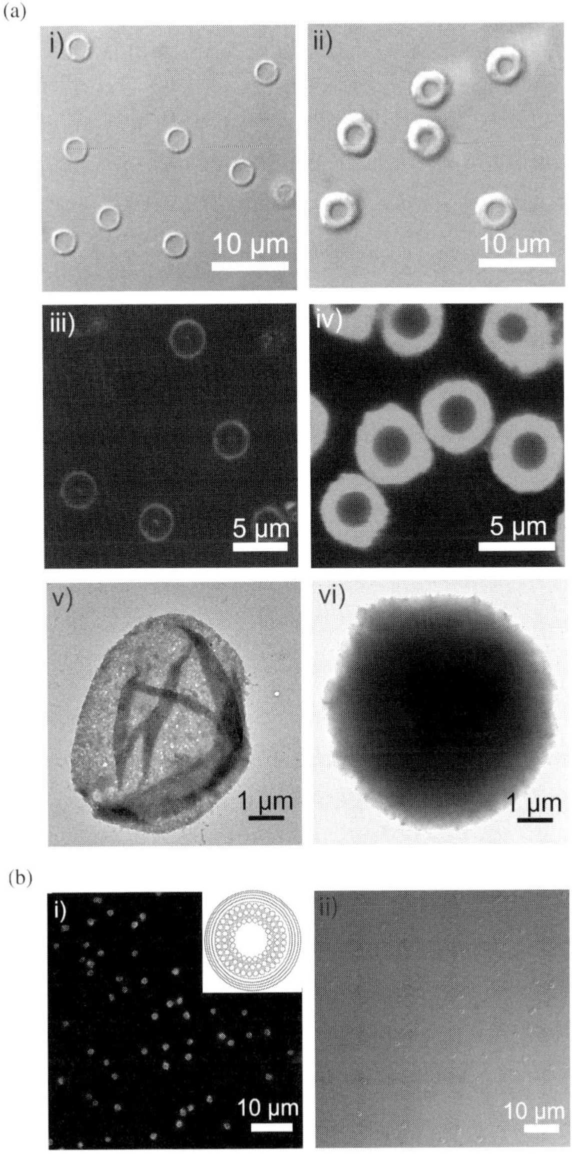

Fig. 1. (a) Differential interference contrast (DIC) microscopy (i, ii), confocal laser scanning microscopy (iii, iv), and transmission electron microscopy (v, vi) images showing capsosomes with one (i, iii, v) and eight (ii, iv, vi) liposome deposition steps. Reproduced with permission from Ref. 48. (b) Fluorescence microscopy (i) and DIC (ii) image of capsosomes containing thiocoraline within in the lipid bilayer of the subunits. Reproduced with permission from Ref. 41.

For controlling the position of the liposomes, we take advantage of the pH-dependent activity of the hydrogen bonds between PVP_c and PMA.[49] For the capsosomes' assembly, both PMA_c and PLL_c led to the formation of capsosomes with the liposomal subunits attached to the capsule's membrane. PLL_c is positively charged, thus electrostatically interacting with the PMA of the polymer carrier capsule. Although PMA_c is negatively charged and electrostatic repulsion is expected to detach the liposomes from the polymer membrane, the "free-floating" of the liposome compartments was not observed, possibly due to a physical entanglement of the polymer chains with the building blocks of the polymer membrane. We therefore synthesize PVP_c, a block copolymer which is uncharged and contains the cholesterol units in a short block to minimize the physical entanglement with the units of the carrier capsule. During deposition at pH 4, the hydrogen bonding between PVP_c and PMA facilitates the assembly, but at physiological conditions the hydrogen bonds are disrupted and the liposomal subunits become "free-floating" within the capsules interior.[49]

3.1.3. *Stability of the Liposomal Subunits*

Upon entry into the organism, micro-carriers are not only expected to be instantly recognized as foreign and efficiently removed and digested via unspecific phagocytosis,[50] but they are also exposed to different kind of biomolecules i.e. enzymes which could destroy them. Therefore, in order to obtain carriers with long circulation times, their surfaces need to be modified such that they become resistant to unspecific phagocytosis, and the polymer membrane should exclude the access of harmful biomolecules to the liposomal subunits. Most strategies to address the former case rely on the immobilization of a poly(ethylene glycol) (PEG) corona onto the carriers' surface,[51,52] while in the latter case, more deposited polymer layers, i.e. thicker polymer membranes, are considered,[53] making the control over the permeability possible, but the assembly becomes time consuming and more labor intensive. Therefore, PEGylation of the outer membrane of capsosomes was considered, not only for improving their fouling characteristics, but also to affect the permeability of the polymer carrier capsules, which was probed by the enzymatic degradation of the liposomal subcompartments by phospholipase A_2 (PLA_2).[45] The results showed that while the fouling properties were not improved, the diffusion of PLA_2 across the polymer membrane was hindered when the capsules were coated with PEG molecules in PBS, in PBS with the presence of BSA, and in cell media at 37°C within 48 hr, with at least 60% of the subunits remaining intact. In the absence of the PEG layer, the subunits were entirely destructed by PLA_2.

3.1.4. *Cargo Retention*

Another important aspect of capsosomes is the cargo retention during assembly and storage in order to understand their potential for sustained encapsulated catalysis. To this end, the size-dependent retention of different cargo in capsosomes during assembly was assessed. Fluorescently labeled hydrophilic molecules covering a size range from 400 Da to 10,000 Da were encapsulated within the liposomes and the retention was assessed.[48] The results suggested that the lower limit of the size of cargo for stable encapsulation is between 400 Da and 550 Da. Further, we encapsulated two different-sized enzymes with similar isoelectric point (β-lactamase (M_w = 31,000 Da) and luciferase (M_w = 79,000 Da)) with the aim to study the encapsulation efficiency, the stability of the enzymatic cargo within liposomes, and the long-term stability of the loaded capsosomes.[24,43] During the polymer membrane assembly there was no measurable loss of cargo, and the number of encapsulated liposomes remained constant upon long-term storage (up to 14 d). However, the quantity of enzymatic cargo trapped in the liposomes in the same time frame progressively decreased, with higher loss for the smaller enzyme.

3.2. *Functionality of Capsosomes*

3.2.1. *Encapsulated Catalysis*

With the aim of mimicking metabolic activities, the capsosomes' functionality was verified by performing encapsulated enzymatic reactions.[24,48,49] In a first attempt, we chose the enzyme β-lactamase as the encapsulate since it can convert the substrate nitrocefin (yellow) into its hydrolyzed product (red), which makes visualization and quantification possible in an easy manner. What is more, this assay was also used to estimate the number of liposomal subcompartments per capsosome to be 8×10^3, which corresponds to ~70% of the maximum theoretical number of liposomes that can cover the surface of a 3 μm silica particle.[24]

In the first attempt, the enzymatic reaction was triggered by lysing the liposomes using the surfactant Triton X and making the enzyme β-lactamase accessible to the substrate. The substrate was only converted in the lysed samples, while the intact liposomes prevented the contact between the enzyme and the substrate. Further, we compared the conversion rate of the substrate nitrocefin into its product of capsosomes containing subunits assembled one and eight liposomal deposition steps, with the aim to demonstrate that the amount of loaded liposomes, and thus, the amount of cargo, correlates with the reaction kinetics, which was found to be 2.5× faster in the latter case.[48] This result demonstrated the potential of capsosomes for encapsulated triggered enzymatic catalysis. However, the deconstruction of the subunits is not ideal when sustained conversion is the aim. Therefore, an alternative trigger, i.e. temperature, was considered. Increasing the temperature above the T_m of the liposomes enhances the permeability of the lipid membrane and allows the substrate nitrocefin to cross the lipid membrane and interact with the enzyme β-lactamase, while the enzyme is too large and remains encapsulated.[48] To confirm that capsosomes could be recycled, we replaced the substrate with fresh nitrocefin, and the enzymatic reaction occurred at a similar rate for three subsequent cycles, demonstrating that the temperature trigger neither damages the liposomes nor releases the enzymes.

To ensure that other enzymes than β-lactamase can be used, we explored the possibility of using the protease subtilisin encapsulated within capsosomes containing "free-floating" or "membrane-associated" subcompartments to convert the substrate *N*-CBZ-*L*-phenylalanyl-*L*-arginine 7-amido-4-methylcoumarin (AMC_s) into the product 7-amido-4-methylcoumarin (AMC_p).[49] The conversion, which comprises the cleavage of an amide bond by the protease subtilisin, was followed over time by fluorescence spectroscopy and triggered using temperature. As expected, when capsosomes containing "membrane-associated" liposomes were incubated at 37°C, the reaction occurred, while at room temperature (~25°C) no conversion was observed. Surprisingly, when incubating capsosomes containing "free-floating" subcompartments, 50% conversion of the product was observed after 24 hr of incubation at room temperature. The difference in behavior of these two types of capsosomes might be explained by the two different cholesterol-containing polymers employed during the assembly. The different distribution of cholesterol units, and thus, the different way of anchoring the liposomes, is likely to affect the permeability of the lipid membrane. From the application point of view, this difference could prove valuable and give rise to the possibility of using the capsosomes for complementary applications: while capsosomes assembled with PMA_c require a trigger to initiate the reaction; capsosomes assembled with PVP_c can be used without a trigger, that is, in a continuous manner.

So far, only the conversion of a small molecule by an encapsulated enzyme was demonstrated. While being valuable to characterize the system, it largely lacks biomedical relevance. Further, most biologically relevant reactions are more than single-step conversions but most often multi-step cascade reactions within the same carrier. Although this is well understood, the successful examples of encapsulated multi-step reactions are scarce. The most promising examples rely on the immobilization of different enzymes at different positions of the carrier, i.e. in the void, inside the membrane, and on the surface.[54,55] The triggered cargo release by encapsulated enzymatic catalysis was used to bring the platform capsosomes a step closer to a real potential medical application. Capsosomes containing the enzyme glutathione reductase entrapped within the liposomal subunits and a peptide (a wild-type antigen used in vaccination termed KP9) conjugated to a polymer carrier via a disulfide bonding as a co-encapsulate within the polymer carrier capsules were assembled.[47] The idea is that glutathione reductase converts glutathione disulfide (GSSG), naturally occurring in the body, into its reduced form, glutathione (GSH), which will be able to cleave the disulfide bond that links the peptide to the polymer.

By doing so, the peptide is released since it is too small to remain trapped within the polymer carrier capsule in its free form (Fig. 2(a)). First, the activity of the entrapped enzyme was demonstrated using Ellman's reagent as indicator. Then, the release of the fluorescently

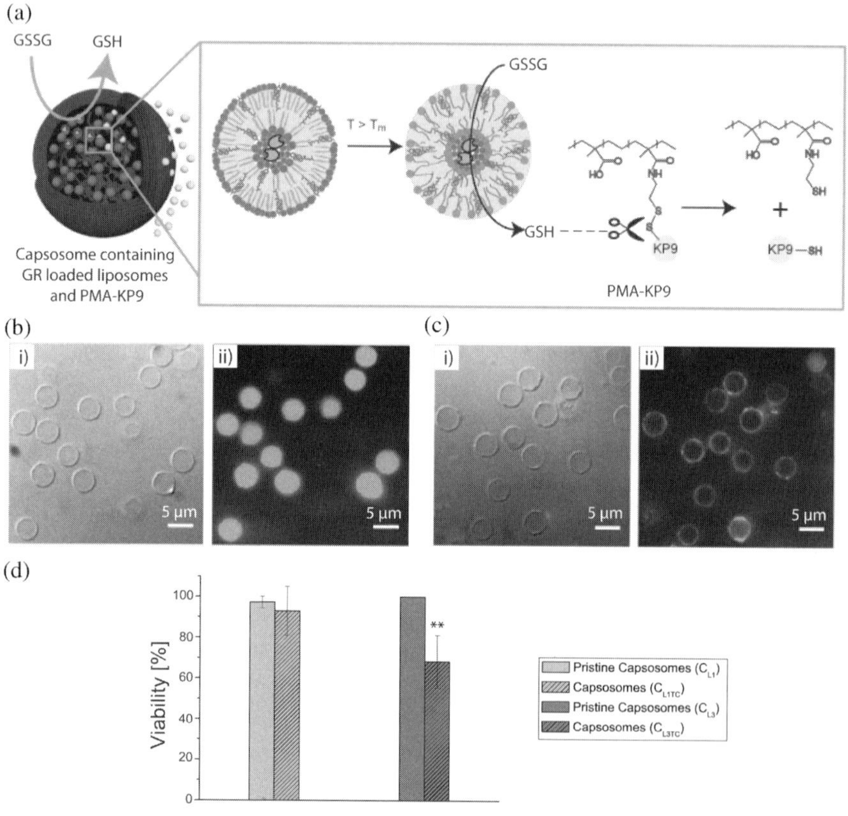

Fig. 2. (a) Release of encapsulated oligopeptides triggered by glutathione reductase. DIC (i) and fluorescence microscopy (ii) images of capsosomes incubated at 37°C, before (b, i and b, ii) and after (c, i and c, ii) the glutathione reductase-induced cleavage and subsequent release of $KP9_{488}$. Reproduced with permission from Ref. 47 (d) Cell viability assay of 1 µm-sized capsosomes incubated with A549 cells for 48 hr. The cell viability was only significantly reduced when exposed to capsosomes containing three layers of TC-loaded liposomes (C_{L3TC}) ($p < 0.01$). Capsosomes containing one layer of TC-loaded liposomes (C_{L1TC}) and pristine capsosomes with one or three liposome layers (C_{L1} and C_{L3}) have a negligible effect on the cell viability. Reproduced with permission from Ref. 41.

labeled peptide due to the enzymatic conversion of GSSG to GSH was monitored using flow cytometry and fluorescence microscopy (Figs. 2(b) and 2(c)). Upon increasing the temperature to 37 °C (above the T_m of the glutathione reductase loaded liposomes), GSSG can cross the lipid membrane and the enzyme is able to convert it into GSH, which, in turn, cleaves the disulfide bond, and a 50% decrease in fluorescence intensity of the capsosomes was observed within 24 hr at 37 °C due to the releases the KP9 peptide.[47] These results demonstrate the potential of capsosomes to continuously generate GSH, an important cellular antioxidant, while simultaneously releasing a small molecule therapeutic at human body temperature, which is, to the best of our knowledge, the first multi-step reaction conducted within the same subcompartmentalized carrier.

3.2.2. *Intracellular Drug Delivery*

Intracellular drug delivery was never the core application considered for capsosomes, largely due to the rather large size of the vehicles, which excludes the intravenous administration pathway. Downscaling would be required, which is not really an option since the liposome loading efficiency for carrier capsules below 1 μm is getting very low. However, if other administration pathways are considered, assessing the interaction of capsosomes with cell becomes valid, especially for the delivery of hydrophobic drugs. In this context, the ability of capsosomes to encapsulate small lipophilic cargo in the lipid membrane of the subunits becomes interesting and was confirmed by the encapsulation of thiocoraline (TC) (Fig. 1(b)), an antitumor cyclic thiodepsipeptide,[42] and taxol (TX),[41] a mitotic inhibitor used in cancer chemotherapy. The use of capsosomes offered several advantages over liposomes or polymer capsules such as: an increased stability of the liposomes, a higher deliverable payload by increasing the amount of liposome layers within one carrier capsule, and the ability to spatially separate different compounds for co-delivery. ~1 μm capsosomes were assembled and the amount of encapsulated TC was confirmed to be dependent on the number of liposome deposition steps, i.e. the amount of encapsulated depsipeptide was the expected three times higher for capsosomes containing three in comparison to one layer of liposomes, demonstrating the easy control over the loading. In order to confirm that the engineered properties of the capsosomes can be translated into biology, the viability of cells upon exposure to capsosomes loaded with the cytotoxic compounds TC or TX was assessed. First, pristine capsosomes did not exhibit any intrinsic toxicity,[41,42] which makes them ideal candidates for biomedical applications. The drug-loaded capsosomes showed a significant decrease in cell viability for both colon cancer cells[42] and human lung carcinoma cells (Fig. 2(d)).[41] In the latter case, the cargo-concentration dependent cytotoxicity was confirmed, showing that the designed capsosomes have the potential to inflict a pre-defined biological response.[41]

3.3. *Next Generation Capsosomes*

Although capsosomes have been demonstrated to be potentially efficient as artificial cells or as drug-delivery systems, there are still unaccomplished challenges, e.g. the number of available triggers to activate the encapsulated catalysis. In order to address this issue, we have developed the assembly of a new class of carrier/reactor vessel containing subcompartments of two different types: liposomes and polymeric capsules.[46] The aim is to enlarge the number of potential triggers by using two conceptually different subunits, i.e. polymer capsules can be triggered chemically, which is a difficult task to accomplish for liposomes. Further, other limitations such as the low loading capacity of liposomes, or the lack of permeability towards small solutes, e.g. substrates for biomedical processes or analytes for diagnostic tests, could be overcome. We encapsulated liposomes and 300-nm-diameter sized PMA capsules into a larger polymer carrier vessel and demonstrated control over the spatial position of the subcompartments by having the liposomes "membrane associated" or "free-floating" within the interior. Further, the selective degradation of the 300-nm-diameter subcompartments was shown, while the enzyme-loaded liposomes remained intact and were able to be employed in an enzymatic reaction.

4. Conclusions and Future Perspectives

Subcompartmentalized systems have the potential to have a considerable impact on therapeutic cell mimicry and, in particular, for enzyme replacement therapy. However, to date, the field is still in its infancy and only few systems, which combine both an advanced assembly and functionality, have been reported. The most established system so far is the platform capsosomes. Advanced assembly from control over number and position of the subunits, permeability, degradability, and cargo encapsulation and retention has been demonstrated. Further, the functionality of this platform has been shown by performing trigger-induced encapsulated enzymatic catalysis. The absence of inherent cytotoxicity makes capsosomes a promising approach for biomedical applications. Although the reported results are promising, a lot of possibilities have yet to be explored: increasing the type of building blocks in the assembly will facilitate the implementation of additional activity, e.g. membrane proteins as specific gates; combining different types of subunits will also allow for the conduction of enzymatic cascade reactions where different enzymatic activity has to be triggered subsequently; also, the administration pathway has to be considered and adjustments in the design have to be made. Aspects of biological cells such as shape, softness, and surface functionality, which are essential for the application, have to be identified and mimicked. Most crucially, all the reports are only covering proof-of-concept ideas performed *in vitro* at best. The next challenge would be the identification of an actual medical issue and the assessment of the performance of a subcompartmentalized system in a potential treatment in order to drive this idea a step closer to clinics.

References

1. Orive G, Hernandez RM, Gascon AR, *et al.* (2003) *Nat Med* **9:** 104.
2. Zhang Y, Ruder WC, LeDuc PR. (2008) *Trends Biotechnol* **26:** 14.
3. Chang TMS. (2005) *Nat Rev Drug Discov* **4:** 221.
4. Chang TMS. (1965) *Science* **146:** 524.
5. Chang TMS. (1969) *Can J Physiol Pharmacol* **47:** 1043.
6. Chang TMS, Malave N. (1970) *T Am Soc Art Int Org* **16:** 141.

7. Chang TMS. (1975) *Kidney Int* 7: S387.
8. Chang TMS, Poznansk MJ. (1968) *Nature* **218:** 243.
9. Bourget L, Chang TMS. (1986) *Biochim Biophys Acta* **883:** 432.
10. Chang TMS. (1989) *Biomater Artif Cells Artif Organs* **17:** 611.
11. Lvov Y, Antipov AA, Mamedov A, *et al.* (2001) *Nano Lett* **1:** 125.
12. Peyratout CS, Dahne L. (2004) *Angew Chem Int Ed* **43:** 3762.
13. Price AD, Zelikin AN, Wang YJ, Caruso F. (2009) *Angew Chem Int Ed* **48:** 329.
14. Price AD, Zelikin AN, Wark KL, Caruso F. (2010) *Adv Mater* **22:** 720.
15. Tiourina OP, Antipov AA, Sukhorukov GB, *et al.* (2001) *Macromol Biosci* **1:** 209.
16. Kisak ET, Coldren B, Evans CA, *et al.* (2004) *Curr Med Chem* **11:** 199.
17. Chiu HC, Lin YW, Huang YF, *et al.* (2008) *Angew Chem Int Ed* **47:** 1875.
18. De Geest BG, De Koker S, Immesoete K, *et al.* (2008) *Adv Mater* **20:** 3687.
19. Kulygin O, Price AD, Chong SF, *et al.* (2010) *Small* **6:** 1558.
20. Fu Z, Ochsner MA, de Hoog HPM, *et al.* (2011) *Chem Commun* **47:** 2862.
21. Driever CD, Mulet X, Johnston APR, *et al.* (2011) *Soft Matter* 7: 4257.
22. Lomas H, Johnston APR, Such GK, *et al.* (2011) *Small* 7: 2109.
23. Fakhrullin RF, Paunov VN. (2009) *Chem Commun* **45:** 2511.
24. Städler B, Chandrawati R, Price AD, *et al.* (2009) *Angew Chem Int Ed* **48:** 4359.
25. Andresen TL, Jensen SS, Jorgensen K. (2005) *Prog Lipid Res* **44:** 68.
26. Drummond DC, Zignani M, Leroux JC. (2000) *Prog Lipid Res* **39:** 409.
27. Torchilin VP. (2005) *Nat Rev Drug Discov* **4:** 145.
28. Quinn JF, Johnston APR, Such GK, *et al.* (2007) *Chem Soc Rev* **36:** 707.
29. Städler B, Chandrawati R, Goldie K, Caruso F. (2009) *Langmuir* **25:** 6725.
30. Zelikin AN, Price AD, Stadler B. (2010) *Small* **6:** 2201.
31. Becker AL, Orlotti NI, Folini M, *et al.* (2011) *ACS Nano* **5:** 1335.
32. Zelikin AN. (2010) *ACS Nano* **4:** 2494.
33. Sexton A, Whitney PG, Chong SF, *et al.* (2009) *ACS Nano* **3:** 3391.
34. De Rose R, Zelikin AN, Johnston APR, *et al.* (2008) *Adv Mater* **20:** 4698.
35. Chong SF, Sexton A, De Rose R, *et al.* (2009) *Biomaterials* **30:** 5178.
36. Sivakumar S, Bansal V, Cortez C, *et al.* (2009) *Adv Mater* **21:** 1820.
37. Yan Y, Johnston APR, Dodds SJ, *et al.* (2010) *ACS Nano* **4:** 2928.
38. Sexton A, Whitney PG, Chong SF, *et al.* (2009) *ACS Nano* **3:** 3391.
39. Chong SF, Lee JH, Zelikin AN, Caruso F. (2011) *Langmuir* **27:** 1724.
40. Zelikin AN, Quinn JF, Caruso F. (2006) *Biomacromolecules* 7: 27.
41. Hosta-Rigau L, Chandrawati R, Saveriades E, *et al.* (2010) *Biomacromolecules* **11:** 3548.
42. Hosta-Rigau L, Stadler B, Yan Y, *et al.* (2010) *Adv Funct Mat* **20:** 59.
43. Chandrawati R, Städler B, Postma A, *et al.* (2009) *Biomaterials* **30:** 5988.
44. Chong SF, Chandrawati R, Stadler B, *et al.* (2009) *Small* **5:** 2601.
45. Chandrawati R, Chong SF, Zelikin AN, *et al.* (2011) *Soft Matter.*
46. Hosta-Rigau L, Kulygin O, Städler B, Caruso F. (2013) *Small* early view. DOI: 10.1002/smll.201300125.

47. Chandrawati R, Odermatt PD, Chong SF, *et al.* (2011) *Nano Lett.* DOI: dx.doi.org/10.1021/nl202906.
48. Chandrawati R, Hosta-Rigau L, Vanderstraaten D, *et al.* (2010) *ACS Nano* **4:** 1351.
49. Hosta-Rigau L, Chung SF, Postma A, *et al.* (2011) *Adv Mater* **23:** 4082.
50. Juliano RL. (1988) *Adv Drug Deliv Rev* **2:** 31.
51. Heuberger R, Sukhorukov G, Voros J, *et al.* (2005) *Adv Funct Mat* **15:** 357.
52. Wattendorf U, Merkle HP. (2008) *J Pharm Sci* **97:** 4655.
53. Becker AL, Zelikin AN, Johnston APR, Caruso F. (2009) *Langmuir* **25:** 14079.
54. Zhang L, Shi J, Jiang Z, *et al.* (2011) *Green Chem* **13:** 300.
55. Baumler H, Georgieva R. (2010) *Biomacromolecules* **11:** 1480.
56. Wang YJ, Hosta-Rigau L, Lomas H, Caruso F. (2011) *Phys Chem Chem Phys* **13:** 4782.

Chapter 13

Soft Magnetic Nanomaterials Towards Nanomedicine: Magnetism Principles, Preparation, Characterization and Potential Applications

G. W. Qin*, S. Li, Y. P. Ren, F. Darain† and K. Dimitrov†

Abstract

Soft magnetic nanomaterials (magnetic nanoparticles and nanorods) have been paid increasing attention in the past decades due to their own unique characteristics that provide many exciting opportunities in the field of nanomedicine. In this chapter, the general magnetism principles of these nanomaterials in both static and AC magnetic fields are reviewed, and the magnetic properties required for various medical applications are also highlighted here. Much effort is focused on reviewing recent progress of synthesis and characterization of diverse magnetic nanomaterials, merits and demerits of preparation techniques, and their surface modifications. Specially, more attention is paid to our recent work on soft magnetic nanomaterials prepared by some novel techniques involving size-tunable Fe_3O_4, NiFe and CoNi

* Key Lab for Anisotropy and Texture of Materials (MOE), Northeastern University, Shenyang 110819, China.

† Australian Institute for Bioengineering and Nanotechnology, The University of Queensland, QLD 4072, Australia.

magnetic nanoparticles, Co-C core–shell nanostructures, and biodegradable Fe-B-P magnetic nanoparticles and nanobars. Their unique magnetic properties are presented, and potential biomedical applications are discussed in detail.

1. Introduction

The development of biomedical science and technology aims to detect disease at the earliest possible time and to get treatment at the fixed place and in the right dose. In the past decade, magnetic nanomaterials (magnetic nanoparticles (NPs) and nanorods), especially soft magnetic ones, have been paid much attention due to their own unique characteristics that provide many exciting opportunities in nanomedicine.[1–3] Generally speaking, soft magnetic nanomaterials not only have quite low remanent magnetization and coercivity to allow monodisperse characteristics in aqueous solution and physiological environment against aggregation, but they also have very high magnetic susceptibility and sufficiently high saturation magnetization to easily respond to external magnetic fields. These unique features make them useful in diagnostics, bio-sensing and medical imaging if they have high saturation magnetization and/or are properly decorated by labeled biomolecules, and feasible in disease treatment including targeted drug delivery, triggered medicine release and carried radionuclide atoms, and generating heat by external magnetic field. The various applications are not only relevant to magnetic properties of these nanomaterials, but also to their biocompatibility, morphology, size, and size distribution.[2,3]

Several particular characteristics of magnetic nanomaterials make them attractive in nanomedicine.[1–3] The first one is their size, from a few nanometers up to tens of nanometers in diameter, offering high specific surface area not only for higher bio-sensitivity, larger loading capacity for drug that is attached on the surface, but also higher leaking ability from blood vessels to tumor issue, even infiltration ability towards the targeted cells. The second one is their higher stability of colloids in physiological environment by proper surface modifications, making them less risky in terms of blood vessel jamming and longer half-life in the blood stream. The third one is their magnetic characteristic, implying these nanomaterials could be manipulated by external magnetic field to a fixed place, e.g. for hyperthermia towards tumor tissue.

During the past decades, many magnetic nanomaterials, including oxides, metals and alloys, core–shell nanostructures, have been extensively investigated from basic research to the development of viable pharmaceutical products.[3,4] However, only some of them are in clinical trials and few of them have reached the market (e.g. magnetic nanooxides in magnetic resonance imaging, MRI). Reasonable design and synthesis of magnetic nanomaterials with the right magnetic properties and tunable size and size distribution, and proper surface modifications as well as their cytotoxicity evaluation are still great challenges for a specific medical task at present time. Moreover, how to endue one magnetic nanomaterial with triple functions, that is, medical diagnosis, treatment and bio-degradation after use, should be seriously considered in the near future.

In this chapter, therefore, we first introduce the general magnetism of soft magnetic nanomaterials in both static and AC magnetic fields, and then review recent preparation progress of various magnetic nanomaterials, mainly focusing on our recent works on soft magnetic nanomaterials prepared by some novel techniques involving magnetic NPs and nanobars (size-tunable Fe_3O_4, NiFe and CoNi magnetic NPs, Co-C core–shell NPs and biodegradable Fe-B-P magnetic NPs and nanobars). Their unique magnetic properties will be presented and potential biomedical applications will be discussed in detail. Finally, perspective remarks will be addressed.

2. Magnetism of Soft Magnetic NPs for Nanomedicine

Since medical diagnostics and disease treatment using magnetic nanomaterials require their different magnetic properties, it is necessary to understand their general magnetic features, especially depending on size, size distribution, magnetization, magnetic anisotropy and susceptibility, and thus it will facilitate to design suitable magnetic nanomaterial for a specific application. Generally speaking, the re-orientation and growth of spontaneously magnetized domains within a given magnetic particle depends on both intrinsic properties such as saturation magnetization, effective magnetic anisotropy (magnetocrystalline, shape and surface anisotropies) and microstructural features of the particle such as defects, impurities, size, etc.[1,2] In most cases, it is not possible to predict strictly what the hysteresis loop looks like at a static magnetic field or magnetization behavior at a radio frequency AC magnetic field. However, their general magnetization features can be obtained by magnetism theory, as summarized below.

2.1 *Magnetism in a Static Magnetic Field*

Soft magnetic nanomaterials with typical sizes less than several hundred nanometers are paid increasing attention in the field of nanomedicine due to high specific surface, saturation magnetization and susceptibility, low coercivity, and small size allowing passing blood vessels and tissues. Therefore, it is of importance to understand the size-dependent magnetic properties of these NPs, as summarized in Fig. 1. With decreasing particle size, magnetic domain changes from a multiple domain to single domain (or pseudo-single domain) at a critical size R_c, the remanent magnetization and coercivity increase significantly, together with the squareness of the magnetic hysteresis loop. However, the magnetic susceptibility usually decreases in this case, which is generally not adopted by magnetic hyperthermia treatment. With further decrease of particles to the other smaller critical size R_{sp}, where the effective magnetic anisotropy energy is comparable to the thermal agitation, the remanent magnetization and coercivity of these magnetic ferromagnetic particles go to zero and they become paramagnetic ones, but respond strongly with a high magnetization and thus high susceptibility under a small external magnetic field, called superparamagnetism.[5]

As illustrated above, two critical sizes, single domain radius R_c and superparamagnetic radius R_{sp}, are important to understand size-dependent magnetic nanomaterials, and thus to design suitable ones for a specific application. Therefore, we outline the general magnetism principles here for these calculations.

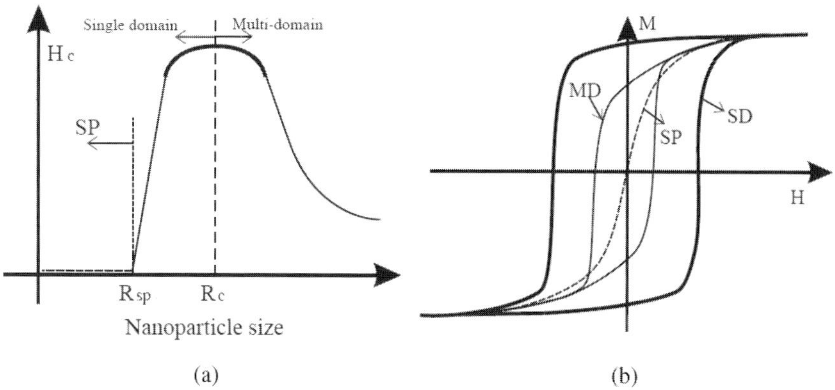

Fig. 1. The sketch of magnetic nanoparticle size-dependent coercivity (a), where R_c is the critical radius of the multiple domain (MD) to the single domain (SD), and R_{sp} is the superparamagnetic (SP) critical radius. (b) The corresponding magnetic hysteresis loops of three kinds of magnetic NPs with various sizes.

Calculation of Critical Radius of Single Domain

In ferromagnetic bulks, they usually form multiple domains to decrease demagnetization energy, and thus to have a lowest energy of system. With decrease of bulk size to microns, even nanometers, the domain wall energy E_γ increases and it makes the increase of system energy. So there is a critical size R_c in which the system has a lowest energy, beyond which the magnetic nanoparticle has multiple domains.[5] Since the magnetic domain structure has different styles depending on the magnetic anisotropy energy of nanomaterials, as shown in Fig. 2, there are different critical sizes for nanomaterials with different magnetic anisotropies.

In the case of the magnetic anisotropy constant $K \gg 0$, the magnetic domain structure exhibits as either Fig. 2(a) (90° domain) or Fig. 2(b) (180° domain). Assuming the particle radius R, saturation magnetization M_s, magnetic permeability in vacuum μ_0, then the demagnetization energy E_d for the case of 90° domain (Fig. 2(a)) is expressed by[5]

$$E_d = \frac{1}{6}\mu_0 M_s^2 \times \frac{4}{3}\pi R^3 = \frac{2}{9}\pi\mu_0 M_s^2 R^3. \quad (1)$$

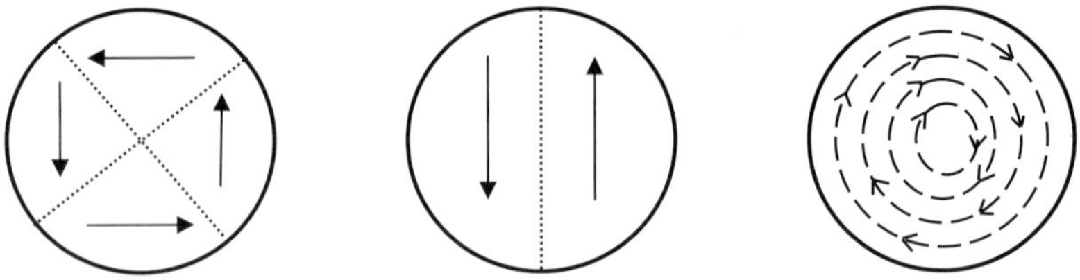

Fig. 2. The magnetic domain structures for the high magnetic anisotropy NPs: (a) 90° domain, (b) 180° domain and (c) low magnetic anisotropy NPs.

The corresponding domain wall energy E_γ is expressed by[5]

$$E_\gamma = 2\pi R^2 \gamma_w, \qquad (2)$$

where γ_w is the domain wall density. It can be known that demagnetization energy $E_d \sim R^3$ while domain wall energy $E_\gamma \sim R^2$, that is, demagnetization energy increases (or decreases) more rapidly than the domain wall energy when the particle size increases (or decreases). When the demagnetization energy is equal to the domain energy, it gives the critical radius R_c, that is,

$$R_c = \frac{9\gamma_w}{\mu_0 M_s^2}. \qquad (3)$$

By the same token, we can get the critical size of domain wall structured NP for 180° domain in the case of uniaxial magnetization, which is the same as Eq. (3).

However, when the magnetic anisotropy of ferromagnetic nanomaterials is small, the magnetization aligns many homocentric spheres in a 3-D NP, as shown in Fig. 2(c), rather than forming any domain walls. Assuming the lattice parameter a, exchange integral constant A_1, the magnetization angle change φ of neighboring ferromagnetic atoms along an arbitrary circle, then the exchange energy of a pair of atoms E_{atoms} is expressed by[5]

$$E_{atoms} = aA_1 \cos\varphi. \qquad (4)$$

The total exchange energy $E_{particle}$ of the nanoparticle is integrated as[5]

$$E_{particle} = 4\pi A_1 R\left[\ln\left(\frac{2R}{a}\right) - 1\right]. \tag{5}$$

At the critical size R_c, the exchange energy should be equal to the demagnetization energy (Eq. (1)), then the critical size of the single domain for weak magnetic anisotropy nanomaterial is determined by Eq. (6).[5]

$$\frac{R_c^2}{\left[\ln\left(\frac{2R_c}{a}\right) - 1\right]} = \frac{18A_1}{\mu_0 M_s^2}. \tag{6}$$

In many ferromagnetic materials, the magnetic exchange integral constant A_1 is at the order of magnitude of 10^{-11} Jm^{-1}.[5]

Calculation of Critical Radius of Superparamagnetic Nanoparticle

As one NP size continues to decrease within the single domain range, the other critical size threshold is reached, at which remanent magnetization and coercivity go to zero.[1,5] When this happens, the nanoparticle becomes superparamagnetic. For superparamagnetic particles, the net magnetic moment will average to zero in zero magnetic field and at a temperature of $T > 0$ K. When an external magnetic field is applied, there will be a large net statistical alignment of magnetic moments. Therefore, the superparamagnetic NPs have a much higher susceptibility value than common paramagnetic materials.

The physical principle of superparamagnetism is that the thermal agitation energy K_BT is sufficient to overcome the anisotropy energy KV due to the decrease of particle size for the NP with a volume V at a temperature T, resulting in a spontaneous reversal of magnetization and random distribution. In absence of magnetic field, there is a finite probability for the magnetization to flip and reverse its direction at a given temperature T. The mean time between two flips is called the

Néel relaxation time t_N and is given by the following Néel–Arrhenius equation[6]

$$t_N f_0 = \exp\left(\frac{KV}{K_B T}\right), \tag{7}$$

where K_B is the Boltzmann constant, f_0 is the attempt frequency (typically $10^9 \sim 10^{11}$ Hz), and K is the effective anisotropy constant. For a typical Néel relaxation time of one second, one can get

$$KV = (20 \sim 25)K_B T. \tag{8}$$

Usually $KV = 25K_BT$ is set for a criteria of superparamagnetism, then $V = \frac{25K_BT}{K}$ for the critical volume of superparamagnetic NP. In the case of spherical NP, the critical radius R_{sp} for superparamagnetic NP,

$$R_{sp} = \left(\frac{75K_BT}{4\pi K}\right)^{1/3}. \tag{9}$$

In a real experiment, when the measurement time t_m is much longer than t_N, the NP magnetization will flip several times during the measurement, then the measured magnetization will average to zero. When the measurement time is much shorter than t_N, the magnetization will not flip during the measurement, so the measured magnetization will be what the instantaneous magnetization is at the beginning of the measurement.[6,7] In the former case, the NP will appear to be in the superparamagnetic state, whereas in the latter case it will appear to be "blocked" in its initial state. The state of the NP (superparamagnetic or blocked) depends on the measurement time. A transition between superparamagnetism and blocked state happens when the measurement time is equal to the Néel relaxation time. In experiments, the measurement time is kept constant but the temperature is varied, so the transition between superparamagnetism and blocked state is seen as a function of the temperature. The temperature for which $t_m = t_N$ is called the blocking temperature T_B, typically by[6,7]

$$T_B = \frac{KV}{25K_B}. \tag{10}$$

In the presence of an external magnetic field H, stable monodispersive magnetic NPs will assemble along the magnetic field, together with an extra Zeeman energy added. In this case, Eq. (9) can be rewritten as the following[8]

$$R_{sp} = \left[\frac{75 K_B T}{4\pi (K + MH)} \right]^{1/3}. \tag{11}$$

From Eq. (11), one can know that the critical size of superparamagnetic nanoparticle will decrease depending on the external magnetic field, which is of essence for the application of superparamagnetic NPs in magnetic hyperthermia. For example, for magnetite NP, its saturation magnetization is ~92 emu/g (4.77×10^5 $\mathrm{Am^{-1}}$) and magnetic anisotropy constant is 1.1×10^4 $\mathrm{Jm^{-3}}$,[9] the superparamagnetic critical radius at zero magnetic field is calculated to be ~13 nm. However, in presence of 1T magnetic field, the critical radius decreases to ~8 nm.

When an assembly of superparamagnetic NPs is under an external magnetic field, their magnetic moments tend to align along the applied field, leading to a large net magnetization. The magnetization curve of the assembly is a reversible S-shaped increasing function. This function is quite complicated in a real case, but one can obtain analytical expressions in the simple case as follows,[7] where all the particles are identical in their anisotropy energy, composition and size.

In the case of temperature T, $T_B < T < KV/10K_B$,

$$M(H) = nm \tanh\left(\frac{\mu_0 H m}{K_B T} \right), \tag{12}$$

$$\chi = \frac{n \mu_0 m^2}{K_B T}, \tag{13}$$

where n is the density of NPs in the sample, m is the magnetic moment of a nanoparticle, and χ is the magnetic susceptibility of the sample.

In the case of $T > KV/10K_B$,

$$M(H) = nmL\left(\frac{\mu_0 Hm}{K_B T}\right), \tag{14}$$

$$\chi = \frac{n\mu_0 m^2}{3K_B T}, \tag{15}$$

where $L(x)$ is the Langevin function, $L(x) = 1/\tanh(x) - 1/x$. The equations above always give an over-estimated magnetization at a given magnetic field. Bandow *et al.* considered the size distribution and surface magnetization contribution and thus modified these equations for the magnetization of superparamagnetic NPs, which were more close to that of the real case.[10]

2.2 *Magnetism in an AC Magnetic Field*

The section above discusses the general magnetism of magnetic NPs in a static magnetic field, however, the application of these NPs is usually under a weak radio frequency field, such as magnetic resonance imaging (MRI) and hyperthermia. It is thus necessary to understand magnetization behavior of these magnetic NPs and physical principles of these methods in order to design and synthesize suitable magnetic nanomaterials for a specific task.

The general idea of hyperthermia using magnetic NPs is to homogeneously disperse them throughout the targeted tumor tissue by surface modification of the NPs with a specific tumor-targeting ligand such as monoclonal antibodies, or by any other biomolecules and driven by an external magnetic field.[1,2] And then a weak magnetic field alternating with high frequency is applied, and thus the magnetization of the particles is continuously reversed, which translates into a conversion from magnetic to thermal energy by different physical mechanisms depending on characteristics of ferromagnetic or superparamagnetic NPs.[1–3] This heat is released to the immediately surrounding diseased tissue, whereby, if the temperature can be maintained at 42°C for 30 min or more, the cancer tissue is destroyed while the neighboring healthy tissue is less affected.[1–3] Gilchrist *et al.*[11] first proposed the hyperthermia in 1957 when they heated various tissue samples with 20–100 nm size particles of γ-Fe_2O_3 exposed to a 1.2 MHz magnetic field. Since then, there have been numerous publications describing a variety of schemes using different types of magnetic materials, different field strengths and frequencies and different methods of encapsulation and delivery of the particles.[12–16]

In general, the released heat using magnetic NPs depends on their magnetic properties, i.e. ferromagnetic or superparamagnetic ones. For the ferromagnetic NPs, they have hysteretic properties when an AC field with a frequency f is applied, resulting in the magnetically induced heating. The specific heat P_{FM} generated is given by Eq. (16)[1]

$$P_{FM} = \mu_0 f \int H \mathrm{d}M. \tag{16}$$

From Eq. (16), the generated specific heat is proportional to applied frequency f and area of magnetic hysteresis loop. Atkinson *et al.* concluded that exposure to the magnetic field where the product $H \cdot f$ does not exceed 4.85×10^8 A^{-1} m^{-1} s^{-1} is safe and tolerable.[17] Here, it should be noted that hard magnetic NPs are not adopted by this application since the applied AC magnetic field is usually weak, up to ~15 kAm^{-1}, with a frequency of 0.05 ~ 1.5 MHz.[1] Under this circumstance, the hard magnetic NPs are hardly magnetized to be saturated. As a consequence, the soft magnetic NPs with proper coercivities and large squareness of magnetic loop are generally considered to conform to the applied field. Since the applied frequency is relatively low and particle size is small, according to the skin-depth principle, the eddy current effect to generate heat, usually happening in ferromagnetic bulks, does not occurs in this case.[5]

As to superparamagnetic NPs, the AC magnetic field induced heat follows the other principle by both rotation of atomic magnetic moments within each nanoparticle (Néel relaxation) and physical friction between NPs and surrounding media (Brownian rotation).[1–4] The Brownian rotation induced heat mostly happens at lower frequency while the Néel relaxation induced heat is generally considered at higher frequency. The theoretical work on AC magnetic field induced heat of superparamagnetic NPs has been well done by Rosensweig,[18] where the generated heat density P_{sp} for the monodispersive NPs is expressed by

$$P_{sp} = \mu_0 \pi \chi H^2 f \frac{2\pi ft}{1+(2\pi ft)^2}, \tag{17}$$

where H is the applied field, χ is the equilibrium susceptibility and can be approximately expressed by Eq. (15) at ambient environment for most superparamagnetic NPs. t is the effective relaxation time containing both Brownian relaxation time t_B and Néel relaxation time t_N (Eq. (7)), and expressed by[18]

$$\frac{1}{t} = \frac{1}{t_B} + \frac{1}{t_N}, \tag{18}$$

$$t_B = \frac{3\eta V_H}{K_B T}, \tag{19}$$

where η is the viscosity coefficient of the matrix fluid and V_H is the hydrodynamic volume of the particle which is a little larger than the real volume V.

For polydispersive NPs, Rosensweig treated the heat density by considering particle size distribution following a Gaussian function $g(R)$, then the averaged heat density $\bar{P}$ is expressed by[18]

$$\bar{P} = \int_0^\infty Pg(R)\mathrm{d}R. \tag{20}$$

And the corresponding temperature rise of an adiabatic colloid is $\Delta T = \bar{P}\Delta t/c$, where Δt is the applied time and c is the specific heat of nanoparticle colloid. From the Rosensweig's calculation results (Fig. 3), several useful conclusions are addressed on how the particle size, size distribution, viscosity of fluid and effective magnetic anisotropy affect the heating rate, which are very important for design and synthesis of superparamagnetic NPs in the future: (1) there is a proper size corresponding to the maximum heating rate depending on effective anisotropy of nanomaterials; (2) narrower size distribution results in a larger heating rate; (3) a relative smaller viscosity of medium gives a

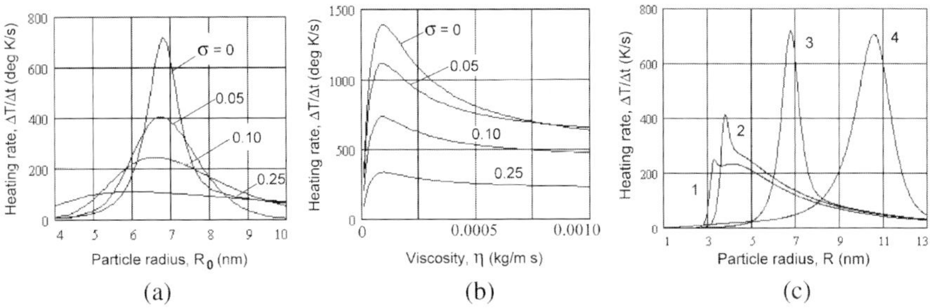

Fig. 3. Temperature rise rate in polydisperse ferrofluid based on magnetite: (a) Tetradecane carrier, $f = 300\,\text{kHz}$; (b) H = 0.06T, f = 900 kHz, R = 7 nm. (c) Comparative heating rates for various magnetic NPs. 1-Barium ferrite, 2-Cobalt ferrite, 3-Magnetite, 4-Maghemite. ($\phi = 0.071$, $\eta = 0.00235\,\text{kgm}^{-1}\text{s}^{-1}$, $f = 300\,\text{kHz}$, $H = 0.09\,\text{T}$, surfactant layer thickness 2 nm). Figure reproduced with permission from Ref. 18. Copyright 2002 Elsevier.

higher heating rate; (4) soft magnetic NPs have a higher heating rate than hard ones.

The MRI technique using superparamagnetic NPs is to enhance decaying in transverse magnetization via interactions with the adjacent proton nuclei, and thus to provide darker contrast to illustrate various soft tissues.[1,2] Generally speaking, MRI images are generated from nuclear magnetic resonance signals received from hydrogen nuclei located in different physiological environments. When a soft tissue or fluid containing atomic nuclei possessing a non-zero spin (e.g. H, C, N, P) is placed in a strong static magnetic field, their nuclear spins orient along the magnetic field *H*, for example, the effective proton moments $2 \times 10^{14}\ \text{mm}^{-3}$ for water containing ^{1}H.[1] When a high radio frequency (RF) field is applied in a plane perpendicular to the static field *H*, this signal can be captured by making use of resonant absorption with the protons precessing by the Larmor precession frequency $f = \frac{1}{2\pi}\gamma H$.[1] For ^{1}H protons the gyromagnetic ratio $\gamma = 2.67 \times 10^8\ \text{rad s}^{-1}\ \text{T}^{-1}$, the Larmor precession frequency corresponds to a radio frequency field with $f = 63.7\,\text{MHz}$ under a typical static field of 1.5 T. By applying RF pulses at the appropriate resonant frequency, MRI scanners perturb the thermal equilibrium distribution of spin states in the samples, resulting in the reorientation of the net nuclear magnetization in a new direction, typically by 90°. After this perturbation, "relaxation" occurs as the induced orientation regresses to its original equilibrium state by the magnetic moment (m_z) along the static field H and in the plane (m_{xy}) perpendicular to H by the following expressings as[1]

$$m_z = m(1 - e^{-t/T_1}), \tag{21}$$

$$m_{xy} = m\sin(2\pi ft + \alpha)e^{-t/T_2}, \tag{22}$$

where m is the net magnetic moment, T_1 and T_2 are the spin-lattice (or longitudinal) and spin-spin (or transverse) relaxation times, respectively; α is the phase constant. The detailed principle is shown in Fig. 4. The spin-lattice relaxation reflects a loss of energy from the system to its surroundings and is primarily a measure of the dipolar coupling of the proton moments to their surroundings. The corresponding contrast is called T_1-contrast, and the typical T_1-contrast agent is the Gd chelate due to its large magnetic moment (due to seven unpaired electrons) and relatively long electron spin relaxation time (10^{-9} sec) in the commercial magnetic field strength, which

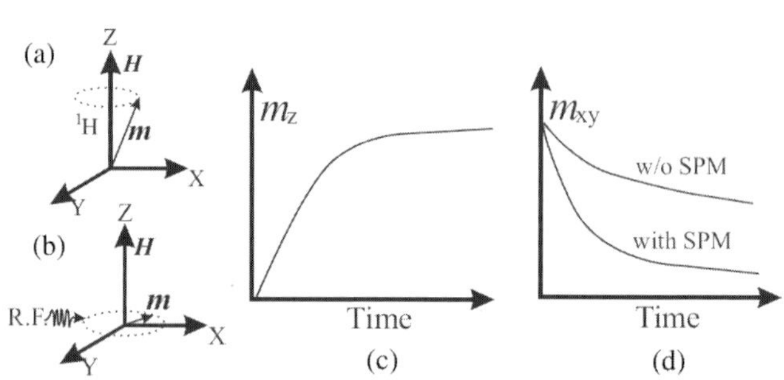

Fig. 4. Sketch of magnetic resonance for a large ensemble of protons with net magnetic moment $\boldsymbol{m}$ in the presence of an external magnetic field $\boldsymbol{H}$. (a) the net moment precesses around $\boldsymbol{H}$ at the characteristic Larmor frequency; (b) an external RF field is applied normally to $\boldsymbol{H}$, the resonantly exciting the moment precession into the plane normal to $\boldsymbol{H}$; the net moment relaxes with time along Z-axis (c) and the in-plane (d) when the RF field is removed, where the m_{xy} relaxes more rapidly to give higher T_2^* contrast when the superparamagnetic NPs are used (d).

principally accelerates T_1 relaxation, producing bright contrast.[19] The Gd chelates have been used as T_1 agent over decades but recently have been identified to possibly induce nephrogenic systemic fibrosis or nephrogenic fibrosing dermopathy in patients with kidney failure.[19] More recently, MnO NPs have paid much attention due to their higher T_1 contrast and lower toxicity.[20] For more detailed information on the T_1 contrast agent, refer to a recent review article by Sharma *et al.*[19]

The paramagnetic T_1 contrast agents obtain the excess energy after proton excitation, and thus lead to a faster recovery of m_z. In contrast, when the superparamagnetic NPs are introduced as T_2 agent, this results in increased field-fluctuations or local variations in the magnetic susceptibility of the system, and thus in a faster dephasing of the in-plane spins. The T_2 relaxation will be treated as the effective one T_2^*, as expressed by Eq. (23).[1,21] It can be known that introduction of the superparamagnetic NPs induces a less effective T_2 relaxation time, and thus a darker negative T_2 contrast. As a consequence, T_2 agents need to have a much larger magnetic moment, higher susceptibility than T_1 agents.

$$\frac{1}{T_2^*} = \frac{1}{T_2} + \frac{\gamma \Delta H}{2}. \tag{23}$$

3. Preparation and Characterization of Soft Magnetic Nanomaterials

Soft magnetic nanomaterials, due to their unique magnetic properties, can be used as MRI agents in diagnostics, as heat mediators in hyperthermia treatment, and as magnetic guidance in drug delivery application.[1–4] In the past decade, magnetic nanomaterials, including magnetic nanooxides, nanoalloys and core–shell nanostructures, have been extensively researched from basic research to clinical trials. Among them, the selection of magnetic nanomaterials for a specific application and in particular, their size and shape controls during chemical synthesis should be seriously considered since they are also related to magnetic properties. Magnetic NPs with a size range of a few to tens nanometers are particularly interesting due to their high specific surfaces, stability in aqueous or physiological environment, and infiltration ability from the blood vessels into tumor tissues. In particular, they possess high surface areas to facilitate both drug encapsulation and functionalization with a large number of targeting biomolecules for targeted cells and tissues.

As discussed in Section 2, it is essential to tune the magnetic properties of NPs and their sizes for a specific task. There have been many reports in the literature on the synthesis methods of magnetic nanooxides, nanoalloys and core–shell nanostructures.[22–54] In this section, the progress of chemical synthesis of these nanomaterials will be reviewed, in particular, our recent results in the past five years.

3.1 *Magnetic Nanooxides*

Magnetic nanooxides, mainly iron oxides such as maghemite (γ-Fe_2O_3),[22] magnetite (Fe_3O_4),[23] and spinel-type structure, such as MFe_2O_4 (M = Mn, Co, Zn)[24,25] have received much attention due to their good chemical stability in physiological environment and superior biocompatibility, and used as both heat mediators and MRI contrast agents. Among them, the maghemite and magnetite have relatively lower saturation magnetization than MFe_2O_4, ~76 emu/g for maghemite[22] and ~92 emu/g for magnetite.[23] Since there is a magnetic disorder layer on the surface of magnetic nanoparticle, the saturation magnetization drastically decreases with the decrease in size; for example, the M_s value of maghemite decreases to ~50 emu/g when the nanoparticle size is ~16 nm and becomes nonmagnetic one when the size is down to ~3 nm.[22] Therefore, it is becoming an important issue to synthesize magnetic nanooxides with a high M_s value. More recently, Lee *et al.* synthesized 12 nm-sized Ni ferrite particles with a M_s value of ~85 emu/g, higher M_s values were achieved by Co and Fe ferrites and the highest M_s value was for the 12 nm-sized $MnFe_2O_4$ NPs (~110 emu/g), which showed a better MRI T_2 contrast.[24]

Several methods have been developed to synthesize iron oxide NPs.[22–30] The simplest, cheapest and most environmentally-friendly procedure is the co-precipitation method, which involves the simultaneous precipitation of Fe^{2+} and Fe^{3+} ions in basic aqueous media.[23,26] The reaction temperature of the co-precipitation method is limited by the boiling point of water, and iron oxide NPs synthesized under these conditions exhibit usually a low degree of crystallinity and large polydispersity. In order to narrow the size distribution, an improved route is proposed by the water-in-oil microemulsion method.[27] However, this processing has quite a low yield, and the produced products have poor crystallinity. Hydrothermal synthesis technique is an alternative method for the preparation of highly crystalline iron oxide NPs, in which the aqueous medium containing iron salts is sealed in a Teflon container and heated above the boiling temperature of water. But the NPs prepared by this processing are usually polydispersive.[28] More recently, significant achievements have been obtained for the colloidal synthesis of nanocrystals in high-boiling point organic solvents, resulting in good crystallinity and fine control of size and size distribution of iron oxide NPs.[29] However, this route requires environmentally non-benign organic chemicals in the synthesis and thus is not regarded as the green approach. On the other hand, a post surface manipulation step is required in order to render the NPs to be water-soluble, biocompatible and target-specific due to the presence of a hydrophobic coating on the surface of the NPs.[2,29]

Considering the characteristics of the hydrothermal processing for good crystallinity of the iron oxides and as an environmentally benign route, we studied the effect of a capping agent with different capping functional groups on size of magnetite nanoparticle in order to improve the size distribution.[30] Three different capping agents, urea containing $-NH_2$, PEG 20k containing –OH and 3-aminopropanol containing both $-NH_2$ and –OH, were used and the reactant mixture was placed in a 120 ml Teflon-sealed autoclave and heated up to 200°C for 10 hr. Figure 5 shows the typical morphologies of three products, where the morphology is dependent on the added capping molecules.[30]

In the case of 3-aminopropanol, the average size of magnetite nanocrystals is about 10 nm, as shown in the TEM image (Fig. 5(a)). Figure 5(b) shows the SEM image of the magnetite obtained with PEG. The product is composed of two kinds of NPs with bimodal

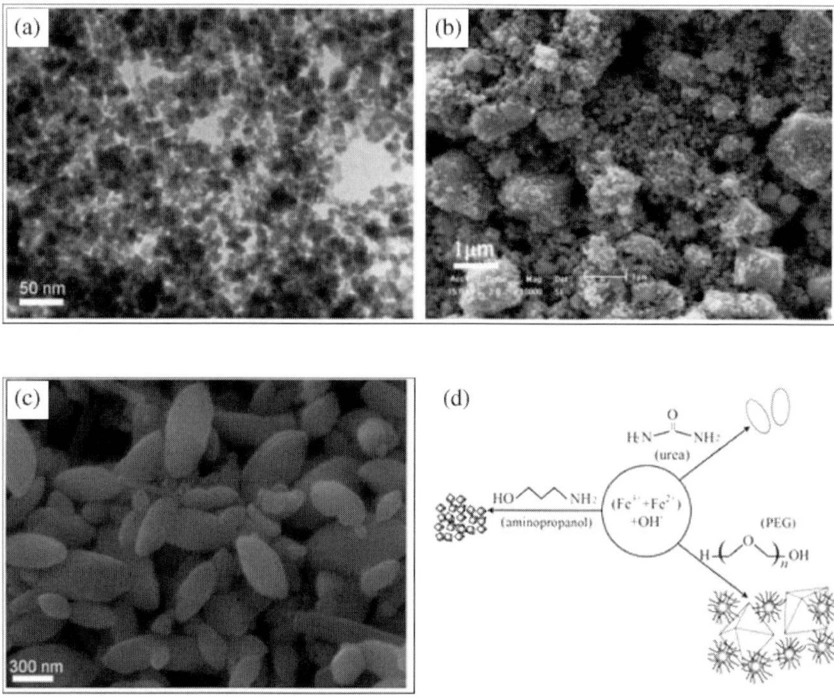

Fig. 5. Magnetite synthesized at 200°C for 10 hr with (a) 3-aminopropanol transmission electron microscope (TEM); (b) polyethylene glycol (PEG) scanning electron microscope (SEM); and (c) urea (SEM) as capping molecules; (d) Schematic illustration of the magnetite particle formation process. Figure reproduced with permission from Ref. 30. Copyright 2009 John Wiley and Sons.

size distributions: one is NPs with diameters below 75 nm; the other is tetrahedral and octahedral particles with an edge length of about 300 nm. The Fe_3O_4 particle is ellipsoidal with a large size of several hundred nanometers with urea (Fig. 5(c)). The morphology difference is expected to result from confined preferential crystal growth caused by added capping molecules and synthesis conditions. Figure 5(d) is a sketch of the magnetite particle formation process. For the detailed discussion, please refer to the literature.[30]

Figure 6(a) shows a typical XRD pattern of synthesized Fe_3O_4 products.[30] All the diffraction peaks can be indexed to the cubic-structured

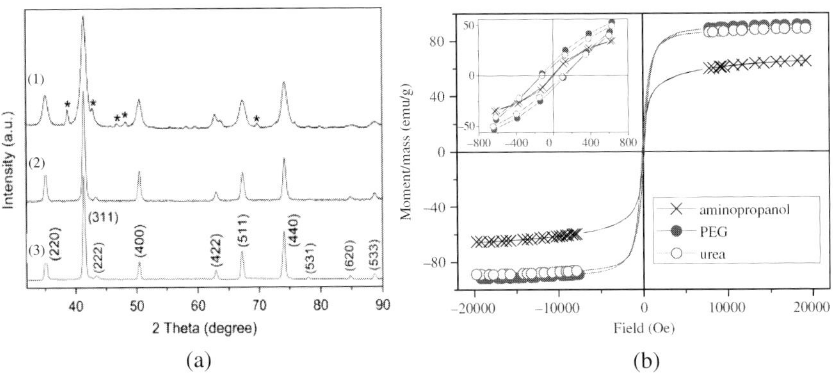

Fig. 6. X-ray diffraction patterns of magnetite synthesized with three surfactants: (a) (1) 3-aminopropanol; (2) polyethylene glycol (PEG); (3) urea. The small peaks marked by asterisks on the curve (1) are assigned to goethite. (b) Magnetic hysteresis loops of as-synthesized magnetite products. Inset: low field region. Figure reproduced with permission from Ref. 30. Copyright 2009 John Wiley and Sons.

magnetite. No obvious impurity phase can be detected from spectra of magnetite obtained with PEG or urea, but small peaks attributed to goethite can be found in the XRD spectrum of the NPs with the aminopropanol. It is noted that the XRD peaks by the aminopropanol are broader than that by PEG or urea as capping agent, suggesting that the crystal size of particles synthesized with 3-aminopropanol is evidently smaller than that synthesized with PEG or urea. The estimated crystal diameter from the width broadening of the XRD peaks with the Scherrer equation is 11.5 nm, which is in good agreement with TEM result. It should be noted that the primary crystal size values calculated from the Scherrer equation for magnetite synthesized with urea and PEG 20k are 28.8 nm and 23.7 nm, which are obviously smaller than that shown in SEM images, implying that the particles may be aggregated from primary particles.

Magnetic properties of the prepared Fe_3O_4 particles are shown in Fig. 6(b). The saturation magnetizations (M_s) of the powder synthesized with the PEG and urea are 91.7 and 89.1 emu/g, respectively, which are close to that of the corresponding bulk Fe_3O_4 (92 emu/g).[23]

The decrease of M_s (65.3 emu/g) of the Fe_3O_4 synthesized with aminopropanol is mostly attributed to the impurity phase of some goethite, and also the existence of the capping molecules on the surface, especially for the particle size of below 10 nm. The inset of Fig. 6(b) shows the coercivity of Fe_3O_4 prepared with aminopropanol approaches to 0 Oe, suggesting superparamagnetism at room temperature due to size effect. The superparamagnetic feature and still sufficiently high M_s (65.3 emu/g) as well as monodisperse characteristic in water suggest its potential application in hyperthermia therapy toward tumors and as MRI contrast agent. Furthermore, a high magnetic complex susceptibility is expected for this sub-10 nm Fe_3O_4 due to its high M_s and zero coercivity and would thus result in an enhanced hyperthermia effect. Actually, the saturation magnetization of this sub-10 nm Fe_3O_4 could be further increased by improving the purity of this phase, that is, to decrease or even remove the goethite via simply changing the initial Fe^{3+}/Fe^{2+} precursor concentrations. These results above suggest that the proper selection of surfactants is an effective way to control the size of magnetic nanooxide by the hydrothermal processing.

3.2 *Magnetic Nanometals and Nanoalloys*

Metallic magnetic NPs usually have higher magnetization than metal oxides, and are thus receiving increasing attention. Preparation of magnetic nanometals and nanoalloys is by either thermal decomposition[31,32] or reduction[33,34] in an organic solvent. Fatty acids, long chain amines, and thiols are generally used as surfactants to stabilize the as-prepared particles from oxidation and agglomeration and also to tune size and size distribution. Monodisperse ε-Co NPs with a controllable size of 3–17 nm have been prepared by rapid pyrolysis of dicobalt octacarbonyl ($Co_2(CO)_8$) in the presence of a surfactant mixture containing oleic acid, lauric acid, and trioctylphosphine.[31] By the same token, iron NPs with a tunable size of 6–19 nm and M_s over 132 emu/g have been obtained by using $Fe(CO)_5$ as precursor and oleic acid and oleylamine as capping agents with/without Pt as nuclei.[32]

An alternative way to prepare magnetic metallic or alloyed NPs with narrow size distribution is to use suitable surfactants and a strong reducing agent such as $NaBH_4$ or warm reducing agents such as alcohols to reduce the dissolved metallic salts in organic solvent, and the synthesis of FePt NPs are a typically successful example.[33] We have reported a series of magnetic nanoalloys, CoNi and NiFe by using the polyol method, where the propylene glycol (PG) or ethylene glycol (EG) has been used that plays triple roles as solvent, reducing agent, and capping agent, and Fe-, Co-, and Ni-salt are dissolved in PG or EG and heated at a high temperature.[34,35] An example is shown in Fig. 7; the synthesized $Ni_{80}Fe_{20}$ NPs are of Face-Centered Cubic (FCC) structure, having a mean size of ~440 nm with a standard deviation less than 10% and very homogenous composition in each nanoparticle.[34] In order to control their sizes down to tens of nanometers, the nuclei of Pt is introduced, similar to the preparation of Fe NPs.[32] Depending on the amount of Pt salt, the $Ni_{80}Fe_{20}$ NP size can be well tuned in the range of ~20–~440 nm.[34]

For the as-prepared NiFe NPs with various sizes, their coercivities decrease with decreasing size, which is different from the conventionally prepared ferromagnetic NPs by a similar processing.[36] After

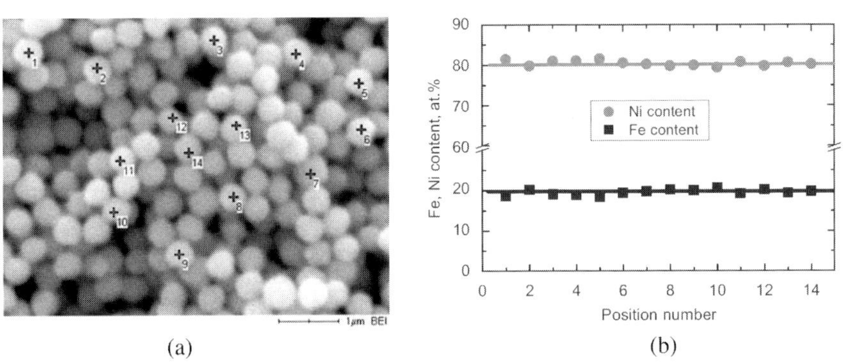

Fig. 7. SEM-EDX examination of the NiFe NPs: (a) SEM observation, and numbers are the detected points for the composition analysis; (b) the composition results at these points, showing the NiFe NPs contain homogeneous Ni and Fe contents inside of 20 at%Fe and 80 at%Ni. Figure reproduced with permission from Ref. 34. Copyright 2009 Elsevier.

annealing of these NiFe NPs in vacuum at 280°C for 5 hr, there is little change in their sizes. In contrast, their M_s values drastically increase and the coercive forces increase with decreasing NP size up to ~50 nm, and then decrease, as shown in Fig. 8. The size-dependent coercivity behavior can be interpreted by magnetic single domain theory, which indicates the critical size of single domain for $Ni_{80}Fe_{20}$ is ~50 nm. The difference in coercivity between as-prepared and annealed NiFe NPs with various sizes indicates a different surface effect. For a detailed discussion, please see the literature.[35] By the same token, a series of CoNi and FeNi alloy NPs with controllable compositions, sizes and size distribution can be prepared.[37]

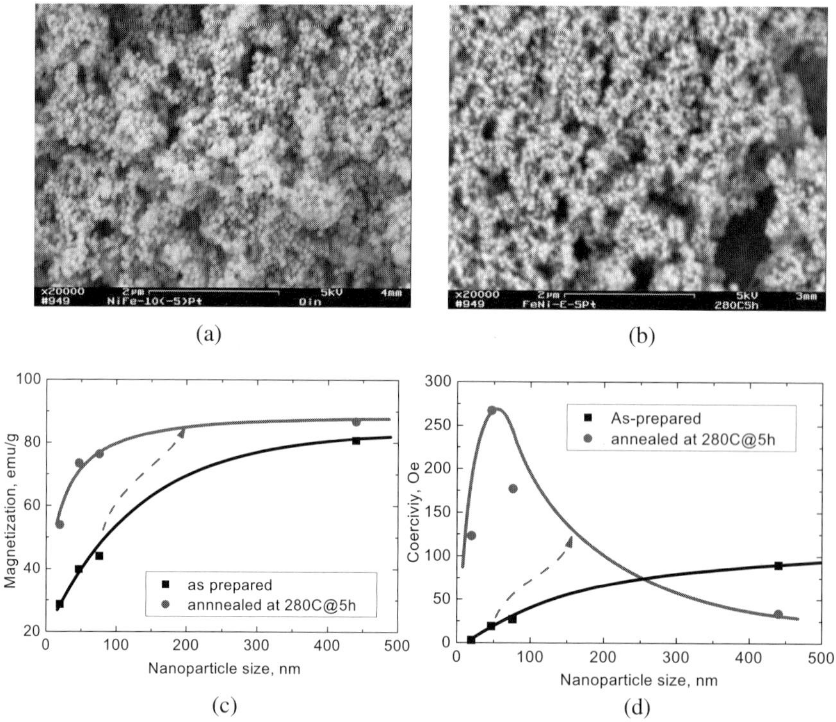

Fig. 8. The NiFe NPs with various sizes annealed at 280°C for 5hr: (a) SEM, as-prepared ~76 nm NPs; (b) SEM, annealed ~76 nm NPs; (c) saturation magnetization and (d) coercivity dependency of NiFe NPs before and after annealing.[34,35]

For many applications in nanomedicine, magnetic properties of NPs such as higher M_s and χ as well as lower H_c are required. However, the three magnetic properties are in trilemma in most cases. That is, the sizes of magnetic NPs are usually in the range of ten ~ tens of nanometers in order to keep their good monodispersion in physiological environment. For the magnetic crystalline NPs, in this case, their magnetic anisotropies are not so small that high H_c and thus low χ are usually resulted. For example, the magnetic anisotropy of $Ni_{80}Fe_{20}$ permalloy bulk is small, but their NPs have high coercity larger than 10 Oe and low χ when their sizes are larger than 10 nm, as shown in Fig. 8.[35] Under usual circumstances, the magnetic anisotropies of Fe, Co, Ni and their alloys are larger than that of permalloy. In contrast, the magnetic amorphous structure should be favorable to decrease magnetic anisotropy, and thus to decrease H_c and finally increase χ. On the other hand, element toxicity should be considered when one prepares magnetic NPs. For example, Co is considered to be a possible carcinogen to humans, and excesses can cause polycythemia, bone marrow hyperplasia, pancreatic failure or congestive heart failure and cardiomyopathy, and large doses interfere with iron absorption.[38] Ni, in particular, has been shown to be highly allergenic.[38]

Here, we propose the preparation of Fe-P-B amorphous NPs with high saturation magnetization, low coercivity and thus high susceptibility, in which all three elements have nutritional value to body. Fe-B-P NPs are prepared by wet chemical reduction at room temperature, using $FeCl_2$, NH_4Cl, sodium citrate and sodium hypophosphite in a basic aqueous solution. At pH 9, the prepared Fe-P-B NPs have a mean size of ~200 nm with M_s of 126.1 emu/g and H_c of 13.8 Oe. When a suitable amount of urea is introduced, the size can be down to ~100 nm and a lower coercivity ~5.0 Oe has been obtained while M_s has no change, as shown in Fig. 9. More interestingly, when an external magnetic field of 4000 Oe is used, the Fe-P-B nanobars with a diameter of ~200 nm and a length of several microns are obtained (Fig. 9(c)). All of these Fe-P-B NPs are of typical amorphous structure (Fig. 9(d)). The degraded magnetization of Fe-P-B NPs with a

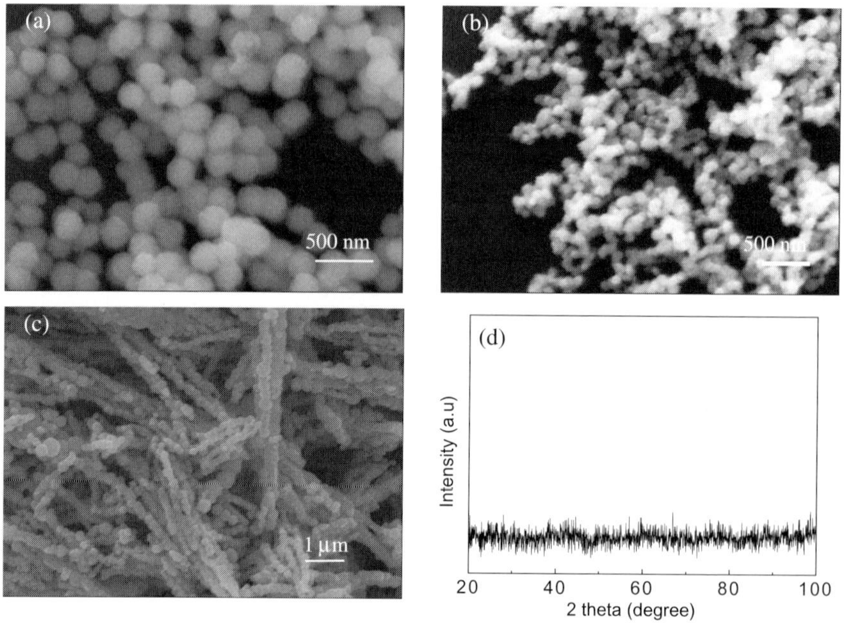

Fig. 9. Prepared Fe-P-B NPs (a) without any surfactant, (b) with urea as surfactant, (c) prepared Fe-P-B nanobars at a magnetic field of 4000 Oe, and (d) typical XRD pattern of these Fe-P-B nanomaterials.

mean size of ~250 nm prepared above have been evaluated by dispersing them in an acellular simulated body fluid at room temperature with a degraded rate of ~8 emu/g per day in the first week. Biodegradation behavior of these Fe-P-B nanomaterials suggests a new trend after they are used as medical diagnosis and treatment.

3.3 *Magnetic Core–Shell Nanostructures*

Magnetic NPs of iron, cobalt, nickel and their alloys show promise for practical nanomedicine applications. However, these NPs are chemically unstable in the ambient environment and easily oxidized, which limits their utility. To prevent the NPs from oxidation, one way is to coat the NPs with another inert layer, that is, making a core–shell nanostructure. The core–shell nanostructure is capable of maintaining favorable magnetic properties of magnetic cores while protecting the NPs from oxidation. Materials such as metal-oxide, inorganic compounds and noble metals are commonly used as the coater for magnetic NPs.

So far, much work has been reported on coating an Au shell on Fe NPs by wet chemical method.[39–42] However, if the Au shell is not densely coated, the Fe/Au core/shell nanostructure cannot prevent the oxidation in air or aqueous solution, and their saturation magnetizations degrade with time.[42] A successful approach has been proposed to prepare well-coated Fe/Au core–shell nanostructure by a laser ablation technique, and these core–shell NPs can be stable up to four months at room temperature.[43] Similar work has also been done on Co/Au, Co/Pd, Co/Pt, and Co/Cu core–shell nanostructures using the redox transmetalation reaction technique.[44]

An alternative way is to coat metallic oxide or SiO_2 on magnetic NPs.[45,46] Qiang *et al.* proposed creating $Fe@Fe_3O_4$ (or Fe_2O_3) core–shell nanoclusters using a nanocluster deposition system; these are superparamagnetic at room temperature for size less than 15 nm, and then become ferromagnetic when the cluster size increases. The specific magnetic moment of core–shell nanoclusters is size dependent and increases rapidly from ~80 emu/g at the cluster size of around 3 nm to over 200 emu/g up to the size of 100 nm.[45] Silica coating has the advantages of exceptional stability, especially in aqueous media and acting simultaneously as a dispersing agent for electrostatic colloids. Since the surfaces of magnetic cores are usually hydrophobic, their surfaces should be modified to be adopted to silica coating in aqueous solution, typically based on a microemulsion synthesis, where micelles or inverse micelles are used to confine and control the coating of silica on the core of the inner nanoparticle, and finally get water-soluble magnetic core–shell NPs.[46,47]

More recently, carbon coating is being paid increasing attention due to the formation of covalent bonds and the binding of functional groups, as is the case for the solid supports for Merrifield synthesis,[48] which cannot be well done on the shells of silica and other metal oxides due to hydrolysis.[49] Carbon coating has also good oxidation resistance in ambient environment.[49–52] Many approaches have been proposed so far to prepare magnetic-core/carbon-shell nanostructures, including the arc discharge method in an inert atmosphere,[50,51] flame spray pyrolysis of the precursor (e.g. cobalt carboxylate),[49] and more recently, hydrothermal method.[52] The major problem with the synthesis of carbon-encapsulated nanocrystals by using the conventional arc method is the production of large amounts of unwanted carbonaceous debris (graphite flakes, carbon nanotubes, amorphous carbon) and micron-sized metallic particles. The side products make the application of the carbon-encapsulated nanocrystals difficult.[53] Flame spray pyrolysis is a facile way to prepare large-scale (> 30 g/h) carbon-coated nanomagnets with high air and thermal stabilities. For example, the prepared carbon-coated Co NPs have a mean size of ~50 nm with a standard deviation of ~40%.[49] Lee *et al.* reported carbon-encapsulated FeCo core–shell NPs synthesized by a hydrothermal method, simply by mixing Fe/Co with sucrose as a carbon source and then by annealing at high temperature. They got the carbon coated FeCo NPs with a mean size of 7 nm which showed a good monodispersivity in water and high T_2 contrast in MRI images.[52]

Here we summarize our proposed approaches to synthesize carbon coated magnetic NPs. One approach is to revise the arc discharge method in ethanol solution with the assistance of an ultrasonic vibrating magnetic anode rather than in an inert atmosphere. This method provides a much lower input power (0.08~0.16 kW) and no need for any vacuum equipment. Figure 10(a) shows the sketch of the experimental apparatus. The as-prepared carbon coated cobalt NPs have a mean size of ~8.3 nm and 75% of the NPs have a diameter less than 10 nm, as shown in Fig. 10(b). The prepared NPs contain some larger cobalt carbides, and the as-prepared ones have a M_s value of ~24 emu/g and H_c value of 64.6 Oe. When the as-prepared sample is annealed at 600°C in an inert atmosphere, cobalt carbide undergoes decomposition and M_s increases up to ~51 emu/g, together with an increased size of ~16.8 nm.[53]

The other approach to preparing carbon coated cobalt NPs is to mix glucose and cobalt acetate aqueous solution in a suitable molar ratio, typically 5:1, and then dry at 70°C, finally annealing at 700°C

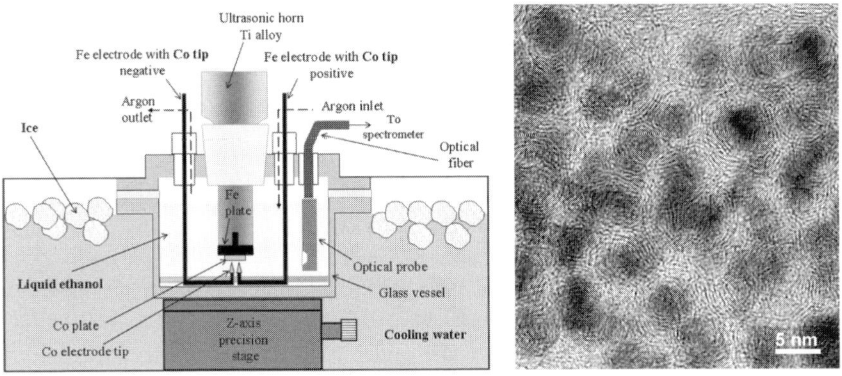

(a) (b)

Fig. 10. Sketch of the experimental apparatus (a) and the as-prepared carbon coated cobalt NPs with a mean size of ~8.3 nm (b). (Figure reproduced with permission from Ref. 53. Copyright 2007 Elsevier.)

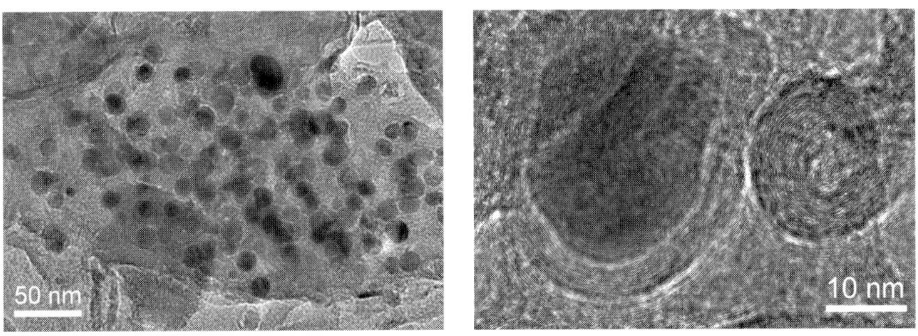

(a) (b)

Fig. 11. TEM image (a) of carbon coated Co NPs, prepared by annealing the mixture of glucose and cobalt acetate by a molar ratio of 5:1 at 700°C for 1 hr; (b) higher magnification.

for one hour under N_2 atmosphere. The obtained Co NPs have a diameter of typically 10–20 nm with a carbon shell of ~2–3 nm in thickness, as shown in Fig. 11. The M_s value is ~50 emu/g, but a much high coercivity of ~500 Oe is achieved due to the presence of HCP structured cobalt. When the annealing temperature is lower than 700°C and/or the glucose amount is not sufficient, the amount of CoO will be present. By designing the appropriate annealing processing, the Co/C core/shell nanostructures with much higher magnetization could be expected to further improve in the future.

4. Surface Modification of NP Surfaces Towards Potential Medical Applications

It is the first step to prepare magnetic NPs with desired magnetic properties for a specific medical or biological application. However, the next and most important issue, which needs to be accomplished after their preparation, is to provide the NPs with a sufficient degree of stability in aqueous or physiological environments, and a specific targeting ability.[1–4] In most cases, the NP surfaces of magnetic metals and alloys are hydrophobic, which results in particle aggregation and precipitation from aqueous solution. If applied directly to a living organism, the aggregation of these magnetic NPs would pose an immediate threat to health by clogging up small blood vessels and obstructing a free blood circulation. On the other hand, even though NP agglomeration could be prevented, unprotected inorganic NPs are prone to opsonization, a defense process initiated by the innate immune system, in which certain serum proteins attach to the nanoparticle surface and initiate phagocytosis.[54–56] Therefore, it is necessary to modify the nanoparticle surface with some specific biomolecules to both assure a high colloidal stability in physiological environments and prevent opsonization by serum proteins.

How to immobilize specific biomolecules on magnetic NPs is becoming an essential issue, which is important for diverse medical applications, such as drug delivery, bio-sensing, tumor cell targeting, disease diagnosis, etc.[57,58] For example, for cancer treatment by hyperthermia, the key issue is the fact that cancer cells overexpress certain kinds of receptors compared to healthy cells. Therefore, magnetic NPs carrying ligands, such as proteins, peptides and aptamers, which will specifically bind to these receptors, will preferably accumulate in tumor tissue and will eventually be taken up by the cancer cells.[59,60] Excellent work has been done so far on the immobilization of various biomolecules on magnetic NPs by covalent bonding using thiol chemistry and carbodiimide activation by random amide bond formation.[57–65]

However, these immobilization processes are time consuming, and a facile way to accomplish this objective is still a great challenge. Considering the experimental evidence by Lee *et al.*, who reported that NiO shell in the Ni/NiO core/shell NPs have a high affinity for polyhistidine,[66] we selected $Ni_{80}Fe_{20}$ permalloy NPs (~71 nm in diameter) as magnetic carriers by immobilizing the immunoglobulin G (IgG) protein, since a thin layer of NiO spontaneously forms after the permalloy NPs are prepared,[67] while providing superior magnetic properties.[34,35]

Human serum IgG is biotinylated with sulfo-NHS-LC Biotin on ice for 2 hr, and then biotinylated IgG is purified. Thereafter, His-tagged streptavidin is immobilized on $Ni_{80}Fe_{20}$ NPs, and finally treated with biotinylated human IgG. The sketch of the procedure of immobilizating antibody onto the surface of $Ni_{80}Fe_{20}$ NPs is shown in Fig. 12(a). The $Ni_{80}Fe_{20}$ NPs functionalized with His-tagged streptavidin are confirmed by the attenuated total reflection-FTIR (Fig. 12(b)), where three new peaks at 1642, 1535, and 1456 cm^{-1} can be assigned as the characteristic peaks for protein amide I, amide II, and amide III, respectively.[62,68] The peak at 3019 cm^{-1} is one of the characteristic peaks of histidine.[69] The presence of these new peaks in Fig. 12(b) indicates that His-tagged streptavidin is successfully anchored on the $Ni_{80}Fe_{20}$ NPs.

In order to exploit diverse medical applications in the future, human serum IgG is chosen as a model protein. Binding of IgG to $Ni_{80}Fe_{20}$ NPs was analyzed with denaturing protein gel electrophoresis. To denature proteins, all samples are treated with lithium dodecyl sulfate (LDS) sample buffer and heated at 70°C for 10 min. Using this technique, any proteins bound to the NPs are expected to dissociate and enter the gel. The migration of proteins depends on the mass/charge ratio of denatured proteins. A series of samples were prepared, as detailed description in Fig. 12(c). A pronounced band appeared for the His-tagged streptavidin bound $Ni_{80}Fe_{20}$ NPs in Lane 3 at exactly the same mobility as the standard His-tagged streptavidin (Lane 4). This result indicates that His-tagged streptavidin is attached on

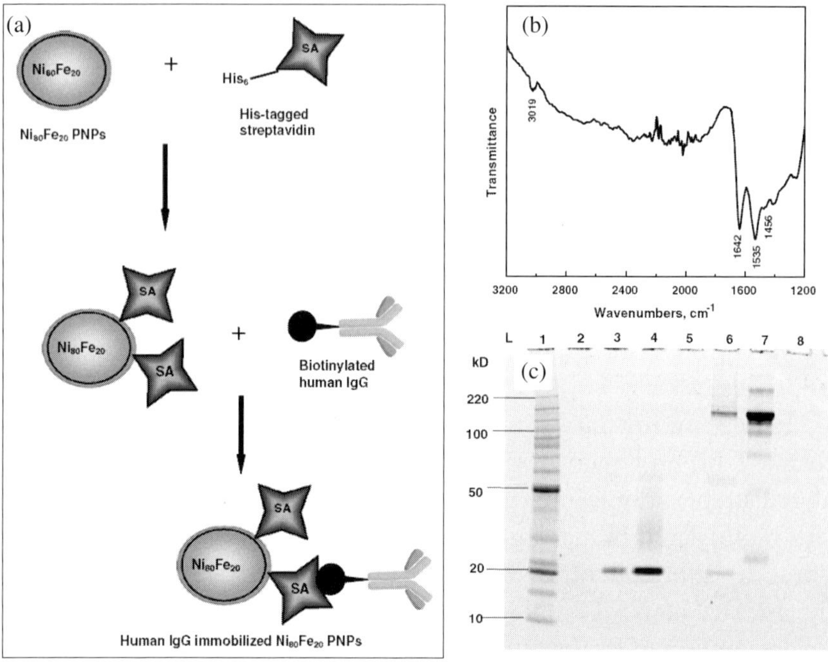

Fig. 12. Coupling human serum IgG onto $Ni_{80}Fe_{20}$ permalloy NPs via streptavidin–biotin bridge: (a) sketch of the modification steps; (b) ATR-FTIR spectra of His-tagged streptavidin functionalized $Ni_{80}Fe_{20}$ permalloy NPs; (c) Immobilization of human IgG on $Ni_{80}Fe_{20}$ permalloy NPs. Lane 1: molecular weight marker (BenchMarkTM), Lane 2: supernatant without $Ni_{80}Fe_{20}$ NPs after His-tagged streptavidin addition, Lane 3: $Ni_{80}Fe_{20}$ permalloy NPs after incubation with His-tagged streptavidin, Lane 4: His-tagged streptavidin standard, Lane 5: supernatant without $Ni_{80}Fe_{20}$ permalloy NPs after His-tagged streptavidin and biotinylated human IgG addition, Lane 6: $Ni_{80}Fe_{20}$ permalloy NPs after incubation with His-tagged streptavidin and biotinylated human IgG, Lane 7: biotinylated human IgG standard, Lane 8: free $Ni_{80}Fe_{20}$ permalloy NPs (without His-tagged streptavidin as a control) after incubation with biotinylated human IgG. Figure reproduced with permission from Ref. 67. Copyright 2011 Springer.

$Ni_{80}Fe_{20}$ NPs and then released from the NPs after heating with denaturing reagent. On the other hand, the supernatant of the mixture without $Ni_{80}Fe_{20}$ NPs (control 1) is clear after intensive washing and no corresponding band is observed, as shown in Lane 2. Two major bands in Lane 6 assigned as His-tagged streptavidin and biotinylated human IgG (same position as standard in Lane 7), respectively, are observed for human IgG-functionalized $Ni_{80}Fe_{20}$ NPs, whereas no band appeared for the supernatant without $Ni_{80}Fe_{20}$ NPs (control 2) in Lane 5. And there is no non-specific protein adsorption on to $Ni_{80}Fe_{20}$ NPs (Lane 8). These results suggest that both streptavidin and IgG are strongly bound on $Ni_{80}Fe_{20}$ NPs by high affinity interactions and the supernatant did not contain any protein after washing.[67] This approach may open a new avenue for anchoring proteins and antibodies to $Ni_{80}Fe_{20}$ NPs or other magnetic NPs by an oriented assembly of proteins onto them. The described bio-functionalized $Ni_{80}Fe_{20}$ permalloy NPs system can also find potential applications in clinical diagnostics for highly sensitive and multiplexed detection of protein or nucleic acid biomarkers associated with specific diseases.

5. Conclusions and Perspective Remarks

Soft magnetic NPs are important platforms for multimodal bio-imaging, bio-sensing and targeted drug delivery since their unique magnetic properties provide specific functions. Sufficiently high saturation magnetization of these NPs gives high T_2-contrast of MRI and also high heating efficiency of hyperthermia. High magnetic susceptibility of superparamagnetic NPs endues facile triggering by external magnetic field and also enhanced hyperthermia. And quite low remanent magnetization together with suitable bio-surfactants encourages less aggregation in aqueous and physiological environments. The recent developments in the synthesis and functionalization of magnetic NPs permit the use of those particles in diagnosis and therapy.

The next generation of soft magnetic NPs is expected to contain multi-functionalities, such as executing medical diagnosis and therapy together using the same NPs, as termed "theragnostics." It is necessary to tune magnetic properties by designing suitable NPs including composition, structure, size, size distribution, etc. On the other hand, little attention has been paid so far to the degradation behavior of magnetic NPs after their administration. It will be more interesting and feasible to have these magnetic NPs bio-degraded in physiological environment after diagnosis and/or therapy. In this case, biocompatible elements should be considered when the magnetic NPs are designed to tailor their magnetic properties. The other important issue is to systematically evaluate the cytotoxicity of magnetic NPs in a physiological environment. Finally, the bio-chemical and/or bio-physical mechanism should be addressed on uptake or infiltration behaviors of these magnetic NPs through biomembranes of cells and tumor tissues, and their leaking behavior from blood vessels. Despite these numerous challenges, soft magnetic NPs are considered to be the most promising tools to achieve the desired sensitivity and efficacy required for future medical diagnostics and therapeutics, which may lead to an exciting breakthrough in nanomedicine.

Acknowledgments

The authors gratefully acknowledge the contributions from all researchers whose work has been described in this chapter, and the permissions from Elsevier, John Wiley and Sons and Springer to reproduce the related figures. G. W. Qin thanks the Chinese Ministry of Education for financial support (No. NCET-10-0272 and No. N100702001). This work was in part supported by the Innovation Projects Fund National and International Research Alliances Program of Queensland Government, Australia.

References

1. Pankhurst QA, Connolly J, Jones SK, Dobson J. (2003) Applications of magnetic nanoparticles in biomedicine. *J Phys D: Appl Phys* **36:** R167–R181.
2. Figuerola A, Corato RD, Manna L, Pellegrino T. (2010) From iron oxide nanoparticles towards advanced iron-based inorganic materials designed for biomedical applications. *Pharm Res* **62:** 126–143.
3. Schladt TD, Schneider K, Schild H, Tremel W. (2011) Synthesis and bio-functionalization of magnetic nanoparticles for medical diagnosis and treatment. *Dalton Trans* **40:** 6315–6343.
4. Latham AH, Williams ME. (2008) Controlling transport and chemical functionality of magnetic nanoparticles. *Acc Chem Res* **41:** 411–420.

5. Yuan DF, Ma XL. (1999) *Magnetic Physics*, Electronic Industry Press, Beijing.
6. Néel L. (1949). Théorie du traînage magnétique des ferromagnétiques en grains fins avec applications aux terres cuites. *Ann Géophys* **5:** 99–136.
7. http://en.wikipedia.org/wiki/Superparamagnetism.
8. Du YW, Xu MX, Wu J, *et al.* (1992) Magnetic properties of ultrafine nickel particles. *Acta Phys Sin* **41(1):** 149–154.
9. Gu WL. (2008) Critical size of particles in superparamagnetic magnetic agents keeping stable and superparamagnetism in magnetic fields. *Trans Beijing Instit Tech* **28(9):** 806–808.
10. Bandow S, Kimura K, Kon-no K, Kitahara A. (1987) Magnetic properties of magnetite ultrafine particles prepared by W/O microemulsion method. *Jpn J Appl Phys* **26(5):** 713–717.
11. Gilchrist RK, Medal R, Shorey WD, *et al.* (1957) Selective inductive heating of lymph nodes. *Ann Surg* **146:** 596–606.
12. Rand RW, Snow HD, Elliott DG, Snyder M. (1981) Thermomagnetic surgery for cancer. *Appl Biochem Biotechnol* **6:** 265–272.
13. Borrelli NF, Luderer AA, Panzarino JN. (1984) Hysteresis heating for the treatment of tumours. *Phys Med Biol* **29:** 487–494.
14. Chan DCF, Kirpotin DB, Bunn PA Jr. (1993) Synthesis and evaluation of colloidal magnetic iron oxides for the site-specific radiofrequency-induced hyperthermia of cancer. *J Magn Magn Mater* **122:** 374–378.
15. Moroz P, Jones SK, Winter J, Gray BN. (2001) Targeting liver tumors with hyperthermia: ferromagnetic embolization in a rabbit liver tumor model. *J Surg Oncol* **78:** 22–29.
16. Jones SK, Winter JW, Gray BN. (2002) Treatment of experimental rabbit liver tumours by selectively targeted hyperthermia. *Int J Hyperthermia* **18:** 117–128.
17. Atkinson WJ, Brezovich IA, Chakraborty DP. (1984) Usable frequencies in hyperthermia with thermal seeds. *IEEE Trans Biomed Eng* **BME 31:** 70–75.
18. Rosensweig RE. (2002) Heating magnetic fluid with alternating magnetic field. *J Magn Magn Mater* **252:** 370–374.
19. Sharma P, Brown SC, Walter G, *et al.* (2007) Gd nanoparticulates: from magnetic resonance imaging to neutron capture therapy. *Adv Powder Technol* **18:** 663–698.
20. Schladt TD, Schneider K, Shukoor MI, *et al.* (2010) Highly soluble multifunctional MnO nanoparticles for simultaneous optical and MRI imaging and cancer treatment using photodynamic therapy. *J Mater Chem* **20(38):** 8297–8304.
21. Browne M, Semelka RC. (1999) *MRI: Basic Principles and Applications*, Wiley Press, New York.
22. Millan A, Urtizberea A, Silva NJO, *et al.* (2007) Surface effects in maghemite nanoparticles. *J Magn Magn Mater* **312:** L5–L9.
23. Kang YS, Risbud S, Rabolt JF, Stroeve P. (1996) Synthesis and characterization of nanometer-size Fe_3O_4 and gamma-Fe_2O_3 particles. *Chem Mater* **8:** 2209–2211.

24. Lee JH, Huh YM, Jun Y, *et al.* (2007) Artificially engineered magnetic nanoparticles for ultra-sensitive molecular imaging. *Nat Med* **13:** 95–99.
25. Kim DH, Zeng HD, Ng TC, Brazel CS. (2009) T-1 and T-2 relaxivities of succimercoated MFe_2O_4 ($M=Mn^{2+}$, Fe^{2+} and Co^{2+}) inverse spinel ferrites for potential use as phase-contrast agents in medical MRI. *J Magn Magn Mater* **321:** 3899–3904.
26. Ricketts CR, Cox JSG, Fitzmaurice C, Moss GF. (1965) The iron dextran complex. *Nature* **208:** 237–239.
27. Munshi N, De TK, Maitra A. (1997) Size modulation of polymeric nanoparticles under controlled dynamics of microemulsion droplets. *J Colloid Interface Sci* **190:** 387–391.
28. Wang X, Zhuang J, Peng Q, Li YD. (2005) A general strategy for nanocrystal synthesis. *Nature* **437:** 121–124.
29. Park J, An KJ, Hwang YS, *et al.* (2004) Ultra-large-scale syntheses of monodisperse nanocrystals. *Nat Mater* **3:** 891–895.
30. Li S, Qin GW, Pei WL, *et al.* (2009) Capping groups induced size and shape evolution of magnetite particles under hydrothermal condition and their magnetic properties. *J Am Ceram Soc* **92:** 631–635.
31. Puntes VF, Krishnan KM, Alivisatos P. (2001) Synthesis, self-assembly, and magnetic behavior of a two-dimensional superlattice of single-crystal epsilon-Co nanoparticles. *Appl Phys Lett* **78:** 2187–2189.
32. Farrell D, Majetich SA, Wilcoxon JP. (2003) Preparation and characterization of monodisperse Fe nanoparticles. *J Phys Chem B* **107:** 11022–1130.
33. Sun SH, Murray CB, Weller D, *et al.* (2000) Monodisperse FePt nanoparticles and ferromagnetic FePt nanocrystal superlattices. *Science* **287:** 1989–1992.
34. Qin GW, Pei WL, Ren YP, *et al.* (2009) $Ni_{80}Fe_{20}$ permalloy nanoparticles: wet chemical preparation, size control and their dynamic permeability characteristics when composited with Fe micron particles. *J Magn Magn Mater* **321:** 4057–4062.
35. Qin GW, Pei WL, Ren YP, *et al.* (2010) Effect of annealing on magnetic properties of $Ni_{80}Fe_{20}$ permalloy nanoparticles with various sizes prepared by polyol method. *Proc IEEE Nanoeletronics Conf* 1116–1117.
36. Toneguzzo P, Acher O, Viau G, *et al.* (1999) Static and dynamic magnetic properties of fine CoNi and FeCoNi particles synthesized by the polyol process. *IEEE Trans Magn* **35:** 3469–3471.
37. Qin GW, Pei WL, Yao C, *et al.* (2010) Formation mechanism, size control and magnetic properties of CoNi nanoparticles prepared by polyol method. *J Functional Mater* **41:** 131–134. (in Chinese).
38. http://corrosion-doctors.org/Elements-Toxic/Elements.htm
39. Lin J, Zhou WL, Kumbhar A, *et al.* (2001) Gold-coated iron (Fe@Au) nanoparticles: synthesis, characterization, and magnetic field induced self-assembly. *J Solid State Chem* **159:** 26–31.

40. Paulus PM, Bönnemann H, van der Kraan AM, *et al.* (1999) Magnetic properties of nanosized transition metal colloids: the influence of noble metal coating. *Eur Phys J D* **9:** 501–504.
41. Cho SJ, Kauzlarich SM, Olamit J, *et al.* (2004) Characterization and magnetic properties of core–shell structured Fe-Au nanoparticles. *J Appl Phys* **95:** 6803–6806.
42. Cho SJ, Idrobo JC, Olamit J, *et al.* (2005) Growth mechanisms and oxidation resistance of gold-coated iron nanoparticles. *Chem Mater* **17:** 3181–3186.
43. Zhang J, Post M, Veres T, *et al.* (2006) Laser-assisted synthesis of superparamagnetic Fe@Au core–shell nanoparticles. *J Phys Chem B* **110:** 7122–7128.
44. Lee WR, Kim MG, Choi JR, *et al.* (2005) Redoxtransmetalation process as a generalized synthetic strategy for core–shell magnetic nanoparticles. *J Am Chem Soc* **127:** 16090–16097.
45. Qiang Y, Antony J, Sharma A, *et al.* (2006) Iron/iron oxide core-shell nanoclusters for biomedical applications. *J Nanopart Res* **8:** 489–496.
46. Chang SY, Liu L, Asher SA. (1994) Creation of templated complex topological morphologies in colloidal silica. *J Am Chem Soc* **116:** 6745–6747.
47. Yi DK, Selvan ST, Lee SS, *et al.* (2005) Silica-coated nanocomposites of magnetic nanoparticles and quantum dots. *J Am Chem Soc* **127:** 4990–4991.
48. Merrifield RB. (1963) Solid phase peptide pynthesis: I. The synthesis of a tetrapeptide. *J Am Chem Soc* **85:** 2149–2154.
49. Grass RN, Athanassiou EK, Stark WJ. (2007) Covalently functionalized cobalt nanoparticles as a platform for magnetic separations in organic synthesis. *Angew Chem Int Ed* **46:** 4909–4912.
50. McHenry ME, Majetich SA, Artman JO, *et al.* (1994) Superparamagnetism in carbon-coated Co particles produced by the Kratschmer carbon arc process. *Phys Rev B* **49:** 11358–11363.
51. Saito Y, Masuda M. (1995) Crystallographic structure and magnetic properties of Co fine particles encaged in carbon nanocapsules. *Jpn J Appl Phys* **34:** 5594–5598.
52. Lee SJ, Cho JH, Lee C, *et al.* (2011) Synthesis of highly magnetic graphite-encapsulated FeCo nanoparticles using a hydrothermal process. *Nanotechnology* **22:** 375603.
53. Sergiienko R, Shibata E, Zentaro A, *et al.* (2007) Formation and characterization of graphite-encapsulated cobalt nanoparticles synthesized by electric discharge in an ultrasonic cavitation field of liquid ethanol. *Acta Mater* **55:** 3671–3680.
54. Cedervall T, Lynch I, Lindman S, *et al.* (2007) Understanding the nanoparticle protein corona using methods to quantify exchange rates and affinities of proteins for nanoparticles. *Proc Natl Acad Sci USA* **104:** 2050–2055.
55. Bruchez M, Moronne M, Gin P, *et al.* (1998) Semiconductor nanocrystals as fluorescent biological labels. *Science* **281:** 2013–2016.

56. Chan WCW, Nie S. (1998) Quantum dot bioconjugates for ultrasensitive nonisotopic detection. *Science* **281:** 2016–2018.
57. Ma YH, Manolache S, Denes FS, *et al.* (2004) Plasma synthesis of carbon magnetic nanoparticles and immobilization of doxorubicin for targeted drug delivery. *J Biomater Sci Polym Ed* **15:** 1033–1049.
58. Nidumolu BG, Urbina MC, Hormes J, *et al.* (2006) Functionalization of gold and glass surfaces with magnetic nanoparticles using biomolecular interactions. *Biotechnol Prog* **22:** 91–95.
59. Xie J, Xu CJ, Xu ZC, *et al.* (2006) Linking hydrophilic macromolecules to monodisperse magnetite (Fe_3O_4) nanoparticles via trichloro-s-triazine. *Chem Mater* **18:** 5401–5403.
60. Weissleder R, Kelly K, Sun EY, *et al.* (2005) Cell-specific targeting of nanoparticles by multivalent attachment of small molecules. *Nat Biotechnol* **23:** 1418–1423.
61. Sestier C, Da-Silva MF, Sabolovic D, *et al.* (1998) Surface modification of supermagnetic nanoparticles (Ferrofluid) studies with particle electrophoresis: application to the specific targeting of cells. *Electrophoresis* **19:** 1220–1226.
62. Gan ZF, Jiang JS, Yang Y, *et al.* (2008) Immobilization of homing peptide on magnetic nanoparticles and its specificity *in vitro*. *J Biomed Mater Res A* **84A:** 10–18.
63. Lu AH, Salabas EL, Schuth F. (2007) Magnetic nanoparticles: synthesis, protection, functionalization, and application. *Angew Chem Int Ed* **46:** 1222–1244.
64. Dave SR, Gao XH. (2009) Monodisperse magnetic nanoparticles for biodetection, imaging, and drug delivery: a versatile and evolving technology. *Wiley Interdiscip Rev Nanomed Nanotechnol* **1:** 583–609.
65. Veiseh O, Gunn JW, Zhang M. (2010) Design and fabrication of magnetic nanoparticles for targeted drug delivery and imaging. *Adv Drug Deliv Rev* **62:** 284–304.
66. Lee IS, Lee N, Park J, *et al.* (2006) Ni/NiO core/shell nanoparticles for selective binding and magnetic separation of histidine-tagged proteins. *J Am Chem Soc* **128:** 10658–10659.
67. Qin GW, Darain F, Wang H, Dimitrov K. (2011) Surface modification of permalloy ($Ni_{80}Fe_{20}$) nanoparticles for biomedical applications. *J Nanopart Res* **13:** 45–51.
68. Liu Z, Galli F, Janssen KGH, *et al.* (2010) Stable single-walled carbon nanotube-streptavidin complex for biorecognition. *J Phys Chem C* **114:** 4345–4352.
69. Yang M, Yang Y, Qu F, *et al.* (2006) Attachment of nickel hexacyanoferrates nanoparticles on carbon nanotubes: preparation, characterization and bioapplication. *Anal Chim Acta* **571:** 211–217.

Chapter 14

GoldMag Composite Particles and Their Applications in Nanomedicine

Mingli Peng*,†, Xu Chao*,‡, Chao Chen*,‡ and Yali Cui*,‡

Introduction

Magnetic composite nanoparticles consisting of a magnetic core (Fe_3O_4, γ-Fe_2O_3) and other matrices (such as polymers, silica, and metal) have been thoroughly studied based on their special chemical and physical properties and potential applications in biomedical fields.[1–4] In the last 10 years, the gold component was introduced as a functional material to synthesize magnetic nanoparticles, and this new kind of composite particles has received great attention due to its responsiveness to an external magnetic field, chemical stability, biocompatibility, optical properties, as well as reactivity with thiol compounds.[5–10] These nanocomposites have been designed and fabricated in different structures of core/shell, dumbbell-like, assembled, and multi-component complicated by different methods (Fig. 1).

* National Engineering Research Center for Miniaturized Detection Systems, Xi'an, 710069, China.
† The College of Chemistry and Material, Northwest University, Xi'an, 710069, China.
‡ The College of Life Sciences, Northwest University, Xi'an, 710069, China.

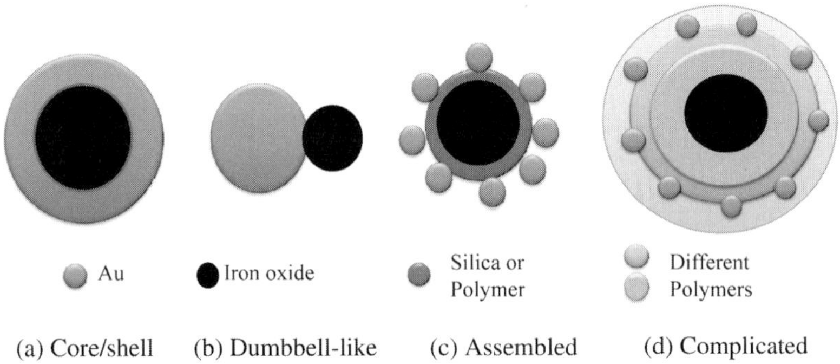

Fig. 1. Gold magnetic composite particles with different structures.

The combination of magnetic nanoparticles and Au in one composite form a unique multifunctional material, and we named them GoldMag particles in 2005.[11] Based on these components, silica or polymer were also used to integrate to form complicated GoldMag composite particles. In this chapter, we focus on the synthesis, characterization and the applications of GoldMag particles in nanomedicine.

1. Synthesis of GoldMag Nanoparticles

We firstly reported the core/shell GoldMag particles in 2001;[5] the composite particles were synthesized by reduction of Au^{3+} onto the surfaces of Fe_3O_4 seeds using hydroxylamine as a reducing agent in aqueous solution. The synthesis method and its property on GoldMag particles with core/shell structure have become a hot topic since 2004.[12,13] Approaches were developed to obtain highly-monodisperse Fe_2O_3/Au or Fe_3O_4/Au nanoparticles in organic phase. One method involved an initial synthesis of Fe_2O_3 and Fe_3O_4 nanoparticles as seeds and a subsequent reduction of $Au(OOCCH_3)_3$ using 1,2-hexadecanediol as reductive, pheyl ether as solvent, oleic acid and oleylamine as capping reagent at 180–190°C. The Fe_2O_3 and Fe_3O_4 nanoparticle seeds were synthesized by thermal deposition of $Fe(CO)_5$ or $Fe(acac)_3$ in high boiling point solvents of ether capped with lauric acid (LA) or oleic acid. The size of Fe-oxide@Au nanoparticles varied in a core size range of 5–15 nm with a shell thickness range of 0.5–2.0 nm by controlling the reaction time and the concentration of Au precursor.[7,14] Another method was thermally activated hetero-interparticle coalescence between gold and magnetic nanoparticles under an encapsulating environment. In the presence of tetraoctylammonium bromide (TOABr), the decanethiols (DTs) capped Au nanoparticles desorbed, coalesced, and re-encapsulated on oleylamine/or oleic acid capped Fe-oxide nanoparticles at elevated temperatures.[15,16] However, the monodispersed GoldMag particles were required as water-based in the biomedical field. Therefore, the phase transfers have been carried out to achieve size control and viability in an aqueous environment. Sun reported the oil–water phase method starts with room temperature coating Au on the surface of Fe_3O_4 nanoparticles by reducing $HAuCl_4$ in a chloroform solution of oleylamine. The Au-coated Fe_3O_4 nanoparticles were then transferred into water by mixing them with sodium citrate and cetyltrimethylammonium bromide (CTAB). The water-soluble core/shell Fe_3O_4/Au nanoparticles served as seeds for the formation of Fe_3O_4/Au nanoparticles with thicker Au coating by simply adding more in the reducing condition.[17] The Li group[18] reported a method using water-based commercial magnetic nanoparticles (EMG304), the silica layer was coated by hydrolysis of TEOS in mixture of water and ethanol, then phosphonium chloride (THPC) gold nanocrystals were attached onto the silica surface to form the THPC gold nanoseeds. Finally, the gold nanoshells were prepared by reduction of K-gold solution with formaldehyde. Another two-step reduced method to synthesize a multi-layer of GoldMag composite particles was developed by the Meyerhoff group.[19] The procedure to coat gold layer on magnetic microparticles started by immersing Dynabead M-500 particles in Sn(II) solution, and oxidized by Ag(I). Then, the beads are immersed in a $Na_3Au(SO_3)_2$ solution, and the Au^+ was reduced to Au^o by formaldehyde.

The dumbbell-like GoldMag nanoparticles were prepared via the decomposition of iron pentacarbonyl, $Fe(CO)_5$, over the surface of the Au nanoseeds followed by oxidation under air in 1-octadecene solvent. The Au nanoseeds can be either synthesized *in situ* by injecting $HAuCl_4$ solution into the reaction mixture or premade in the presence of oleylamine. The dumbbell is formed through epitaxial growth of iron oxide on the Au seeds, and the growth can be affected by the polarity of the solvent, as the use of diphenyl ether results in flower-like Au-Fe_3O_4 nanoparticles.[20,21] A strategy for engineering binary and ternary hybrid nanoparticles based on spontaneous epitaxial nucleation and growth of a second and third component onto seed nanoparticles in high-temperature organic solutions was reported by the Prasad group. The Fe_3O_4-Au particles were prepared by decomposition of iron acetylacetonate ($Fe(acac)_3$) on the surfaces of Au nanoparticles in high-boiling-point solvents for nucleation, then heating to reflux 300°C for growth of gold nanoparticles separately.[22] The third method for synthesis of dumbbell like Au-Fe_3O_4 nanoparticles was reported by the Hyeon group from inexpensive, nontoxic metal-surfactant complexes. Au-oleylamine complex solution obtained from the reaction of oleylamine, and hydrogen tetrachloroaurate(III) hydrate in 1-octadecene was injected into the Fe-oleate complex solution at 140°C, and the resulting mixture was annealed at this temperature for 20 min, subsequently heated to 230°C, this step lasting for 30 min, resulting in a violet colloidal solution.[23]

The assembled GoldMag particles were reported by the Caruntu and O'Connor group using the covalent bond of Au-NH_2 and electrostatic interaction. The 2–3 nm Au particles were attached to 10 nm Fe_3O_4 nanospheres through a simple, two-step synthetic process. First, Fe_3O_4 nanoparticles were prepared and functionalized with (3-aminopropyl) triethoxysilane (APTES), then the negative charged 2–3 nm Au particles attached to the amine-functionalized positive charged Fe_3O_4 nanoparticles in ethanolic dispersion to form the highly stable Fe_3O_4/Au nanocomposites.[24] We first reported the assembled micro-sized GoldMag particles synthesized by the method of self-aggregation of surface-chemically modified Fe_3O_4 nanoparticles with 3-mercaptopropyltriethoxysilane (MPTES) and then assembled colloid gold on their surfaces.[8] The same method was applied to synthesize the smaller size assembled GoldMag particles using the γ-Fe_2O_3/SiO_2 composites as a core and modified with mercaptosilane, and then gold nanoparticles were attached to the surface of the γ-Fe_2O_3/SiO_2 composite *via* Au–S bonds to form γ-$Fe_2O_3/SiO_2/$Au assembled magnetic composites.[25] Cysteine was also used to modify Fe_3O_4 and obtain amine-functionalized particles with thiol group, then the Au nanoparticles synthesized in aqueous solution through the reduction of $HAuCl_4$ by $NaBH_4$ was conjugated to the Fe_3O_4 particles under ultrasonic conditions.[9] Polymers such as polypyrrole (PPy)[26] and polyaniline (PANI)[27] were used as a functional linker to assemble Au nanoparticles by means of electrostatic interactions between the positive charged polymer coating and the negative charged citrate stabilized Au nanoparticles. For PPy linker, first uniform Fe_3O_4 nanospheres with a diameter of 100 nm were rapidly prepared via the microwave solvothermal method. Then, $Fe_3O_4/$PPy composite nanospheres were obtained through oxidative polymerization of pyrrole with $FeCl_3$. Finally, plentiful Au nanoparticles with a diameter of 15 nm were assembled on the surface of $Fe_3O_4/$PPy by mixing the two solutions to form the three component particles.

Besides the three structures mentioned above of GoldMag particles, gold magnetic composite particles with more complicated structured such as multilayer, irregular shape, nanoflower were designed and prepared.[28–31] Mirkin's group first prepared the three-layer composite magnetic nanoparticle ($SiO_2/Fe_3O_4/$Au). They used positively charged amino-modified SiO_2 particles as templates for the electrostatic assembly of negatively charged Fe_3O_4 nanoparticles. Then the SiO_2/Fe_3O_4 particles electrostatically attracted Au nanoparticle seeds that act in a subsequent step as nucleation sites for the formation of a continuous gold shell upon $HAuCl_4$ reduction. These three-layer magnetic nanoparticles can be functionalized with DNA and applied in high selectivity DNA detection.[28] Using magnetic silica spheres as core, exploiting molecular self-assembly (layer-by-layer (LbL) self-assembly technique) and colloidal growth chemistry, the Salgueirino–Maceira group prepared the magnetic silica spheres with three inner polymers shell and outer shell of gold. The silica spheres were synthesized by sol-gel methods with the hydrolysis of TEOS, then the inner three polyelectrolyte film was deposited by the alternate adsorption of poly-(diallyldimethyl ammonium chloride) (PDADMAC), poly(sodium4-styrenesulfonate) (PSS), and PDADMAC onto magnetic silica spheres. 15 nm gold particles were adsorbed as a seed to form the outer layer of gold shell.[29] The Amal group reported the Fe_3O_4/ polyethyleneimine(PEI)/Au_{seed}/PEI/Au multilayer structured particles. First, the 50 nm PEI-functionalized cubic Fe_3O_4 Cores was synthesized, 2 nm colloidal gold particles were added as Au seed, and then the particle surfaces were functionalized again with PEI by heating in a 60°C oven for 1 hr in the presence of a 5 g/L PEI solution. The outer Au shells were next grown by iterative reduction of $HAuCl_4$ onto the Fe_3O_4/PEI/Au_{seed}/PEI.[30] Zhang reported a facile approach to synthesize multilayer magnetic microspheres by layer-by-layer self-assembly. At first, the Fe_3O_4/C magnetic microspheres were synthesized by two-step reactions including solvothermal and hydrothermal reactions. Next, in order to adsorb negatively charged Au nanoparticles onto the surface of the Fe_3O_4/C magnetic microspheres, the surface was then modified with PDDA, which provides a homogeneous distribution of positive charges, then Fe_3O_4/C/Au magnetic microspheres were obtained by a self-assembly approach.[31] Recently, our group reported a facile method to synthesize novel Fe_3O_4/Au/Fe_3O_4 nanoparticles (nanoflowers). The Fe_3O_4 cores were first synthesized using the hydrothermal method, In order to render the Fe_3O_4 cores water-soluble, the as-prepared NPs were oxidized using an oxidant and functionalized using CTAB surfactant. After adding $HAuCl_4$ solution dropwise and hydroxylamine iteratively to allow reduction of Au^{3+} ions on the surface of as-prepared Fe_3O_4 cores, finally, nanoflower-shaped composite particles were formed.[32]

2. Characterization of GoldMag Composite Particles

Characterization of the different structured gold magnetic composite particles is necessary to evaluate whether the particles are appropriate as a specific carrier in nanomedicine. The most important physicochemical properties of GoldMag particles which influence their biocompatibility, distribution and circulation time *in vivo* were composition analysis, size, surface charge, optical absorption, and magnetic behaviors. However, the characterization of gold magnetic composite nanoparticles is critical, since the structure and composition are complicated. Many characterization techniques the same as those used for single nanoparticles are required to apply together. The most significant characterization techniques are used for the measurement of size, shell thickness, elemental and surface analysis, optical properties, and thermal stability among others. Therefore, the usual characterization techniques such as transmission electron microscopy (TEM), dynamic light scattering (DLS), X-ray photoelectron spectroscopy (XPS), thermal gravimetric analysis (TGA), X-ray diffraction (XRD), atomic emission spectroscopy (AES), ultraviolet-visible spectroscopy (UV-Vis), Fourier transform infrared spectroscopy (FTIR), superconducting quantum interference device (SQUID) or Vibrating Sample Magnetometer (VSM) are the ones most often used.

The characterization of core/shell gold magnetic composite nanoparticles is important since these kinds of particle are the most frequently used in nanomedicine. For example, the core/shell GoldMag particles synthesized in aqueous solution as Fe oxide core/Au shell nanoparticles by iterative hydroxylamine reduction[12] were characterized by transmission electron microscopy (TEM), energy dispersive spectroscopy (EDS), UV-visible absorption spectroscopy, and SQUID magnetometry. The sizes and compositions of the Au-coated Fe oxide nanoparticles were shown by TEM and EDS spectra. The surface plasmon resonances were characterized by UV-visible absorption spectroscopy, exhibiting a peak at 520 nm. SQUID magnetometry reveals that a shell of Au has a decreased effect on magnetic behavior. For the monodisperse core/shell nanoparticles prepared in organic phase, besides the characterization used in water phase, the XRD, XPS, FTIR or Surface-Enhanced Raman Scattering (SERS) were also carried out for the detection of the crystal change and the surface compositions, and the UV-visible absorption spectroscopy was measured in hexane solution.[14–16] These gold magnetic composite nanoparticles show great promise in magnetic resonance imaging (MRI) and NIR photothermal therapy; the MR relaxation study and Vis/NIR spectra should have been performed.

High-resolution transmission electron microscopy (HRTEM) with electron diffraction, XRD and magnetization were employed to characterize the dumbbell-like gold magnetic composite nanoparticles. HRTEM images not only show both components crystalline and the interfaces between them, they also showed parallel lattice planes for Au and Fe_3O_4. For assembled and complicated structure gold magnetic composite particles, the surface charge usually known as zeta-potential are the key factor to assemble the gold nanoparticles[24] or for the layer-by-layer self-assembly.[30] For the PEI coating composite particles, the positive value of zeta-potential indicates that increasing the PEI on Fe_3O_4 surfaces decreases the effective particle aggregate size as well as absorption of gold nanoparticles, For the Fe_3O_4/C magnetic microspheres, before modification of PDDA, the zeta-potential was −59.3 mV. However, after the modification, the surface zeta-potential of the magnetic microspheres in aqueous solution (pH 7.0) changed to +2.33 mV, which confirmed that the PDDA molecules were immobilized onto the magnetic microspheres, and the surface can adsorb the negative charge Au nanoparticles to form the multilayer gold magnetic composite particles.[31]

The optical property based on local surface plasmon resonance (SPR) depends on the shape and thickness of gold in nanoparticle, the surface and solution environment of the particles. Figure 2a shows

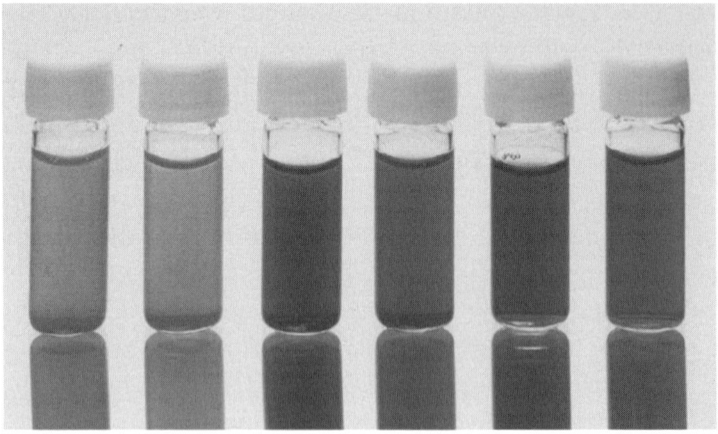

Fig. 2a. Photography of GoldMag nanoparticles with different colors.

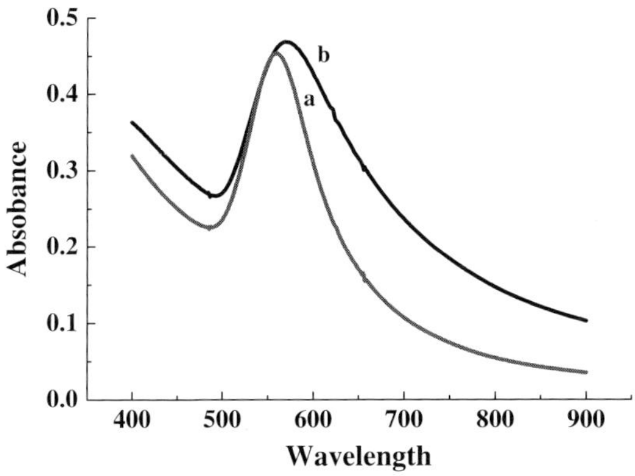

Fig. 2b. UV-Visible spectra of GoldMag suspended in, (a) H_2O λ_{max} = 560 nm, (b) 1×PBS λ_{max} = 570 nm.

that the Fe_3O_4/Au with core/shell structure has different color with increasing average diameter from 30~170 nm. The characteristic peaks of GoldMag nanoparticles with 50 nm diameter suspended in different buffer also differed from 560 nm to 570 nm, giving 10 nm redshifted (Fig. 2b). These versatile optical properties varied with the surface composite of biomolecules could be used to detect target molecules at point-of-care test manner.

3. Modification and Applications of GoldMag Composite Particles in Nanomedicine

The GoldMag advantages of superparamagnetism which can be attracted by an external magnet, and the gold component, such as gold layer or colloid gold assembled on Fe oxide particle surface, provide the chemical functionality with a thiol or amino group. Therefore, biomolecules or anti-tumor drugs can be readily coupled to the surface of GoldMag without the need for additional cross-linkers. On the other hand, thiol ligands such as 11-mercaptoundecanoic acid (MUA) could also introduce the groups such as a carboxyl terminal on its surface based on Au-S chemistry, and the bioconjugation could be carried out by covalent interaction between biomolecules and the carboxyl groups *via* EDC as a cross-linker. The modified GoldMag composite nanoparticles with or without cross-linkers are of great interest for magnetic bimolecular separation,[16,33] bio-sensors for molecular detection,[34,35] contrast agents[36,37] for MRI and magnetically targeted drug delivery.[38] Based on the Au-S specific bond of cysteine or methionine and gold nanoparticles assembled on magnetic particle surface, Okada's group created a variety of peptide tags containing methionine and cysteine and established a novel protein purification system using the "methionine tags."[39] The particles were also used to purify the specific DNA fragment or genomic DNA from whole blood or different tissues.[40,41] Monodispersion and good colloidal stability in buffer of GoldMag nanoparticles are key issues in the application of biomolecule detection. Sometimes the particles need to be modified with surfactant and functional groups to ensure stability in biological media and get a much deeper understanding of surface properties of the composite particle. Irudayaraj's group reported that single strands of oligonucleotides can be immobilized onto amino- and carboxylate-functionalized gold-coated magnetic nanoparticles using the streptavidin-biotin interaction, after carbodiimide activation. The immobilization of streptavidin was more effective when the particles were functionalized by L-aspartic acid because of the availability of two binding sites through its carboxyl and amino groups. They also demonstrated that DNA hybridization can be directly quantified using the microarray scanner.[42] We focus on the introduction of GoldMag nanoparticles used in point-of-care test (POCT) and magnetic drug delivery system using liver cancer as a model.

The driving notion behind POCT is to bring the test conveniently and immediately to the patient. POCT is usually accomplished through the use of transportable, portable, and handheld instruments and related test kits. Cheaper, smaller, faster, and smarter POCT devices and reagents have increased the use of POCT approaches by making it effective for many diseases. Combining the property of surface plasmon resonance of gold nanoparticles and conveniently coupling for biomolecule on its surface, GoldMag can be used as an ideal carrier to carry out POCT. The interaction of biomolecules on

(a) (b)

Fig. 3. (a) GoldMag nanoparticles. (b) Responsiveness to an external magnetic field.

its surface may form the assembly of particles and cause color change of the colloid. Therefore, probes composed of gold magnetic particle and biomolecules could be used to detect target molecules via color change by the naked eye. The responsiveness to an external magnetic field (EMF) makes the coupling process faster during immobilization of biomolecules on its surface. Figure 3 shows the magnetic property of these composite nanoparticles. Figure 4 shows a photo of GoldMag-IgY probe (a) and the aggregate (b) after adding the target protein. The GoldMag-IgY probe could change its color from red to dark purple when the target protein was added to the system. This detection process can be finished in few minutes. The results indicate that this simple and adaptive nanoparticle design may accommodate rapid and visual protein detection.

Lateral-flow immunoassay (LFIA), also known as immunochromatographic tests, have been a popular platform for rapid immunoassays since their introduction in the mid-1980s. LFIA permits a one-step and rapid low-cost analysis as well as a user-friendly format, very short assay time, and no requirement for skilled technicians. The simple, economical and rapid properties of lateral-flow tests make them widely applied in medical tests. Colloidal gold, latex, selenium, carbon and liposome have been used in this system. The

Fig. 4. GoldMag-IgY probe (a) and the aggregate (b) after reaction with the target protein.

most common material used in this strip is colloid gold. However, traditional reporters provide only qualitative and semiquantitative results. Thus, only the signal from top 10 μm layer of the three-dimensional matrix (usually hundred micrometers thick) can be detected by visual interpretation. The signal generated below was inevitably missed, making this method less sensitive. The nanoparticles equipped with the signal of magnetic intensity are not only characterized by rapidity, but also high sensitivity and accuracy. Tang[43] developed a novel magnetic nanogold microsphere-based lateral-flow immunodipstick for rapid detection of aflatoxin B_2 in food. This method is as simple in its operation as a colloidal gold-based strip test but has a threefold higher sensitivity. The results can be obtained within 15 min without expensive equipment. Using GoldMag nanoparticles as a carrier, we established the lateral-flow immunostrips for detection of antibody/antigen or small molecule based on sandwich or competition principle, respectively. Figure 5a shows the strips based on GoldMag nanoparticles for detection of the different concentration of anti-*Treponema pallidum* (TP) antibody. Figure 5b shows the result of magnetic signal of C and T lines in the Magnetic Assay Reader (MAR, Magnabiosciences, USA).

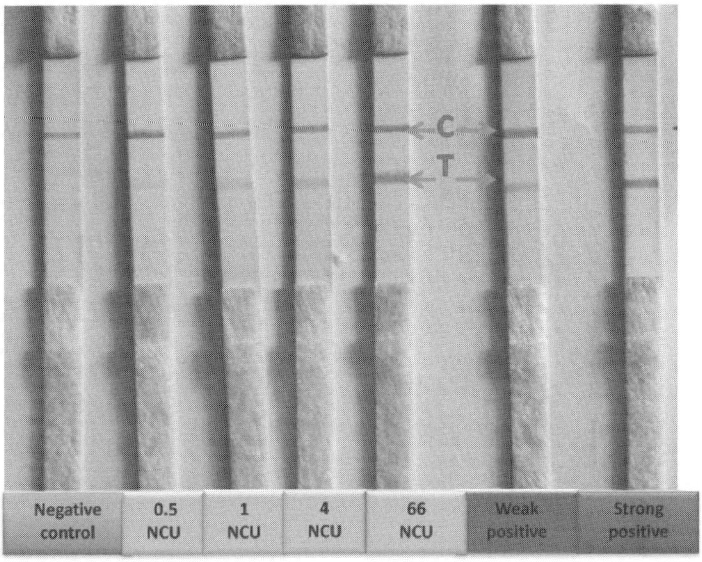

Fig. 5a. Magnetic lateral flow immunoassay in sandwich format for TP antibody detection.

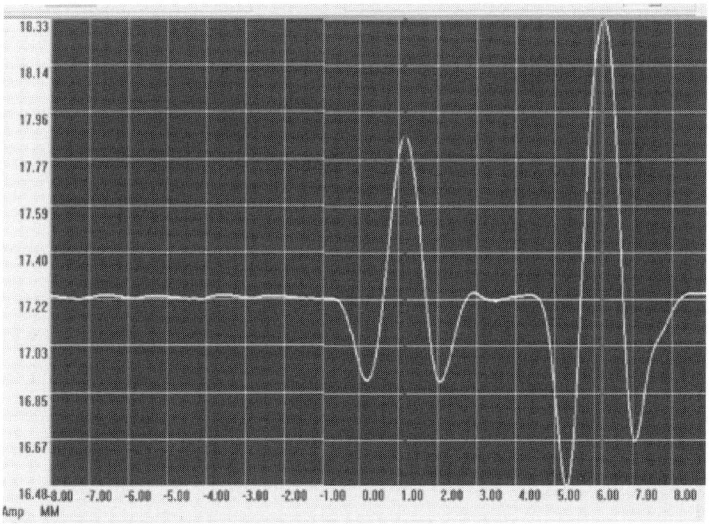

Fig. 5b. Quantitative detection for "C" and "T" line based on magnetic signal.

Conventional chemotherapeutic compounds are distributed non-specifically in the body, thus indiscriminately affecting both normal healthy cells as well as rapidly proliferating cancer cells. For this reason, the desired therapeutic dose to reach the tumor, causes side effects in most cases due to high level toxicity. The lack of targeting specificity of conventional chemotherapeutic agents makes magnetic nanoparticles as attractive drug nanocarriers. We developed polyethylene glycol-modified GoldMag nanoparticles (PGMNPs) as a new magnetically targeted drug carrier combined with the magnetic field for local drug delivery.[38] PGMNPs were synthesized by CTAB and used as the ionic surfactant to disperse the GoldMag nanoparticles, and modified by PEG5000-thiol to stabilize it. Characterization of PGMNPs showed that the average diameter of the PGMNPs was about 50 nm with highly uniform spherical shape. The results of dynamic light scattering showed that the PGMNPs had a slightly broad size distribution in PBS (pH 7.4), but did not aggregate *in vitro*. The kinetic analysis of drug release from DOX-loaded PGMNs showed the maximum doxorubicin (DOX) loading capacity of PGMNs is 107.78 mg/g. The PGMNPs loaded with DOX (DOX-loaded PGMNPs) exhibited a controlled drug-release pattern within the first 4 h and then a slow, sustained release pattern for 4 days. The results of biodistribution of DOX-loaded PGMNPs *in vivo* indicated that the drug was enriched in a restricted area by an external magnetic field and the concentration of the drug was reduced in other organs. The studies of PEG modified GMNPs (PGMNPs) as a drug carrier of the chemotherapy drug of doxorubicin (DOX) *in vivo* to evaluate the therapeutic efficiency of DOX-conjugated PGMNPs (DOX-PGMNPs) using an EMF-guided delivery were also carried out. The relative tumor volumes in DOX-PGMNPs-M (DOX-conjugated PGMNPs using an EMF-guided), DOX-PGMNPs, and DOX groups were 5.46 ± 1.48, 9.22 ± 1.51, and 14.8 ± 1.64, respectively (each $p<0.05$), following treatment for 33 days. The life span of tumor-bearing mice treated with DOX-PGMNPs-M, DOX-PGMNPs, and DOX were 74.8 ± 9.95, 66.1 ± 13.5, and 31.3 ± 3.31 days, respectively (each $p<0.05$) (Fig. 6).

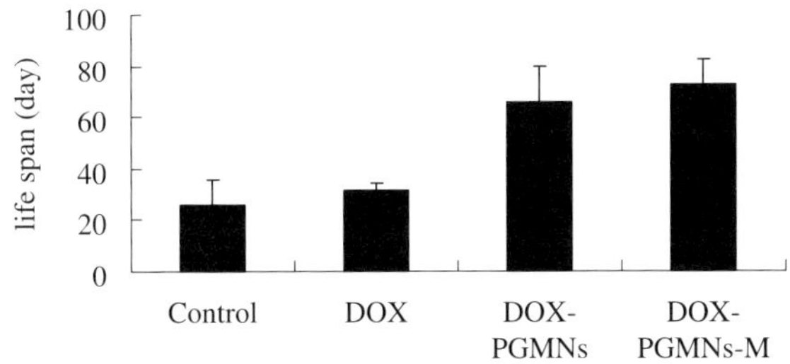

Fig. 6. The life span of H22 tumor-bearing mice model that received different treatments.

The results of anti-tumor effect *in vivo* showed PGMNPs constitute highly attractive nanocarriers and offer a promising perspective for an external magnetic field-guided delivery of anticancer therapeutics and integration in cancer-treating protocols. We also developed GoldMag coated with polysaccharide as an optimized carrier for drug delivery. Therefore, this simple and adaptive nanoparticle design may accommodate chemotherapy for drug delivery optimization and *in vivo* drug-target definition in system biology profiling, increasing the margin of safety in treatment of cancers in the near future.

References

1. Gao L, Zhuang J, Nie L, *et al.* (2007) Intrinsic peroxidase-like activity of ferromagnetic nanoparticles. *Nat Nanotechnol* **2:** 577–583.
2. Lu AH, Salabas EL, Schüth F. (2007) Magnetic nanoparticles: synthesis, protection, functionalization, and application. *Angew Chem Int Ed* **46:** 1222–1244.
3. Jun YW, Huh YM, Choi JS, *et al.* (2005) Nanoscale size effect of magnetic nanocrystals and their utilization for cancer diagnosis via magnetic resonance imaging. *J Am Chem Soc* **127:** 5732–5733.
4. Jeong U, *et al.* (2007) Superparamagnetic colloids: controlled synthesis and Niche applications. *Adv Mater* **19:** 33–605.
5. Cui Y, Hu D, Fang Y, *et al.* (2001) Preparation and mechanism of Fe_3O_4/Au core/shell super-paramagnetic microspheres. *Sci China B* **44:** 404–410.
6. Xu ZH, Hou YL, Sun SH. (2007) Magnetic core/shell Fe_3O_4/Au and Fe_3O_4/Au/Ag nanoparticles with tunable plasmonic properties. *J Am Chem Soc* **129:** 8698–8699.

7. Wang L, Luo J, Maye MM, *et al.* (2005) Iron oxide–gold core–shell nanoparticles and thin film assembly. *J Mater Chem* **15:** 1821–1832.
8. Cui Y, Zhang L, Su J, *et al.* (2006) Synthesis of GoldMag particles with assembled structure and their applications in immunoassay. *Sci China B* **49:** 534–540.
9. Bao J, *et al.* (2007) Bifunctional Au-Fe_3O_4 nanoparticles for protein separation. *ACS Nano* **1**: 293–298.
10. Wang D, Li Y. (2010) One-pot protocol for Au-based hybrid magnetic nanostructures via a noble-metal-induced reduction process. *J Am Chem Soc* **132:** 6280–6281.
11. Cui Y, Wang Y, Hui W, *et al.* (2005). The synthesis of GoldMag nano-particles and its application for antibody immobilization. *Biomed Microdevices* **7:2** 153–156.
12. Lyon JL, Fleming DA, Stone MB, *et al.* (2004) Synthesis of Fe oxide core/Au shell nanoparticles by iterative hydroxylamine seeding. *Nano Lett* **4:** 719–723.
13. Salgueiriño-Maceira V, Correa-Duarte MA, Farle M, *et al.* (2006) Bifunctional gold-coated magnetic silica spheres. *Chem Mater* **18:** 2701–2706.
14. Wang L, Luo J, Fan Q, *et al.* (2005) Monodispersed core-shell Fe_3O_4/Au nanoparticles. *J Phys Chem B* **109:** 21593.
15. Wang L, Park HY, *et al.* (2008) Core@shell nanomaterials: gold-coated magnetic oxide nanoparticles. *J Mater Chem* **18:** 2629–2635.
16. Park HY, Schadt MJ, Wang L, *et al.* (2007) Fabrication of magnetic core@shell Fe Oxide@Au nanoparticles for interfacial bioactivity and bio-separation. *Langmuir* **23:** 9050–9056.
17. Xu Z, Hou Y, Sun S. (2007) Magnetic core/shell Fe_3O_4/Au and Fe_3O_4/Au/Ag nanoparticles with tunable plasmonic properties. *J Am Chem Soc* **129:** 8698–8699.
18. Ji X, Shao R, Elliott AM, *et al.* (2007) Bifunctional gold nanoshells with a superparamagnetic iron oxide-silica core suitable for both MR imaging and photothermal therapy. *J Phys Chem C* **111:** 6245–6251.
19. Zhang H, Meyerhoff ME. (2006) Gold-coated magnetic particles for solid-phase immunoassays: enhancing immobilized antibody binding efficiency and analytical performance. *Anal Chem* **78:** 609–616.
20. Ho D, Sun X, Sun S. (2011) Monodisperse magnetic nanoparticles for theranostic applications. *Accounts of Chem Res* **44:** 875–882.
21. Yu H, Chen M, Rice PM, *et al.* (2005) Dumbbell-like bifunctional Au-Fe_3O_4 nanoparticles. *Nano Lett* **5:** 379–382.
22. Shi W, Zeng H, Sahoo Y, *et al.* (2006) A general approach to binary and ternary hybrid nanocrystals. *Nano Lett* **6:** 875–881.
23. Choi SH, Na HB, Park Y. (2008) Simple and generalized synthesis of oxide-metal heterostructured nanoparticles and their applications in multimodal biomedical probes. *J Am Chem Soc* **130:** 15573–15580.

24. Caruntu D, Cushing BL, Caruntu G, O'Connor CJ. (2005) Attachment of gold nanograins onto colloidal magnetite nanocrystals. *Chem Mater* **17:** 3398–3402.
25. Li ZX, Peng ML, Jin YY, *et al.* (2011) Synthesis of Y-Fe_2O_3/SiO_2/Au magnetic composites for immobilization of bovine serum albumin. *Chinese Sci Bull* **56:** 2911–2915.
26. Zhang H, Zhong X, Xu JJ, Chen HY. (2008) Fe_3O_4/Polypyrrole/Au nanocomposites with core/shell/shell structure: synthesis, characterization, and their electrochemical properties. *Langmuir* **24:** 13748–13752.
27. Xuan S, Wang YXJ, Yu JC, *et al.* (2009) Preparation, characterization, and catalytic activity of core/shell Fe_3O_4@polyaniline@Au nanocomposites. *Langmuir* **25(19):** 11835–11843.
28. Stoeva SI, Huo F, Lee JS, *et al.* (2005) Three-layer composite magnetic nanoparticle probes for DNA. *J Am Chem Soc* **127:** 15362–15363.
29. Salgueirino-Maceira V, Correa-Duarte MA, Farle M. (2006) Bifunctional gold-coated magnetic silica spheres. *Chem Mater* **18:** 2701–2706.
30. Goon IY, Lai MH, Lim M. (2009) Fabrication and dispersion of gold-shell-protected magnetite nanoparticles: systematic control using polyethyleneimine. *Chem Mater* **21:** 673–681.
31. Qi W, Zhang HY, Tang J, *et al.* (2010) Facile synthesis of mercaptophenylboronic acid-functionalized core-shell structure Fe_3O_4@C@Au magnetic microspheres for selective enrichment of glycopeptides and glycoproteins. *J Phys Chem C* **114:** 9221–9226.
32. Hui W, Shi F, Yan K. (2012) Fe_3O_4/Au/Fe_3O_4 nanoflowers exhibiting tunable saturation magnetization and enhanced bioconjugation. *Nanoscale* **4:** 747–751.
33. Kinoshita T, Seino S, Mizukoshi Y, *et al.* (2005) Magnetic separation of amino acids by gold/iron-oxide composite nanoparticles synthesized by gamma-ray irradiation. *J Magn Magn Mater* **293:** 106–110.
34. Li S, Liu H, Liu L, *et al.* (2010) A novel automated assay with dual-color hybridization for single-nucleotide polymorphisms genotyping on gold magnetic nanoparticle array. *Anal Biochem* **405:** 141–143.
35. Zhou X, Xu W, Wang Y, *et al.* (2010) Fabrication of cluster/shell Fe_3O_4/Au nanoparticles and application in protein detection via a SERS method. *J Phys Chem C* **114:** 19607–19613.
36. Jin Y, Jia C, Huang S, *et al.* (2010) Multifunctional nanoparticles as coupled contrast agents. *Nat Commun* **1**: 1–8.
37. Larson TA, Bankson J, Aaron J, *et al.* (2007) Hybrid plasmonic magnetic nanoparticles as molecular specific agents for MRI/optical imaging and photothermal therapy of cancer cells. *Nanotechnology* **18:** 325101–325109.
38. Chao X, Guo L, Zhao Y, *et al.* (2011) PEG-modified GoldMag nanoparticles (PGMNs) combined with the magnetic field for local drug delivery. *J Drug Target* **19:** 161–170.

39. Okada Y, Takano TY, Kobayashi N, *et al.* (2011) New protein purification system using gold-magnetic beads and a novel peptide tag. *Bioconjug Chem* **22:** 887–893.
40. Kinoshitaa T, Seinoa S, Mizukoshi Y, *et al.* (2007) Functionalization of magnetic gold/iron-oxide composite nanoparticles with oligonucleotides and magnetic separation of specific target. *J Magn Magn Mater* **311:** 255–258.
41. Zhao M, Zhang XQ, Wang S, *et al.* (2009) A simple method for purification of genomic DNA from whole blood using Fe_3O_4/Au composite particles as a carrier. *Journal of Medical Colleges of PLA* **24:** 239–243.
42. Kouassi GK, Irudayaraj J. (2006) Magnetic and gold-coated magnetic nanoparticles as a DNA sensor. *Anal Chem* **78:** 3234–3241.
43. Tang D, Sauceda JC, Lin Z, *et al.* (2009) Magnetic nanogold microspheres-based lateral-flow immunodipstick for rapid detection of aflatoxin B_2 in food. *Biosens Bioelectron* **25:** 514–518.

Chapter 15

Nanoparticles for Imaging and Therapy: Functionalization, Endocytosis and Characterization

Yu Zhang[*,§], Yi Li[§], Gürer Budak[¶], Mihrimah Ozkan[†,‡] and Cengiz S. Ozkan[*,‡]

1. Introduction

With the explosive growth of nanotechnology, a tremendous variety of nanostructures in different shapes and structures of different composites has been developed.[1–5] Unique electronic, photonic, magnetic, highly controlled and interesting properties are revealed after the size of materials is reduced from bulk to nanoscale.[6] The similar size dimensions of nanomaterials compared to biomolecules such as proteins and DNA offer a great potential to integrate nanotechnology into biotechnology.

*Department of Mechanical Engineering, University of California Riverside, CA 92521, USA.
†Department of Electrical Engineering, University of California Riverside, CA 92521, USA.
‡Materials Science and Eng, Program, University of California Riverside, CA 92521, USA
§Institute of Textiles and Clothing, the Hong Kong Polytechnic University, Hong Kong, China.
¶Nanomedicine & Advanced Technologies Research Center, Ankara, Turkey.

Liposomes, first proposed in 1974 by Gregoriadis *et al.*, are the earliest examples of nanotechnology applied in cancer therapeutics, and diagnostics as nanoscale-targeted magnetic resonance imaging contrast agents.[7] Today, there are a tremendous number of nanoscale devices developed for the early detection, diagnostics and therapeutic strategies in cancer research. Most recently developed nanomedicine and nanodevices include nanoparticles, quantum dots, nanowires, nanotubes, nanocantilevers, nanopores and nanoshells.[8–10] Nanoparticles for drug delivery can be in the form of nanospheres (matrix systems in which drugs are dispersed throughout the particle) and nanocapsules (where the drug is confined in an aqueous or oily cavity surrounded by a single polymeric membrane).[11] According to The National Cancer Institute, nanotechnology has proven to be very useful in cancer therapy, allowing effective and cancer-specific delivery of therapeutic and imaging agents by avoidance of bio-barriers, biomarker-based targeting and the increase of therapeutic efficacy.

Nanoparticles constitute one of the most important forms of nanoscale components that have proven to have a multitude of useful applications in cancer research. A multifunctional nanoparticle is illustrated in Fig. 1.4 by Ferrari in 2005.[8] This ideal multifunctional nanoparticle is designed to have the ability to carry therapeutic agents, bimolecular targeting agents, imaging contrast enhancement agents, permeation enhancer agents as well as a suitable coating to avoid bio-barriers. Although the number of different types of nanoparticles is increasing rapidly, most can be classified into organic and inorganic naoparticles. Organic nanoparticles, which contain organic molecules as a major building material, include liposomes, micelles, polymersomes, dendrimers, emulsions, viruses and other polymers. Inorganic nanoparticles, including metal and metal oxide nanoparticles, quantum dots and nanoshells, usually consist of a central core that defines the optical, magnetic, and electronic properties of the nanoparticle, with a protective and/or a functional coating on the surface.[12,13] A summary of biomedical nanoparticles is listed in Table 1.

Table 1. A Summary of Biomedical Nanoparticles

Name	Role
Lipid-based nanoparticles (Liposomes)	Increase the solubility of hydrophobic chemotherapeutics limit drug toxicity; Coated with tumor-specific antibodies for tumor targeting
Dendrimers	Highly branched, monodisperse macromolecules; Holding therapeutics substances such as DNA in their cavities; MRI contrast agents; Targeted drug delivery and gene transfection against tumors, bacteria, and viruses
Polymer nanoparticles	Poly(d,l-lactic-co-glycolic acid) (PLGA), polystyrene, polyethylene glycol (PEG) covered or PEGylated, poly-vinyl-chloride (PVC) nanoparticles, PLA nanoparticles East incorporation of the drug within the nanoparticles, Stabilize the drugs *in vivo* and no side effects Biodegradable Targeted drug delivery Sustained drug release
Virus based nanoparticles	Cowpea mosaic virus (CPMV), cowpea chlorotic mottle virus (CCMV), Flock House virus (FHV), etc. Benign to humans and other animals; Highly organized nanoblock structure, chemically addressable sites on the protein shell (the capsid) surface, pH and thermally stability; Vaccines, vascular imaging, and targeted drug delivery
Quantum dots	Commonly consist of a CdSe or CdTe core enclosed within a ZnS shell Highly fluorescent and photobleach resistant Optical detection of genes and proteins in animal models and cell assays Tumor and lymph node visualization

(*Continued*)

Table 1. (*Continued*)

Name	Role
Nanoshells	A nonconducting nanoparticle (silica) core with a thin layer of metal (gold) shell Strong absorption at near-infrared light Deep tissue tumor cell thermal ablation (photothermal cancer therapy) Tumor-specific imaging
Magnetic nanoparticles	Typically iron oxide nanoparticles, magnetite Fe_3O_4 and maghemite (γ-Fe_2O_3) Non-toxic, biocompatible MRI contrast agents Hyperthermia therapy (heat under an oscilating magnetic field up to around 43°C to kill cancer cells whereas the normal cells can survive) Targeted drug delivery

2. Overview of Biomedical Nanoparticles

Liposomes

Liposomes are one of the nanoscale lipid-based vehicles. Liposomes are nontoxic and biodegradable because of their natural composition. The unique ability of liposomes to entrap drugs both in an aqueous and a lipid phase make them used most to increase the solubility of hydrophobic chemotherapeutics and for limiting drug toxicity.[9] Liposomes coated with tumor-specific antibodies have been used for tumor targeting.[14,15] But there are several problems that limit their biomedical applications. These problems include nonspecific uptake by the macrophages of the reticuloendothelial system (RES), fast removal from blood circulation and instability.[16,17] Recently, to overcome the limitations, researchers have made some progresses on synthesizing advanced liposomes. Liposomes made of novel lipid polymers significantly increased their membrane stability and bioavailability.[17,18] Liposomes coated with PEG also have increased bioavailability significantly by reducing nonspecific RES uptake.[17,19]

Dendrimers

Dendrimers were discovered in the early 1980s by Tomalia and co-workers as a new class of polymeric materials.[20] Dendrimers have unique properties, such as their high degree of branching, globular architecture and widely recognized mimicry of protein size, and mutivalency, which make them promising nanoscale scaffolding and nanocontainers for drug delivery.[21,22] Dendrimers have also been used in MRI as contrast agents, and have aided visualization of various pathological processes.[23,24] By simplifying and optimizing the synthesis of dendrimers such as the "lego" and "click" approaches, dendrimers provide a large variety of structures while reducing the cost of their production.[25] Modified dendrimers have been shown to act as targeted drug delivery and gene transfection against tumors, bacteria, and viruses.[26]

Virus Based Nanoparticles

Viruses are well-defined natural molecular assemblies and containers with a variety of shapes, sizes, stabilities, dynamic properties, and chemical reactivities, which could act as multifunctional nanoscale devices.[27,28] The viruses that have been explored for nanotechnology extensively are the cowpea mosaic virus (CPMV), the cowpea chlorotic mottle virus (CCMV), MS2 Bacteriophage and other several icosahedral viruses such as the insect nodavirus Flock House virus (FHV), papillomaviruses, polyomaviruses, and the Hepatitis B virus.[27] The cowpea mosaic virus (CPMV) is an icosahedron with a spherical average diameter of 28.4 nm which is built from 60 copies of two asymmetric protein units assembled around a single-stranded bipartite RNA genome.[29,30] Its highly organized nanoblock structure, chemically addressable sites on the protein shell (the capsid) surface, and pH and thermally stability over a wide range of conditions make CPMV an excellent choice in biomedical applications including vaccines, vascular imaging, and targeted drug delivery.[28,30–32] For CPMV's application in cancer research, Fig. 1.5 shows the intravital visualization of human fibrosarcoma-mediated tumor angiogenesis using fluorescent CPMV-A555 by Lewis *et al.* in 2006, which contributed to the identification of arterial and venous vessels and to the monitoring of the neovascularization of the tumor microenvironment.[29]

Polymeric Nanoparticles

Polymeric nanoparticles have several advantages, such as the ability to stabilize the drugs *in vivo*, easy incorporation of the drugs, and no side effects for the biodegradable polymers.[33] Biodegradable polymeric nanoparticles, especially Poly d,l-lactic-co-glycolic acid (PLGA), polystyrene, polyethylene glycol (PEG) covered or PEGylated, polyvinyl-chloride (PVC) nanoparticles and Poly d,l-lactide (PLA) based nanoparticles, have been widely studied for therapeutic applications, especially in drug delivery. They could also be functionalized with cancer marker targeting molecules, which enable specifically targeting the cancer cells and then releasing the drugs to the cancer.[18,34]

Quantum Dots

Quantum dots are novel semiconductor nanocrystals, and are one of the most widely used nanoparticles in cancer research.[35] Quantum dots have wide absorption ranges and their fluorescence emission properties are size and composition dependent.[36] Compared to their organic molecular counterparts, quantum dots are not only able to provide single excitation for multicolor images, but they are also highly fluorescent and photobleach resistant.[37] Quantum dots have been used in cancer research to image cancer biomarkers in blood assays and as contrast agents for tumor and lymph node visualizations.[38–40] But one major concern for the use of quantum dots *in vivo* is the potential human cytotoxicity of the semiconductor material.

Nanoshells

Nanoshells, which contain a non-conducting dielectric core with a thin layer of metallic shell, are another kind of nanoparticle for cancer imaging and therapy applications.[41] The absorption band of core-shell particles can been tuned by adjusting the ratio of the thickness of the metal shell to the diameter of the silica core, thus enabling both strong scattering and absorption efficiency.[42] Nanoshells that absorb near-infrared light are very useful because they can easily penetrate several centimeters of human tissue.[43] The absorption of light by the nanoshells will create an intense heat that could kill cancer cells. Therefore they have been used as dual imaging/therapy contrast agents, and photothermal cancer therapy *in vivo*.[44,45]

Magnetic Nanoparticles

Magnetic nanoparticles, especially iron oxide nanoparticles, including magnetite Fe_3O_4 and maghemite (γ-Fe_2O_3), are non-toxic and biocompatible.[46,47] Iron oxide nanoparticles have been widely studied for their potential applications in a variety of biomedical fields for their unique physical, chemical and magnetic properties.[48,49] Superparamagnetic iron oxide nanoparticles play an important role as MRI contrast agents to better differentiate healthy and pathological tissues.[50,51] In hyperthermia therapy, after magnetic nanoparticles are embedded around a tumor site, an oscillating magnetic field will heat them up to around 43 °C. Cancer cells are destroyed, whereas the normal cells can survive at higher temperatures.[52–54] Another promising application of colloidal magnetic nanoparticles is to act as carriers for site-specific delivery of drugs.[13,55,56]

3. A Surface Charge Study on Cellular Uptake Behaviors of F3 Peptide Conjugated Iron Oxide Nanoparticles

3.1. *Introduction*

Nanoparticles have been proven to play a very important role in current cancer research. Cellular interaction with nanoparticles is of great interest for researchers. In this study, we investigate the cellular uptake behaviors of F3 peptide conjugated dextran coated iron oxide nanoparticles with normal breast epithelial cells (MCF10A) and cancer breast epithelial cells (MCF7) by cell surface zeta potential measurements. Iron oxide nanoparticles are the most commonly investigated nanoparticles for current cancer diagnosis and therapeutics based on their unique chemical and magnetic properties.[48,49] Their applications include magnetic resonance imaging (MRI),[57–60] targeted drug delivery[13,55,56] and magnetic intracellular hyperthermia.[54,61,62] A wide variety of magnetic iron oxide nanoparticles studies have been reported to date, differing in size and type of coating material used, such as dextran, starch, albumin, silicones and poly(ethyleneglycol) (PEG).[48,51,63] The polymeric coating of nanoparticles can not only improve the stability of the colloid suspensions, but also avoid degradation of the fluid with time.[64] Moreover, for biomedical applications, if the adsorbed polymer layer completely covers the particles, the protection of the iron oxide core against contact with blood proteins and phagocytosis-associated receptors is enabled.[65] Dextran is one of the most common coatings, because it has already being used in clinical trials[66] and proven to be long circulating with no measurable reported toxicity.[63,67,68] Dextran is a polymer $(C_6H_{10}O_5)_n$ of anhydroglucose having mainly alpha-D(1-6) linkages with some unusual 1,3 glucosidic linkages at branching points. F3 peptide was first discovered in a screening procedure that used a phage-displayed cDNA library and combined *ex vivo* screening on cell suspensions prepared from mouse bone marrow cells and *in vivo* screening for homing to HL-60 human leukemia xenografts tumors.[69] It is a 31-amino acid sequence of the N-terminal fragment of human high-mobility group protein 2 (HMGN2). F3 peptide, reported to have cell-penetrating properties, is promising for drug targeting and gene therapy applications, since it is taken up by the cells and it is able to carry a payload into the target cell nucleus.[69–75] It was demonstrated in previous studies that FITC-labeled F3 peptide accumulated on the cell surface and then translocated to the nucleus of breast cancer cells both *in vitro* and in xenograft studies.[69,73]

The zeta (ξ) potential, which is the electrostatic potential that exists at the shear plane of a particle, is related to both surface charge and the local environment of the particle. Zeta potential is commonly used as an important parameter in colloid science to understand the colloid electrostatic interactions.[76] Zeta potential has been used in cell biology to study cell adhesion, activation and agglutination based on cell surface charge properties.[77–79] Altankov *et al.* stated that the zeta potential might be a critical parameter for cellular interaction.[80] We have reported the surface charge responses by zeta potential measurements of normal breast epithelial cells (MCF10A) and cancer breast epithelial cells (MCF7) after incubation with untreated iron oxide nanoparticles and cowpea mosaic virus (CPMV) nanoparticles. Distinct patterns between normal and cancer breast cells are observed and indicate that zeta potential measurements could be a feasible tool in describing the interaction between cells and nanoparticles.[81] Here, we observe the significant and sensitive changes in zeta potential measurements of normal breast epithelial cells (MCF10A) and cancer breast epithelial cells (MCF7) after incubation with F3 peptide conjugated dextran coated iron oxide nanopaticles. Our results suggest that zeta potential has a great potential to be a valuable biological characteristic in describing the various interactions between different cells and specific nanoparticles.

3.2. *Theory and Analysis*

The potential distribution around the double layer region is expressed with the Poisson–Boltzmann equation[76]:

$$n_i(x) = n_i^0 \exp\left[\frac{-z_i e\psi}{kT}\right], \tag{1}$$

where n_i is the number of ions of type i per unit volume in the double layer region. $n_i = n_i^0$ when $\psi = 0$. z_i is the valence of ion i. ψ is the potential. Using the Debye–Huckel approximation, Eq. (1) is solved. The potential distribution in the double layer is:

$$\psi = \frac{1}{4\pi\varepsilon_0} \cdot \frac{Q}{D(1+\kappa a)} \frac{e^{-\kappa(r-a)}}{r}, \tag{2}$$

where Q is cell surface charge, a is the radius of cell, $\kappa = \left(\frac{e^2 n_i^0 z_i^2}{\varepsilon kT}\right)^{1/2}$, ε is the permittivity.

The potential at the shipping plane surface $r = a + 1/\kappa$ gives the zeta potential ζ. For the cell, as the radius a is much larger than the double layer length $1/\kappa$, the zeta potential can be simplified to:

$$\zeta = \frac{Qe^{-1}}{4\pi\varepsilon_0 Da^2} \cdot \frac{1}{\kappa}. \tag{3}$$

When nanoparticles are adsorbed at the cell surface, they may influence the zeta potential in two ways. The adsorption characteristics of the ions present will be influenced, and the plane of shear will be shifted from the particle surface. The additional free energy is incorporated into the Boltzmann equation[76] and the modified Poisson–Boltzmann equation with nanoparticle adsorption is given below, where $\beta(x)$ measures how the presence of nanoparticles affects the free energy of the ions at a distance from the cell surface.

$$n_i(x) = n_i^0 \exp\left[\frac{-z_i e\psi}{kT} + \beta(x)\right]. \tag{4}$$

Equation (4) is then solved in the Debye–Huckel approximation with $\beta(x) = \beta$ (a constant). The assumption is the uniform distribution of nanoparticles in the cell interface.

Zeta potential is the potential at the shifted shipping plane surface $r = a + 1/\kappa_\beta$, where $\kappa_\beta = \kappa \exp\left(\frac{-\beta}{2}\right)$ is the modified Debye–Huckel parameter.

$$\zeta_\beta = \frac{Qe^{-1}}{4\pi\varepsilon_0 Da^2} \cdot \frac{1}{\kappa_\beta}. \tag{5}$$

Comparing Eqs. (3) and (5), the relative zeta potential $Z = \zeta_\beta / \zeta_0$ is:

$$Z = \exp\left(\frac{\beta}{2}\right). \tag{6}$$

Nanoparticle uptake by cells is considered to be a two-step process: first is the binding of nanoparticles to the cell surface, followed by the internalization of nanoparticles by the specific endocytosis pathway. In our previous study,[81] when nanoparticles bind onto the cell surface with the same sign of zeta potential which increases the free energy of the ions at a distance from the cell surface, $\beta^{s}_{\text{binding}}$ is positive, i.e. $Z > 1$. In this study, as the F3 peptide conjugated iron oxide nanoparticles has measured positive zeta potential, which has the opposite sign of zeta potential compared to cells. When nanoparticles bind onto the cell surface with opposite sign of zeta potential which decreased the free energy of the ions at a distance from the cell surface, $\beta^{0}_{\text{binding}}$ is negative, i.e. $Z < 1$. And β_{int} is positive, i.e. $Z > 1$, during the internalization of nanoparticles by the endocytosis pathway, the position of the plane of shear is shifted and the negative surface charge is increased compared to the binding process. The relative zeta potential of cells during the adsorption of nanoparticles is expressed as:

$$Z = \exp\left(m_{\text{binding}} \cdot \frac{\beta_{\text{binding}}}{2} + m_{\text{int}} \cdot \frac{\beta_{\text{int}}}{2} \right), \tag{15.7}$$

where m_{binding} is the total mass of nanoparticles binding to the cell surface; m_{int} is the total mass of nanoparticles internalized within the cell. Langmuir adsorption could be employed to model the binding and internalization processes.[81,82]

The simulation of cell zeta potential change with the same-sign zeta potential nanoparticles has been discussed in last chapter. In this model, cell zeta potential change after incubation with nanoparticles with the opposite sign of zeta potential is explored under the effect of the binding and internalization process. Cells generally have negative zeta potential and the nanoparticles have positive zeta potential in this model. Here, the relative zeta potential Z before nanoparticle incubation is normalized to 1. In Fig. 1, when the binding effect dominates first and the internalization effect plays a more important role in the following stage, the relative zeta potential Z will first decrease with time and then increase. Accordingly, the real negative cell zeta potential $-Z$ will first increase with time and then decrease. In Fig. 2, when the internalization effect dominates during the whole adsorption process, the relative zeta potential Z will increase and then reach a stable stage. Accordingly, the real negative cell zeta potential $-Z$ will decrease to a stable stage.

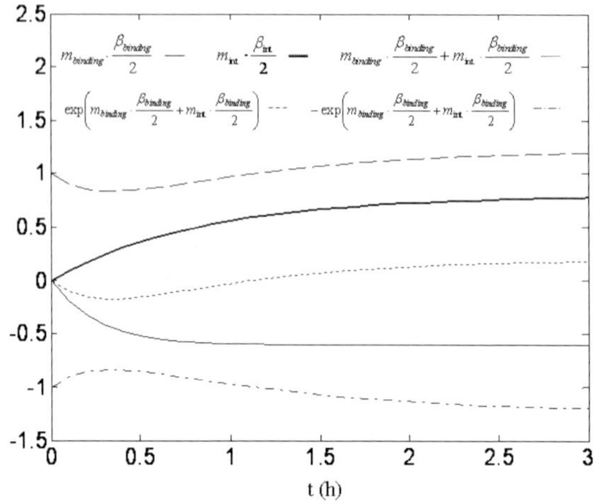

Fig. 1. Theoretical evaluation of cell zeta potential change vs. time after incubation with cationic nanoparticles. The binding effect dominates in the first stage, and then internalization effect dominates.

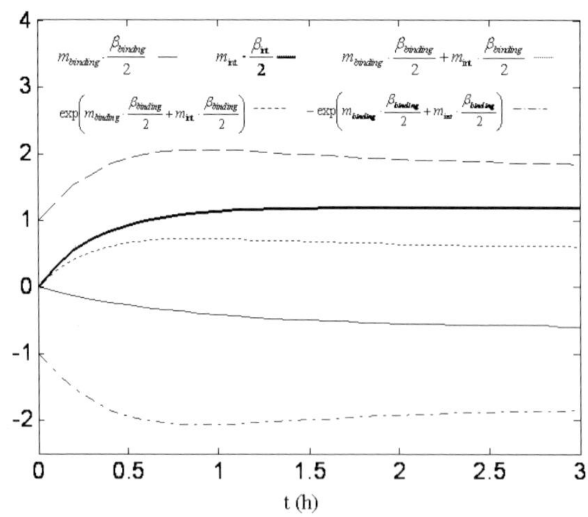

Fig. 2. Theoretical evaluation of cell zeta potential change vs. time after incubation with cationic nanoparticles. The internalization effect dominates all the time.

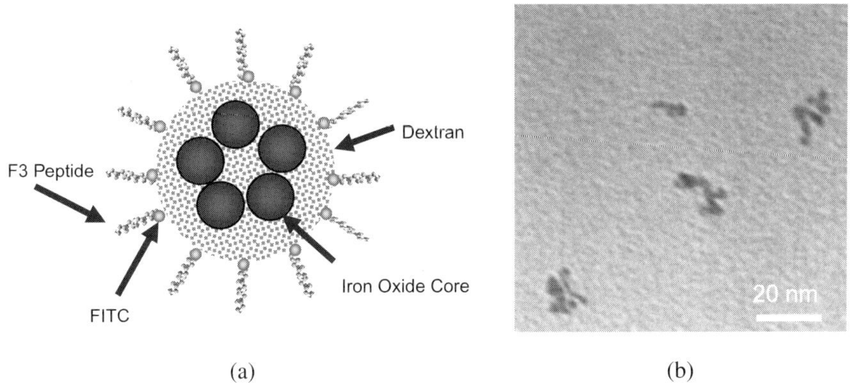

Fig. 3. (a) The schematic illustration of the structure for F3 peptide conjugated dextran coated iron oxide nanoparticles; (b) Transmission Electron Micrograph of iron oxide cores of the nanoparticles.

3.3. *Results and Discussion*

TEM Images of F3 Peptide Conjugated Dextran Coated Iron Oxide Nanoparticles

Figure 3(a) shows the schematic illustration of one F3 peptide conjugated dextran coated iron oxide nanoparticle. Several iron oxide cores are embedded in dextran coating and F3 peptides are then conjugated on the surface. FITC attached to F3 peptide acts as a probe for nanoparticles and is used to track the cellular uptake behavior of the nanoparticles. The Transmission Electron Micrograph of the iron oxide nanoparticles is shown in Fig. 3(b), which shows only the iron oxide cores of the nanoparticles. The average hydrodynamic diameter of compound nanoparticles was determined by DLS to be 50.3 nm. And their zeta potential value measured at pH=7.4 in Hepes buffer is $+10.59 \pm 0.14$ mV.

Zeta Potential Measurements for Normal and Cancer Cells

Figures 4 and 5 show the experimental results of zeta potential measurements at pH 7.4 for MCF7 cancer breast epithelial cells and MCF10A normal breast epithelial cells after incubation with F3

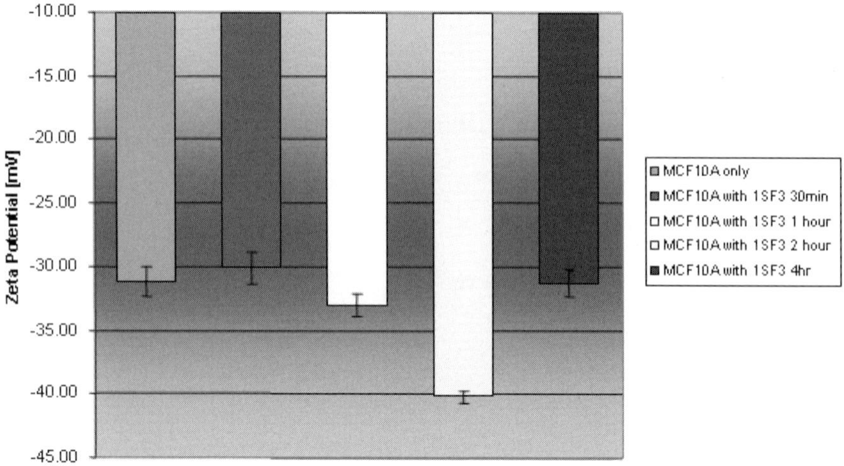

Fig. 4. Comparison of zeta potential values of MCF10A cells alone and MCF10A cells after incubation with F3 peptide conjugated iron oxide nanoparticles for 30 min, 1 hr, 2 hr and 4 hr at 37°C. All the zeta potential values were measured using a ZetaPALS system in Hepes buffer (Concentration 40 mM, pH 7.4). *Note*: On the bar graph, 1SF3 stands for F3 peptide conjugated iron oxide nanoparticles, and the error bar indicates the standard error for 10 runs of zeta potential measurements.

peptide conjugated dextran coated iron oxide nanoparticles for 30 min, 1 hr, 2 hr and 4 hr. The zeta potential of MCF10A normal cells alone was –31.16 ± 1.12 mV and the zeta potential of MCF7 cancer cells alone was –20.32 ± 2.43 mV. And the zeta potential of F3 peptide conjugated dextran coated iron oxide nanoparticles has an opposite sign to cells and was measured to be +10.59 ± 0.14 mV. The results of zeta potential measurements of MCF7 and MCF10A cells after incubation with F3 peptide conjugated iron oxide nanoparticles turn out to follow very different trends. Figure 4 shows that the zeta potential value of MCF10A cells after incubated with F3 peptide conjugated iron oxide nanoparticles slightly increased from –31.16 ± 1.12 mV to –30.11 ± 1.25 mV after 30 min of incubation, dropped to –33.01 ± 0.89 mV after 1 hr of incubation, further dropped to –40.23 ± 0.46 mV after 2 hr of incubation and then increased to –31.27 ± 1.01 mV after 4 hr of incubation. Figure 5 shows that for

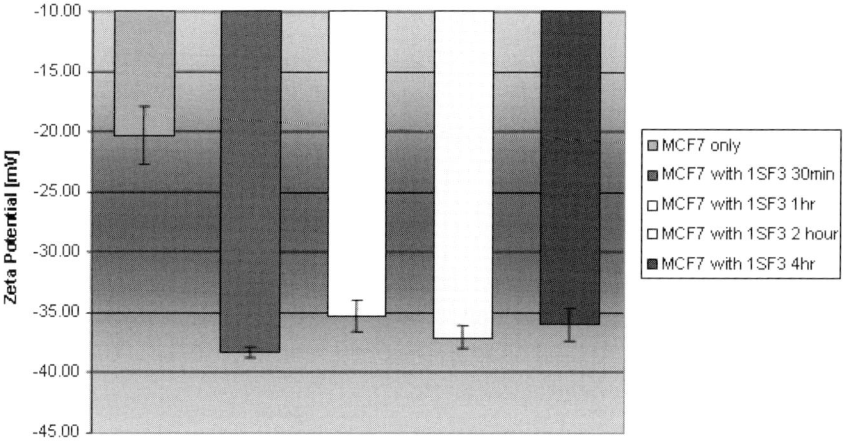

Fig. 5. Comparison of zeta potential values of MCF7 cells alone and MCF7 cells after incubation with F3 peptide conjugated iron oxide nanoparticles for 30 min, 1 h, 2 hr and 4 hr at 37°C; All the zeta potential values were measured using a ZetaPALS system in Hepes buffer (Ionic strength 40 mM, pH 7.4). *Note*: On the bar graph, 1SF3 stands for F3 peptide conjugated iron oxide nanoparticles, and the error bar indicates the standard error for 10 runs of zeta potential measurements.

MCF7 cells, the change of zeta potential value follows a different pattern after they were incubated with F3 peptide conjugated iron oxide nanoparticles. The zeta potential value dropped significantly from -20.32 ± 2.43 mV to -38.29 ± 0.45 mV after the first 30 min of incubation and slightly increased to -35.24 ± 1.29 mV after 1 h of incubation. The zeta potential value dropped again to -37.08 ± 0.97 mV after 2 hr of incubation and then slightly increased back to -35.99 ± 1.37 mV.

The nanoparticle uptake by cells could be considered as a two-step process: first, a binding process, then followed by an internalization process. The zeta potential change after nanoparticle incubation has been explained and simulated in theory and analysis. The attachment of cationic nanoparticles on the cell plasma membrane will cause the zeta potential to become intuitively more positive. The internalization process will result in the change of zeta potential to an opposite way mainly due to the vesicular transport based cell endocytosis.

Confocal microscope analysis of F3 peptide conjugated iron oxide nanoparticles after incubation with MCF7 cancer cells and MCF10A normal cells.

For the F3 peptide conjugated iron oxide nanoparticles in this study, the tumor-homing peptide F3 conjugated on the nanoparticle surface specifically bound to nucleolin which is overexpressed on the surface of cultured breast carcinoma cells and these carcinoma cells bound, internalized conjugated F3 peptide and transported it into the nucleus.[73–75] Nucleolin was confirmed to be a receptor molecule and acted as a shuttle protein that traffics between membrane, cytoplasm and nucleus.[73–75] The nucleolin polypeptide consists of a negatively charged NH_2-terminal domain, an RNA-binding domain, and a COOH–terminal domain rich in RGG motifs, and the main functions of nucleolin relate to rRNA maturation and ribosome assembly[83,84] Although nucleolin was originally known as a nuclear and cytoplasmic protein, a number of studies confirm that it can also be expressed at the cell surface.[85–88] Cell surface nucleolin expression is a novel angiogenesis marker and provides a tool for studying tumor angiogenesis, including the contribution of precursor cells to this process and for targeting drugs into tumors. We believe that the F3 peptide conjugated iron oxide nanoparticles are also uptaken by MCF7 cancer breast cells through the overexpressed nucleolin receptors on the surface of breast carcinoma cells. Figure 6 shows the confocal microscope analysis of MCF7 cancer cells after incubation with 40 μg Fe/ml FITC attached F3 peptide conjugated iron oxide nanoparticles for different time lengths of 30 min, 1 h, 2 hr and 4 hr. Single DAPI channel, single FITC channel, overlaid image and cross section image obtained from confocal *Z* series images are showed. The nanoparticles are visualized in green by attached FITC and nuclei are visualized in blue by DAPI counterstaining. From the confocal images, we observed that F3 peptide conjugated iron oxide nanoparticles are actively uptaken by MCF7 cancer breast epithelial cells. Most importantly, although F3 peptide conjugated iron oxide nanoparticles was only incubated with MCF7 cells for 30 min, nanoparticles have already been internalized and significant amount of F3 peptide conjugated dextran coated iron oxide nanoparticles were delivered into the nuclei after

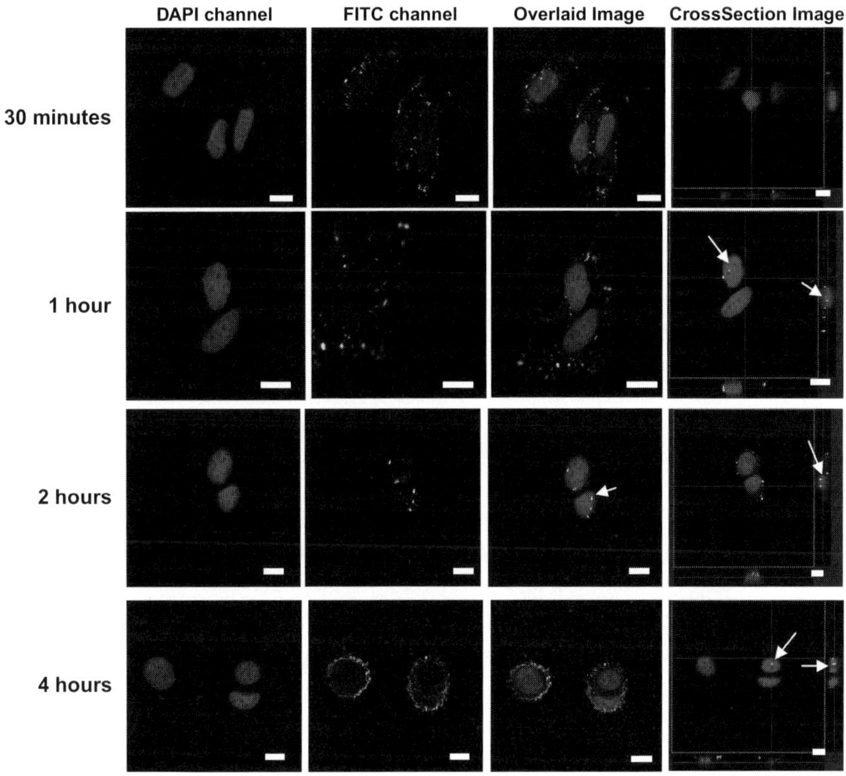

Fig. 6. Confocal microscope analysis of MCF7 cells after incubation with 40 μg Fe/ml F3 peptide conjugated iron oxide nanoparticles after 30 min, 1 h, 2 hr and 4 hr. Leica SP2 Confocal Microscope with a 63*x* objective was used for imaging. Single DAPI channel, single FITC channel, overlaid image and cross section image obtained from Confocal Z series images are showed. For cross section image, large panel is *x–y* sections, small panel at the bottom is *x–z* sections and small panel at right is *y–z* sections. The nanoparticles are visualized in green by FITC and nuclei are visualized in blue by DAPI counterstaining. The bar represents 10 μm. The white arrows point out some of the F3 peptide conjugated iron oxide nanoparticles that are delivered in the nuclei of MCF7 cells.

1 h incubation. We believe that overexpressed nucleolin on the surface of cancer breast cells are responsible for the fast internalization and a significant amount delivery of F3 peptide conjugated iron oxide nanoparticles into the nuclei of MCF7 cells.

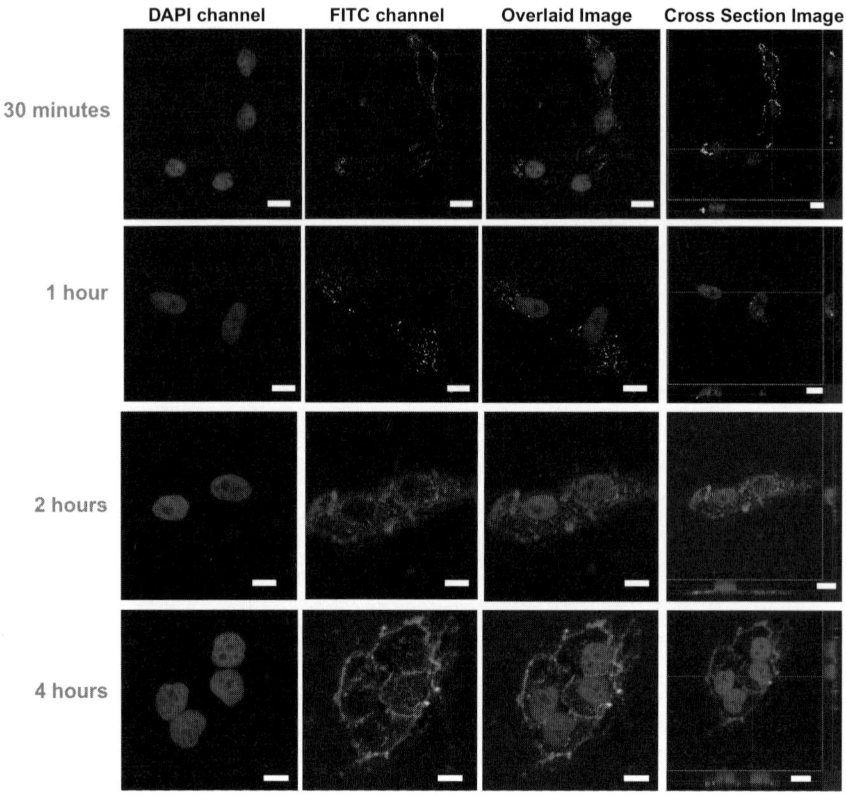

Fig. 7. Confocal microscope analysis of MCF10A normal breast cells after incubation with 40 μg Fe/ml F3 peptide conjugated iron oxide nanoparticles after 30 min, 1 h, 2 hr and 4 hr. Leica SP2 Confocal Microscope with a 63× objective was used for imaging. Single DAPI channel, single FITC channel, overlaid image and cross section image obtained from Confocal Z series images are showed. For cross section image, large panel is *x–y* sections, small panel at the bottom is *x–z* sections and small panel at right is *y–z* sections. The nanoparticles are visualized in green by FITC and nuclei are visualized in blue by DAPI counterstaining. The bar represents 10 μm.

Figure 7 shows the confocal microscope analysis of FITC attached F3 peptide conjugated iron oxide nanoparticles after incubation with MCF10A normal breast epithelial cells for different time lengths of 30 min, 1 h, 2 hr and 4 hr. Internalization of F3 peptide conjugated iron oxide nanoparticles are also observed for MCF10A cells over time. Although the F3 peptide uptake pathway for MCF10A cells is not clear, F3 is rich in basic amino acids and heparan sulfate proteoglycans may act as nonspecific receptors for cationic macromolecules.[75,89] Compared to MCF7 cancer breast cells, not a significant number of F3 conjugated iron oxide nanoparticles were delivered into the nuclei of MCF10A normal cells, which indicates the lack of nucleolin both on the cell surface and in cytoplasm as a shuttle protein compared to cancer breast cells. Untreated iron oxide nanoparticles did not enter cell nucleus from the previous study we did,[81] and also indicated that the F3 peptide determines the cellular uptake behavior for MCF10A and MCF7 cells.

Figure 8 reveals the changes of zeta potential values for both MCF10A normal cells and MCF7 cancer cells after incubation with F3

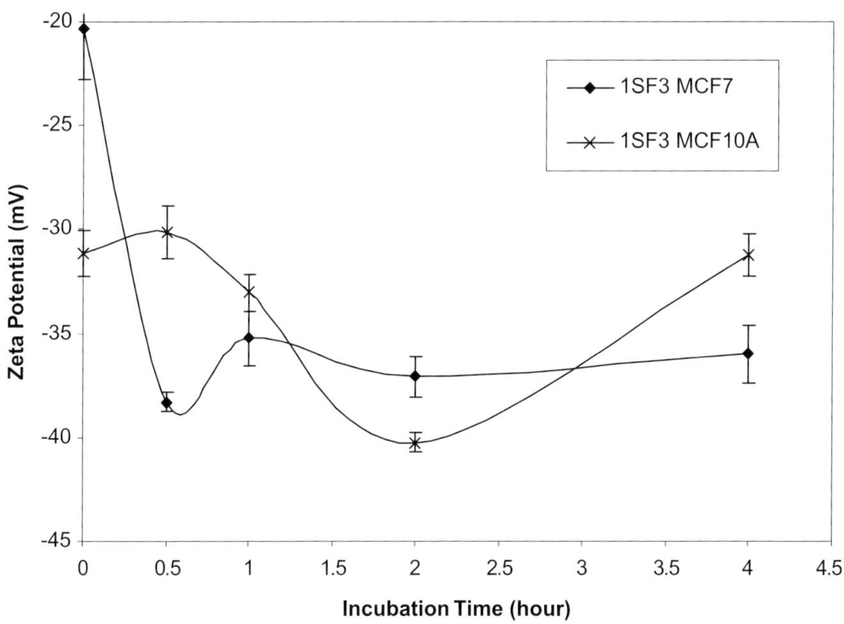

Fig. 8. Summary of zeta potential change for MCF10A normal breast cells and MCF7 cancer breast cells after incubation with F3 peptide conjugated iron oxide nanopaticles for 30 min, 1 h, 2 hr and 4 hr. 1SF3 stands for F3 peptide conjugated iron oxide nanoparticles.

peptide conjugated iron oxide nanoparticles over time. For MCF10A cells, the first increase of measured zeta potential value after 30 min incubation with F3 peptide conjugated iron oxide nanoparticles indicates the binding of cationic nanoparticles on the cell surface. The measured zeta potential values of MCF10A cells continuously decreased after incubation with F3 peptide conjugated dextran coated iron oxide nanoparticles after 1 h and 2 hr, which indicates that the internalization effect of F3 peptide conjugated iron oxide nanoparticles starts to dominate the change of zeta potential value. After 4 hr, the zeta potential of MCF10A cells increased. The possible reason might be that the excytosis effect starts to dominate, which indicates the escape of cationic F3 peptide conjugated iron oxide nanoparticles. The change of zeta potential of MCF10A normal breast cells basically follows the simulated zeta potential change pattern after incubation with the opposite sign of nanoparticles that we discussed before. The binding effect dominates first, and the internalization effect plays a more important role in the following stage.

In the case of MCF7 cells, only after 30 min incubation with F3 peptide conjugated iron oxide nanoparticles, the measured zeta potential value dropped significantly, which indicates that MCF7 cells internalized the nanoparticles quickly due to the overexpressed nucleolin receptors. We could not observe the binding effect that should cause the increase of zeta potential, but the drop of zeta potential due to internalization effect. For MCF10A cells, they uptake nanoparticles slower than MCF7 cells and the increase of zeta potential due to binding effect could be observed. As F3 peptide tends to bind much faster with overexpressed nucleolin on MCF7 cancer breast cells than non-specific binding in the case of MCF10A normal cells, the zeta potential of MCF7 cells drops much faster than MCF10A cells. And the much smaller fluctuation of zeta potential for MCF7 cancer cells indicates less escape of nanoparticles in MCF7 cells than in MCF10A cells after incubation with nanoparticles. The possible reasons might be the overexpressed nucleolin in the cell cytoplasm and the significant amount of nuclei delivery of F3 conjugated nanoparticles for MCF7 cells. The zeta potential change of MCF7 cells basically follows the simulated zeta potential change pattern after incubation with opposite sign of nanoparticles when the internalization effect dominates during the whole process.

3.4. *Conclusion*

In summary, we investigated the different cellular uptake behaviors of F3 peptide conjugated dextran coated iron oxide nanoparticles in terms of cell surface zeta potential change after incubation with cancer breast epithelial cells (MCF7) and normal breast epithelial cells (MCF10A). The tumor-homing F3 peptide bound with overexpressed nucleolin receptors on the cell surface and inside the cytoplasm of breast carcinoma cells, which enables fast internalization compared to normal breast cells and also acts as a shuttle to deliver F3 peptide conjugated nanoparticles into the nuclei of cancer breast cells. The distinct zeta potential change patterns of MCF7 cancer breast cells and MCF10A normal breast cells after incubation with F3 conjugated nanoparticles demonstrate a sensitive and applicable method for describing the interaction between nanoparticles and biological cells. This suggests a great potential for the surface zeta potential measurements to be a promising approach in studying cellular interaction of nanomaterials. The surface zeta potential measurement opens the door for creating a combinatorial label-free approach to quickly identify and monitor the dynamic cell responses in biomedical research.

References

1. Alivisatos AP. (1996) Semiconductor clusters, nanocrystals, and quantum dots. *Science* **271(5251):** 933–937.
2. Jin RC, *et al.* (2001) Photoinduced conversion of silver nanospheres to nanoprisms. *Science* **294(5548):** 1901–1903.
3. Sun YG, Xia YN. (2002) Shape-controlled synthesis of gold and silver nanoparticles. *Science* **298(5601):** 2176–2179.
4. Couvreur P, Vauthier C. (2006) Nanotechnology: intelligent design to treat complex disease. *Pharm Res* **23(7):** 1417–1450.
5. Thurn KT, *et al.* (2007) Nanoparticles for applications in cellular imaging. *Nanoscale Res Lett* **2(9):** 430–441.

6. Edelstein AS, Cammarata RC. (1996) *Nanoparticles: Synthesis, Properties and Applications.* Inst Phys, Bristo l.
7. Gregoria G, *et al.* (1974) Drug-carrier potential of liposomes in cancer chemotherapy. *Lancet* **1(7870):** 1313–1316.
8. Ferrari M. (2005) Cancer nanotechnology: opportunities and challenges. *Nat Rev Cancer* **5(3):** 161–171.
9. Cuenca AG, *et al.* (2006) Emerging implications of nanotechnology on cancer diagnostics and therapeutics. *Cancer* **107(3):** 459–466.
10. Peer D, *et al.* (2007) Nanocarriers as an emerging platform for cancer therapy. *Nat Nanotechnol* **2(12):** 751–760.
11. Brigger I, Dubernet C, Couvreur P. (2002) Nanoparticles in cancer therapy and diagnosis. *Adv Drug Deliv Rev* **54(5):** 631–651.
12. Yezhelyev MV, *et al.* (2006) Emerging use of nanoparticles in diagnosis and treatment of breast cancer. *Lancet Oncology* **7(8):** 657–667.
13. Gupta AK, Wells S. (2004) Surface-modified superparamagnetic nanoparticles for drug delivery: preparation, characterization, and cytotoxicity studies. *IEEE Trans Nanobioscience* **3(1):** 66–73.
14. Torchilin VP, *et al.* (2003) Immunomicelles: targeted pharmaceutical carriers for poorly soluble drugs. *Proc Natl Acad Sci USA* **100(10):** 6039–6044.
15. Ahmad I, *et al.* (1993) Antibody-targeted delivery of doxorubicin entrapped in sterically stabilized liposomes can eradicate lung-cancer in mice. *Cancer Res* **53(7):** 1484–1488.
16. Sahoo SK, Labhasetwar V. (2003) Nanotech approaches to delivery and imaging drug. *Drug Discov Today* **8(24):** 1112–1120.
17. Torchilin VP. (2005) Recent advances with liposomes as pharmaceutical carriers. *Nat Rev Drug Discov* **4(2):** 145–160.
18. Xu JP, *et al.* (2005) Novel biomimetic polymersomes as polymer therapeutics for drug delivery. *Journal of Controlled Release* **107(3):** 502–512.
19. Blume G, Cevc G. (1990) Liposomes for the sustained drug release *in vivo. Biochim Biophysi Acta* **1029(1):** 91–97.
20. Tomalia DA, *et al.* (1985) A New Class of Polymers — Starburst-Dendritic Macromolecules. *Polymer J* **17(1):** 117–132.
21. Gillies ER, Frechet JMJ. (2005) Dendrimers and dendritic polymers in drug delivery. *Drug Discov Today* **10(1):** 35–43.
22. Klajnert B, Bryszewska M. (2001) Dendrimers: properties and applications. *Acta Biochim Pol* **48(1):** 199–208.
23. Stiriba SE, Frey H, Haag R. (2002) Dendritic polymers in biomedical applications: from potential to clinical use in diagnostics and therapy. *Angew Chem Int Ed* **41(8):** 1329–1334.
24. Sato N, *et al.* (2001) Pharmacokinetics and enhancement patterns of macromolecular MR contrast agents with various sizes of polyamidoamine dendrimer cores. *Magn Reson Med* **46(6):** 1169–1173.

25. Svenson S, Tomalia DA. (2005) Commentary — Dendrimers in biomedical applications — reflections on the field. *Adv Drug Deliv Rev* **57(15):** 2106–2129.
26. Patri AK, Kukowska-Latallo JF, Baker JR. (2005) Targeted drug delivery with dendrimers: comparison of the release kinetics of covalently conjugated drug and non-covalent drug inclusion complex. *Adv Drug Deliv Rev* **57(15):** 2203–2214.
27. Singh P, Gonzalez MJ, Manchester M. (2006) Viruses and their uses in nanotechnology. *Drug Dev Res* **67(1):** 23–41.
28. Manchester M, Singh P. (2006) Virus-based nanoparticles (VNPs): platform technologies for diagnostic imaging. *Adv Drug Deliv Rev* **58(14):** 1505–1522.
29. Lewis JD, *et al.* (2006) Viral nanoparticles as tools for intravital vascular imaging. *Nat Med* **12(3):** 354–360.
30. Blum AS, *et al.* (2004) Cowpea mosaic virus as a scaffold for 3-D patterning of gold nanoparticles. *Nano Lett* **4(5):** 867–870.
31. Koudelka KJ, *et al.* (2007) Interaction between a 54-kilodalton mammalian cell surface protein and cowpea mosaic virus. *J Virol* **81(4):** 1632–1640.
32. Lomonossoff GP, Hamilton WDO. (1999) Cowpea mosaic virus-based vaccines. In: *Plant Biotechnology*, pp. 177–189.
33. Moghimi SM, Hunter AC, Murray JC. (2001) Long-circulating and target-specific nanoparticles: theory to practice. *Pharmacol Rev* **53(2):** 283–318.
34. Vasir JK, Labhasetwar V. (2007) Biodegradable nanoparticles for cytosolic delivery of therapeutics. *Adv Drug Deliv Rev* **59(8):** 718–728.
35. Seydel C. (2003) Quantum dots get wet. *Science* **300(5616):** 80–81.
36. Bruchez M, *et al.* (1998) Semiconductor nanocrystals as fluorescent biological labels. *Science* **281(5385):** 2013–2016.
37. Qu LH, Peng XG. (2002) Control of photoluminescence properties of CdSe nanocrystals in growth. *J Am Chem Soc* **124(9):** 2049–2055.
38. Wu XY, *et al.* (2003) Immunofluorescent labeling of cancer marker Her2 and other cellular targets with semiconductor quantum dots. *Nat Biotechnol* **21(1):** 41–46.
39. Smith AM, *et al.* (2006) Multicolor quantum dots for molecular diagnostics of cancer. *Expert Rev Mol Diagn* **6(2):** 231–244.
40. Parungo CP, *et al.* (2005) Intraoperative identification of esophageal sentinel lymph nodes with near-infrared fluorescence imaging. *J Thorac and Cardiovasc Surg* **129(4):** 844–850.
41. Loo C, *et al.* (2004) Nanoshell-enabled photonics-based imaging and therapy of cancer. *Technol Cancer Res Treat* **3(1):** 33–40.
42. Oldenburg SJ, *et al.* (1998) Nanoengineering of optical resonances. *Chem Phys Lett* **288(2–4):** 243–247.
43. National Cancer Institute, *Cancer Imaging*. 2008. http://www.cancer.gov/

44. Hirsch LR, *et al.* (2003) Nanoshell-mediated near-infrared thermal therapy of tumors under magnetic resonance guidance. *Proc Natl Acad Sci USA* **100(23):** 13549–13554.
45. Gobin AM, *et al.* (2007) Near-infrared resonant nanoshells for combined optical imaging and photothermal cancer therapy. *Nano Lett* 7(7): 1929–1934.
46. Perez JM, *et al.* (2002) DNA-based magnetic nanoparticle assembly acts as a magnetic relaxation nanoswitch allowing screening of DNA-cleaving agents. *Journal of the American Chemical Society* **124(12):** 2856–2857.
47. Dyal A, *et al.* (2003) Activity of *Candida rugosa* lipase immobilized on gamma-Fe_2O_3 magnetic nanoparticles. *J Am Chem Soc* **125(7):** 1684–1685.
48. Gupta AK, Gupta M. (2005) Synthesis and surface engineering of iron oxide nanoparticles for biomedical applications. *Biomaterials* **26(18):** 3995–4021.
49. Berry CC, Curtis ASG. (2003) Functionalisation of magnetic nanoparticles for applications in biomedicine. *J Phy D:Appl Phys* **36(13):** R198–R206.
50. Lee H, *et al.* (2006) Antibiofouling polymer-coated superparamagnetic iron oxide nanoparticles as potential magnetic resonance contrast agents for *in vivo* cancer imaging. *J Am Chem Soc* **128(22):** 7383–7389.
51. Babes L, *et al.* (1999) Synthesis of iron oxide nanoparticles used as MRI contrast agents: a parametric study. *J Colloid Interface Sci* **212(2):** 474–482.
52. Hergt R, *et al.* (1998) Physical limits of hyperthermia using magnetite fine particles. *IEEE Trans Magn* **34(5):** 3745–3754.
53. Pardoe H, *et al.* (2003) A magnetic resonance imaging based method for measurement of tissue iron concentration in liver arterially embolized with ferrimagnetic particles designed for magnetic hyperthermia treatment of tumors. *Magn Reson Imaging* **21(5):** 483–488.
54. Sonvico F, *et al.* (2005) Folate-conjugated iron oxide nanoparticles for solid tumor targeting as potential specific magnetic hyperthermia mediators: synthesis, physicochemical characterization, and *in vitro* experiments. *Bioconjug Chem* **16(5):** 1181–1188.
55. Alexiou C, *et al.* (2000) Locoregional cancer treatment with magnetic drug targeting. *Cancer Res* **60(23):** 6641–6648.
56. Jain TK, *et al.* (2005) Iron oxide nanoparticles for sustained delivery of anticancer agents. *Mol Pharm* **2:** 194–205.
57. Kubaska S, *et al.* (2001), Dual contrast enhanced magnetic resonance imaging of the liver with superparamagnetic iron oxide followed by gadolinium for lesion detection and characterization. *Clin Radiol* **56(5):** 410–415.
58. Atanasijevic T, *et al.* (2006) Calcium-sensitive MRI contrast agents based on superparamagnetic iron oxide nanoparticles and calmodulin. *Proc Natl Acad Sci USA* **103(40):** 14707–14712.
59. Smirnov P, *et al.* (2006) *In vivo* cellular imaging of lymphocyte tracking by MRI: A tumor model approach to cell-based anticancer therapy. *Magn Reson Med* **56(3):** 498–508.

60. Cengelli F, *et al.* (2006) Interaction of functionalized superparamagnetic iron oxide nanoparticles with brain structures. *J Pharmacol Exp Ther* **318(1):** 108–116.
61. Hilger I, *et al.* (2002) Heating potential of iron oxides for therapeutic purposes in interventional radiology. *Acad Radiol* **9(2):** 198–202.
62. Babincova M, *et al.* (2000) Selective treatment of neoplastic cells using ferritin-mediated electromagnetic hyperthermia. *Med Hypotheses* **54(2):** 177–179.
63. Lacava LM, *et al.* (2002) Use of magnetic resonance to study biodistribution of dextran-coated magnetic fluid intravenously administered in mice. *J Magn Magn Mater* **252(1–3):** 367–369.
64. Hiemenz PC, Rajagopalan R. (1997) *Principles of Colloid and Surface Chemistry.* Marcel Dekker Inc, New York.
65. Weissleder R, *et al.* (1995) Long-circulating iron-oxides for Mr-imaging. *Adv Drug Deliv Rev* **16(2–3):** 321–334.
66. Harisinghani MG, *et al.* (1997) Differentiation of liver hemangiomas from metastases and hepatocellular carcinoma at MR imaging enhanced with blood-pool contrast agent Code-7227. *Radiology* **202(3):** 687–691.
67. Moore A, *et al.* (2000) Tumoral distribution of long-circulating dextran-coated iron oxide nanoparticles in a rodent model. *Radiology* **214(2):** 568–574.
68. Berry CC, *et al.* (2004) Cell response to dextran-derivatised iron oxide nanoparticles post internalisation. *Biomaterials* **25(23):** 5405–5413.
69. Porkka K, *et al.* (2002) A fragment of the HMGN2 protein homes to the nuclei of tumor cells and tumor endothelial cells *in vivo. Proc Natl Acad Sci USA* **99(11):** 7444–7449.
70. Akerman ME, *et al.* (2002) Nanocrystal targeting *in vivo. Proc Natl Acad Sci USA* **99(20):** 12617–12621.
71. Ruoslahti E. (2004) Vascular zip codes in angiogenesis and metastasis. *Biochem Soc Trans* **32:** 397–402.
72. Ruoslahti E, Duza T, Zhang L. (2005) Vascular homing peptides with cell-penetrating properties. *Curr Pharm Des* **11(28):** 3655–3660.
73. Christian S, *et al.* (2003) Nucleolin expressed at the cell surface is a marker of endothelial cells in angiogenic blood vessels. *J Cell Biol* **163(4):** 871–878.
74. Enback J, Laakkonen P. (2007) Tumour-homing peptides: tools for targeting, imaging and destruction. *Biochem Soc Trans* **35:** 780–783.
75. Reddy GR, *et al.* (2006) Vascular targeted nanoparticles for imaging and treatment of brain tumors. *Clin Cancer Res* **12(22):** 6677–6686.
76. Hunter RJ. (1981) *Zeta Potential in Colloid Science Principles and Applications.* Academic Press Inc.
77. Veronesi B, *et al.* (2002) The surface charge of visible particulate matter predicts biological activation in human bronchial epithelial cells. *Toxicol Appl Pharmacol* **178(3):** 144–154.

78. Fontes A, *et al.* (2006) Studying red blood cell agglutination by measuring electrical and mechanical properties with a double optical tweezers. *Microsc Microanal* **12:** 1758–1759.
79. Lin DQ, Zhong LN, Yao SJ. (2006) Zeta potential as a diagnostic tool to evaluate the biomass electrostatic adhesion during ion-exchange expanded bed application. *Biotechnol Bioeng* **95(1):** 185–191.
80. Altankov G, Richau K, Groth T. (2003) The role of surface zeta potential and substratum chemistry for regulation of dermal fibroblasts interaction. *Materialwiss Werkstofftech* **34(12):** 1120–1128.
81. Zhang Y, *et al.* (2008) Zeta potential: a surface electrical characteristic to probe the interaction of nanoparticles with normal and cancer human breast epithelial cells. *Biomed Microdevices* **10(2):** 321–328.
82. Wilhelm C, *et al.* (2002) Interaction of anionic superparamagnetic nanoparticles with cells: kinetic analyses of membrane adsorption and subsequent internalization. *Langmuir* **18(21):** 8148–8155.
83. Ginisty H, *et al.* (1999) Structure and functions of nucleolin. *J Cell Sci* **112(6):** 761–772.
84. Srivastava M, Pollard HB. (1999) Molecular dissection of nucleolin's role in growth and cell proliferation: new insights. *Faseb J* **13(14):** 1911–1922.
85. Deng JS, Ballou B, Hofmeister JK. (1996) Internalization of anti-nucleolin antibody into viable HEp-2 cells. *Mol Biol Rep* **23(3–4):** 191–195.
86. Larrucea S, *et al.* (1998) Cellular adhesion mediated by factor J, a complement inhibitor — Evidence for nucleolin involvement. *J Biol Chem* **273(48):** 31718–31725.
87. Said EA, *et al.* (2002) The anti-HIV cytokine midkine binds the cell surface-expressed nucleolin as a low affinity receptor. *J Biol Chem* **277(40):** 37492–37502.
88. Sinclair JF, O'Brien AD. (2002) Cell surface-localized nucleolin is a eukaryotic receptor for the adhesin intimin-gamma of enterohemorrhagic *Escherichia coli* O157 : H7. *J Biol Chem* **277(4):** 2876–2885.
89. Khalil IA, *et al.* (2006) Uptake pathways and subsequent intracellular trafficking in nonviral gene delivery. *Pharmacol Rev* **58(1):** 32–45.

Chapter 16

Noscapines: Novel Carrier Systems to Target Tumor Cells

Ramesh Chandra*, J. Madan†, Prashant Singh†, Pradeep Kumar†, V. Tomar†, Sujata K. Dass‡ and Ankush Chandra§

Abstract

Noscapine, a plant-derived, non-toxic, antitussive alkaloid which has a tubulin binding anticancer activity, has been identified to inhibit tumor growth in nude mice bearing human xenografts of hematopoietic, breast, lung, ovarian, brain, and prostate origin. The clinical trials of noscapine and its analogs as an anticancer agent are undergoing. The analogs appear to have higher tubulin binding activity than noscapine and affect tubulin polymerization differently than noscapine, and these analogs appear to be active and able to arrest cell cycle progression at mitosis using concentrations much lower than noscapine in inhibiting the proliferation of various human cancer cells, including those that are resistant to Paclitaxel and epothilone. Noscapine potentiated the

*Corresponding author: Prof. Ramesh Chandra Founder Director, Dr. B.R. Ambedkar Centre for Biomedical Research, University of Delhi, Delhi-110007, India. E-mail: acbrdu@hotmail.com

†Dr. B.R. Ambedkar Centre for Biomedical Research, University of Delhi, Delhi-110007.

‡V.P. Chest Institute, University of Delhi, Delhi-110007, India.

§Massachusetts General Hospital, Harvard Medical School, Boston, 55 Fruit St., MA-02114, USA.

anticancer activity of Doxorubicin in a synergistic manner against Triple Negative Breast Cancer (TNBC) via inactivation of NF-KB and anti-angiogenic pathways, while stimulating apoptosis, suggesting the potential benefit for the use of oral noscapine and doxorubicin combination therapy for the treatment of more aggressive TNBC.

The efficacy of the drug can be significantly affected by the drug delivery systems, such as nanoparticles (NPs). The recent chemotherapeutic work on drug delivery using NPs as carriers for the targeted treatment of cancer as compared to direct drug delivery through a carrier can increase the efficacy of a drug but decrease side effects by utilizing the enhanced permeability and retention (EPR) effect and tumor-specific targeting. The research for efficient and safe transport vehicles (carriers) to achieve better drug availability at the target site has been a challenging and exciting area. The colloidal NPs, including the biodegradable polymer and liposome systems bio-conjugating with antitumor drugs, have been an area of importance. These NPs, with enlarged surface area to volume ratio can overcome non-cellular and cellular base mechanisms of resistance and increase the selectivity of drugs towards cancer cells and reduce their toxicity for normal tissues. To improve the biodistribution of the noscapine cancer drug, NPs have been designed for optimal size and surface characteristics to increase their circulation time in the blood stream. They are also able to carry their loaded active drug to cancer cells by selectively using the unique patho-physiology of tumors, such as their enhanced permeability and retention effect and the tumor microenvironment. NPs have the ability to accumulate in cells without being recognized by P-glycoprotein, one of the main mediators of Multi Drug Resistance (MDR), resulting in the increased intracellular concentration of drugs. Our group has constructed noscapine-enveloped gelatin NPs and Poly(ethylene glycol) grafted gelatin NPs as well as inclusion complex of NOS in β-cyclodextrin (β-CD) and evaluated their physicochemical characteristics. The Fe_3O_4 NPs were also used to prepare noscapine-loaded magnetic Polymeric NPs (NMNP) which encapsulate the noscapine on the surface of the polymer matrix, whereas the molecular weight of the polymer plays a crucial role in the capacity of drug loading on the polymer surface. The enhanced noscapine delivery using μPAR-targeted optical-MR imaging trackable NPs offer a great potential for image directed targeted delivery of noscapine for the management of prostate cancer. Human Serum Albumin NPs (150–300 nm) as efficient noscapine drug delivery systems have also been developed for potential use in breast cancer.

1. Introduction

Noscapine is a novel tubulin binding, antiangiogenic, anticancer drug that causes cell cycle arrest and induces apoptosis in cancer cells both *in vitro* and *in vivo*.[1–4] Based on the GLOBOCAN 2008 estimates, about 12.7 million cancer cases and 7.6 million cancer deaths are estimated to have occurred in 2008; of these, 56% of the cases and 64% of the deaths occurred in the economically developing world. Breast cancer is the most frequently diagnosed cancer and the leading cause of cancer death among females, accounting for 23% of the total cancer cases and 14% of the cancer deaths. Lung cancer is the leading cancer site in males, comprising 17% of the total new cancer cases and 23% of the total cancer deaths. Breast cancer is now also the leading cause of cancer death among females in economically developing countries, a shift from the previous decade during which the most common cause of cancer death was cervical cancer. The bioactive compounds are emerging as potential future drugs in addition to those already under research and development, and a major challenge for efficient drug delivery and targeting. Thus, Newman and Cragg[5] have shown that 63% of anticancer drugs introduced over the last 25 years are natural products or can be traced back to a natural products source, and similar observations have been made by many others. A recent review by Butler lists 79 natural products or natural product analogs that entered clinical trial as anticancer agents in the 2005–2007 timeframe.[6]

The very slow progress in the efficacy of the treatment of severe diseases suggested a need for a multidisciplinary approach to the delivery of therapeutic drugs to the targeted tissues. New ideas on controlling the pharmacokinetics, pharmacodynamics, non-specific toxicity, immunogenicity, bio-recognition and efficacy of drugs are being generated. These new strategies are called as drug delivery systems (DDS), and are based on interdisciplinary approaches that combined bio-conjugate chemistry, pharmaceutics, polymer science and molecular biology. Targeting is the ability to direct the drug-loaded system to the site of interest. Two major mechanisms can be distinguished for addressing the desired sites for drug release: (i) passive and (ii) active targeting. An example of passive targeting is the preferential accumulation of chemotherapeutic agents in solid tumors as a result of the enhanced vascular permeability of tumor tissues compared with healthy tissue. A strategy that could allow active targeting involves the surface functionalization of drug carriers with ligands that are selectively recognized by receptors on the surface of the cells of interest. Since ligand–receptor interactions can be highly selective, this could allow a more precise targeting of the site of interest. When developing these formulations, the goal is to obtain systems with optimized drug loading and release properties, long shelf-life and low toxicity. The incorporated drug participates in the microstructure of the system and may even influence it due to molecular interactions, especially if the drug possesses amphiphilic and/or mesogenic properties.

Noscapine, a phthalide isoquinoline alkaloid derived from opium, has been used as an oral antitussive agent and has shown minimal toxic effects in animals and humans. It is a naturally occurring tubulin-binding agent currently undergoing clinical trials for anticancer therapy. During last 25 years, researchers have appreciated the potential benefits of nanotechnology in providing vast improvements in drug delivery and drug targeting. Improving delivery techniques that minimize toxicity and improve efficacy offers great potential benefits to patients and opens up new markets for pharmaceutical and drug delivery companies. Other approaches to drug delivery are focused on crossing particular physical barriers, such as the blood–brain barrier, in order to better target the drug and improve its effectiveness; or on finding alternative and acceptable routes for the delivery of protein drugs other than via the gastro-intestinal tract, where degradation can occur.

Microtubule-inhibitor-based chemotherapy is a major therapeutic approach for the treatment of cancer which may be used alone or combined with other forms of therapy. In this series, our group is continuously working on noscapine and its derivatives for better future prospects.[7,8] However, their low aqueous solubility, short oral and plasma half-life, low oral bioavailability, and high therapeutic dose produce hurdles in their extensive commercialization. Nanotherapeutics is rapidly progressing in the direction of solving several limitations of conventional drug delivery systems. Nonspecific targeting of cancer chemotherapy leads to the damaging of rapidly proliferating normal cells. The hydrophobic nature of chemotherapeutics leads to poor aqueous solubility and low bioavailability, which can be overcome by colloidal drug delivery systems.

Chemotherapy-induced side effects caused by some current chemotherapy drugs can be so severe that they can lead to cessation of the treatment plan for the patient. Therefore, development of drug systems that could selectively deliver drug molecules to the tumor site without a concurrent increase in its level in the normal healthy cells of the tissues is currently one of the most active areas of investigation in cancer research. Several different possibilities have been explored to reduce the toxicity and increase anti-tumor efficacy of anti-cancer drugs using nano-devices. Encapsulating noscapine in NPs will help to increase its efficacy and lowers any side effects. In this manuscript, we have discussed the work done in our and collaborating labs, various types of NPs of noscapine and their efficacy when used to target various types of tumors.[9–11]

2. Mechanism of Action of Noscapine

Among the various mechanisms of action of natural products, that of interaction with the cellular protein tubulin is one of the most important, and over 25% of the new clinical candidates listed by Butler operate by this general mechanism. Two major classes of anticancer drugs owe their effectiveness to this mechanism: the first class is that of the tubulin polymerization inhibitors, and the second is that of tubulin polymerization promoters. The cellular protein tubulin is a crucial protein for cellular replication. The cell cycle involves the replication of DNA and the packaging of the resulting replicated chromosomes into two daughter cells. The separation of the daughter chromosomes in mitosis is brought about by microtubules, which are formed by the polymerization of α- and β-tubulin. The microtubules radiate in cells from centrosomes and from the poles of mitotic spindles, and are formed by attachment of GTP-tubulin to the growing end of an existing microtubule. Microtubules undergo rapid assembly and disassembly in cells, and this property enables those associated with the mitotic spindle to generate a large collection of structures and thus to produce those structures which will interact constructively with the centromeres of daughter chromosomes and generate the necessary aligned chromosomes at metaphase. The normal functioning of tubulin assembly and disassembly is thus crucial to cell division, and any interference with this process will disrupt cell division and cause cell death by apoptosis. Although the most dramatic effect of the tubulin-interactive drugs is that of changing the extent of microtubule polymer mass, either decreasing it for the tubulin polymerization inhibitors, or increasing it for the tubulin polymerization promoters, cancer cell growth can be inhibited at concentrations significantly lower than those necessary to exert these macroscopic effects. This fact can be explained by the observation that cell growth inhibition at low concentrations is caused by the suppression of microtubule dynamics. The structure of the tubulin heterodimer has been solved by electron diffraction. Both the vinca alkaloids and the taxane drugs bind to β-tubulin, but at different locations on the protein; the vinca alkaloids bind to β-tubulin between amino acids 175 and 213, while paclitaxel binds both to an N-terminal unit of β-tubulin14 as well as to the region bounded by amino acids 217–231.[12,13] Colchicine, which is not a clinically used drug for cancer but which has been studied extensively, binds between the two subunits.[14] The epothilones also bind at the paclitaxel site.[15] Noscapine attenuates microtubule dynamics just enough to activate the mitotic checkpoints to stop cell cycle and do not alter the steady state monomer/polymer ratio of tubulin.[16]

2.1. *Anti-Angiogenic Effect of Noscapine*

Angiogenesis, necessary for tumor growth, involves cell proliferation and directed migration. Thus, there is clearly a crucial role of cytoskeletal microtubule (MT) dynamics in angiogenesis, linking perturbations of MT dynamics to inhibition of tumor angiogenesis. The process includes the action of various growth factors and cytokines, in particular vascular endothelial growth factor (VEGF). For example, nitric oxide causes local vasodilation and metalloproteinases, a kind of protease that degrades the local basement membrane and the local matrix, and also mobilizes further growth factors from the matrix. Further, endothelial cells migrate out, forming a solid capillary sprout followed by stabilization of the endothelial layer through cell-to-cell binding by adherence proteins and integrin binding of cells to the matrix.[17] Over expression of hypoxia-inducible factor-1 (HIF-1) is a common feature in solid malignancies, thus HIF-1 and its signaling pathway are well-known targets for cancer chemotherapy.[18] Noscapine down regulated hypoxia mediated HIF-1α expression in human glioma cells, concomitantly with reduced secretion of the potent angiogenic cytokine, VEGF.[19]

2.2. *Physicochemical, Pharmacokinetic and Pharmacodynamic Limitations of Tubilin-binding Anticancer Agents*

Beside the major benefits of noscapine as a novel resistance-free anticancer drug, it is also associated with several limitations, like other cancer chemotherapeutic drugs.

(a) *Limited aqueous solubility*: Noscapine, being an alkaloid, is hydrophobic in nature and possesses limited solubility in aqueous phase[20]; (b) *Ionization at acidic pH*: Noscapine ionizes at acidic pH in the stomach, which may lead to low bioavailability due to slow absorption from the intestine; (c) *Lack of selectivity of anticancer drugs*: Most chemotherapeutics lack selectivity toward cancerous cells, causing significant damage to rapidly proliferating normal cells; (d) *Limited oral bioavailability:* Noscapine is reported to have low oral bioavailability both in humans and mice, below 44%[20,21]; (e) *Short oral and plasma half-life*: Noscapine presents short oral and plasma half-life of below 2 hr, which poses hurdles in the design of controlled release formulations[20]; (f) *Rapid elimination from body tissues*:

Noscapine rapidly eliminates from body tissues with first order kinetic, therefore requiring multiple injections for successive chemotherapy.[22]

2.3. *Novel Potent Analogues of Noscapine*

Our group has synthesized and reported various potent analogues of noscapine. We showed that 9-chloronoscapine is more potent than 9-bromonoscapine when tested on U87 glioma cells, however, it did not show a significant difference at low concentration.[23] Moreover, 9-bromonoscapine was also tested on various cancer cell lines and possesses 5- to 10-fold higher anticancer activity in comparison to parent compound, Noscapine, in preclinical testing.[24–26] Moreover, a cyclic ether fluorinated Noscapine analog showed potent antiproliferative and anticancer activity in both hormone-responsive (MCF-7) and hormone non-responsive (MDA-MB-231) breast cancer cells.[27] 9-aminonoscapine also showed promising results and it binds to tubulin at a site overlapping with colchicine-binding site or close to it. Thus, 9-aminonoscapine has better anti-tumor activity than noscapine.[28] Researchers have validated more potent antitumor noscapinoid, 9-azido-noscapine, and reduced 9-azido-noscapine with IC_{50} value of 1.2 to 56.0 μM in human acute lymphoblastic leukemia cells.[29]

The lead compound noscapine is currently undergoing phase I/II clinical trials for cancer treatment at the University of Southern California, USA. EM011 is potently effective against vinblastine-sensitive line CEM. Although noscapine appears to be a weak inhibitor of microtubule polymerization, its low cost and ready availability allow for further exploration of this natural product.

2.4. *Noscapine: Combination Therapy*

Current approach in cancer chemotherapy involves a mixture of drugs and techniques that designed to selectively target cancer cells. Therefore, noscapine, being a novel anticancer drug, was also investigated with a number of drugs for better cancer chemotherapy. Researchers showed that noscapine and doxorubicin combination

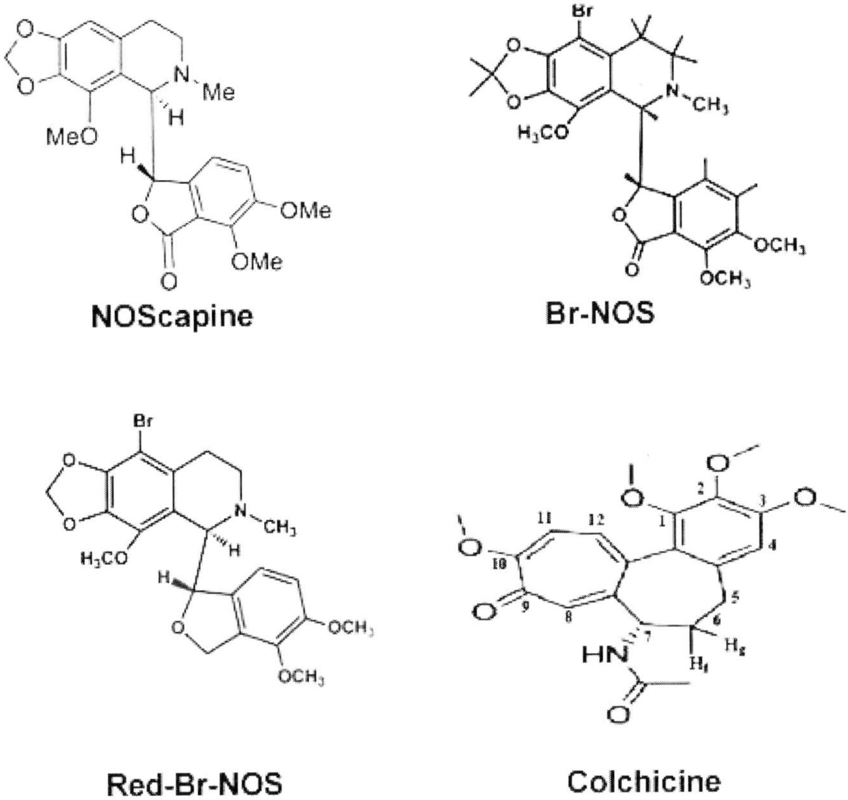

Fig. 1. Molecular structure of Noscapine, 9-bromonoscapine (Br-nos), reduced 9-bromonoscapine (red-Br-nos) and colchicines.

treatment significantly increased the extent of apoptosis in breast cancer cells.[30] This combination showed decreased expression of NF-KB pathway proteins, VEGF, cell survival, and increased expression of apoptotic and growth inhibitory proteins compared to single-agent treatment and control groups. These findings indicated the potential use of oral noscapine and doxorubicin combination therapy for treatment of more aggressive tumors. Further, noscapine was combined with cisplatin and tested on A549 and H460 lung cancer cells and *in vivo* in murine xenograft model.[31] The combination showed the higher percentage of apoptotic non-small cell lung cancer cells and increased expression of p53, p21, caspase-3, cleaved caspase-3, cleaved-PARP, Bax, and decreased expression of Bcl-2 and surviving proteins compared with treatment with either agent. Moreover, docetaxel (25 nM) in combination with 9-bromonoscapine caused an additive increase in proapoptotic activity in prostate cancer cells.[32] But the researchers did not observe the synergistic effect in all combinations. Combination of noscapine with imatinib mesylate showed antagonist effect in T98G human GBM cells *in vitro*.[33]

3. Nanoscale-based Drug Delivery Systems for Cancer Chemotherapeutics

Nanotechnology literally means technology performed on a nanoscale. The nanoscale/nanocavities are ultrafine vesicles in the size of nanometer from 1 nm to 1000 nm. Nanomedicine is an important area in nanotechnology which refers to highly specific medical intervention at the molecular scale for diagnosis, prevention and treatment of diseases. Nanotherapeutics are a rapidly progressing field, and are utilized to solve several limitations of conventional drug delivery system such as nonspecific biodistribution, lack of targeting, lack of aqueous solubility, poor oral bioavailability, and low therapeutic indices. Some important technological advantages of nanotherapeutic drug delivery systems (NDDS) are (a) NDDS provides longer shelf-life; (b) both hydrophilic and hydrophobic substances can be incorporated in NDDS; (c) NDDS can be administrated orally, nasally, parenterally, intraocularly, etc.; (d) NDDS improve the biodistribution of cancer drugs. Whereas optimal size and surface characteristics of NPs increases the circulation time of the drug, (e) NDDS provides controlled and sustained release of the drug both during the transportation and at the site of action, and (f) NDDS enhances drug concentration in tumor network either by enhanced permeability and retention effect (EPR) or by endocytosis mechanism.[34]

3.1. *Passive Targeting*

Size of nanovesicles and behavior of tumor tissue vasculature plays a significant role in passive targeting. A solid tumor consists of tumor parenchyma and stroma, which in turn consist of vasculature and other supporting cells. Due to the increased metabolic requirements of growing tumor cells, pre-existing blood vessels are subjected to angiogenic pressure, which leads to the development of new capillaries to the tumor in a process called angiogenesis. Scanning electron microscopic studies revealed that the formed tumor capillaries are highly irregular and show gross architectural changes. Normal tissue vasculatures are lined by tight endothelial cells thereby preventing entry of nanovesicles, whereas tumor tissue vasculatures are leaky (gaps as large as 200 to 2000 nm between adjacent endothelial cells) and hyper permeable. This defective vascular architecture induces an EPR and permits accumulation of nanovesicles in the tumor interstitial space. Accumulation of nanovesicles in tumor tissues depends on interstitial fluid pressure, which is higher in tumor tissues than in benign tumor and normal tissues. In particular, interstitial pressure would be higher at the center diminishing towards the periphery, which is responsible for the induction of drug outflow from the cells that may lead to drug redistribution in some portions of the cancer tissue. Accumulation of nanovesicles in tumor tissues also depends on size, surface character, circulation half-life of the nanovesicles and the degree of angiogenesis of the tumor. Selected delivery systems to achieve passive targeting are liposomes, polymeric NPs, nanocrystals, inorganic NPs, micelles, dendrimers, etc.

3.2. *Active Targeting*

Paul Ehrlich coined the term "magic bullet," which is an idealized package that would target and deliver drugs to a specific place in the body. This idea of active targeting was proposed even before a rational targeting ligand was discovered. Active targeting involves conjugation of targeting molecules (like antibodies, ligands, peptides, nucleic acids, etc.) on the surface of nanovesicles with receptors overexpressed on a tumor cell surface. Tumor targeting molecules on the nanovesicles bind to cells through an endosome-dependent mechanism which bypasses the drug efflux pump, leading to high intracellular concentration.[35,36]

4. Nanoparticles (NPs) for Drug Delivery in Cancer

NPs hold great promise for improving cancer treatment. For example, they can guide drugs directly to tumors, increasing effectiveness and reducing side effects. However, significant challenges need to be overcome before these engineering marvels make it to the clinic. On the engineering side, it is difficult to make anything that small, around 100 nanometers (a nanometer is one billionth of a meter). Researchers must also generate particles that are uniform in size and shape and, once they have done their job, these particles must break down safely in the body. On the treatment side, NPs share the same obstacles as all potential treatments — cancer is wily. Because the disease evolves so rapidly, it finds ways to escape treatments, leading to drug resistance. So even the perfect nanoparticle containing a single treatment might not be effective in the long run. NPs, by using passive and active targeting strategies, can enhance the intracellular concentration of drugs on cancer cells while avoiding toxicity in normal cells. Furthermore, when NPs bind to specific receptors and then enter the cell, they are usually enveloped by endosomes via receptor-mediated endocytosis, thereby bypassing the recognition of P-glycoproteins, one of the main drug-resistance mechanisms. Although NPs offer many advantages as a drug-carrier system, there are still many limitations to be solved, such as poor oral bioavailability, instability in circulation, inadequate tissue distribution and toxicity. NPs applied as drug delivery systems are submicro sized particles (3–200 nm), devices or systems that can be made using variety of material including polymers (polymeric NPs, micelles or dendrimers) and even oganometallic compounds (nanotubes).[37]

4.1. *Two-faced NPs and Cancer*

Named for the Roman god with two faces, Janus particles are spherical and have two very different sides — one that loves water and one that hates it. This conflicted relationship with water comes into play when attaching therapies. The two distinct sides can hold different agents. "These particles give us a lot of versatility," according to

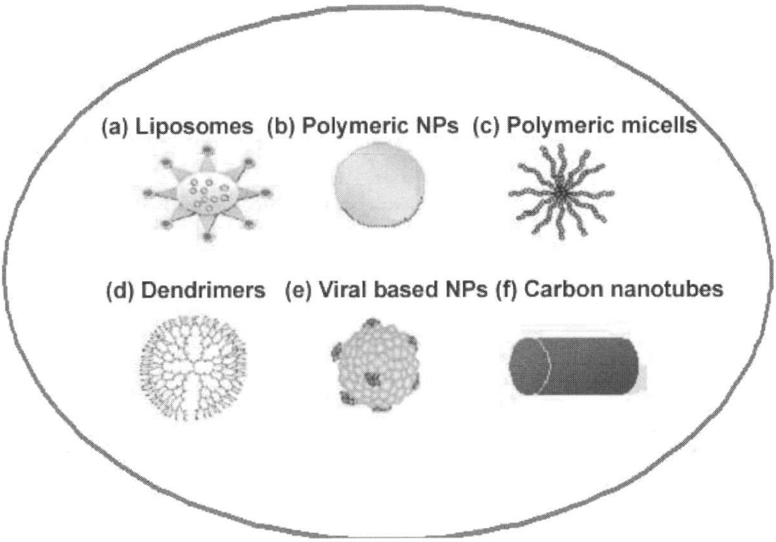

Fig. 2. Types of nano carriers for drug delivery. (a) Liposomes: self-assembling structures composed of lipid bilayers in which an aqueous volume is entirely enclosed by a membranous lipid bilayer; (b) polymeric NPs, in which drugs are conjugated to or encapsulated in polymers; (c) polymeric micelles: amphiphilic block copolymers that form nanosized core/shell structures in aqueous solution. The hydrophobic core region serves as a reservoir for hydrophobic drugs, whereas the hydrophilic shell region stabilizes the hydrophobic core and renders the polymers water-soluble; (d) dendrimers: synthetic polymeric macromolecules of nanometer dimensions, composed of multiple highly branched monomers that emerge radially from the central core; (e) viral-based NPs: in general structure are the protein cages, which are multivalent, self-assembled structures; (f) carbon nanotubes: carbon cylinders composed of benzene rings.

Dr. Jeffrey Smith, whose lab is working to perfect Janus particles. "For example, they can create a theranostic, a particle with a therapy on one side and a diagnostic imaging agent on the other." Janus particles have been around for a few years, but they have been difficult to make in the quantity and uniformity needed for clinical trials. The Smith laboratory may have solved these problems. "These particles are stunningly uniform," according to Dr. Smith. In addition, the particles are made of a polymer called PLGA, which is already approved by the Food and Drug Administration and breaks down into harmless glycolic acid and lactic acid in the bloodstream. Janus particles could be a great boon to cancer treatment. For example, clinicians could create particles with chemically different treatments on each face, allowing them to combine treatments that normally could not be combined. In addition, some treatments known for their toxicity are significantly less toxic when delivered by NPs. Dr. Smith is also creating particles with a treatment on one side and a gene or piece of RNA on the other. The genetic material would make the cancer cell more vulnerable to the treatment. Ultimately, Dr. Smith would like to make particles with multiple faces, allowing him to create a nano-cocktail to attack cancer. "You just cannot be effective without attacking from multiple angles." "More faces mean more therapeutics and hopefully more success."[38]

4.2. *Noscapine Nanotherapeutics*

Drug substances are considered highly soluble when the largest dose of the drug is soluble in < 250 mL of water throughout the physiological pH range from 1–8. Most anticancer drugs show poor aqueous solubility. Chemotherapeutics with poor aqueous solubility, both from plant sources and synthetic, often demonstrate decreased bioavailability, increased chance of food effect, more frequent incomplete release from the dosage form and higher interpatient variability. However, administration of poor-water-soluble drugs through the systemic route requires solvents like Cremophor EL, Tween (polysorbate)-80, etc. which in turn lead to severe adverse effects, including acute hypersensitivity reactions, fluid retention and peripheral neuropathy. There are two basic approaches to overcome the poor water solubility and poor bioavailability problems: (a) increase saturation solubility (by complex formation) and (b) increase dissolution velocity (dissolution velocity can be increased by increasing the surface area of the drug powder, i.e. nanonisation).

5. Noscapine-loaded Magnetic Polymeric NPs

Among the current attractive schemes of drug targeting, a promising one employs magnetically guided NPs.[39,40] This technology is based on the encapsulation of a drug with magnetic NPs or magnetic NPs coupled to a polymeric carrier. The potential advantage with magnetically guided drug NPs is that the strength of the magnetic field enables the NPs to escape uptake by the reticular endothelial system and multiple drug-resistance pumps, which are common factors for poor drug performance. Therefore, in recent years the scientific community has been seeking to exploit the intrinsic properties of magnetic lead to cessation of the treatment plan for the patient. Therefore, researchers have focused on tailoring magnetic NPs for drug delivery and diagnosis.[41–45] NPs can act at the tissue or cellular level. The latter implies that they can be endocytosed or phagocytosed (i.e. by dendritic cells, macrophages) resulting in internalization of the NPs. In this process, the NPs can reach beyond the cytoplasmatic membrane and, in some cases, also beyond the nuclear membrane (i.e. transfection applications). Tumor targeting with magnetic NPs may use passive or active strategies. Passive targeting occurs as a result of extravasations of the NPs at the diseased site (tumor) where the microvasculature is hyper permeable and leaky, a process aided by tumor-limited lymphatic drainage. Combined, these factors lead to the selective accumulation of NPs in tumor tissue, a phenomenon known as enhanced permeation and retention (EPR).[43]

Another major advantage of using magnetic particles coupled to a biodegradable polymeric carrier is that the therapeutic drug concentrations are maintained for sustained periods of time. The polymeric matrix prevents drug degradation and also allows precise control over the release kinetics of the drug from the NPs. Moreover, the duration of drug levels released from the NPs can be easily modulated by altering formulation parameters such as drug–polymer ratio, or polymer molecular weight and composition. Super paramagnetic iron oxide NPs (Fe_3O_4 NPs), which are commonly used to confer the magnetic property to any polymeric drug delivery systems, are also used as popular contrast agents for clinical diagnostics with magnetic resonance imaging (MRI). Therefore, most magnetic drug delivery systems supersede other targeted drug delivery systems because of their dual property as a drug delivery system and an *in vivo* imaging system.

Noscapine has recently been found to bind to tubulin and alter its conformation, assembly properties, and microtubule dynamic. The analog EM105 (chloro) is more potent and regresses breast cancer xenografts in nude mice without significant toxicity. Noscapine and its analogs are thus interesting lead compounds. The delivery of noscapine therapies directly to the site of the tumor would ultimately allow higher concentrations of the drug to be delivered, and prolong circulation time *in vivo* to enhance the therapeutic outcome of this drug. Therefore, we designed magnetic-based polymeric NPs for the site-directed delivery of noscapine to invasive tumors and synthesized Fe_3O_4 NPs with an average size of 10 ± 2.5 nm. These Fe_3O_4 NPs were used to prepare noscapine-loaded magnetic polymeric NPs (NMNP) with an average size of 252 ± 6.3 nm. FT-IR spectroscopy showed the encapsulation of noscapine on the surface of the polymer matrix. The encapsulation of the Fe_3O_4 NPs on the surface of the polymer was confirmed by elemental analysis, mass and other techniques. We studied the drug loading efficiency of polylactide acid (PLLA) and poly(l-lactide acid-co-gylocolide) (PLGA) polymeric systems of various molecular weights. Our findings revealed that the molecular weight of the polymer plays a crucial role in the capacity of the drug loading on the polymer surface. Using a constant amount of polymer and Fe_3O_4 NPs, both PLLA and PLGA at lower molecule weights showed higher loading efficiencies for the drug on their surfaces.

We hypothesize that creating magnetically guided noscapine NPs will present a new method for administering this drug with high targeting efficacy, which will significantly increase its therapeutic index. While reports have not conclusively provided a reason for the remnant cancerous mass upon noscapine treatment, it may be attributed to the rapid elimination of the drug and/or its widespread distribution into non-targeted organs and tissues. Currently, clinical data (available on pcref.com) reveal that prostate cancer patients who consume off-label noscapine are on a regimen of about 1000 mg/day of the drug to achieve a lowering of PSA (prostate specific antigen) levels. Although no detectable toxicity has been so far reported with such enormously high doses of noscapine, consuming such large quantities of drug can present long-term side effects due to accumulation in healthy tissues.

Thus, this system will ensure the delivery of a concentrated dose of noscapine directly into the tumor, which will possibly eliminate the need for the patient to consume large quantities of the drug. This will have tremendous positive impact, both physiological and economical, on cancer patients. Another advantage of this system is the possibility to use MRI to visualize the distribution of drug-loaded NPs in the tumor and its surrounding healthy tissues. This is very important because it leads to advantages such as: (i) pre-imaging of the distribution of drug-free NPs (a tracer dose) to enable good prediction of tumor targeting; (ii) monitoring during and after administration of the drug-loaded NPs leading to more efficacious treatment.[46]

5. Cyclodextrin-based Nanomicells to Enhance Aqueous Solubility and Bioavailability

Drugs are considered highly soluble when the therapeutic dose of a drug is freely soluble in 250 ml of the dissolution medium throughout the physiological pH (1–8). Most anticancer drugs show poor aqueous solubility. Moreover, administration of poor water-soluble drugs through the parental route requires solvents like Cremophor EL,

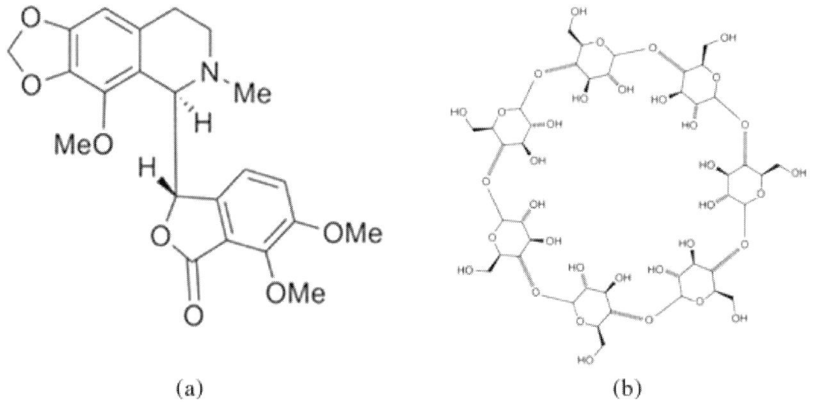

Fig. 3. Schematic representation of the chemical structures of (a) NOS, a phthalideisoquinoline alkaloid obtained from opium; (b) β-cyclodextrin (β-CD).

polysorbate-80 that lead to severe adverse effects, including acute hypersensitivity reactions, fluid retention and peripheral neuropathy. Cyclodextrins were used to enhance aqueous solubility and stability of anticancer drugs in dissolution medium. Researchers studied the inclusion complexation behavior of paclitaxel with a series of oligo (ethylenediamino) bridged bis(β-cyclodextrin) possessing bridge chains in different lengths of 1–4 to improve the water solubility and bioavailability of paclitaxel.[47,48] The cyclodextrin inclusion complex of anticancer drug alters the solubility and bioavailability and can be utilized to deliver cancer therapeutics with low solubility and bioavailability. Noscapine, due to its short half-life and low bioavailability, requires high therapeutic doses of the drug (ED_{50}–300 mg/kg bodyweight) for induction of anticancer activity. Therefore, we synthesized the noscapine-β-cyclodextrin inclusion complex with *Kc* value of 0.454 mM^{-1} at pH 7.4, indicating the formation of a stable complex with low affinity.

We also explored the molecular mechanism in the inclusion mode with spectroscopic techniques and correlated this with the molecular modeling. Our data proved that O–CH_2–O moiety of noscapine entered in the β-cyclodextrin cavity and formed the inclusion complex in 1:1 ratio. Noscapine requires basic environment to remain in unionized form and also for its absorption. The dissolution profile of noscapine and

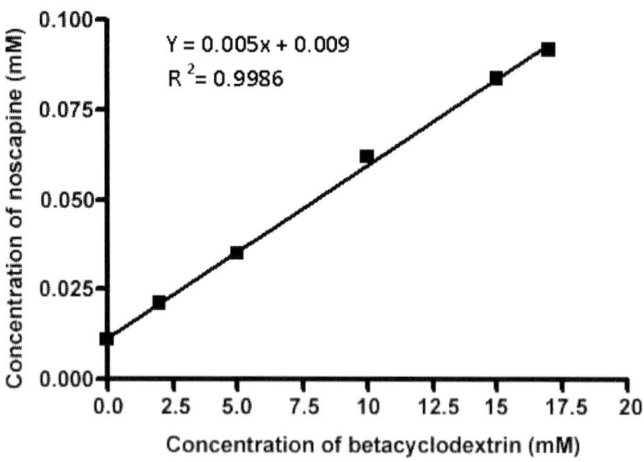

Fig. 4. Phase-solubility diagram of NOS in phosphate buffer saline (PBS, pH 7.4, 37°C) β-CD solution. An A_L type curve was obtained showing the linear increase in solubility of NOS with respect to concentration of β-CD (n = 3).

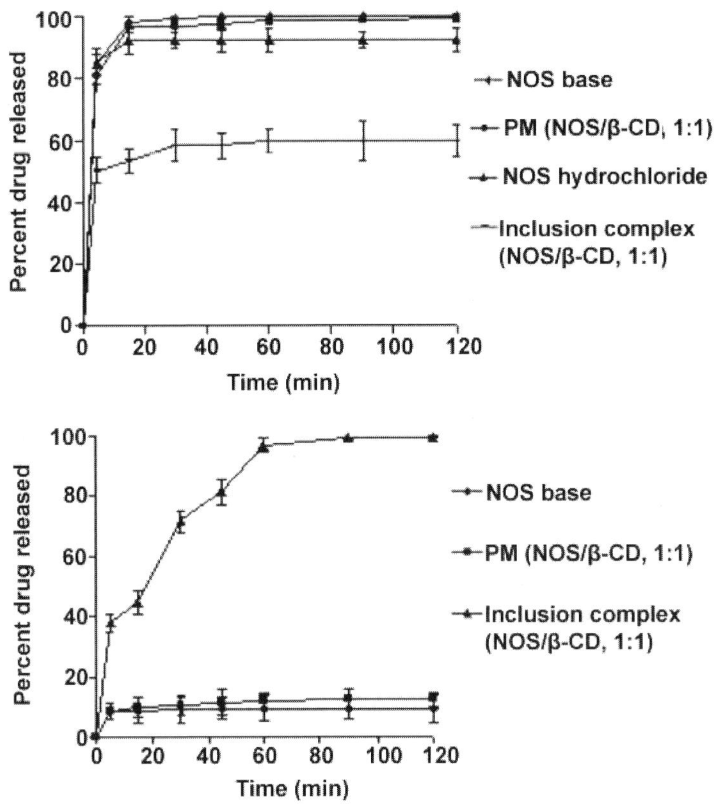

Fig. 5. (a) *In vitro* release of NOS base, PM of NOS/β-CD, NOS hydrochloride and NOS-β-CD inclusion complex in simulated gastric fluid at pH 1.2. Data are presented as mean ± S.D. for $n = 3$; $P < 0.05$ *vs.* NOS-β-CD and NOS base, PM or NOS hydrochloride; NOS hydrochloride and NOS base. (b) Dissolution profile in PBS at pH 7.4. Results show that NOS-β-CD inclusion complex release high percentage of drug at pH 7.4 in comparison with NOS base and PM. Data are presented as mean ± S.D. for $n = 3$; $P > 0.05$ *vs.* NOS β-CD and NOS base or PM; $P < 0.05$ *vs.* NOS base and PM.

noscapine-β-cyclodextrin inclusion complex demonstrated that noscapine rapidly releases from the inclusion complex in comparison with noscapine. Moreover, the noscapine-β-cyclodextrin inclusion complex gave shorter t_{max} and higher AUC and C_{max} than did noscapine. Pharmacokinetic analysis disclosed the enhanced bioavailability of 1.87-fold in absolute bioavailability, indicating that the enhanced oral bioavailability of noscapine in the inclusion complex might be due to increased absorption rate of noscapine through intestinal pH. However, the $t_{1/2}$ value of noscapine was not significantly changed in inclusion complex, indicating that inclusion complex allowed fast absorption in the initial phase, thereby improving bioavailability.[20]

6. Noscapine Nanotherapeutics Enhance Plasma Half-Life and Therapeutic Index

Nanoparticle-mediated anticancer drug delivery to solid tumors can be an effective approach to enhance the therapeutic concentration of

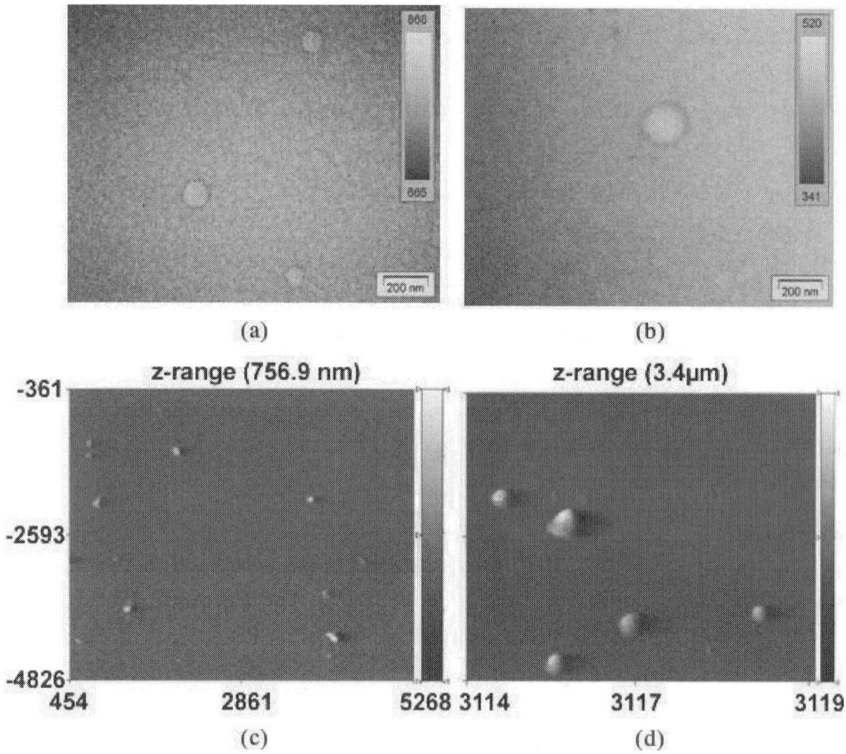

Fig. 6. Transmission electron micrograph of (a) noscapine-loaded gelatin nanoparticles (NOS-GN); (b) noscapine-loaded poly(ethylene glycol)-grafted gelatin nanoparticles (NOS-PEG-GN). The scale bar represents a distance of 200 nm. Atomic force microscopy of (c) NOS-GN (d) NOS-PEG-GN.

the drug in tumor cells, usually with a permeability cut-off less than 600 nm. However, intravenously injected nanoparticles are rapidly cleared from the circulation by a process of opsonization, which is initiated by complement activation and preferential uptake of the nanoparticles by the organs of RES. Therefore, we designed and synthesized noscapine encapsulated poly(ethylene glycol)-grafted gelatin nanoparticles and gelatin nanoparticles.

A relatively smaller elimination rate constant was observed in PEGylated formulation in non-compartmental pharmacokinetic analysis compared with noscapine and gelatin nanoparticles that was equivalent to enhanced bioavailability of PEGylated formulation of noscapine. Notable in these results are the approximately 1.46-fold increase in the AUC_{last} for the PEGylated formulation compared with gelatin nanoparticles and the 13.40-fold increase over the noscapine.

This also coupled with the higher cytotoxicity and low IC_{50} value of PEGylated formulation of noscapine in MCF-7 cells, which might be attributed to greater cellular uptake of PEGylated formulation of noscapine in cancer cells compared with gelatin nanoparticles and noscapine.[45]

7. Noscapine Delivery Using μPAR-Targeted Optical-MR Imaging

Despite noscapine's nontoxic attributes, significant elimination of the disease has not been achieved, perhaps since the bioavailability of noscapine to tumors saturates at an oral dose of 300 mg/kg body weight. To enable the selective and specific delivery of noscapine to prostate cancer cells, a multifunctional nanoscale delivery vehicle have been engineered that takes advantage of urokinase plasminogen activator receptor (μPAR) overexpression in prostate cancer compared to normal prostate epithelia and can be tracked by magnetic resonance imaging (MRI) and near-infrared (NIR) imaging. Specifically, the human-type 135 amino-acid amino-terminal fragment (hATF) of urokinase plasminogen activator (μPA), a high-affinity natural ligand for μPAR, has been employed. Noscapine was efficiently adsorbed onto the amphiphilic polymer coating of μPAR-targeted NPs (NPs).

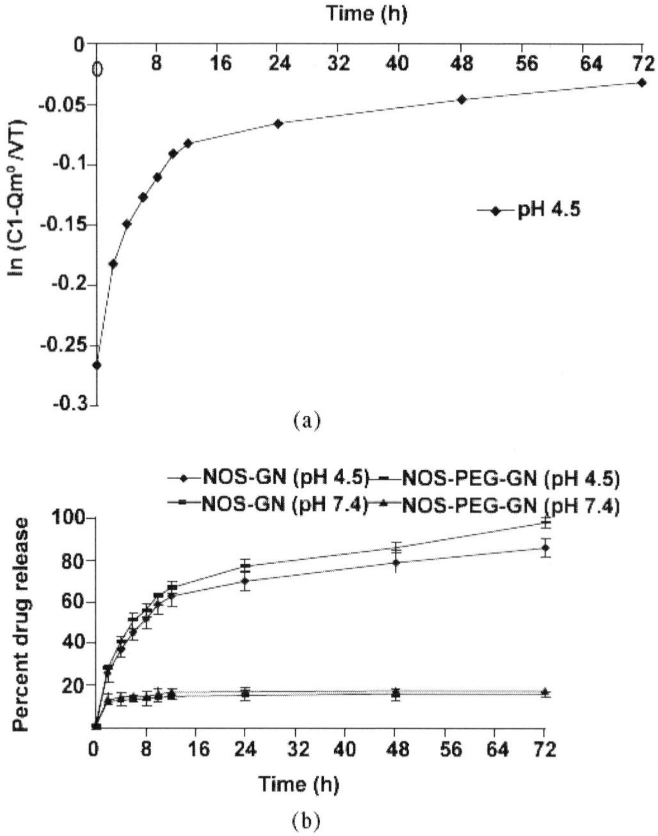

Fig. 7. (a) Release kinetic profile of noscapine (NOS) from gelatin nanoparticles (GN) in different phases with release rate constants; (b) *in vitro* release kinetics of NOS from GN (GN1) and poly(ethylene glycol)-grafted gelatin nanoparticles (PEG-GN) (GN8) at pH 4.5 and 7.4. Results documented that PEG-GN releases 97.28 ± 1.1% of NOS whereas GN releases 85.1 ± 4.2% of NOS at pH 4.5. This release is significantly ($P < 0.05$) higher than the NOS released (PEG-GN: 17.1 ± 1.2%; GN: 15.3 ± 2.7%) at pH 7.4. Values are shown as mean ± standard deviation for $n = 3$.

Noscapine-loaded NPs were uniformly compact-sized, stable at physiological pH and efficiently released the drug at pH 4–5 within a span of 4 hr. It has been demonstrated that these μPAR-targeted NPs were capable of binding to the receptor and were internalized by PC-3 cells. μPAR-targeted noscapine-loaded NPs enhanced intracellular noscapine accumulation as evident by the ~6-fold stronger inhibitory

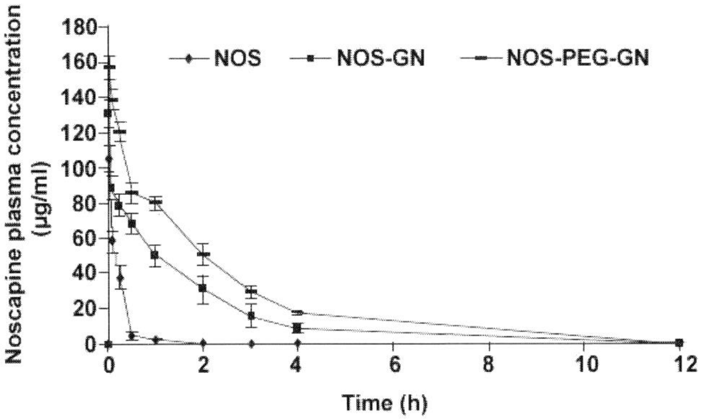

Fig. 8. Comparative plasma concentration–time profile curves of noscapine (NOS), noscapine-loaded gelatin nanoparticles (NOS-GN) (GN1), and noscapine-loaded poly(ethylene glycol)-grafted gelatin nanoparticles (NOS-PEG)-GN (GN8) at the dose of 50 mg/kg injected intravenously in female Swiss albino mice. Values are shown as mean ± standard deviation for $n = 3$.

effect on PC-3 growth compared to free noscapine. In addition, noscapine-loaded iron oxide NPs maintained their T2 MRI contrast effect upon internalization into tumor cells owing to their significant susceptibility effect in cells. Thus, this data provides compelling evidence that these optically and magnetic resonance imaging (MRI)-trackable µPAR-targeted NPs may offer a great potential for image-directed targeted delivery of noscapine for the management of prostate cancer.[49]

8. Noscapine-loaded Human Serum Albumin NPs

NPs made of human serum albumin (HSA) represent a promising strategy for targeted drug delivery to tumor cells by enhancing the drug's bioavailability and distribution and reducing the body's response towards drug resistance. Satya Prakash *et al.* have reported for the first time the incorporation and delivery of noscapine-loaded HSA NPs to tumor cells. The NPs were designed and optimized to achieve a particle size in the range of 150–300 nm with a

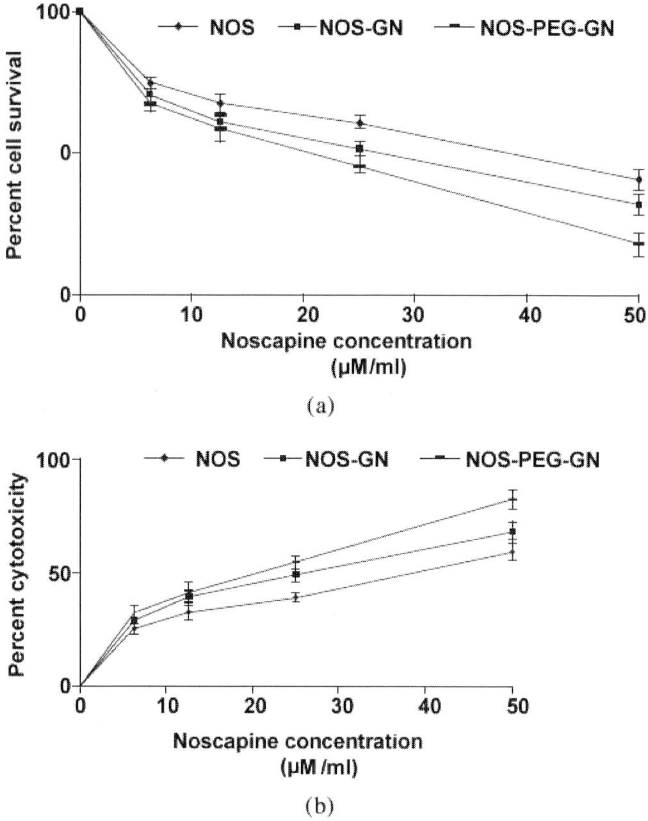

Fig. 9. (a) Percentage cell viability of human breast cancer cell line (MCF-7) cells incubated with serial dilutions of noscapine (NOS), noscapine-loaded gelatin nanoparticles (NOS-GN) (GN1), and noscapine-loaded poly(ethylene glycol)-grafted gelatin nanoparticles (NOS-PEG-GN) (GN8); (b) Cell cytotoxicity of MCF-7 cells incubated with serial dilutions of NOS, NOS-GN (GN1), and NOS-PEG-GN (GN8). The IC_{50} value of NOS-PEG-GN was equivalent to 20.8 μmol/l, which was significantly ($P < 0.05$) lower than IC_{50} value of NOS-GN (26.3 μmol/l) and NOS (40.5 μmol/l). Values are shown as mean ± standard deviation for $n = 4$.

drug-loading efficiency of 85%–96%. The NPs were evaluated *in vitro* for their anticancer activity and efficacy on breast cancer cells.

Drug-free HSA NPs and noscapine-loaded NPs cross-linked with glutaraldehyde were prepared using a pH-coacervation method. 100 mg of HSA was dissolved in 2 mL of water or NaCl. Noscapine

Table 1. Comparative Pharmacokinetic Parameters of Noscapine, Noscapine-Loaded Gelatin Nanoparticles, and Noscapine-Loaded Poly(Ethylene Glycol)-Grafted Gelatin Nanoparticles at the Dose of 50 mg/kg Injected Intravenously in Female Swiss Albino Mice

Parameters	NOS[a]	NOS-GN (GN1)[a]	NOS-PEG-GN (GN8)[a]
$-k_e$ (/h)	0.622 ± 0.02	0.276 ± 0.01	0.06 ± 0.01
$t_{1/2}$ (h)	1.11 ± 0.08	2.51 ± 0.10	11.55 ± 0.16
AUC_{last} (h µg/ml)	22.29 ± 4.58	203.94 ± 9.57	298.90 ± 7.78
AUC_{inf} (h µg/ml)	25.82 ± 1.36	235.38 ± 4.22	592.23 ± 3.21
AUMC (h µg/ml)	26.94 ± 4.59	501.04 ± 19.87	888.69 ± 15.47
MRT (/h)	1.20 ± 0.05	2.45 ± 0.15	2.97 ± 0.25
Vd (I)	3.11 ± 0.7	0.768 ± 0.009	1.4 ± 0.008
CL_{total} (I/h)	1.93 ± 0.11	0.212 ± 0.015	0.084 ± 0.004

AUC_{last}, area under the curve up to the last sampling point; AUC_{inf}, area under the curve with extrapolation to infinity; CL_{total}, total body clearance; MRT, mean residence time; NOS-GN, noscapine-loaded gelatin nanoparticles; NOS-PEG-GN, noscapine-loaded poly(ethylene glycol)-grafted gelatin nanoparticles; Vd, volume of distribution.

[a]Values are shown as mean ± standard deviation for $n \geq 3$.

at a concentration of 5–30 mg/mL was incubated with solution for 4–8 hr at room temperature. The pH was adjusted to 8 by the addition of 1 M NaOH. NPs were formed by adding 8 mL of ethanol drop-wise at a constant rate of 1 ml/min under constant magnetic stirring. The particles were stabilized by cross-linking with 100 µL of 8% glutaraldehyde solution. The cross-linking was performed for at least 24 hr under constant magnetic stirring at room temperature. Drug loading was evaluated following an indirect method of collecting the supernatant of purified particles. The quantity of unloaded free drug present in the supernatant was measured by spectrophotometer, leading to the quantification of the percentage of drug loaded into the NPs.

Investigations were also carried out on noscapine-loaded human serum albumin nanoparticles for enhanced delivery of noscapine in breast cancer (SK-BR-3) cells. The efficacy of albumin nanoparticles was found to be concentration dependent. Noscapine nanoparticles with concentrations of 5 mg/mL had the lowest efficacy, with cell viability of 60% and less than 35% at 10 and 15 mg/mL after 144 hr. The drug free nanoparticles had little effect on cell viability of cancer cells.[50] Our group has also exploited other polymeric noscapine nanoparticles for better patient compliance.[51–53]

Conclusion

Recent research in the area of Noscapine drug delivery system is actively aimed toward achieving specific and targeted delivery of anti-cancer agents through new avenues of directing drugs to tumors as well as to new types of drugs. It has been proven that the versatility of NPs, including polymer and lipidic NPs, makes them very suitable as potential delivery carriers due to their small size and enlarged surface area to volume ratio. During the last decade, large numbers of inorganic NPs have been studied as anti-cancer drug delivery carriers because of their versatile physico-chemical properties. NPs are readily available, and can be easily functionalized, as they possess good bio-compatibility and low cyto-toxicities. Better understanding of the interaction between NPS and bio-molecules and the driving forces in the transfection process would lead to efficient cell-specific targeted deliveries. It is also stressed that magnetic-nano hybrid particles could be the next-generation drug delivery carriers for successful chemo-therapy as this method enhances the transfection efficiency by using external magnetic fields. A major problem in cancer chemotherapy is multidrug resistance, which can be effectively circumvented via mesoporous silica NPs, solid lipid NPs, polymeric NPs and magnetic NPs. Poor aqueous solubility and low bioavailability of cancer chemo-therapeutic can be effectively overcome by NPs, albumin-based NPs, liposomal formulation, polymeric micelles, cyclodextrin and chitosan based NPs. The significant challenge is target delivery to a specific cell, wherein the difficulty lies in the recognition of the targeted cells among thousands of untargeted cells as well as suitable availability of NPs during the delivery process. In conclusion, the key to cancer therapy is to treat as early as possible, which requires superior detection and targeting methods combined with nanotechnology. Therefore, Noscapine NPs have a strong possibility of working as the best possible chemotherapeutic agents for the treatment of various types of tumors.

Acknowledgments

The authors thank Dr. B.R. Ambedkar Center for Biomedical Research for providing financial and infrastructural facilities to carry out the Noscapine research, and also coworkers, particularly Prof. Harish C. Joshi, Dr. Ritu Aneja and Dr. Anju Katyal for their continuing collaboration.

References

1. Landen JW, Lang R, McMohan SJ, *et al.* (2002) Noscapine alters microtubule dynamics in living cells and inhibits the progression of melanoma. *Clin Cancer Res* **62:** 4109–4114.
2. Zhou J, Gupta K, Aggarwal S, *et al.* (2003) Brominated derivatives of noscapine are potent microtubule-interfering agents that perturb mitosis and inhibit cell proliferation. *Mol Pharmacol* **63:** 799–807.
3. Aneja R, Vangapandu SN, Lopus M, *et al.* (2006) Development of a novel nitro-derivative of noscapine for the potential treatment of drug resistant ovarian cancer and T-cell lymphoma. *Mol Pharmacol* **69:** 1801–1809.
4. Aneja R, Zhou J, Zhou B, *et al.* (2006) Treatment of harmone-refractory breast cancer: apoptosis and regression of human tumours implanted in mice. *Mol Cancer Ther* **5:** 2366–2377.
5. Newman DJ, Cragg GM. (2007) Natural products as sources of new drugs over the last 25 years. *J Nat Prod* **70:** 461–477.
6. Butler MS. (2008) Natural products to drugs: natural product-derived compounds in clinical trials. *Nat Prod Rep* **25:** 475–516.
7. Alisaraie L, Tuszynski JA. (2011) Determination of noscapine's localization and interaction with the tubulin-α/β heterodimer. *Chem Biol Drug Des* **78:** 535–546.
8. Barken I, Geller J, Rogosnitzky M. (2010) Prophylactic noscapine therapy inhibits human prostate cancer progression and metastasis in a mouse model. *Anticancer Res* **30:** 399–401.

9. Portney NG, Ozkan M. (2006) Nano-oncology: drug delivery, imaging, and sensing. *Anal Bioanal Chem* **384(3):** 620–630.
10. West JL, Halas NJ. (2000) Applications of nanotechnology to biotechnology commentary. *Curr Opin Biotechnolo* **11(2):** 215–217.
11. Vogelson C. (2001) Advances in drug delivery syatems. Available from: http://pubs.acs.org/subscribe/journal/mdd/v04/i04/html/MDD04Feature Vogelson.html
12. Stanton RA, Gernert KM, Nettles JH, Aneja R. (2011) Drugs that target dynamic microtubules: a new molecular perspective. *Med Res Rev* **31:** 443–481.
13. Canta A, Chiorazzi A, Cavaletti G. (2009) Tubulin: a target for antineoplastic drugs into the cancer cells but also in the peripheral nervous system. *Curr Med Chem* **16:** 1315–1324.
14. Ozçelik B, Kartal M, Orhan I. (2011) Cytotoxicity, antiviral and antimicrobial activities of alkaloids, flavonoids, and phenolic acids. *Pharm Biol* **49:** 396–402.
15. Jordan MA. (2002) Mechanism of action of antitumour drugs that interact with microtubules and tubulin. *Curr Med Chem Anticancer Agents* **2:** 1–17.
16. Aneja R, Zhou J, Zhou B, *et al.* (2006) Treatment of hormone-refractory breast cancer: apoptosis and regression of human tumours implanted in mice. *Mol Cancer Ther* **5:** 2366–2377.
17. Lechertier T, Hodivala-Dilke K. (2011) Focal adhesion kinase and tumour angiogenesis. *J Pathol* [in press].
18. Newcomb EW, Lukyanov Y, Schnee T, *et al.* (2006) Noscapine inhibits hypoxia-mediated HIF-1α expression and angiogenesis *in vitro*: a novel function for an old drug. *Int J Oncol* **28:** 1121–1130.
19. Newcomb EW, Lukyanov Y, Alonso-Basanta M, *et al.* (2008) Antiangiogenic effects of noscapine enhance radio response for GL261 tumours. *Int J Radiat Oncol Biol Phys* **71:** 1477–1484.
20. Madan J, Dhiman N, Parmar VK, *et al.* (2010) Inclusion complexes of noscapine in beta-cyclodextrin offer better solubility and improved pharmacokinetics. *Cancer Chemother Pharmacol* **65:** 537–548.
21. Dahlström B, Mellstrand T, Löfdahl CG, Johansson M. (1982) Pharmacokinetic properties of noscapine. *Eur J Clin Pharmacol* **22:** 535–539.
22. Nayak KP, Brochmann-Hanssen E, Way EL. (1965) Biological disposition of noscapine. I. Kinetics of metabolism, urinary excretion, and organ distribution. *J Pharm Sci* **54:** 191–194.
23. Verma AK, Bansal S, Singh J, *et al.* (2006) Synthesis and *in vitro* cytotoxicity of haloderivatives of noscapine. *Bioorg Med Chem* **14:** 6733–6736.
24. Karna P, Sharp SM, Yates C, *et al.* (2009) EM011 activates a survivin-dependent apoptotic program in human non-small cell lung cancer cells. *Mol Cancer* **8:** 93–105.
25. Karna P, Rida PC, Pannu V, *et al.* (2011) A novel microtubule-modulating noscapinoid triggers apoptosis by inducing spindle multipolarity via centrosome amplification and declustering. *Cell Death Differ* **18:** 632–644.

26. Aneja R, Miyagi T, Karna P, *et al.* (2010) A novel microtubule-modulating agent induces mitochondrially driven caspase-dependent apoptosis via mitotic checkpoint activation in human prostate cancer cells. *Eur J Cancer* **46:** 1668–1678.
27. Mishra RC, Karna P, Gundala SR, *et al.* (2011) Second generation benzofuranone ring substituted noscapine analogs: synthesis and biological evaluation. *Biochem Pharmacol* **82:** 110–121.
28. Naik PK, Chatterji BP, Vangapandu SN, *et al.* (2011) Rational design, synthesis and biological evaluations of amino-noscapine: a high affinity tubulin-binding noscapinoid. *J Comput Aided Mol Des* **25:** 443–454.
29. Santoshi S, Naik PK, Joshi HC. (2011) Rational design of novel anti-microtubule agent (9-azido-noscapine) from quantitative structure activity relationship (QSAR) evaluation of noscapinoids. *J Biomol Screen* **16:** 1047–1058.
30. Chougule MB, Patel AR, Jackson T, Singh M. (2011) Antitumour activity of noscapine in combination with doxorubicin in triple negative breast cancer. *PLoS One* **6:** e17733.
31. Su W, Huang L, Ao Q, *et al.* (2011) Noscapine sensitizes chemoresistant ovarian cancer cells to cisplatin through inhibition of HIF-1α. *Cancer Lett* **305:** 94–99.
32. Pannu V, Karna P, Sajja HK, *et al.* (2011) Synergistic antimicrotubule therapy for prostate cancer. *Biochem Pharmacol* **81:** 478–487.
33. Erguven M, Bilir A, Yazihan N, *et al.* (2011) Decreased therapeutic effects of noscapine combined with imatinib mesylate on human glioblastoma *in vitro* and the effect of midkine. *Cancer Cell Int* **11:** 18.
34. Irache JM, Esparza I, Gamazo C, *et al.* (2011) Nanomedicine: novel approaches in human and veterinary therapeutics. *Vet Parasitol* **180:** 47–71.
35. Cho K, Wang Xu, Nie S, *et al.* (2008) *Cancer Res* **14:** 1310.
36. Cammas-Marion S. (2009) Nanotherapeutics: drug delivery concepts in nanoscience, *Targeting Approaches*, Rennes Cedex, France, Pan Stanford Publishing USA, p. 67.
37. Zhang J, Lan QC, Post M, *et al.* (2006) Design of nanoparticles as drug carrier for cancer therapy. Review. *Cancer Genomics & Proteomics* **3:** 147–158.
38. Smith J. Beaker, A medical research blog by Sanford-Burnham Medical Research Institute, La Jolla, California, USA.
39. Alexiou C, Arnold W, Klein RJ, *et al.* (2000) Cancer treatment with magnetic drug targeting. *Cancer Res* **60:** 6641–6648.
40. Lubbe AS, Bergemann C, Brock J, McClure DG. (1999) Physiological aspects in magnetic drug-targeting. *J Magn Magn Mater* **194:** 149–155.
41. Novakova AA, Lanchinskaya VY, Volkov AV, *et al.* (2003) Magnetic properties of polymer nanocomposites containing iron oxide nanoparticles, *J Magn Magn Mater* **258–259:** 354–357.
42. Dandamudi S, Campbell RB. (2007) The drug loading cytotoxicity and tumor vascular targeting characteristics of magnetite in magnetic drug targeting. *Biomaterials* **28:** 4673–4683.

43. Arruebo M, Fenandez-Pacheco R, Ibarra MR, Santamaria J. (2007) Magnetic nanoparticles for drug delivery. *Nano Today* **2:** 22–32.
44. Hu FX, Neoh KG, Kang ET. (2006) Synthesis and *in vitro* anti-cancer evaluation of tamoxifen-loaded magnetite/PLLAcomposit enanocomposites. *Biomaterials* **27:** 5725–5733.
45. Fonseca C, Simoes S, Gasper R. (2002) Paclitaxel-loaded PLGAnanoparticles: preparation, physiochemical characterization and *in vitro* anti-tumoral activity. *J Control Release* **83:** 273–286.
46. Abdalla OM, Aneja R, Dean D, *et al.* (2010) Synthesis and characterization of noscapine loaded magnetic polymeric nanoparticles. *J Magn Magn Mater* **322:** 190–196.
47. Zhang A, Anyarambhatla G, Ma L, *et al.* (2005) Development and characterization of a novel cremophor EL free liposome based paclitaxel (LEP-ETU) formulation. *Eur J Pharm Biopharm* **59:** 177–187.
48. Liu Yu, Chen GS, Chen Y, *et al.* (2004) Inclusion complex of paclitaxel and oligo (ethylenediamino) bridged bis (β-cyclodextrin)s: solubilization and antitumor activity. *Bioorg Med Chem* **12:** 5767–5775.
49. Abdalla MO, Karna P, Sajja HK, *et al.* (2011) Enhanced noscapine delivery using μPAR-targeted optical-MR imaging trackable nanoparticles for prostate cancer therapy. *J Control Release* **149(3):** 314–322.
50. Sebak S, Mirzaei M, Malhotra M, *et al.* (2010) Human serum albumin nanoparticles as an efficient noscapine drug delivery system for potential use in breast cancer: preparation and *in vitro* analysis. *Int J Nanomedicine* **5:** 525–532.
51. Chandra R, Madan J, Singh P, *et al.* (2012) Implications of nanoscale based drug delivery systems in delivery and targeting tubulin binding agent, noscapin, in cancer cells. *Curr Drug Metab* **3(10):** 1476–1483.
52. Madan J, Pandey RS, Jain V, *et al.* (2013) Poly(ethylene)-glycol conjugated solid lipid nanoparticles of noscapin improve biological halflife, brain delivery and efficacy in globlastoma cells. *Nanomedicine* **9(4):** 492–503.
53. Jain V, Jain B, Tiwari P, *et al.* (2013) Nanosolvated microtubule modulating chemotherapeutics: a case-to-case study. *Anticancer Drugs* **29(9):** 327–336.

Chapter 17

High-Throughput Methods for Miniaturization of Implantable Artificial Cells

Maryam Mobed-Miremadi*

Abstract

Using three different fabrication methods, namely Electrostatic Spraying, On-Chip Synthesis and Inkjet Bio-Printing, miniaturization of alginate-based microcapsules ranging from 20–100 μm has been achieved. With a limited amount of published cell and drug-delivery data, the projected therapeutic applications differ by miniaturization technique, dependent in turn on the starting alginate concentration. The capsules miniaturized using the On-Chip Synthesis method would be ideal for implantation in tumors in non-invasive drug delivery applications. As for the other two methods with documented cell encapsulation feasibility, the benefits would be in the areas of programmed cell delivery, metabolic disorders and wider bio-membrane porosity studies. The effect of miniaturization on the strength, porosity and permeability of alginate capsules using Inkjet Bio-Printing with a throughput of 1.8×10^6 microcapsules/hr are the following: (1) Strength has been enhanced by a higher degree of cross-linking, thickening the cross-linking solution or

*Department of Biomedical, Chemical and Materials Engineering, Davidson's College of Engineering, San Jose State University, San Jose CA 95192-0082. E-mail: maryam.mobed-miremadi@sjsu.edu

subsequent coating steps as shown by results of spectrophotometric measurements; (2) Neither porosity nor the reflection coefficients have been affected for blue dextran (MW = 2000 kDa) and FITC dextran (MW = 4 kDa), as shown by results of Atomic Force Microscopy measurements of pore sizes; and (3) Permeability has been enhanced as a result of miniaturization when comparing standard 1000 μm macrocapsules (1.5% MV alginate cross-linked with 1.5% $CaCl_2$) to 30 μm minicapsules (0.5% LV alginate cross-linked with 15% $CaCl_2$ and coated with 0.5% chitosan) as shown by results of the Fluorescence Microscopy measurements. In light of these preliminary findings, the hypothesis of diffusion across the membrane being inversely correlated to the surface to volume ratio (S/V) has been confirmed for artificial cell miniaturization, and the membrane strength and porosity can be tuned for a wide array of promising applications in cutting-edge areas of nanomedicine.

Introduction

Over the past 50 years, microencapsulation has been used to treat many disabling metabolic disorders like phenylketonuria, kidney failure and diabetes mell it us affecting 1%,[1] 7.8%[2] and 10%[3] of the American population, respectively. Bio-printing is used in many bioengineering applications, namely microencapsulation of cells[4] and enzymes, and nanodispensing in genomics.[5] Depending on the desired print volume and precision, the driving force for the generation of the drop(s) falls into multiple categories. Embryonic stem cells have been encapsulated in alginate microcapsules for enhanced tissue regeneration studies in bioreactors.[6] By designing the nutrient media in the microcapsule core and the membrane thickness, one can control differentiation until the bio-membrane degrades and re-absorbs as waste.[7] Bioencapsulation of implantable liver cells (non-differentiating hepatocytes) in Alginate-Polylysine-Alginate (APA) membranes has been widely conducted using atomization, producing capsule diameters between 200 μm and 1000 μm and membrane thicknesses between 5 μm and 10 μm.[4]

The MWCO, expressed in terms of Stokes radius (a), is the maximum molecular weight that is allowed through the passage of the microcapsule pores. For transplantation, the MWCO calculated using Stokes radius (a), given by Eq. 1, must be high enough to allow the passage of nutrients but low enough to reject antibodies and other immune molecules. In addition to the diffusivity coefficient and in the absence of electrostatic and hydrophobic interactions, the reflection coefficient (σ) which is a function of the Stokes radius (a) and the pore radius (R_p) of the membrane, is used to characterize "sieving" through ideal cylindrical pores of radius (R_p).[8] The knowledge of σ given by Eq. 2 is critical for immuno-isolation membrane design in terms of filtration selectivity. As σ approaches unity, the membrane becomes more impermeable to the solute.

$$a = \left(\frac{3MW}{4\pi \rho N_A} \right)^{1/3} \quad (1)$$

$$\sigma = 1 - \left(1 - \frac{a}{R_P}\right)^2 \left[2 - \left(1 - \frac{a}{R_P}\right)^2\right]\left[1 - \frac{2}{3}\left(\frac{a}{R_P}\right)^2 - 0.165\left(\frac{a}{R_P}\right)^3\right] \quad (2)$$

a	*Stokes' radius (cm)*
MW	*Molecular weight (Da)*
ρ	*Solute density* ($g.cm^3$)
N_A	*Avogadro's number*
σ	*Reflection corfficient*
R_P	*Pore radius (cm)*

With the advances in stem cell technology and cancer therapy, the demand for reliable and high-throughput production of smaller (<200 microns) microcapsules is increasing because of the following reasons: (1) Diffusion of healing nutrients across the membrane is inversely proportional to the microcapsule radius and proportional to the surface to volume ratio (S/V)[9]; (2) The reduction in microcapsule size allows more choices of implantation target sites for therapy[10]; (3) Using small cell-loaded microcapsules improves cell viability post-cryopreservation[11]; (4) Encapsulating a single cell vs. multiple allows conducting single cell vs. population studies in an appropriate stem cell niche.[12] Recently, high-throughput production of artificial cells has been enabled by the following cutting-edge methodologies for miniaturized droplet generation: Electrostatic Spraying,[13] Inkjet Bio-printing[14,15,16] and On-Chip Synthesis.[17] With the implementation of a robust miniaturization technique, capsule strength and porosity should not be impacted. Systematic nutrient flux determinations are hindered by lack of experimental data on membrane topography, thus preventing the full therapeutic potential of the miniaturized artificial cells to be reached. In this chapter, an overview of the fabrication methods of miniaturization and the ramifications of miniaturization in terms of strength, porosity and permeability will be discussed.

Fabrication Methods for Artificial Cell Miniaturization

As shown in Fig. 1, there are three main methods of miniaturization documented in the literature. A brief description of each technique and its potential therapeutic application is discussed in this section.

Electrostatic Spraying

Using electrostatic spraying, alginate microcapsules are prepared using an electrostatic spray device (Fig. 1(a)).[13] This is a modification of the

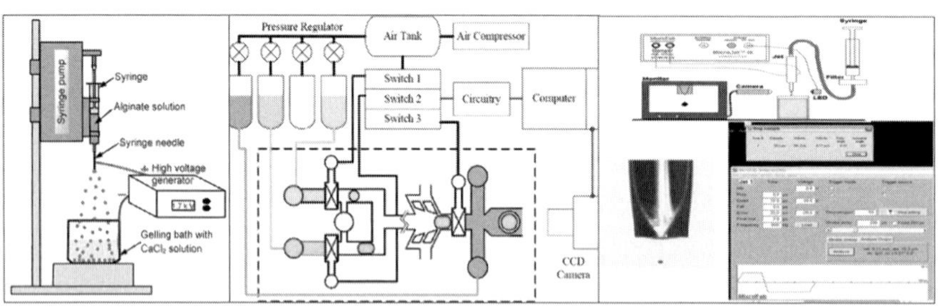

Fig. 1. Overview of miniaturization methods for alginate-based artificial cells (*from left to right*). (a) Shown on the left is an image of the Electrospraying set-up (with permission from Ref. 13); (b) Shown in the middle is an image of the On-Chip Synthesis Microfluidic set-up (with permission from Ref. 17); (c) Shown on the right is an image of the Inkjet Printing Equipment and software interface (with permission from Ref. 16).

standard electrostatic spraying technique,[18] where barium is used instead of calcium as a cross-linking cation. The microcapsules are generated by extruding 1.2% (w/v) sodium alginate solution into a cross-linking solution where the ionotropic gelation occurs. The solution is pushed through a blunt syringe needle using a syringe pump. The inner diameters and wall thickness of the syringe needle used are 150 μm and 75 μm, respectively. During extrusion, droplets of sodium alginate are formed on the tip of the needle and sprayed into the calcium chloride solution gelling bath as a result of multiple forces such as the driven force of syringe pump, electrostatic force due to the high electrostatic potential applied between the gelling bath and the needle, the surface tension, and gravity.[19] Using the mouse mesenchymal stromal cell line (C3H10T1/2), a cell loading of 5 cells/capsule with an average capsule size of 100 μm has been reported.[13] To obtain small cell loaded microcapsules with good morphology, the optimal range off low rate is 1.5–3 mL/hr and the optimal cell density is (3–5) $\times 10^6$ cells/mL. For the same study, a survival rate of 89% has been reported. Neither throughput nor the rheological properties of the solution alginate solution have been documented.

On-Chip Synthesis

Recently, it has been demonstrated that on demand, controlled encapsulation and subsequent coating steps can be achieved on an integrated micro fluidic system (Fig. 1(b)). A PDMS molding and bonding process is used to fabricate the proposed micro fluidic device. A three-layered PDMS micro-device with pneumatically actuated diaphragm valves constructed on top of specially designed fluidic-channels is utilized to control the droplet size ranging from 53–151 μm.[17] Small sodium alginate droplets are metered, trapped, and then drawn into relatively large $CaCl_2$ droplets, while they crosslink at the interface. The size of the droplet is a function of valve opening times. Optimal sphericity is obtained at a concentration of 3% (w/v) because of the uniform interfacial stress throughout the solidification process. In addition, entrapment and transfer of the resulting capsules can also be performed on the same micro fluidic system to coat the beads with poly-L-lysine (PLL) and liquefy the beads using sodium citrate. Using this new technique, neither a specific cell/drug encapsulation nor a throughput for the miniaturized alginate cell production has been reported. The rheological properties of the alginate solution have not been documented as well.

Inkjet Bio-Printing

The use of inkjet bio-printing of alginate beads was initially reported for 3D patterning in tissue engineering.[20] Recent studies encompass waveform and cross-linking solution optimization for tuning capsule sphericity and strength.[14,15,16] Miniaturized microcapsules ranging from 25–100 μm are fabricated using an integrated set-up presented in Fig. 1(c). The apparatus (Microfab Jet Lab III) consists of a CCD camera, a control unit, a print-head, a triggering unit, a fluid delivery unit, and a PC equipped with proprietary software to tune the bio-ink formulation to the jetting paramaters. Printheads can be chosen based on the aperture size ranging from 50–100 μm aperture and are rated for a specific dynamic viscosity (μ) and surface tension (γ). After inputting the jetting variable settings, droplet generation begins with the triggering box sending electrical signals to the ink-jet control unit and to the CCD camera control PC, simultaneously. The ink-jet engine fires through the inkjet head to the 15% $CaCl_2$ solution. The CCD camera captures a picture of the droplet exiting the print-head. The imaging software has a motion analysis feature that captures droplet size and velocity. Typically, low-viscosity alginate is used with concentrations ranging from 0.5%–1.0%. Dyes and bacterial cells have been encapsulated in the inkjet miniaturized cells.[14,21] The reported throughput for a 120 min pilot run using the optimal waveform parameters and 0.5% alginate ink is 1.8×10^6 microcapsules/hr, averaging 30–40 μm in diameter.[16] Comparative images of bacterial cells encapsulated in 500 μm "macrocapsules" and 35 μm "microcapsules" are presented in Fig. 2. The miniaturized cells were prepared using a printhead aperture of 50 μm.

Clearly, the use of the micro fluidic device is the most advantageous when stronger capsules are required due to the ability to process

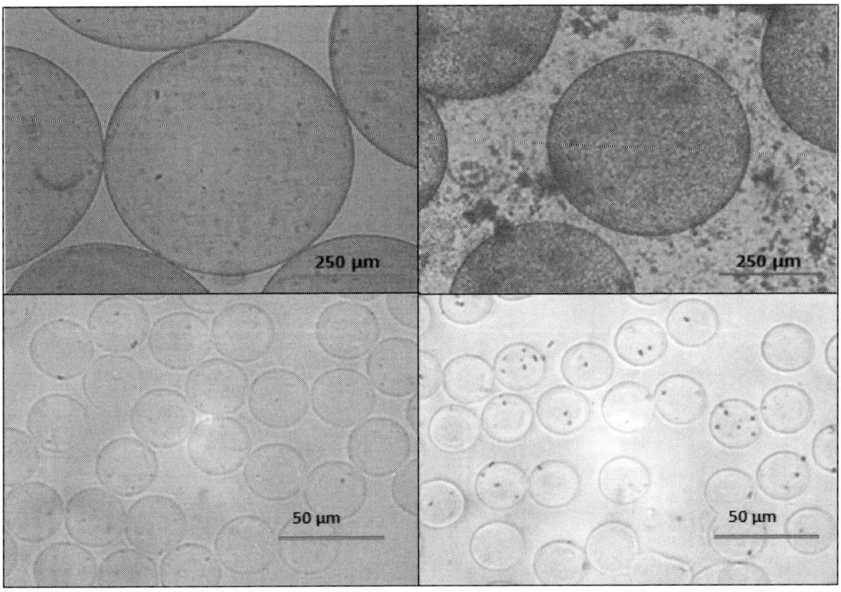

Fig. 2. Viability of *E. coli DH5* in alginate-based artificial cells. (a) Shown on top is bacterial growth from day (0) (left) to day (5) (right) in macrocapsules; (b) Shown on the bottom is bacterial growth from day (0) (left) to day (5) (right) in microcapsules generated using inkjet technology (Ref. 21).

higher alginate concentrations at 3% (w/v) as compared to 1% (w/v) associated with the other two methods. Future applications such as implantation in tumors for non-invasive drug delivery would be an example.[22] The use of lower alginate concentrations associated with the other two methods could be programmed cell delivery,[7] porosity studies for general bio-membrane design and toxin removal amongst others. For the remainder of this chapter, recent progress on artificial cells made using inkjet printing will be presented. This method has been chosen because of its track record of reliability and high throughput[23] as well as the commercial availability of the prototype.

Miniaturized Bio-Membrane Characterization

Due to the smaller artificial cell sizes and lower alginate viscosities, high centrifugation speeds are required to separate the miniaturized capsules. Lower alginate viscosities can in turn lead to leakage of the microencapsulated content for non-spherical capsules with tails. In order to compensate for the changes in the alginate formulation, either higher concentrations of cross-linkers are used or the receiving cross-linking solution is modified. Membrane topography in turn, characterized by porosity and ultimately the calculation of the reflection coefficient could be affected, thus membrane permeability would be altered. In this section, membrane strength is quantified in terms of leakage detected by UV-VIS spectrophotometry, pore sizes have been measured using Atomic Force Microscopy (AFM) and membrane permeability has been examined using Fluorescence Microscopy (FM).

Strength Characterization

Results of studies focusing on the formulation of the receiving cross-linking solutions and strength characterization are presented below. For analytical purposes the terms release rates and leakage are used interchangeably.

In the first study, the release kinetics of rhodamine 6G dye (MW = 479 Da, $a = 0.57\,\text{nm}$) from calcium alginate microcapsules were measured.[14] Time to initial burst release, release rate and percentage release were measured for different input parameter combinations using a factorial design. Surface to volume ratio (S/V) of the microcapsule had a significant effect on the release kinetics, the range of microcapsule sizes varying from 20–30 µm. Typically, faster release kinetics were observed for smaller microcapsule sizes (20 µm) based on their higher surface to volume ratio. Higher concentrations of alginate and $CaCl_2$ resulted in higher cross-linking of the microcapsules and vice versa.

In a second study, it was proven that the shape of the formed microcapsules can be influenced by increasing the viscosity of the receiving solution and can vary from elongated to spherical to flattened, respectively.[15] The receiving $CaCl_2$ solution was thickened using various concentrations of glycerol with viscosities ranging from 2–100 mPa.s. The release rate of two model substances methylene blue (MW = 320 Da, $a = 0.5\,\text{nm}$) and vitamin B12 (MW = 1355 Da, $a = 0.82\,\text{nm}$) from calcium alginate beads was also investigated and the dependence of the effective diffusivity of each compound on the alginate concentration was obtained. For an alginate concentration of 0.5% (w/v), membrane diffusivities (D_{eff}) of $6.64 \times 10^{-6}\,m^2 \cdot s^{-1}$ and $1.9 \times 10^{-7}\,m^2 \cdot s^{-1}$ have been determined by modeling the transient kinetics of release using Fick's second law for methylene blue and vitamin B12, respectively. Inward membrane diffusivities for 1.5% (w/v) medium-viscosity alginate cross-linked with 1.5% $CaCl_2$ for creatinine (MW = 113 Da, $a = 0.36$ nm) have been determined to be $2.5 \times 10^{-14}\ m^2 \cdot s^{-1}$.[23] Since the Stokes radii of creatinine and the above-mentioned dyes are close (<1 nm), it can be extrapolated that the molecules have the same diffusive properties and hence miniaturization has enhanced diffusivity by a factor of 10^7.

In a third study, it was proven that the shape of the formed microcapsules can be tuned by varying the concentration of $CaCl_2$ and adding chitosan to receiving solution while using the previously-determined optimal waveform parameters.[21,24] An L_{16} (five factors, four levels) Taguchi design combining the jetting and receiving solution parameters was designed. Blue dextran (MW = 2000 kDa, $a = 9.3$ nm) was pre-mixed with 0.5% (w/v) low-viscosity (LV) alginate prior to jetting. The turbidity of the receiving solution indicative of the precipitation of chitosan at higher calcium chloride concentrations was also monitored. It was found that below a $CaCl_2$ of 2% no capsules are formed. Best results were obtained for a chitosan concentration of 0.5% (w/v) mixed with 10–15% $CaCl_2$ (w/v) chloride characterized by a non-turbid receiving solution viscosity of 4 mPa.s. The miniaturized capsules averaging 40 μm in size with an aspect ratio of 1.06 were centrifuged at 10,000 g and re-examined for damage. All re-suspended capsules from the concentrated pellet were intact. In addition no leakage of blue dextran was detected, indicating the impermeability of the membrane to the high molecular weight solute.

Porosity Characterization

Atomic Force Microscopy

Pore size and porosity measurements streamline the therapeutic application of the microcapsules. AFM is a quantitative surface characterization tool for the estimation of the pore size diameter.[25] SEM is a quantitative surface characterization tool used for pore location and pore structure (closed, open, linked). AFM and Cryo-SEM have been used to measure the pore size of impermeable alginate membranes[7,18] with pore sizes reported in the micron range.

In this study, pore sizes in the nanometer range, in alignment with the MWCO order of magnitude, have been reported for submerged cross-linked and immobilized alginate films in saline using AFM in contact mode. Since the sample is hydrated, it has been assumed that film surfaces mimic membrane topography in solution. The working hypothesis behind this comparison is that higher alginate concentrations result in a higher degree of cross-linking.[14] The instrument used was an Agilent 5500 connected to a 100 μm scanner with resolution in the x and y directions of 20–30 nm. The probes used were Pyrex Nitride triangular probes manufactured by Nano World (Part # PNP-TR-20). Data reduction was carried out using Gwyddion software, which is a modular program for SPM (scanning probe microscopy), data visualization and statistical analysis. Sample membrane topography for the 0.5% (w/v) alginate is presented in Fig. 3.

For a scan area of 0.1 μm^2, the ranges of the pore diameters reported are 8–15 nm and 5–16 nm for the 1.5% (w/v) and the 0.5% (w/v) alginate, respectively. These numbers are still higher than the observed MWCO of 67 kDa (a = 2.98 nm) in terms of the Stokes radius for at least the standard 1.5% (w/v) macrocapsule membrane.[4] This discrepancy could be due to the AFM probe geometry (inadequate aspect ratio) and the unknown distribution of closed vs. open pores. In future experiments, the use of carbon nanotube (CNT) probes will be considered. The work done by Nguyen *et al.* indicates that the minimum measurement is to 2–5 nm,[26] within the range of the Stoke radii for the alginate membranes. With knowledge of the experimentally-measured pore radii (R_p), the range of reflection coefficients was in turn calculated using Eq. 2. As shown in Table 1, at the lower limits of R_p both micro and microcapsule membranes are impermeable to albumin. Meanwhile, transport is uninhibited for the 4 kDa dextran solute across all pore sizes. Hence, the working hypothesis of degree of cross-linking being proportional to the alginate concentration is refuted for both solutes.

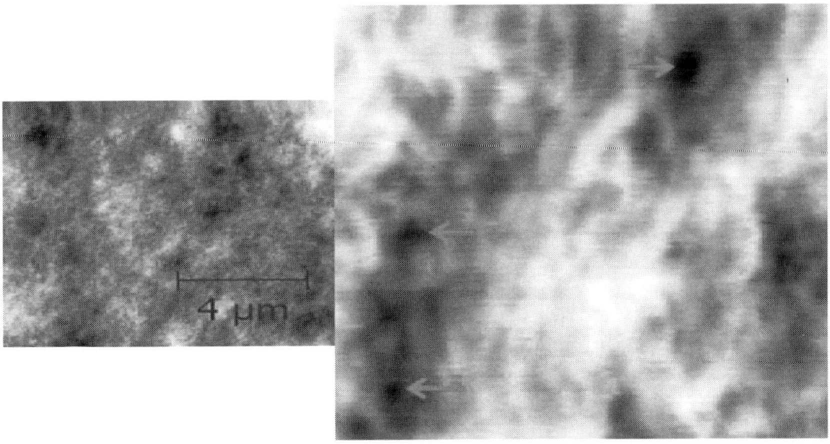

Fig. 3. AFM Imaging of 0.5% (w/v) cross-linked alginate film for a scan area of 0.1 μm^2. (a) Shown on the left is an image of the pores; (b) shown on the right is the magnified membrane topography where the pores are indicated by red arrows.

Table 1. Pore Size Ranges and Reflection Coefficients for Immobilized Alginate Films

Compound/ MW (kDa)	a (nm)	R_{pmin} (1.5%) (nm)/σ	R_{pmax} (1.5%) (nm)/σ	R_{pmin} (0.5%) (nm)/σ	R_{pmax} (0.5%) (nm)/σ
Albumin/67	2.98	4/0.930	7.5/0.475	2.5/1.000	8/0.432
Dextran/4	1.17	4/0.294	7.5/0.098	2.5/0.591	8/0.087

Permeability Characterization

Fluorescence Microscopy

Fluorescence microscopy using FITC-labeled dextran molecules has been previously used to visualize diffusion through alginate membranes.[27] In this study, fluorescence microscopy has been used to assess whether miniaturization enhances permeability with a working hypothesis that the diffusive flux across the membrane is inversely proportional to the radius of the capsule or (S/V). The formulation for the miniaturized artificial cells (microcapsules) was comprised of 0.5% (w/v) low-viscosity alginate cross-linked with 15% $CaCl_2$ and coated with low MW chitosan 0.5% (w/v) as compared to the standard macrocapsules made with 1.5% (w/v) medium-viscosity alginate cross-linked with 1.5% $CaCl_2$. Empty capsules were incubated in FITC dextran (1 mg/mL) solution for 12 hr. The solutions were centrifuged and washed followed by suspension in 0.9% NaCl solution. An inverted microscope (Nikon Epiphot Model 200) equipped with a fluorescence filter (Nikon, 96302 C-FL B2, excitation wavelength 470 nm, emission wavelength 515 nm) was used. Outward diffusion of the dextran molecules using fluorescent intensity for an area of interest (ROI) as a metric was monitored and data analysis was conducted using the NIS Elements software. Intensity was in turn converted to concentration. For the 500 kDa dextran (Sigma-Aldrich, FD500S), no changes in intensity were observed during the course of the 30 min study and there was no evidence of permeation post the 12 hr incubation period. Both types of capsules were impermeable to the molecule as shown in Fig. 4.

In contrast, for the 4 kDa marker (Sigma-Aldrich #FD4), evidence of permeation was observed (Fig. 4), and the net change in concentration was recorded as a function of time. As shown in Fig. 5, after 6 min, a concentration decrease of 81% for the micro and 38% for the macrocapsules were measured indicative of faster solute equilibration through the membrane for the miniaturized artificial cells. Theoretically the observed (S/V) ratio is proportional to the inverse of the spherical radius of the artificial cell, and diffusive flux should have been enhanced by a factor of 33. Future efforts will include more frequent and extended sampling in order to determine the diffusivity coefficient using Fick's second law for comparative flux quantitation. In addition, this comparative study needs to be extended to incorporate a wider range of alginate concentrations, molecular weights, cross-linked concentrations and fabrication methods in order to calculate the net effect of miniaturization in terms of diffusivity coefficients. As previously stated, blue dextran does not leak from the miniaturized capsules. This is in agreement with this coarse MWCO using the 500 kDa marker with future plans to incorporate FITC-Albumin (MW = 67 kDa) in the finer MWCO studies.

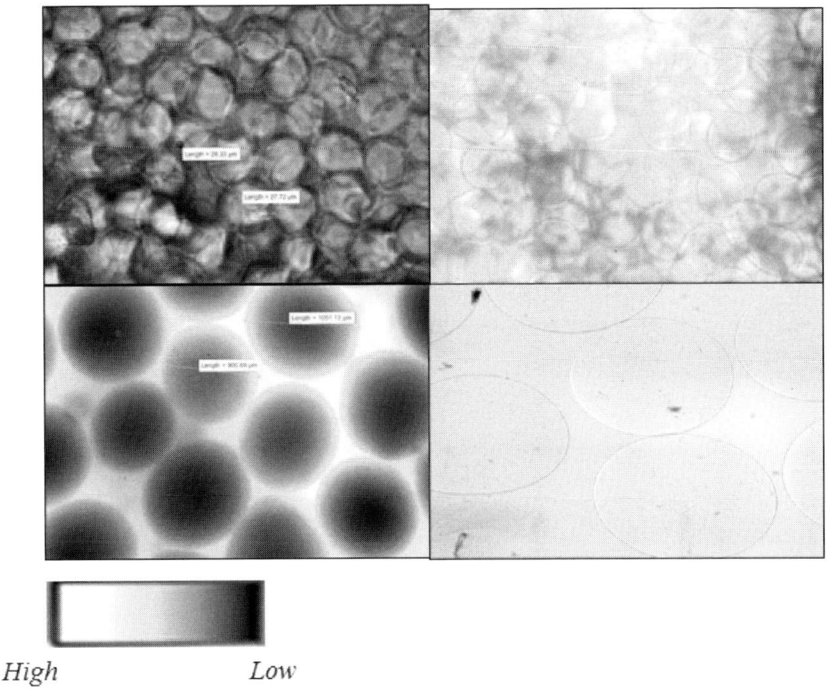

Fig. 4. Membrane Permeability Visualization for microcapsules (≅30 μm) and macrocapsules (≅ 1000 μm) using Fluorescence Microscopy. (a) Shown on the top left and bottom left is the lack of permeability of microcapsules and macrocapsules, respectively to 500 kDa FITC dextran after a 12 hr incubation period; (b) Shown on the top right and bottom right the saturation of the microcapsules and macrocapsules, respectively with 4 kDa FITC dextran after a 12 hr incubation period.

Conclusion and Future Directions

In summary, a limited amount of metabolic or drug delivery-related data has been published on the miniaturization techniques for producing alginate-based artificial cells. High viability of bacterial and mammalian cell encapsulation has been proven using the methods of Electro spraying and Inkjet Bio-Printing, respectively. The aforementioned methods are limited by the use of lower alginate viscosity formulations. Nano-indentation experiments need to be conducted

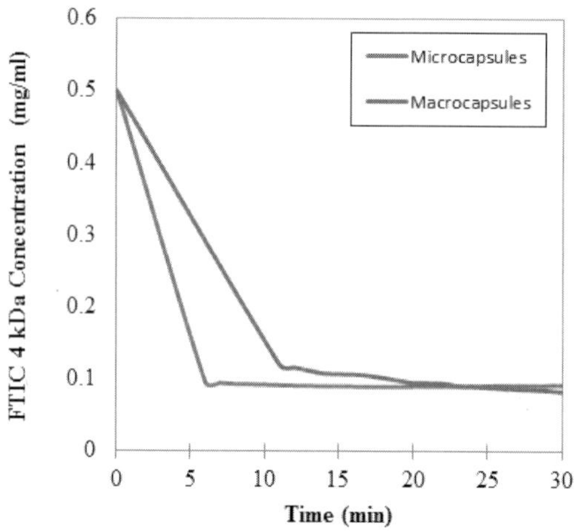

Fig. 5. Comparative concentration profile for the outward diffusion of FITC 4 kDa FITC dextran.

in conjunction with the reported leakage experiments to measure the stiffness and strength of the capsules for implantation purposes. Meanwhile, no cell viability data using On-Chip Synthesis approach is available, however high-viscosity alginate formulations (up to 3% (w/v)) can be successfully processed to make stronger capsules ideal for implantation. In order to estimate the net effect of miniaturization, simultaneous studies using multiple MW markers need to be conducted between the diffusive and sieving properties of the 200–1000 μm range macrocapsules fabricated using different alginate, cross-linking and coating agent concentrations, and 20–100 μm range miniaturized microcapsules using the same formulations. For small molecules less than 4 kDa ($a < 1$ nm), the extrapolated effect of miniaturization and the projected hypothesis of (S/V) enhancement have been confirmed, with a gain of $10^7\ m^2 \cdot s^{-1}$ in terms of diffusivity coefficient. This is in agreement with the results of the permeability study using Fluorescence Microscopy indicating that miniaturization has enhanced solute equilibration time and thus membrane permeability. These preliminary findings have promising implications for contributing to societal needs in terms of hemodialysis treatment and design of other immune-isolation membranes as well as the cutting-edge stem cell and cancer therapies: (1) Miniaturized alginate-based artificial cells could be used for urea and other lower molecular weight toxins as an alternative to hollow fiber and MEMS fabricated bio-membranes; (2) based on the current metabolic flux data on stem cells encapsulated in macrocapsules, the microenvironment and the dosing can be scale-up or down to compensate for the effects of miniaturization.

Acknowledgments

The author would like to acknowledge the support of Davidson's College of Engineering Faculty Development Grants, the CSU Mini-Grants and the C-SUPERB Joint Venture Grant "*Bio-Printing of Mammalian Cells*" for funding this effort. The author also wishes to thank the following individuals for invaluable discussions and timely contributions: Ms. Yiming Shan, Ms. Cheryl Simpliciano, Ms. Behrokh Asi, Mr. Ricardo Duque, and Dr. Michel Goedert.

References

1. Visakorpi JK, Palo J, Renkonen OV. (2008) The incidence of PKU in Finland. *Acta Pædiatrica* **60(6):** 666–668.
2. Fissell WH, Fleichman AJ, Humes AJ, Roy S. (2007) Development of continuous implantable renal replacement: past and future. *Transl Res* **150:** 327–336.
3. http://diabetes.niddk.nih.gov/DM/PUBS/statistics/#allagesAccessedon 11/15/2011. Accessed 7/16/2013.
4. Abba MC, Hu Y, Sun H, *et al.* (2005) Gene expression signature of estrogen receptor a status in breast cancer. *BMC Genomics* [online]. Available at: http://www.biomedcentral.com/1471–2164/6/37. Accessed on 11/15/2011. Accessed 7/16/2013.
5. Chang TMS. (2007) *Artificial Cells: Biotechnology, Nanomedicine, Regenerative Medicine, Blood Substitutes, Bioencapsulation, Cell/Stem Cell Therapy.* Singapore: World Scientific Publishing Co. Pte. Ltd, London: Imperial College Press, London [online] Available at: www.artcell.mcgill.cafor free access or download. Accessed on 11/15/2011. Accessed 7/16/2013.
6. Hwang YS, Cho J, Tay F, *et al.* (2009) The use of murine embryonic stem cells, alginate encapsulation, and rotary microgravity bioreactor in bone tissue engineering. *Biomaterials* **30(4):** 499–507.

7. Wang N, Adams G, Buttery L, *et al.* (2009) Alginate encapsulation technology supports embryonic stem cells differentiation into insulin-producing cells. *J Biotechnol* **144(4):** 304–312.
8. Fournier RL. (2011) *Basic Transport Phenomena in Biomedical Engineering*, 3rd edn. Taylor & Francis, Philadelphia.
9. Uludag H, De Vos P, Tresco PA. (2000) Technology of mammalian cell encapsulation. *Adv Drug Deliv Rev* **42(1–2):** 29–64.
10. Wilson JT, Chaikof EL. (2008) Challenges and emerging technologies in the immunoisolation of cells and tissues. *Adv Drug Deliv Rev* **60(2):** 124–145.
11. He X, Park EY, Fowler A, *et al.* (2008) Vitrification by ultra-fast cooling at a low concentration of cryoprotectants in a quartz micro-capillary: a study using murine embryonic stem cells. *Cryobiology* **56(3):** 223–232.
12. Karoubi G, Ormiston ML, Stewart DJ, Courtman DW. (2009) Single-cell hydrogel encapsulation for enhanced survival of human marrow stromal cells. *Biomaterials* **30(29):** 5445–5455.
13. Wujie Z, Xiaoming H. (2009) Encapsulation of living cells in small (~100 um) alginate microcapsules by electrostatic spraying: a parametric study. *J Biomech Eng* **131(7):** 074515–(1–6).
14. Desai L, Perkins J, Harrison BS, Sankar J. (2010) Understanding release kinetics of biopolymer drug delivery microcapsules for biomedical applications. *Mater Sci Eng B* **B168:** 127–131.
15. Dohnal J, Štěpánek F. (2010) Inkjet fabrication and characterization of calcium alginate microcapsules. *Powder Technol* **200:** 254–259.
16. Mobed-Miremadi M, Acks E, Polasaward S, Chen D. (2011) High throughput miniaturization of artificial cells. *Art Cells Blood Substitutes & Biotech* **39(5):** 310–316.
17. Chang SC, Su YC. (2011) On-demand micro-encapsulation utilizing on-chip synthesis of semi-permeable alginate-PLL capsules. *Microfluids Nanofluids* **10:** 1165–1174.
18. Zimmermann U, Zimmermann H, Wahlisch F, *et al.* (2007) Physical and biological properties of barium cross-linked alginate membranes. *Biomaterials* **28(7):** 1327–1345.
19. Strand BL, Gaserod O, Kulseng B, *et al.* (2002) Alginate-polylysine-alginate microcapsules: effect of size reduction on capsule properties. *J Microencapsul* **19(5):** 615–630.
20. Nakamura M, Nishiyama Y, Henmi C. (2008) 3D Micro-fabrication by Inkjet biofabrication for 3D tissue engineering. *Micro-NanoMechatronics and Human Science* 451–456.
21. Shan Y, Wong J, Deol S, *et al.* (2011) High throughput encapsulation of E. Coli DH5 using inkjet bio-printing. *Proceedings of the 23rd CSU Annual Biotechnology Symposium* [online]. Available at: http://csuperb.org/oars/abstracts.php?p=author&o=14. Accessed 11/15/2011. Accessed 7/16/2013.

22. Sridhar-Keralapura M., Thirumalai S., Mobed-Miremadi M. (2013) Structural Changes and Imaging Signatures of Acoustically Sensitive Microcapsules under Ultrasound. *Ultrasonics* (doi.org/10.1016/j.ultras.2013.02.001)
23. Djomehri S, Mobed-Miremadi M, Keralapura M (2013). Modeling Diffusivity Through Alginate-Based Microfibers: A Comparison of Numerical And Analytical Models Based On Empirical Spectrophotometric Data. *J Membra Sci Technol* (doi.org/10.6000/1929-6037.2013.02.01.8)
24. Hsueh YS, Iyer G, Sarwary M. (2011) Optimization and Quality Improvement of Bioinkjet Printing. Masters Project Report [online]. Available at: http://www.engr.sjsu.edu/ges/media/pdf/mse_prj_rpts/fall2011/OPTIMIZATION%20AND%20QUALITY%20IMPROVEMENT%20OF%20BIOINKJET%20PRINTING.pdf. Accessed on 7/16/2013.
25. Butt HJ, Capella B, Kappi M. (2005) Force measurements with the atomic force microscope: technique, interpretation and applications. *Surf Sci Rep* **59(1–6):** 1–152.
26. Nguyen C, So C, Stevens RM, *et al.* (2004) High lateral resolution imaging with sharpened tip of multi-walled carbon nanotube scanning probe. *J Phys Chem B* **108(9):** 2816–2821.
27. Chen H, Ouyang W, Lawuyi B, Prakash S. (2006), Genipin cross-linked alginate-chitosan microcapsules: membrane characterization and optimization of cross-linking reaction *Biomacromolecules* 7(7): 2091–2098.

Chapter 18

Amphiphilic Core–Shell Nanoparticles Containing Hairy Polyethyleneimine Shells as Effective Nanocarriers for Gene and siRNA Delivery

Pei Li*, Yuen Shan Siu, Kin Man Ho and Wei Li

Abstract

Gene therapy is a promising treatment for a variety of genetic disorders. The success of gene therapy greatly depends on suitable gene carriers that should provide safe, efficient and targeted delivery. Polyethyleneimine (PEI) is considered an effective non-viral gene carrier because of its high complexing abilities with both DNA and siRNA, as well as excellent transfection efficiency. However, the use of PEI for gene delivery has achieved only limited success owing to its cytotoxicity and broad size distribution of its polyplexes. In this chapter, we discuss our studies using PEI-based core–shell nanoparticles as effective nanocarriers for gene delivery. Two types of core–shell nanoparticles have been synthesized and evaluated. (1) Amphiphilic core–shell nanoparticles consisting of biocompatible poly(methyl methacrylate) (PMMA) cores and hairy PEI shells. The amphiphilic nanoparticles have an average diameter of around 150 nm with

* Corresponding author: Department of Applied Biology and Chemical Technology, The Hong Kong Polytechnic University, Hung Hom, Kowloon, Hong Kong, P. R. China. E-mail: bcpeili@polyu.edu.hk

uniform size distribution. (2) Core–shell nanogel particles containing biodegradable gelatin cores and PEI shells with an average diameter of around 200 nm and narrow size distribution. Properties of these two types of nanocarriers with respect to their abilities to condense, protect and release nucleic acids have been examined. Their cytotoxicities, cellular uptake abilities and gene delivery efficiencies have also been studied. Furthermore, the efficiency of the gelatin-PEI nanogel as the siRNA nanocarrier for down-regulation of human argininosuccinate synthetase 1 (ASS1) gene in HeLa cells is demonstrated.

Introduction

Gene therapy is a promising approach to treat a variety of genetic disorders.[1,2] This approach is based on the introduction of functional genes (e.g. gene segments, siRNA) to alter defective gene expression and restore normal metabolism, cellular and physiological responses of patients.[3,4] Up till now, the development of safe and efficient gene delivery carriers for clinical use is still a major challenge in human gene therapy. In the past two decades, viral vectors have been widely applied in clinical protocols. However, their applications are seriously hampered by safety issues (i.e. the possibility of inducing severe immune responses and the provocation of mutagenesis), difficulty in mass production and limitation in gene loading capacity.[5]

Synthetic non-viral vectors have been rapidly developed because of progress made in material science and nanotechnology in the last decade. These non-viral vectors are considered as alternatives to overcome the adverse effects of viral vectors. Among the various types of non-viral vectors, cationic polymers such as polylysine (PLL),[6,7] polyethyleneimine (PEI),[8–11] polyamidoamine dendrimer (PAMAM)[12–15] and chitosan[16–18] have received much attention owing to their advantageous properties over the viral and cationic liposome-based vectors. For examples, they do not induce any specific immune responses. They are also easily prepared and chemically modified. Among the various types of cationic polymers, linear and branched PEI are considered to be one of the most promising types of materials for gene delivery because of their

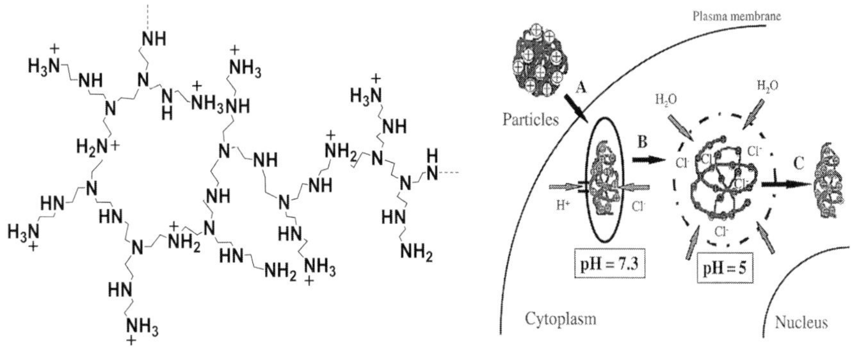

Fig. 1. (a) Chemical structure of branched polyethyleneimine; (b) Schematic diagram of proton sponge effect of the PEI/DNA complex. (Figure taken from Ref. 20.)

unique chemical and structural properties (Fig. 1(a)).[19] The PEI molecules possess high cationic charge density, thus they can effectively condense nucleic acids into nano-sized particles through strong electrostatic interaction. The resultant polyplexes can protect nucleic acids against enzymatic degradation. The cationic PEI can also facilitate interaction between the polyplexes and cell surface, thus enhancing cellular uptake efficiency. After cellular internalization, the high amine density of PEI can assist the endosomal escape of the polyplexes from lysosomal compartments via the well-known proton sponge effect (Fig. 1(b)).[10] The gene cargos are then released and expressed in a variety of mammalian cells. Thus, PEI often possesses high transfection efficiency.

Despite the above mentioned advantages of the PEI-based vectors, the use of PEI vectors has achieved only limited success, possibly due to the broad particle size distribution of the resultant polyplexes and the cytotoxicity issue.[21] To address these problems, various approaches to modify PEI have been made, such as conjugating PEI with biodegradable polymer,[22] modifying PEI to contain carboxylic acid and succinylated amino groups.[23] However, these PEI modification methods usually involve multi-step syntheses and tedious purification processes. They may also alter the unique chemical properties of PEI such as its sponge effect, thus reducing its transfection efficiency.

In this chapter, we describe our work using PEI-based core–shell nanoparticles to tackle the problems associated with broad size distribution and cytotoxicity. Two types of core–shell nanoparticles have been designed and synthesized. (1) An amphiphilic core–shell nanoparticle consisting of a biocompatible poly(methyl methacrylate) (PMMA) core and a hairy PEI shell. The nanoparticles have an average diameter of around 150 nm with uniform size distribution. (2) A core–shell nanogel particle containing biodegradable gelatin core and PEI shell with an average diameter of around 200 nm and narrow size distribution. Properties of these nanocarriers with respect to their abilities to condense, protect and release nucleic acids as well as their cytotoxicities have been studied. Their cellular uptake abilities and gene delivery efficiencies have also been investigated.

Amphiphilic PMMA-PEI Core-Shell Nanoparticles for DNA Delivery

In recent years, our group has developed a simple and versatile route to prepare well-defined amphiphilic core–shell particles in water based on the redox initiation between alkyl hydroperoxide and amino-containing water-soluble polymer.[24] Based on this new synthetic method, we have prepared amphiphilic nanoparticles that are composed of biocompatible PMMA cores and thick PEI shells, as well as applied them as nanocarriers for DNA delivery.[24–26] The hard PMMA core acts as a support for DNA molecules that are encapsulated into the soft PEI shell, thus forming homogenous complexed nanoparticle. Another function of the PMMA core is to reduce PEI cytotoxicity which is caused by the multiple attachment of cationic PEI onto the cell surface.[27] Hence, immobilizing the PEI onto the PMMA solid core could restrict its interaction with cell surfaces, resulting in reduction of cytotoxicity. Since the PEI molecule contains high amino content and positive charges, the highly extended PEI shell in water is able to provide a large surface area to encapsulate DNA molecules and facilitate cellular uptake of the complexed nanoparticles.

Synthesis of PMMA-PEI Core–Shell Nanoparticles

The PMMA-PEI nanoparticles have been synthesized according to our graft polymerization method.[24] The nanoparticles comprised 29% PEI and 71% PMMA, and displayed positive surface charges of around +40 mV as measured in a 1 mM of NaCl solution. The number- and volume-average diameters of the nanoparticles were 135 nm and 148 nm, respectively, indicating narrow particle size distribution [polydispersity index (D_v/D_n) = 1.10]. SEM image shows that the nanoparticles were of spherical shape and highly uniform size (Fig. 2(a)). Through selective staining of the PEI shells with a diluted phosphotungstic acid (PTA) solution, the core–shell nanostructure of the particle was clearly revealed in the TEM image (Fig. 2(b)).

Formation of Nanoparticle/DNA Complexes

DNA condensation on the gene carrier is the first step in gene delivery. For a cationic gene carrier, the interaction between the carrier and the negatively charged DNA molecule is simply based on electrostatic interaction. We have examined the ability of the core–shell

(a) (b)

Fig. 2. Morphology of the PMMA-PEI core–shell nanoparticles: (a) SEM image of the highly uniform nanoparticles; (b) TEM image of a nanoparticle that shows a well-defined PMMA core (lighter part) and thick PEI shell (darker region). (Figure taken from Ref. 25.)

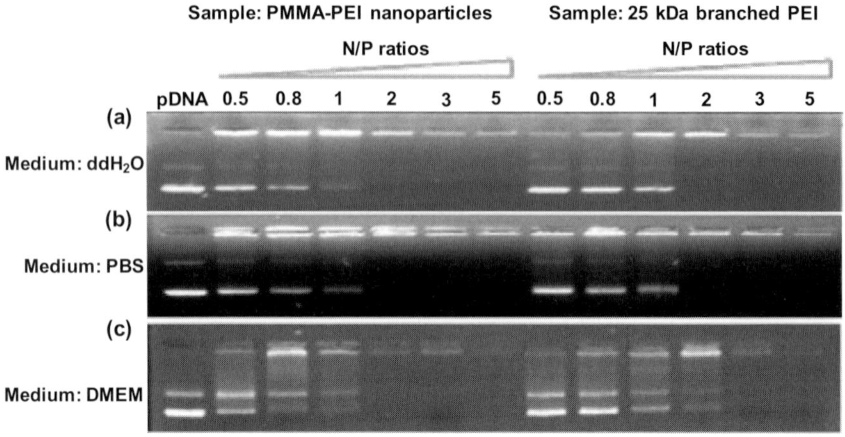

Fig. 3. Agarose gel retardation study of nanoparticle/pDNA and PEI/pDNA complexes at various N/P ratios (ranging from 0.5 to 5) in different media: (a) ddH_2O; (b) PBS; (c) DMEM. Conditions: various amounts of nanoparticles (ranging from 0 to 806 μg) and PEI polymers (ranging from 0 to 196 μg) were mixed with 0.3 μg pGL3-control plasmid in different media. After incubation at room temperature for 20 min, the complexes were analyzed by 1% agarose gel electrophoresis.

nanoparticle to condense plasmid DNA (pDNA) and compared the results with the 25 kDa branched PEI polymer in different media based on agarose gel retardation study. Both PMMA-PEI nanoparticle and PEI polymer could completely condense 0.3 μg pDNA in double deionized water (ddH_2O) and phosphate buffered saline (PBS) at the N/P ratio of 2 (Figs. 3(a) and (b)). In the case of Dulbecco's modified Eagle medium (DMEM), which contains several amino acids and vitamins, complete pDNA retardation was also achieved at the N/P ratio of 2 for nanoparticle, but at the N/P ratio of 3 for the native PEI. We have further investigated the complexing ability of the nanoparticles through varying the molecular weight of PEI and the thickness of PEI shell.[25] Results suggest that even though the nanoparticles with higher molecular weight of PEI (750 kDa) could bind to pDNA more effectively, the nanoparticles that contain a low molecular weight of PEI (25 kDa) are more suitable for use in gene delivery due to their lower cytotoxicity.

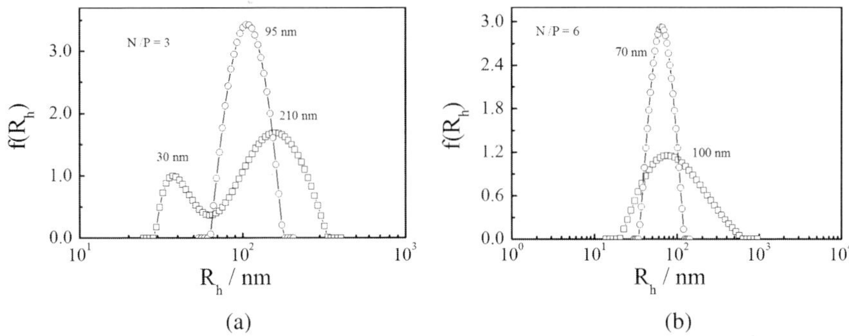

Fig. 4. Hydrodynamic radius (R_h) distributions of PEI/pDNA complexes (□); nanoparticle/pDNA complexes (○) measured by laser light scattering at 17.5° in the deionized Milli-Q water. (a) N/P ratio = 3; (b) N/P ratio = 6.

We have also investigated the particle size and size distribution of the nanoparticle/pDNA and PEI/pDNA complexes at N/P ratios of 3 and 6 using laser light scattering (LLS) determination. Figure 4(a) shows that PEI/pDNA complexes at the N/P ratio of 3 resulted in a bi-model size distribution, while the nanoparticle/pDNA complexes gave a much narrower size distribution with a hydrodynamic radius (R_h) of 95 nm. Increasing the N/P ratio up to 6 could produce pDNA complexes with a narrower size distribution (Fig. 4(b)). The LLS results suggest that nanoparticle/pDNA complexes could form more homogeneous and uniform complexed nanoparticles than the native PEI, especially at lower N/P ratios.

Protection Ability of the Nanoparticles and their In Vitro Cytotoxicity

Study of the protection ability of the nanoparticle/pDNA complexes showed that the nanoparticles could effectively protect the condensed pDNA molecules against enzymatic degradation.[25] The pDNA could also be released from the complexed nanoparticles in the presence of excess aspartic acid. These results clearly indicate that pDNA molecules remain intact after complexation and could be easily disassembled by polyanions.

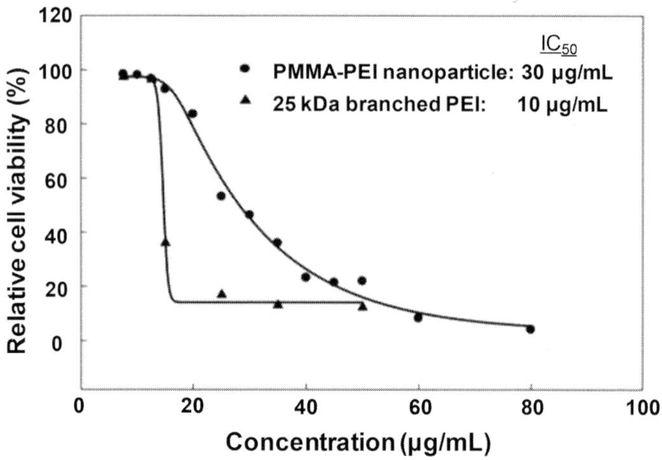

Fig. 5. The percentage of viable HeLa cells after treatment with various concentrations of native PEI (▲) and PMMA-PEI nanoparticles (●). (Figure taken from Ref. 25.)

It is known that PEI toxicity is attributed to the multiple attachment of cationic PEI onto the cell surface.[27] In our system, we proposed that immobilizing the PEI molecules onto the PMMA support could restrict the mobility of the PEI chain, thus reducing PEI interaction with the cell surface. We have evaluated the cytotoxicity of the PMMA-PEI nanoparticle in HeLa cells, and compared the results with that of the native PEI. Figure 5 shows the relative cell viability as a function of PEI concentration. Based on the 3-(4,5-dimethylthiazole-2-yl)-2,5-diphenyltetrazolium bromide (MTT) assay, IC_{50} concentrations of the PMMA-PEI nanoparticles and the native PEI are 30 and 10 μg/mL, respectively. This result indicates that covalent linkage of the PEI molecules to the PMMA particle surface can effectively reduce cytotoxicity of PEI by threefold.

Cellular Uptake and Intracellular Distribution of the Nanoparticle/pDNA Complexes

One of the prerequisites of an efficient gene delivery system is the ability to deliver genetic materials into cells and release them at the site of action. Thus the cellular uptake efficiency of the PMMA-PEI core–shell nanoparticles has been studied through labeling the nanoparticles with fluorescein isothiocyanate (FITC), followed by flow cytometry determination. Figure 6(a) shows that almost 100% of the treated cells contain FITC fluorescent signals. This result indicates that the amphiphilic nanoparticles can be efficiently taken up by the HeLa cells. The cellular uptake of the nanoparticles and nanoparticle/pDNA complexes was also investigated with confocal laser scanning microscopy. Figure 6(b) illustrates that both the nanoparticles and nanoparticle/pDNA complexes can be effectively internalized into the HeLa cells. The *in vitro* time series experiments also indicate that both nanoparticle/pDNA complexes and PMMA-PEI nanoparticles were distributed in the cytoplasmic region of cells after transfection for 10 min and 60 min, respectively. Nuclear localization was also observed in both samples after transfection for 20 min and 60 min, in which the nucleoli in nuclear region was lightened up by FITC.[28]

Dual fluorescent labeling study through labeling pDNA with tetramethyl-rhodamine (TM-rhodamine) and the nanoparticle with FITC confirmed that the pDNA remained bound to the nanoparticle even after 4 hr transfection (Fig. 6(c)). All these results suggest that PMMA-PEI nanoparticles can effectively deliver the pDNA to the nuclear region of HeLa, where the transcription takes place.

In Vitro Transfection Efficiency

The transfection efficiency of the nanoparticle/pDNA complexes was optimized through studying various transfection time and N/P ratios (Fig. 7(a)). It was found that the pDNA delivery ability of the PMMA-PEI nanoparticle was much more efficient than that of the native PEI polymer in the N/P ratios ranging from 3 to 15 (Fig. 7(b)). The optimal N/P ratio was 5:1 with transfection time of 36 hr. It was also found that increasing the N/P ratio would result in a decrease of transfection efficiency. This effect may be attributed to the increase of PEI cytotoxicity. We have also investigated the transfection efficiencies of the PEI-based complexes in different cell lines.[28] Results demonstrated that the nanoparticles were able to transfect both human

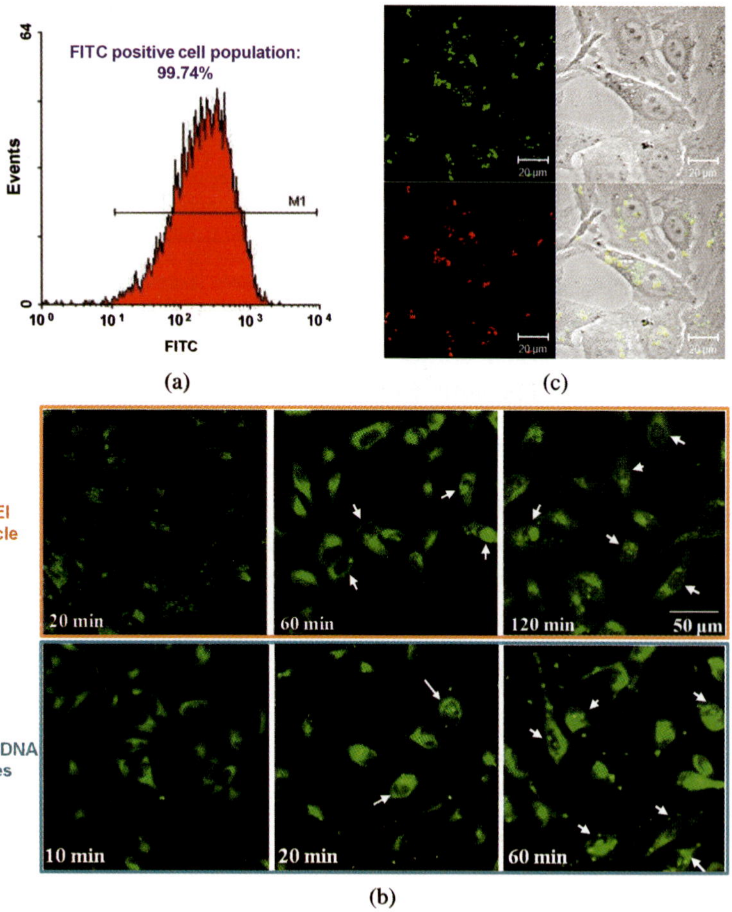

Fig. 6. Flow cytometric analysis of cellular uptake of FITC-labeled PMMA-PEI nanoparticles. (a) Region of gate M1 demarcates the fluorescent positive cell population. (b) Cellular internalization and localization of FITC-labeled PMMA-PEI nanoparticle or nanoparticle/pDNA complexes at various transfection time. Cells with nuclear transportation are indicated by arrows. (c) Microscopic image of HeLa cells after 4 hr of transfection with dual fluorescent labeled nanoparticle/pDNA complexes. The yellow signals illustrate the red signals (pDNA) are co-localized with the green signals (nanoparticle), indicating the presence of pDNA with the nanoparticles (Figures taken from Ref. 28.)

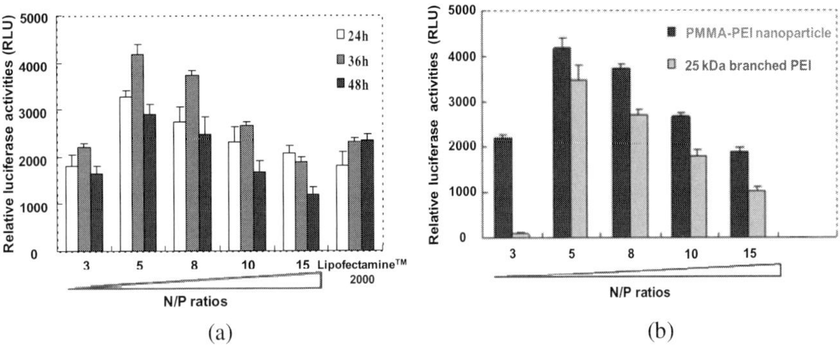

Fig. 7. (a) Effect of different N/P ratios and transfection time on the PMMA-PEI nanoparticle-mediated transfection efficiency in HeLa cells. HeLa cells were seeded at 4×10^5 cells per well in a 6- well plate and 2 μg of pGL3-Control plasmid was used for the preparation of each complex. (b) Comparison between PMMA-PEI nanoparticle and PEI polymer mediated transfection at various N/P ratios in HeLa cells. (Figure taken from Ref. 28.)

(MCF-7, BEL 7404) and non-human cell lines (C6, CHO-K1), and the transfection efficiency was cell-dependent. Among the tested cell lines, nanoparticle-mediated transfection displayed gene expression levels comparable to, or even better than the Lipofectamine™ 2000, a commercially available transfection agent.

To further enhance the transfection efficiency, we have incorporated a targeting agent, high mobility group box 1 (HMGB1) protein,[28] through simple electrostatic complexation with the nanoparticle/pDNA complexes. Results indicated that the expression level of the firefly luciferase gene encoded by the pDNA was significantly enhanced by up to eightfold with the presence of HMGB1. All the above results suggest that the PMMA-PEI nanoparticle is an efficient and promising nanocarrier for gene delivery.

Gelatin-PEI Core–Shell Nanogel for siRNA Delivery

Our previous work has demonstrated that the PMMA-PEI nanoparticles are promising nanocarriers for pDNA delivery. The PEI-based core–shell nanoparticles are able to complex pDNA at low N/P ratios with homogeneous size distribution, and can protect the pDNA from enzymatic degradation. The nanoparticle/pDNA complexes can also be effectively internalized by the cells, resulting in high transfection efficiency. Finally, the nanoparticles are three times less toxic than the native PEI. Despite the advantages of the PMMA-PEI nanoparticles, the non-biodegradable PMMA core may hinder the clinical use of the nanoparticles. Thus a new type of core–shell nanoparticle containing a biodegradable core is designed and synthesized.

Synthesis and Characterization of Gelatin-PEI Core–Shell Nanogel Particle

The core–shell nanogel particles were synthesized using a two-stage reaction[29]: (1) preparation of highly uniform gelatin nanoparticles through appropriate treatments of gelatin solution; and (2) conjugation of branched PEI to the preformed gelatin nanoparticles, followed by generation of core–shell nanostructure (gelatin core and PEI shell) through repeated cycles of desolvation and drying of the gelatin-PEI nanogels in an ethanol/water mixture. SEM image illustrates that the nanogels are mostly spherical and highly uniform in size (Fig. 8(a)). Statistical analysis of the nanogel size based on the SEM image gives an average diameter of 200 ± 40 nm. TEM image shows that the nanogel particle has a well-defined core–shell nanostructure with a hairy PEI outer layer in water (Fig. 8(a), inset). The gelatin-PEI nanogel particles display high positive surface charges (> +40 mV) at neutral pH as shown in Fig. 8(b). The actual composition of the gelatin-PEI core–shell nanogel particles, determined based on a gravimetry and a trinitrobenzenesulfonic acid (TNBS) method, is 57% gelatin and 43% PEI, respectively.

Formation of Nanogel/siRNA Complexes and their Cytotoxicity

To demonstrate the effectiveness of the gelatin-PEI core–shell nanogel as a potential nanocarrier, we used siRNA molecules that target

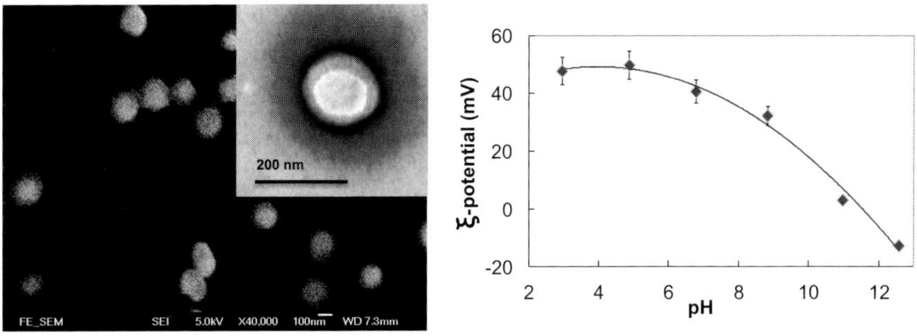

(a) (b)

Fig. 8. Morphology and surface property of the gelatin-PEI core–shell nanogel particles: (a) SEM image of the nanogels (inset: TEM image showing core–shell nanostructure); (b) *zeta*-Potential profile of the nanogels as measured in a 1 mM NaCl solution. (Figures taken from Ref. 29.)

human argininosuccinate synthetase 1 (ASS1) gene as a model study. ASS1 is a key enzyme in the urea cycle and is involved in arginine synthesis. It has been recently reported that some tumors which have high arginine requirements (e.g. melanomas and hepatocellular carcinomas) and lack expression for ASS1 gene are sensitive to arginine-depleting enzymes such as arginine deiminase (ADI).[30] Although ADI treatment was proven to have great success in anticancer therapy,[31,32] it is only effective for ASS1-deficient tumors. Tumors with normal ASS1 gene expression (e.g. human cervix adenocarcinoma) are resistant to ADI treatment.[33] Therefore, delivery of the ASS1-siRNA using our nanocarriers to down-regulate the ASS1 gene in the ASS1-positive cancer cells may resolve the limitation of the ADI treatment. We have evaluated the gelatin-PEI core–shell nanogel for its ability to condense ASS1-siRNA and determined the optimal complexation ratio between the nanogels and ASS1-siRNA using gel retardation assay. Figure 9(a) shows that the migration of ASS1-siRNA was fully retarded at the N/P ratio of 5, indicating a complete complex formation between the nanogels and ASS1-siRNA molecules.

The cytotoxicity of the gelatin-PEI core–shell nanogels was evaluated in HeLa cells using MTT assay. The IC_{50} concentration of the

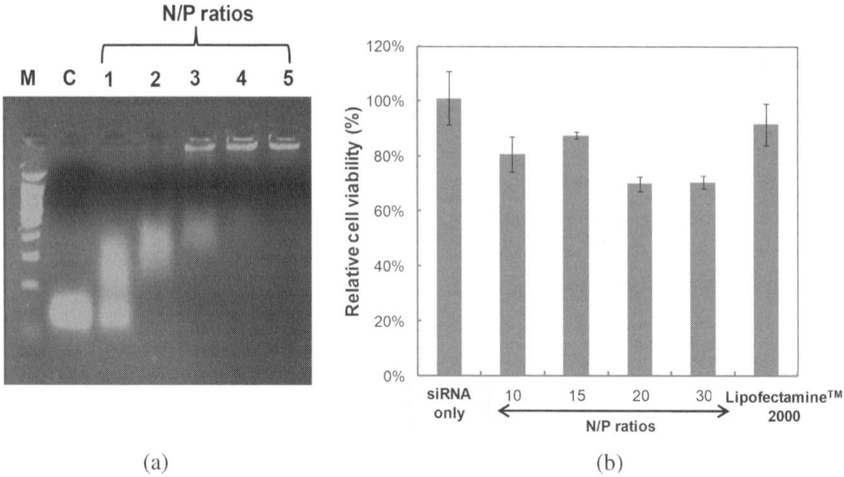

Fig. 9. (a) Gel retardation assay of the nanogel/siRNA complexes with various N/P ratios ranging from 1 to 5. M and C represent 100 bp DNA marker and naked siRNA, respectively. (b) Relative cell viabilities at various N/P ratios of nanogel/siRNA complexes and Lipofectamine™ 2000/siRNA lipoplexes. The concentration of siRNA used was 50 pmol/mL. (Figures taken from Ref. 29.)

nanogels was 33 μg/mL. We have also studied the cytotoxicity of the nanogel/siRNA complexes as a function of N/P ratios between 10 and 30 (Fig. 9(b)). At the N/P ratios of 10 and 15, their cell viabilities were above 80% with similar cytotoxicity (based on the statistical test of one-way ANOVA, $P \geq 0.05$). Increasing the N/P ratios to 20 and 30 resulted only in a slight reduction in cell viability from 80 to 70%. These results suggest that the nanogel/siRNA complexes have low cytotoxicity, thus are suitable nanocarriers for siRNA delivery.

In Vitro Cellular Uptake of the Nanogel/siRNA Complexes

To evaluate the delivery efficiency of the nanogel/siRNA complexes, the nanogels and the siRNA molecules were separately labeled with FITC and Alexa Fluor® Red. Their cellular uptake level and internalization were then studied using flow cytometry and confocal laser scanning microscopy, respectively. The cellular uptake efficiency of the nanogels was first investigated through counting the number of HeLa cells that contained FITC signals. As shown in Fig. 10(a), in the absence of the nanogels, there were no Alexa Fluor® Red positive cells, similar to the control (i.e. without siRNA). When the cells were treated with nanogel/siRNA complexes at the N/P ratio of 10 (which is equivalent to 4.6 μg/mL gelatin-PEI nanogels), the cellular uptake level of the Alexa Fluor® Red-labeled siRNA increased considerably to 40.5%. Further increasing the complexing N/P ratio to 30 (equivalent to 13.7 μg/mL of gelatin-PEI nanogels) could significantly enhance the siRNA uptake efficiency to 83.6%. These results demonstrate that nanogel/siRNA complexes could effectively deliver the siRNA into the cells.

The cellular uptake of the nanogel/siRNA complexes was further confirmed by the confocal laser scanning microscopy (CLSM). Figure 10(b) illustrates that HeLa cells treated with only Alexa Fluor® Red-labeled siRNA display negligible red fluorescence. When the siRNA was complexed onto the gelatin-PEI nanogels, the treated cells showed an increase of red fluorescence with the increase of N/P ratios from 10 to 30. These results clearly indicate that the nanogel/siRNA complexes could be effectively internalized into the cells. The cellular distributions of the delivered siRNA and the nanogel particles were also studied using dual fluorescent complexes (FITC-labeled gelatin-PEI nanogel and Alexa Fluor® Red-labeled siRNA) formed at the N/P ratio of 30 (Fig. 10(c)). The CLSM image of HeLa cells shows that the siRNA and nanogels were co-localized in the cytoplasmic region, suggesting that the nanogels could effectively carry and deliver the siRNA to the site of function, i.e. the cytoplasm of HeLa cells.

In Vitro Endogenous Gene Silencing Studies

The gene silencing activity of the delivered siRNA was studied by semi-quantitative reverse transcription–polymerase chain reaction (RT-PCR), using ASS1 as the target gene. The ASS1 gene knockdown efficiency was expressed as a percentage by comparing the relative ASS1-mRNA expression level of the treated group with the untreated cells. Figure 10(d) shows the effect of N/P ratios of 10, 15 and 30 on gene

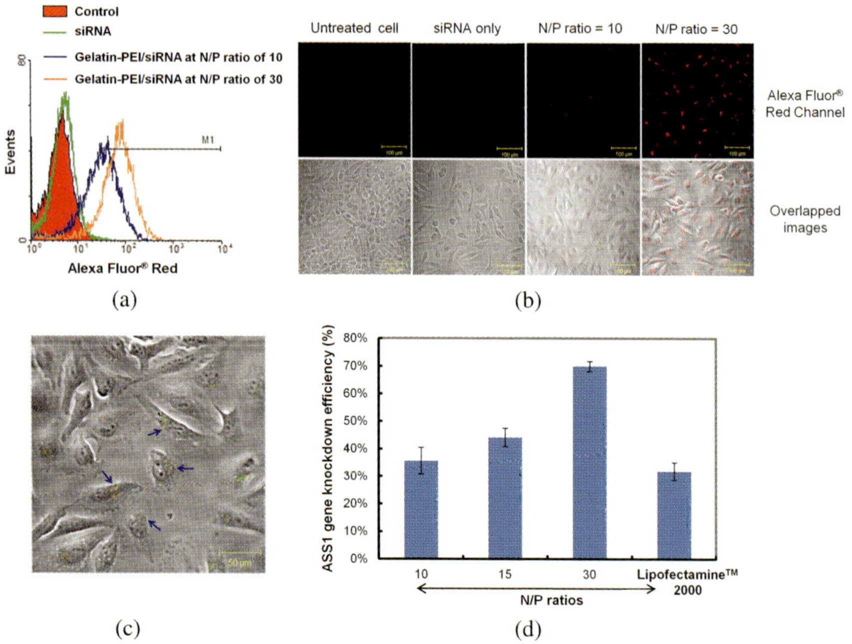

Fig. 10. (a) Flow cytometric histogram of the Alexa Fluor® Red fluorescence intensity of HeLa cells which were treated with nanogel/siRNA complexes (the concentration of siRNA was 50 pmol/mL). (b) CLSM images of HeLa cells treated with different N/P ratios of the nanogel/siRNA complexes. *Upper row*: red fluorescence channel images; *Lower row*: overlapping differential inference (DIC) and red fluorescence channel images (The original magnification was 20×). (c) Confocal laser scanning microscopic image with overlapping DIC and fluorescence channels for HeLa cells being treated with nanogel/siRNA complexes at N/P ratio of 30. (d) ASS1 gene knockdown efficiency after transfected with different N/P ratios of the nanogel/siRNA complexes and Lipofectamine™ 2000/siRNA lipoplexes. (Figure taken from Ref. 29.)

silencing efficiency. For comparison, Lipofectamine™ 2000 was used as the positive control. It was found that as the N/P ratios increased from 10 to 30, ASS1 gene knockdown efficiency increased significantly from 36.0 ± 4.8% to 70.0 ± 3.2%. This effect is probably attributed to the increasing cellular uptake events with increasing N/P ratio, as evidenced by the flow cytometry and CLSM data. Furthermore, the gene silencing efficiency of siRNA delivered with the nanogels is considerably higher than that of Lipofectamine™ 2000 (32.0% ± 8.5%). Thus, it is believed that gelatin-PEI core–shell nanogel particles are an effective type of nanocarrier for non-viral siRNA delivery.

Conclusion

We have described two types of PEI-based core–shell nanocarriers, namely PMMA-PEI nanoparticle and gelatin-PEI nanogel, for their potential applications in pDNA and siRNA deliveries. These novel types of amphiphilic core–shell nanocarriers have the following advantageous features: (1) they are able to complex and protect pDNA and siRNA; (2) they are three- to four-times less toxic than the native 25 kDa PEI polymer; (3) they have higher gene delivery efficiencies owing to their higher cellular uptake ability; (4) they are able to deliver the genetic materials to the site of function. Our results demonstrate that the core–shell nanoparticle containing hairy PEI layer and solid core is a unique nano-structural design for safer and more efficient gene delivery. The examples shown in this chapter only represent our work using PEI-based core–shell nanocarriers for pDNA and siRNA deliveries. We believe that this kind of nanocarrier will find many important applications in the delivery of other desirable biomolecules into cells.

Acknowledgments

We gratefully acknowledge the financial support from The Hong Kong Polytechnic University Niche Areas Funding, Research Grants Control of Hong Kong S.A.R. and Lo Ka Chung Centre for Natural Anti-Cancer Drug Development.

References

1. Liu F, Huang L. (2002) Development of non-viral vectors for systemic gene delivery. *J Control Release* **78:** 259–266.
2. Lundstrom K, Boulikas T. (2003) Viral and non-viral vectors in gene therapy: technology development and clinical trials. *Technol Cancer Res T* **2:** 471–486.

3. Segura T, Shea LD. (2001) Materials for non-viral gene delivery. *Ann Rev Mater Res* **31:** 25–46.
4. Thomas M, Klibanov AM. (2003) Non-viral gene therapy: polycation-mediated DNA delivery. *App Microbiol Biotechnol* **62:** 27–34.
5. Kay MA, Liu D, Hoogerbrugge PM. (1997) Gene therapy. *Proc Natl Acad Sci USA* **94:** 12744–12746.
6. Jeon E, Kim HD, Kim JS. (2003) Pluronic-grafted poly-(L)-lysine as a new synthetic gene carrier. *J Biomed Mater Res* **66A:** 854–859.
7. Cho KC, Kim SH, Jeong JH, Park TG. (2005) Folate receptor-mediated gene delivery using folate-poly(ethylene glycol)-poly(L-lysine) conjugate. *Macromol Biosci* **5:** 512–519.
8. Lungwitz U, Breunig M, Blunk T, Göpferich A. (2005) Polyethylenimine-based non-viral gene delivery systems. *Eur J Pharm Biopharm* **60:** 247–266.
9. Günther M, Lipka J, Malek A, *et al.* (2011) Polyethylenimines for RNAi-mediated gene targeting *in vivo* and siRNA delivery to the lung. *Eur J Pharm Biopharm* 77: 438–449.
10. Godbey WT, Wu KK, Mikos AG. (1999) Poly(ethylenimine) and its role in gene delivery. *J Control Release* **60:** 149–160.
11. Wang CF, Lin YX, Jiang T, *et al.* (2009) Polyethylenimine-grafted polycarbonates as biodegradable polycations for gene delivery. *Biomaterials* **30:** 4824–4832.
12. Kitchens KM, El-Sayed MEH, Ghandehari H. (2005) Transepithelial and endothelial transport of poly(amidoamine) dendrimers. *Adv Drug Deliv Rev* **57**: 2163–2176.
13. Zhang XQ, Wang XL, Huang SW, *et al.* (2005) *In vitro* gene delivery using polyamidoamine dendrimers with a trimesyl core. *Biomacromolecules* **6:** 341–350.
14. Navarro G, Tros de Ilarduya C. (2009) Activated and non-activated PAMAM dendrimers for gene delivery *in vitro* and *in vivo*. *Nanomed-Nanotechnol* **5:** 287–297.
15. Marvaniya HM, Parikh PK, Patel VR, *et al.* (2010) Dendrimer nanocarriers as versatile vectors in gene delivery. *J Chem Pharm Res* **2:** 97–108.
16. Mao S, Sun W, Kissel T. (2010) Chitosan-based formulations for delivery of DNA and siRNA. *Adv Drug Deliv Rev* **62:** 12–27.
17. Lavertu M, Methot S, Tran-Khanh N, Buschmann MD. (2006) High efficiency gene transfer using chitosan/DNA nanoparticles with specific combinations of molecular weight and degree of deacetylation. *Biomaterials* **27:** 4815–4824.
18. Strand SP, Lelu S, Reitan NK, *et al.* (2010) Molecular design of chitosan gene delivery systems with an optimized balance between polyplex stability and polyplex unpacking. *Biomaterials* **31:** 975–987.
19. Neu M, Fischer D, Kissel T. (2005) Recent advances in rational gene transfer vector design based on poly(ethylene imine) and its derivatives. *J Gene Med* **7:** 992–1009.
20. Kichler A, Behr JP, Erbacher P. (1999) Polyethylenimines: a family of potent polymers for nucleic acid delivery. In: Huang L, Hung MC, Wagner E (eds.), *Nonviral Vectors for Gene Therapy*. Academic Press, San Diego; London, pp. 191–206.

21. Kunath K, von Harpe A, Fischer D, *et al.* (2003) Low-molecular-weight polyethylenimine as a non-viral vector for DNA delivery: comparison of physicochemical properties, transfection efficiency and *in vivo* distribution with high-molecular-weight polyethylenimine. *J Control Release* **89:** 113–125.
22. Wen Y, Pan S, Luo X, *et al.* (2009) A biodegradable low molecular weight polyethylenimine derivative as low toxicity and efficient gene vector. *Bioconjug Chem* **20:** 322–332.
23. Zintchenko A, Philipp A, Dehshahri A, Wagner E. (2008) Simple modifications of branched PEI lead to highly efficient siRNA carriers with low toxicity. *Bioconjug Chem* **19:** 1448–1455.
24. Li P, Zhu J, Sunintaboon P, Harris FW. (2002) New route to amphiphilic core-shell polymer nanospheres: graft copolymerization of methyl methacrylate from water-soluble polymer chains containing amino groups. *Langmuir* **18:** 8641–8646.
25. Zhu J, Tang A, Law LP, *et al.* (2005) Amphiphilic core-shell nanoparticles with poly(ethylenimine) shells as potential gene delivery carriers. *Bioconjug Chem* **16:** 139–146.
26. Feng M, Lee D, Li P. (2006) Intracellular uptake and release of poly(ethyleneimine)-co-poly(methyl methacrylate) nanoparticle/pDNA complexes for gene delivery. *Int J Pharm* **311:** 209–214.
27. Choksakulnimitr S, Masuda S, Tokuda H, *et al.* (1995) *In vitro* cytotoxicity of macromolecules in different cell culture systems. *J Control Release* **34:** 233–241.
28. Siu YS, Li L, Leung MF, *et al.* (2012) Polyethylenimine-based amphiphilic core-shell nanoparticles: study of gene delivery and intracellular trafficking. *Biointerphases* DOI: 10.1007/s13758-011-0016-4.
29. Mimi H, Ho KM, Siu YS, *et al.* (2012) Polyethyleneimine-based core-shell nanogels: a promising siRNA carrier for argininosuccinate synthetase mRNA knockdown in HeLa cells. *J Control Release* **158**: 123–130.
30. Ensor CM, Holtsberg FW, Bomalaski JS, Clark MA. (2002) Pegylated arginine deiminase (ADI-SS $PEG_{20,000\ mw}$) inhibits human melanomas and hepatocellular carcinomas *in vitro* and *in vivo*. *Cancer Res* **62:** 5443–5450.
31. Ascierto PA, Scala S, Castello G, *et al.* (2005) Pegylated arginine deiminase treatment of patients with metastatic melanoma: results from phase I and II studies. *J Clin Oncol* **23:** 7660–7668.
32. Izzo F, Marra P, Beneduce G, *et al.* (2004) Pegylated arginine deiminase treatment of patients with unresectable hepatocellular carcinoma: results from phase I/II studies. *J Clin Oncol* **22:** 1815–1822.
33. Shen LJ, Lin WC, Beloussow K, Shen WC. (2003) Resistance to the anti-proliferative activity of recombinant arginine deiminase in cell culture correlates with the endogenous enzyme, argininosuccinate synthetase. *Cancer Lett* **191:** 165–170.

Chapter 19

Nano-Chaperones: Nanoparticles Acting as Artificial Chaperones

S. Boridy* and D. Maysinger*,†

1. Introduction

Protein misfolding and subsequent aggregation are regarded as early and causal events in the pathogenesis of neurodegenerative diseases.[1–3] Alzheimer's (AD), Parkinson's (PD), and Huntington's disease (HD) are prime examples wherein the improper folding and subsequent oligomerization of amyloid-β, α-synuclein, and Huntington, respectively, generate toxic disease-initiating soluble and insoluble aggregates. Until recently, larger, insoluble aggregates were thought to be the most toxic species.[4–6] However, in cases such as Alzheimer's disease, it was revealed that soluble oligomers, as small as dimers and trimers, are now regarded as the main neurotoxic species, impairing neuronal signaling and activating pro-inflammatory processes, ultimately leading to the formation of neural lesions. The cell's main defense mechanism against such insults is to mount an unfolded protein response which constitutes in large part the upregulation of chaperone proteins, responsible for sequestering misfolded proteins, promoting proteasome-mediated degradation and chaperone-mediated autophagy

* Department of Pharmacology and Therapeutics, McGill University, 3655 Promenade Sir-William-Osler, Montreal, Quebec, H3G 1Y6, Canada.
† Corresponding author. dusica.maysinger@mcgill.ca

of protein aggregates, and ensuring the stability of nascent polypeptide chains and general protein homeostasis[7–10] (summarized in Fig. 1). While it is still not entirely understood why toxic misfolded proteins accumulate in neurodegenerative disease states, in addition to the exact mechanism by which they mediate their harmful effects, the reduced capacity of the nervous system to mount a stress response is thought to be a critical component, especially in aging.

In the last decade, several nanoparticle-based systems, termed "*nano-chaperones*," were shown to exhibit "chaperone-like" activity based on their ability to bind to and promote the proper folding of protein enzymes and RNAses.[11] In particular, certain nanomaterials have the capacity to bind protein oligomers and aggregates and limit their toxicity *in vitro*, pointing towards the potential application of novel nanomedicine-based therapeutic interventions to treat protein-misfolding diseases. Examples include polysaccharide-based nanogels, hydrophobized poly(vinyl alcohol) nanoparticles, and silica nanoparticles, among others. Nano-chaperones can be tailored to mediate specific interactions with suspect protein oligomers and aggregates, aiding in their elimination, preventing subsequent toxicity, and slowing disease progression. As such, they may serve to supplement the endogenous molecular chaperone machinery, and can work in conjunction with present therapeutic strategies.[12,13] This combinatorial approach is superior to strategies aimed solely at over-expressing chaperone proteins, taxing the stress response and potentially leading to oncogenic transformation.[14,15]

In this chapter, we provide a concise overview of molecular chaperone function in the central nervous system under normal and pathological states, followed by a description of nanoparticles with chaperone-like protein folding and aggregation-inhibiting activity. We further provide insight into the differences and similarities between the mechanisms and behavior of artificial versus professional chaperones. And finally, we indicate likely directions this field can take towards improving interventions in neurodegenerative disease.

2. Chaperoning Function in Health and Disease

Molecular chaperones serve as the primary cellular defense mechanism against stress induced by protein misfolding and aggregation.[9,10,16–21]

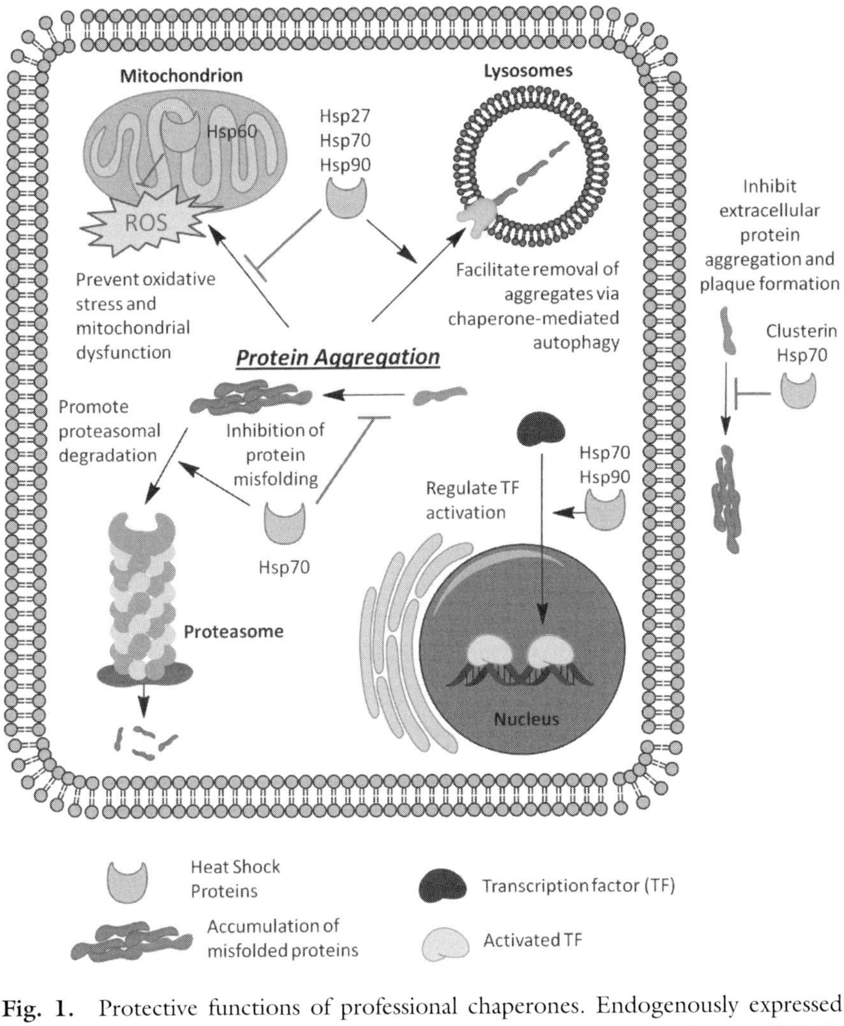

Fig. 1. Protective functions of professional chaperones. Endogenously expressed molecular chaperones serve many protective functions, both intra- and extracellularly. Namely, they are responsible for sequestering and facilitating the folding of nascent and misfolded proteins, promoting the degradation of damaged or aggregated proteins (via proteasomal degradation and chaperone-mediated autophagy), and preventing organelle damage mediated by misfolded and aggregated proteins (mitochondria are shown as one example in this figure). ROS: reactive oxygen species; HSP: heat shock proteins.

An evolutionarily conserved network of both constitutively expressed and inducible variants, chaperone proteins are responsible for the maintenance of cell homeostasis by promoting the proper folding of nascent proteins, preventing unintended protein–protein interactions, and stabilizing signal transducers, such as protein kinases and transcription factors. Chaperone function is indispensable to survival given a cell's susceptibility to macromolecular crowding, and this is especially true for neurons, which have thin, elongated processes that serve as "molecular highways" for the energy-dependent transportation of proteins and organelles to distal synapses. Neurons are also particularly susceptible to protein damage, misfolding and aggregation because as postmitotic cells, they do not have the capacity to dilute damaged cellular matter by cell division.[19] In addition, chaperone function encompasses the direct regulation of signal transduction pathways. For example, on top of stabilizing protein kinase B (PKB/AKT), a pro-survival kinase, they also directly inhibit the function of pro-apoptotic regulators, namely c-Jun N-terminal Kinase (JNK), Bcl-2–associated X protein (BAX), and pro-apoptotic caspases 8 and 9.[22]

The endogenous chaperone machinery constitutes a family of proteins known as heat shock proteins (Hsps), which were discovered serendipitously because of their ability to be induced under stress.[23,24] Notably, Hsps ensure the elimination of damaged or misfolded proteins, in addition to protein aggregates.[25] Due to their significant overlap in terms of function, they are classified according to their approximate molecular weight: Hsp27, Hsp60, Hsp70 and Hsp90 are known to play significant roles in cellular biogenesis. Hsp60 and above are energy-dependent chaperones, requiring adenosine 5′-triphosphate (ATP) to execute their demanding protein-folding functions. Hsp synthesis is regulated at the transcriptional level by heat shock factors (HSFs), the most prominent being HSF-1, which is a monomer in resting state. Under conditions of stress, Hsp expression is induced following HSF-1 homo-trimerization and subsequent binding to heat shock elements (HSEs) within the regulatory regions of target Hsp genes.[8,12,26] Under physiological conditions, the conservation of this response, in addition to proteasomal degradation and chaperone-mediated autophagy (CMA), is sufficient to maintain protein homeostasis and cell survival (Fig. 1).

However, there is an age-related detectable decline in these responses for reasons that are not entirely understood, but are known to involve increased oxidative damage, in addition to the accumulation of oxidized and ubiquitinated proteins and damaged organelles that have been incompletely autophagocytosed.[7,27,28] These mechanisms serve as an explanation for why aging is the most common risk factor for the development of neurodegenerative diseases, such as AD and other debilitating diseases in the elderly. Evidence collected over the last decade has shown that elevation of chaperone expression is protective in several models of neurodegeneration (summarized in Table 1), reviewed in Refs. 9, 19 and 29. This strongly suggests

Table 1. List of Endogenous Chaperones that Play an Integral Role in Neurodegenerative Disease

Family Name	Human Gene	Relevant Co-chaperones	Cellular Distribution of Family Members	Associated Disease	References
Clusterin	CLU	Unknown	Extracellular lipoprotein	AD	86–88
αB-crystallin	CRYAB	N/A	Cytosolic	PD AD ALS	136 137 138
Hsp27	HSPB1	N/A	Cytosolic	AD PolyQ, HD PD	139 140 141
Hsp60	HSPD1	Hsp10	Cytosol (TRiC) mitochondria (Hsp60)	AD	141, 142
Hsp70	HSPA1A	Hsp40, CHIP	Cytosol-nucleus (Hsc70), mitochondria (mtHsp70), ER (GRP78)	AD PD ALS PolyQ	62, 63 143, 144 144 145
Hsp90	HSP90A	Hsp70, cdc37, CHIP, FKBP51/2	Cytosol-nucleus	AD PD	146–148 149

Legend: AD, Alzheimer's disease; PD, Parkinson's disease; ALS, Amyotrophic lateral sclerosis; PolyQ, Polyglutamine expansion diseases; HD, Huntington's disease.

that the cellular chaperone machinery represents an ideal therapeutic target capable of reversing degenerative changes common in aging, especially given the redundancy in function amongst chaperone proteins and because modulators of the response are already known.[7,12,19,25,30]

2.1. *Molecular Chaperones in the Central Nervous System*

2.1.1. *Hsp60*

Hsp60 is the eukaryotic analog of the essential bacterial chaperone, GroEL, and is the main mitochondrial chaperonin in eukaryotic cells. Chaperonins are distinct from other chaperones based on their characteristic "double doughnut" structure; Hsp60 is a 14 subunit molecular machine made-up of two rings, layered one atop of the other, comprising 7 subunits each, with a complementary 7 subunit "lid," denoted hsp10, leading to the entrapment of a single protein sized 60kDa or less.[31,32] With hydrophobic residues at the apex and hydrophilic sequences along the inside walls, Hsp60 binds to exposed hydrophobic sequences of misfolded proteins, engulfing and sequestering them within the cavity (Fig. 3(b)), where it is more energetically favorable for proteins to assume their native fold.[33] Misfolded polypeptide chains may go through several "catch-and-release" cycles before finally achieving their proper fold.[34]

Hsp60's main function is to orchestrate the refolding of misfolded proteins within the mitochondrial matrix and inner membrane. Mitochondria are specific targets of the protective effect mediated by Hsp60, primarily against misfolding of multi-subunit protein complexes within the electron transport chain.[35,36] Interestingly, strong evidence indicates that mitochondrial dysfunction due to oxidative injury acts causally in aging-related neurodegenerative disease pathogenesis[37,38] and, in particular, increased oxidative damage is the earliest detectable event in AD.[39,40] Hsp60 was exclusively shown to relieve the selective inhibition of complex IV mediated by amyloid β in cortical neurons of mice,[41] maintain mitochondrial integrity and ATP generation in neonatal rat cardiac myocytes during ischemic injury,[42] and blunt the pro-apoptotic Bax/cytochrome c/caspase-3 cascade in an animal model of brain stem cell death.[21,43] Further supporting its essential neuroprotective role, a missense mutation in Hsp60 leads to a degenerative nerve disorder, termed *hereditary spastic paraplegia*, caused by defects in cellular transport,[44,45] whereas complete inactivation of the Hsp60-encoding gene, HSPD1, in mice is embryonic lethal.[46]

2.1.2. *Hsp70*

As the main mammalian cytosolic chaperone, Hsp70 is responsible for many cellular functions implicating protein quality control, including prevention of protein aggregation, proper folding of nascent and newly synthesized polypeptides, protein transport and degradation, solubilization of protein aggregates and refolding of misfolded protein, in addition to regulation of intracellular signal transduction pathways.[19,47,48] Eukaryotic Hsp70 chaperones exist intracellularly as monomers, with an ATP hydrolysis domain at the amino terminus and a polypepetide-binding domain at the carboxy-terminus.[49–51] Similarly to the apical domains of Hsp60, Hsp70 binds to short hydrophobic sequences that tend to be exposed in nascent and misfolded cytosolic polypeptide chains. Hydrolysis of ATP is stimulated by association with its J-domain containing co-chaperone, Hsp40, and is required for substrate binding. The primary folding mechanism for Hsp70 constitutes the unfolding of local domains, e.g. β-sheets and α-helices, along the polypeptide chain to overcome kinetic barriers that arise during spontaneous folding.[10,20,52] This same unfolding mechanism is believed to be involved in the disaggregation of large inclusions,[53] such as the plaques found in patients with AD. Unique to this family of chaperones, distinct constitutive and inducible forms exist; Hsc70/Hsp73 is constitutively expressed, while Hsp70/Hsp72 is the inducible form expressed following classic heat shock[54] and following ischemic and oxidative stress.[55] In addition, shuttling of the cognate Hsc70 between the nucleus and cytosol, which plays an essential role in protein transport,[56] is inhibited under stress, leading to nuclear accumulation of Hsc70 and anchoring to nucleoli-resident proteins.[57,58]

Importantly, Hsp70 is found to co-localize with inclusion bodies in several different neurodegenerative diseases involving protein misfolding, such as polyQ diseases and Huntington's disease,[59,60] Alzheimer's disease,[19] and Parkinson's disease.[61] Overexpression of cytosolic Hsc70 in cultured neurons suppresses intracellular Aβ production[62] and facilitates tau ubiquitination, blunting subsequent tau-mediated toxicity *in vitro*.[63,64] In addition, it was recently shown that when a mouse model of AD (i.e. APPsw) was crossed with a mouse overexpressing Hsp70, the progeny exhibited a decrease in age-dependent cognitive impairment compared to transgenic mice alone, and they also displayed decreased levels of circulating Aβ, plaque number, and synaptic loss.[65] Our lab and others have also shown evidence of age-related abnormalities in subcellular distribution and expression of chaperone proteins.[66–68] These findings suggest a neuroprotective role of Hsp70 in brain disorders, where protein misfolding appears to play a major role in disease pathogenesis.

2.1.3. *Hsp90*

Hsp90, similarly to Hsp70, is also an essential cytosolic chaperone. In addition to being responsible for the proper folding of cytosolic proteins, it serves a central role in the regulation of intracellular signal transduction and, importantly, the cytosolic stability of protein kinases, transcription factors, nuclear receptors, and numerous other proteins that lack any consistent functional or conformational features.[69,70] Hsp90 exists as a homodimer in nature, containing a carboxy-terminus dimerization domain and an amino-terminus ATP-binding domain. It mediates protein folding via an interaction with hydrophobic residues on substrate proteins and stabilizes them in an ATP-bound "clamp-like" conformation, which forms due to a transient N-terminal dimerization event. Following ATP-hydrolysis, the homodimer "clamp" returns to an open conformation, leading to the release of the substrate protein. One of the unique roles of Hsp90 is the repression of HSF-1 activity, and therefore serving as an ideal candidate to target in neurodegenerative disease.[12]

2.2. *Therapeutics Targeting the Endogenous Chaperone Machinery*

Based on the neuroprotective functions mediated by the chaperone machinery both intracellularly and extracellularly (Fig. 1), activating the stress protein response via activation of HSFs, in particular HSF-1, is viewed as a promising disease-modifying intervention in neurodegenerative disorders.[12] Given Hsp90's role in modulating the function of HSF-1, it comes as no surprise that ATP-binding pocket Hsp90 inhibitors, such as geldanamycin (Fig. 2(a)) and its less toxic derivative, 17-N-allylamino 17-demethoxygeldanamycin (17-AAG) (Fig. 2(b)), are some of the most potent inducers of chaperone protein expression and are protective in mouse models of neurodegenerative disorders such as PD.[71,72] This approach,

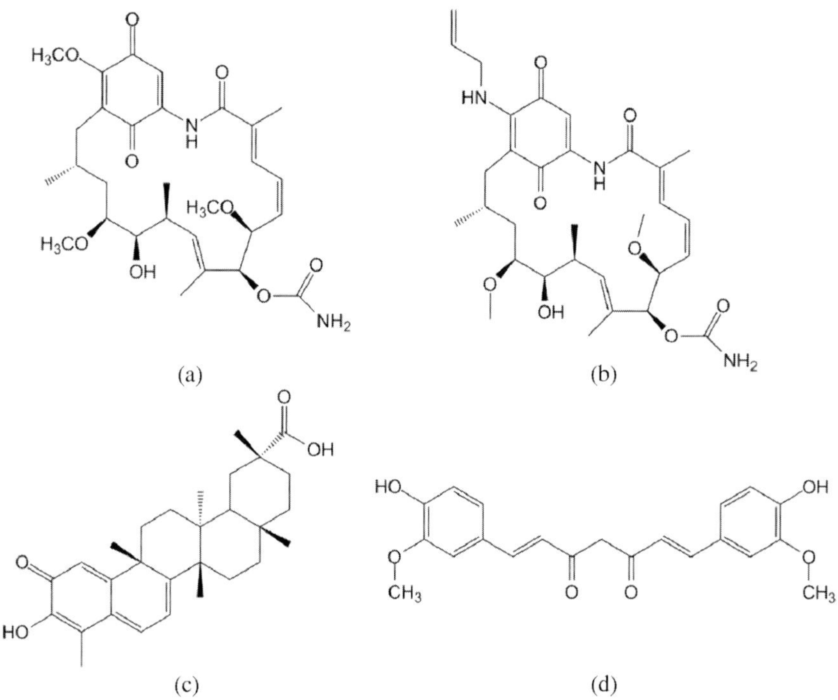

Fig. 2. Structure of HSF-1 activating drugs. (a) Geldanamycin, (b) 17-AAG (Tanespimycin), (c) Celastrol, (d) Curcumin.

however, is a double-edged sword, because it also leads to the down regulation of many protein kinases that regulate differentiation and cell survival. Other well-known chaperone inducers are the natural phenolic compounds, e.g. curcumin[73] and celastrol[74] (Figs. 2(c) and 2(d)). Both of these compounds showed some protective effects in animal models of AD.[75,76] Neuroprotection, induced by curcumin, although clinically relevant, remains to be established with orally administered curcumin in AD patients.[77] Treatment with non-steroidal anti-inflammatory drugs (NSAIDs) *in vitro* leads to HSF-1 DNA-binding, however chaperone expression has only been observed when coupled with heating to temperatures that would normally be sub-optimal for induction of the heat shock response.[78] It should be noted that more robust responses have been observed more recently with sulindac, nabumetone and phenylbutazone in the absence of additional stress.[79] Present strategies are also aimed at indirectly targeting the post-translational modifications of HSF-1 that together regulate its activation state, namely acetylation, sumoylation, and phosphorylation, by directly targeting the pathways that regulate these processes in the context of HSF-1 target gene transactivation. A concise review of these strategies has been provided by Neef *et al.*[12]

The major concern that is still outstanding with regards to the induction of chaperones by therapeutic agents is to what extent the endogenous chaperone machinery can cope with stress in an aging cellular network. With the accumulation of oxidative damage and the initiation of compensatory mechanisms, it is difficult to predict whether the signaling pathways we are studying remain intact under such conditions. Ideally, the therapeutic goal in neurodegeneration is to supplement and/or complement the protein folding and aggregate solubilization functions of chaperones, without affecting the regulation of cell survival and apoptosis mediated by molecular chaperones, as outlined above. Hence, designing nano-sized artificial chaperone systems that can directly sequester and eliminate toxic aggregates may prove to be a cost-effective disease-modifying approach.

3. Artificial Nano-Chaperones to Treat Protein Misfolding Disorders of the Central Nervous System

In light of the essential biological role that endogenous chaperones play in protein-misfolding diseases of the central nervous system, such as Creutzfeldt–Jakob disease (CJD), HD, PD, and AD, it is not surprising that therapeutics aimed at enhancing their expression and function are highly sought after.[12,13,19,30,80] Remarkably, rapid developments in nanomedicine within the last decade have paralleled these advancing efforts, through the development of novel nanomaterials with tailored surface properties capable of interacting with protein substrates and inducing conformational changes in single protein complexes.[11,81,82] Pioneering work by Rozema and Gellman in the early 90s using surfactants, such as cetyl trimethylammonium bromide (CTAB), to prevent the misfolding and aggregation of recombinant proteins for biotechnological applications, was the first known account of artificial chaperones being used for biomedical purposes.[83–85] Their model was inspired by the catch-and-release mechanism of the prokaryotic chaperonin, GroEL,[34] and has served as the driving model for the development of therapeutic nanomaterial-based artificial chaperones.

Due to their size, shape, surface functionalization, and large surface-to-volume ratio, nanoparticles are ideal candidate artificial chaperones both extracellularly, analogous to the way in which clusterin modulates the folding of extracellular Aβ and the clearance of extracellular proteins,[86–88] and intracellularly, analogous to the function of Hsps, such as the eukaryotic GroEL orthologue, Hsp60 (refer to Fig. 1). The interaction of nanoparticles with proteins in human body fluids and cell-derived proteins is well studied, and significant attention is paid to the importance of charge and composition of the surface functionalized layer.[89–92] The following section focuses on several examples of nanomaterials whose composition and surface chemistry have been tailored to not only favor protein interactions, but to also facilitate protein refolding and solubilization of inclusion bodies (summarized in Table 2).

Table 2. List of Nanoparticles with "Chaperone-Like" Activity

Name	Function and Protein Refolding Activity	Stripping Agent	Reference
Cholesteryl-bearing pullulan (CHP) nanogels	Inhibits amyloid-β aggregation, Refolding of CAB, CS, GFP Similar refolding kinetics to GroEL/Hsp60	Methyl-β-cyclodextrin	93, 100, 107, 108, 135, 150–153
2-(10-mercapto-decyl)-malonic acid functionalized gold nanoparticles (AuDA)	Refolding of chymotrypsin, lysozyme, and papain	NaCl solution	112, 113
Hydrophobized poly(vinyl alcohol) micellar nanoparticles (Macry-PVA)	Refolding of CAB	Methyl-β-cyclodextrin	114–117
N-isopropylacrylamide/ N-tert-butylacrylamide (PNIPAAm/BAM) nanoparticles	Temperature-responsive Dextran grafting provides increased amphiphilicity Refolding of β-lactamase	N/A	118–120

Legend: CAB, carbonic anhydrase; CS, citrate synthase; GFP, green fluorescence protein

3.1. *Nanogels*

Nanogels are characterized as "chemically" (via covalent interactions) or "physically" (via hydrogen bonds or hydrophobic interactions) cross-linked polymer networks. When dispersed in aqueous media, swollen nanogels assume soft-nanoparticle architecture and retain the capacity to absorb a considerable volume of water. They are primarily used to encapsulate macromolecules, such as proteins and peptides,[93,94] or siRNA[95] for drug delivery purposes. Over a decade ago, Akiyoshi's group developed one of the first ever characterized nanogel preparations by generating hydrophobized chains of the naturally occurring polysaccharide, pullulan, which were found to spontaneously self-associate when cholesterol moieties were interspersed at a frequency of 1.6 per 100 glucose units within each polymer chain (Fig. 3(a)). This preparation was termed Cholesteryl-bearing pullulans (CHP).[96] The spontaneous self-assembly of the amphiphilic polymers is driven by the intramolecular non-covalent cross linking between hydrophobic cholesterol moieties, resulting in a spherical particle ~30 nm in size, shielding local hydrophobic microenvironments within its cavity.[97]

The activity of these nanoparticles was particularly intriguing because they mimicked that of "professional" molecular chaperones by acting as hosts for guest proteins. In fact, they formed complexes with various proteins and enzymes, binding more specifically with heat denatured proteins than with native proteins,[96,98,99] and as such they served as the first model *nano-chaperone* system. Furthermore, proteins were released by the nanogels in their native fold over time, a process that could be stimulated by adding cyclodextrin to the system. Using an *in vitro* model of protein refolding, carbonic anhydrase (CAB) and citrate synthase (CS) were chemically denatured using guanidium chloride, after which nanogels were added.[100] Following methyl-β-cyclodextrin (m-β-CD) addition to the nanogel-enzyme complex, 60% of the original enzymatic activity of CAB was recovered and close to 70% of CS activity was recovered, while spontaneous folding in the absence of nanogels resulted in a mere 18% and 8% recovery of enzymatic activity, respectively. Reassuringly, nanogels did not bind native CAB, even following 24 hr incubation. More recently, the refolding activity of CHP-nanogels was directly compared with the essential prokaryotic heat shock protein 60 analogue, GroEL.[101] Both chaperone systems were incubated with acid-denatured green fluorescence protein (GFP) and fluorescence recovery was tracked following the addition of co-factors (ATP for GroEL and m-β-CD for CHP). CHP-nanogels recovered more than 90% of native GFP activity, 20% greater than that achieved with spontaneous refolding and equivalent to the recovery achieved with the professional GroEL chaperone system. The mechanism of protein refolding was stipulated to be similar to that of the chaperone, i.e. by GroEL/ES-inspired "catch-and-release" (Fig. 3(b)), allowing the protein several new occasions to refold properly, clearly demonstrating the artificial chaperone-like character of this nanogel-based system.

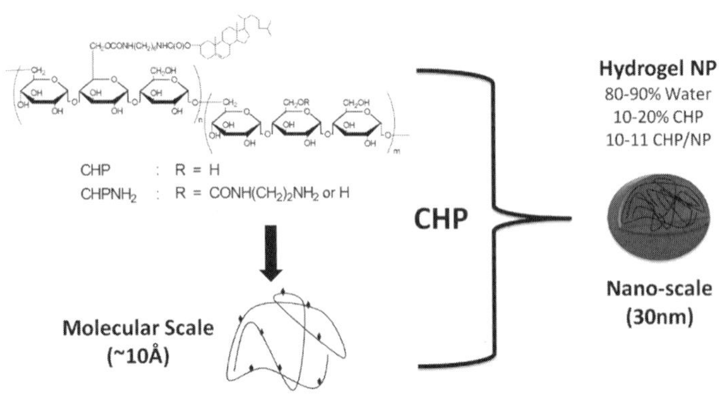

(a)

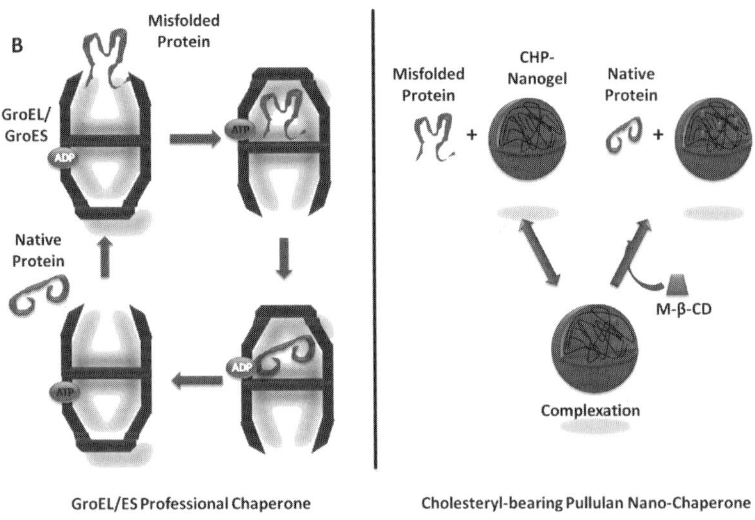

(b)

Fig. 3. Architecture of cholesteryl-bearing pullulan (CHP) nanogels and comparison between catch-and-release reaction cycles of GroEL/ES professional chaperone and CHP-nanogel based nano-chaperone system. (a) Molecular structure of cholesteryl-modified pullulan polysaccharide and assembly into nanogel tertiary structure. (b) Catch-and-release mechanism GroEL/ES vs. that of CHP-nanogels. Left, misfolded protein substrate interacts with hydrophobic residues of GroEL apical domain. Following nucleotide exchange for ATP, GroES traps the substrate inside the hydrophilic cavity, forcing it to refold in a sequestered environment. ATP hydrolysis aids in refolding activity and the cycle repeats in the adjacent cavity.

In an effort to demonstrate the potential therapeutic application of this artificial chaperoning system, we employed CHP nanogels from Akiyoshi's group to test the capacity of this system to prevent protein aggregation using AD as a model. Notably, the hallmark of these disorders is the accumulation of protein aggregates intra- and extracellularly.[102] In AD, these aggregates form plaques which are presently regarded as reservoirs for toxic oligomeric species of the amyloid-β peptide (Aβ).[103–106] Much of this research has focused on the aggregation-prone 42 amino acid variant, $A\beta_{1-42}$, the product of sequential processing of the membrane-bound amyloid precursor protein (APP) by β-secretase, followed by γ-secretase.[102] We first tested whether CHP-nanogels could bind Aβ, either in the monomeric or oligomeric form, using fluorescently tagged derivatives of both Aβ and nanogels and employing the principal of Förster resonance energy transfer (FRET) (Fig. 4(a)). We reported the capacity of nanogels, both cationic (-NH_2 modified pullulan, $CHPNH_2$) and neutral, to interact with and bind Aβ monomers and oligomers *in vitro*.[107,108] Moreover, unlabeled cationic and neutral nanogels prevented Aβ-mediated toxicity in mouse microglia (N9) and primary neural cells, respectively, as evidenced using the 3-(4,5-dimethylthiazol-2-yl)-2,5-diphenyl tetrazolium bromide (MTT) assay[108] (Fig. 4(b)). Our results suggested a mechanism of action whereby CHP-nanogels were capable of sequestering soluble Aβ monomers and oligomers, thereby blocking their cytotoxic activity (summarized in Fig. 4(c)). In addition, we also studied the uptake and intracellular fate of nanogels (i.e. CHP and $CHPNH_2$) in microglia. Microglia, and monocytes in general, can act as carriers for drug delivery systems and can result in specific targeting to the site of injury where microglia tend to accumulate, as is observed in AD at the site of senile plaques.[109] Using fluorescently-labeled nanogels modified with tetramethyl rhodamine isothiocyanate (TRITC), we observed the rapid accumulation of $CHPNH_2$ nanogels within 30 min, whereas uncharged derivatives slowly accumulated over 3 hr. Further studies need to be conducted before it can be confirmed whether microglia have the capacity to transport nanogels to the site of injury; however previous studies illustrating this behavior are encouraging.[110,111]

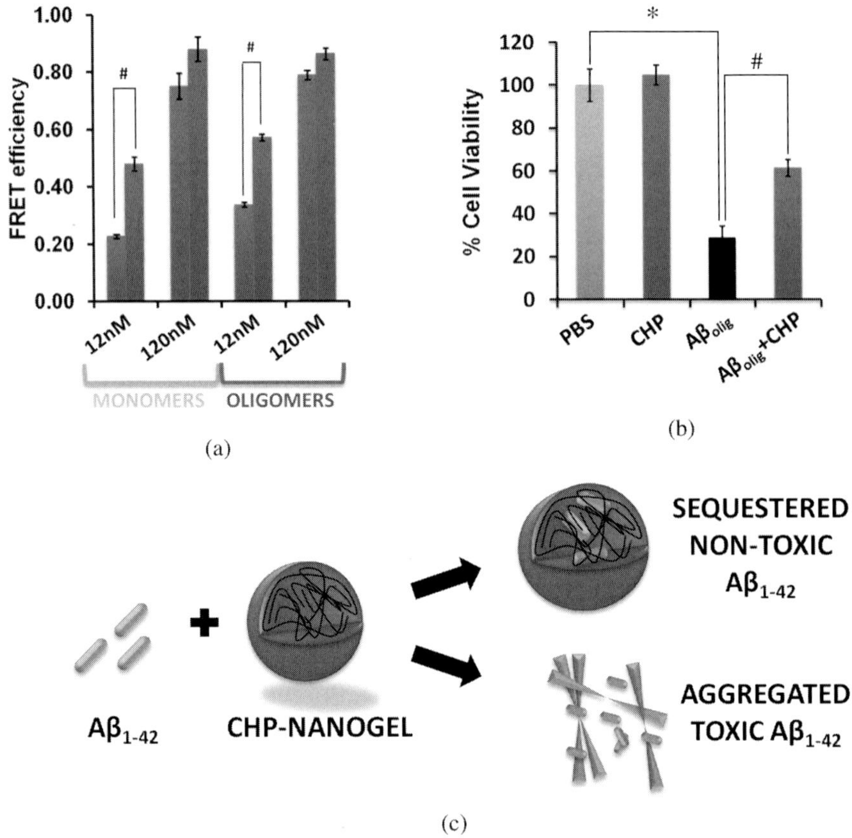

Fig. 4. Cholesteryl-bearing Pullulan (CHP) nanogels bind and inhibit the toxicity of $A\beta_{1\text{-}42}$. (a) Interaction between monomeric/oligomeric FITC-Aβ and TRITC-CHP was quantified using the principal of Förster Resonance Energy Transfer (FRET) *in vitro*. # $p < 0.01$ (b) CHP nanogels protect primary neural cultures from $A\beta_{1\text{-}42}$ mediated toxicity. * $p < 0.01$ vs. PBS # $p < 0.01$ vs. Aβdig (c) Schematic of CHP binding to and sequestering $A\beta_{1\text{-}42}$, preventing aggregation and subsequent toxicity. (Adapted from Boridy *et al.*, *Biomaterials*, 2009.)

3.2. *Functionalized Gold Nanoparticles*

Rotello *et al.* recently synthesized 2 nm core gold nanoparticles functionalized with 2-(10-mercapto-decyl)-malonic acid (AuDA), which exhibit a high surface localized negative charge density. AuDA readily formed a high affinity complex with cationic proteins, e.g. chymotrypsin, based on complementary electrostatic surface recognition, and moreover, upon dissociation of the complex in high ionic strength solution, chymostrypsin protein activity was completely restored.[112] To further test this model, Rotello and colleagues investigated the chaperone-like acitivity of AuDA nanoparticles by incubating them with heat-denatured cationic proteins: α-chymotrypsin, lysozyme, and papain.[113] AuDA nano-chaperones restored the activity of the thermally unfolded enzymes within 1 hr, with an average of ~95% restoration of activity, compared with ~50% restoration with spontaneous refolding in a high salt (NaCl) solution. Ongoing studies to test the application of this nano-chaperone system to target amyloid-type protein misfolding are underway.

3.3. *Hydrophobized Poly(vinyl alcohol) Micellar Nanoparticles*

Originally envisioning the potential use of macrogels for drug delivery and tissue engineering, Cavalieri *et al.* designed amphiphilic poly(vinylalcohol) polymers (PVA) randomly grafted with methacryloyl groups (Macry-PVA) and studied their behavior in aqueous solutions.[114,115] They were shown to spontaneously self-aggregate, similar to nanogels, however in this case forming micellar particles via intra- and interpolymeric association between hydrophobe pendants along the polymer chains, generating nanosized particles with a segregated hydrophobic core and a looping hydrophilic corona.[116] Based on similarities in structural features compared with previously designed artificial chaperones, they tested the chaperone-like activity of Macry-PVA nanoparticles in preventing heat-induced unfolding and aggregation of CAB. CAB is an ideal model of protein unfolding and aggregation because it readily disassembles at 55°C (<5 min), exposing hydrophobic residues, leading to the irreversible aggregation and subsequent loss of enzymatic activity. PVA nanoparticles with increasing methacryloyl substitution degree (Macry5-PVA and Macry25-PVA) were tested and shown to effectively chaperone the refolding of heat-denatured CAB.[117] In particular, the high substitution degree of Macry25-PVA resulted in a presumably stronger interaction with denatured CAB based on the investigation of secondary structure elements using circular dichroism (CD). CD revealed a higher ellipticity in the CAB-Macry25-PVA complex, indicating greater α-helical content in CAB. Moreover, following the administration of m-β-CD, Macry25-PVA complexation resulted in the complete restoration of enzymatic activity.

Cavalieri *et al.* propose a similar model to that proposed by Akiyoshi's group for the artificial chaperoning activity of hydrophobized polymeric complexes: The hydrophobic residues exposed by protein denaturation serve as binding sites for the methacroloyl and cholesterol hydrophobes within the nanogel or micellar core, resulting in (i) prevention of irreversible aggregation, and (ii) controlled unfolding of the protein secondary structure. Refolding is achieved by "stripping" the hydrophobic groups from the protein surface in a controlled manner with m-β-CD and subsequently allowing the recovery of the native-like fold in a semi-sequestered environment, similar in fashion to professional chaperonins, such as GroEL/Hsp60.[31,34,101]

3.4. *N-isopropylacrylamide/N-tert-butylacrylamide (NIPAM/BAM) Nanoparticles*

Given the need for stripping agents in the above examples of artificial chaperone models, it is difficult to imagine using these models *in vivo*, where precise manipulation of protein folding and aggregation depends on the spatial and temporal context. For instance, in AD, insoluble aggregates accumulate extracellularly as well as intracellularly within lysosomes, high pH environments.[102] By designing systems with changeable hydrophobicity using *stimulus-response* polymers (termed SMART polymers), artificial chaperones can turn on or off their chaperoning capacity in response to a localized microenvironment.

Temperature stimulus-responsive polymers poly(Nisopropylacrylamide) (PNIPAAm) were effective in refolding urea-denatured β-lactamase *in vitro*.[118,119] Refolding reached maximal yield (i.e. 100%) when the temperature was raised to 40°C, but was minimally effective at temperatures lower than 20°C, indicating that polymer hydrophobicity and subsequently, affinity for the substrate changed with a changing microenvironment. More recently, dextran-grafted

NIPAM was synthesized to increase the amphiphilicity of the polymer, providing greater sequestration of the substrate and increased refolding yield.[120] Furthermore, pH-responsive Eudragit S-100 polymers, commercially available anionic copolymers based on methacrylic acid and methyl methacrylate, were shown to alter their level of hydrophobicity based on the protonation of carboxyl groups at low pH.[121] In this state, the esterified carboxyl groups lose their charge and hydrophobic interactions predominate. Gupta *et al.* clearly demonstrate the capacity of this polymeric system to prevent aggregation and promote the successful refolding capacity of urea-denatured α-chymostrypsin.[122] It remains to be seen, however, to what extent these systems are effective at limiting protein misfolding and aggregation within cells, where a multitude of protein–polymer interactions can potentially take place.

4. Comparison between the Mechanisms of Professional Chaperones and Artificial Nano-Chaperones

The original inspiration for the development of artificial chaperones was the catch-and-release mechanism of action of the GroEL chaperonin,[31,34,101] which can be used to describe the activity of the majority of the nanoparticle systems described herein (refer to Fig. 3(b)). An important distinction, however, must be made between that of professional chaperones and the artificial systems designed to replicate their activity. Endogenous cellular chaperones, such as the chaperonin Hsp60, recognize and associate with their substrates based on hydrophobic interactions, which occur at the apical surface of the cage-like GroEL/Hsp60 complex. Upon encapsulation, however, the protein encounters a hydrophilically-walled cavity; sequestered from the crowded cytosol, the protein has an opportunity to re-fold in an entropically favorable environment, eliminating the risk of aggregation.[18] Moreover, this entire process is energy-dependent, consuming valuable ATP in the process. The artificial chaperone system, although based on a similar model, is vastly different in mechanism. For nanogels and Macry-PVA nanoparticles, substrate-recognition and active protein *unfolding* are mediated entirely by hydrophobic forces and is an entropically-driven reaction. Moreover, because the process lacks an energy-driven step that alters the energetic state of the complex leading to the release of the substrate, as occurs for Hsp60, Hsp70, or Hsp90, the dissociation requires a "stripping" agent, such as m-β-CD, which masks the hydrophobic residues on polymer chains, releasing the protein substrate and allowing for another opportunity to refold in a semi-sequestered environment. Conceptually, this difference has impeded the use of artificial chaperones to treat protein-misfolding diseases. It should be noted that binding, sequestering, and eliminating toxic misfolded protein aggregates without a requirement for refolding, should, in theory, do away with the need for a stripping agent.

5. Conclusion

Herein, we have summarized recent advances in the development of nanomaterials with protein refolding activities similar to that of professional, endogenously expressed chaperones and have highlighted their application in nanomedicine. The development of artificial chaperones for *in vivo* use could provide a novel therapeutic and preventative approach by reducing or completely eliminating the burden of accumulated misfolded protein aggregates, which lead to age-related and other neurodegenerative diseases, according to prevailing hypotheses. Strategies focused on taxing the endogenous machinery by enhancing the expression and function of heat shock proteins have been costly and have yet to reach their theoretical potential; this is an opportune time to develop novel therapeutics, including nanomaterial-based systems, which could expand the limited choice of disease-modifying interventions.

6. Future Perspective

Presently, studies focusing on the use of artificial chaperones in nanomedicine have made the greatest impact in the biotechnology sector, increasing the efficiency and yield of recombinant protein grown and purified from bacteria, in addition to being an ideal vehicle for protein delivery *in vivo*.[123–125] These attributes of artificial chaperones are mainly due to their natural capacity to bind to protein substrates with high affinity and thereby to provide slow release properties and passive targeting *in vivo*.[126–130]

As investigations begin to broaden, they will extend to the testing of the capacity of various artificial chaperones to bind to and sequester other disease-causing amyloid forming proteins. These include α-synuclein,[131,132] Huntington,[133] and superoxide dismutase,[134] with implications in PD, HD, and ALS respectively. The foundation of a novel and refreshing modality to intervene in disease pathology is emerging. For example, the *in vitro* testing of the binding and toxicity-inhibiting properties of such chaperones, similar to those performed in our lab and others for Aβ,[108,135] can serve as a filtering model for the selection of the best candidates. Another example would utilize mouse models of neurodegenerative disease that primarily recapitulate the amyloid pathology; the local administration, e.g. by intracerebral injection, of promising artificial chaperones would serve as a proof of principle and help to set the stage for preclinical testing within a decade's time. These approaches promise a real glimpse into the novel and effective strategies that nanotechnologies can afford to combat disease.

Acknowledgments

This work was supported by a grant from the Alzheimer's Association USA (IIRG-05-1352) and the Canadian Institutes of Health Research (nos. MOP-142626 MOP-119425). S.B. is supported by a CIHR-funded Vanier Canada Graduate Scholarship.

List of Abbreviations

Alzheimer's disease (AD)
Parkinson's disease (PD)
Huntington's disease (HD)
Amyloid β (Ab)
Heat shock proteins (Hsp)
Heat shock elements (HSE)
Heat shock factors (HSFs)
Carboxy-terminus of Hsc70 interacting protein (CHIP)
Creutzfeldt–Jakob disease (CJD)
Cetyl trimethylammonium bromide (CTAB)
Cholesteryl-bearing pullulans (CHP)
Carbonic anhydrase (CAB) and citrate synthase (CS)
Methyl-β-cyclodextrin (m-β-CD)
Green fluorescence protein (GFP)
Amyloid precursor protein (APP)
Förster resonance energy transfer (FRET)
Tetramethyl rhodamine isothiocyanate (TRITC)
Gold nanoparticles functionalized with 2-(10-mercapto-decyl)-malonic acid (AuDA)
Poly(vinyl-alcohol) polymers (PVA) randomly grafted with methacryloyl groups (Macry-PVA)
Poly(Nisopropylacrylamide) (PNIPAAm)
Adenosine 5′-triphosphate (ATP)
c-Jun N-terminal Kinase (JNK)
Protein Kinase B (PKB/Akt)
Bcl-2–associated X protein (BAX)

References

1. Gupta S, Jie S, Colby DW. (2011) Protein misfolding detected early in the pathogenesis of a transgenic mouse model of Huntington's disease using an amyloid seeding assay. *J Biol Chem.*
2. Ben-Zvi A, Miller EA, Morimoto RI. (2009) Collapse of proteostasis represents an early molecular event in Caenorhabditis elegans aging. *Proc Natl Acad Sci USA* **106(35):** 14914–14919.
3. Ross CA, Poirier MA. (2004) Protein aggregation and neurodegenerative disease. *Nat Med* **10 Suppl:** S10–S17.
4. Katzman R, *et al.* (1988) Clinical, pathological, and neurochemical changes in dementia: a subgroup with preserved mental status and numerous neocortical plaques. *Ann Neurol* **23(2):** 138–144.
5. Terry RD, *et al.* (1981) Some morphometric aspects of the brain in senile dementia of the Alzheimer type. *Ann Neurol* **10(2):** 184–192.
6. Walsh DM, *et al.* (2002) Naturally secreted oligomers of amyloid beta protein potently inhibit hippocampal long-term potentiation *in vivo*. *Nature* **416(6880):** 535–539.
7. Kourtis N, Tavernarakis N. (2011) Cellular stress response pathways and ageing: intricate molecular relationships. *EMBO J* **30(13):** 2520–2531.
8. Sreedhar AS, Csermely P. (2004) Heat shock proteins in the regulation of apoptosis: new strategies in tumor therapy: a comprehensive review. *Pharmacol Ther* **101(3):** 227–257.
9. Voisine C, Pedersen JS, Morimoto RI. (2010) Chaperone networks: tipping the balance in protein folding diseases. *Neurobiol Dis* **40(1):** 12–20.
10. Young JC, *et al.* (2004) Pathways of chaperone-mediated protein folding in the cytosol. *Nat Rev Mol Cell Biol* **5(10):** 781–791.
11. Fei L, Perrett S. (2009) Effect of nanoparticles on protein folding and fibrillogenesis. *Int J Mol Sci* **10(2):** 646–655.
12. Neef DW, Jaeger AM, Thiele DJ. (2011) Heat shock transcription factor 1 as a therapeutic target in neurodegenerative diseases. *Nat Rev Drug Discov* **10(12):** 930–944.
13. Neckers L, Workman P. (2012) Hsp90 molecular chaperone inhibitors: are we there yet? *Clin Cancer Res* **18(1):** 64–76.
14. Dai C, *et al.* (2007) Heat shock factor 1 is a powerful multifaceted modifier of carcinogenesis. *Cell* **130(6):** 1005–1018.
15. Fang F, Chang R, Yang L. (2011) Heat shock factor 1 promotes invasion and metastasis of hepatocellular carcinoma *in vitro* and *in vivo*. *Cancer.*
16. Ali YO, Kitay BM, Zhai RG. (2010) Dealing with misfolded proteins: examining the neuroprotective role of molecular chaperones in neurodegeneration. *Molecules* **15(10):** 6859–6887.

17. Broadley SA, Hartl FU. (2009) The role of molecular chaperones in human misfolding diseases. *FEBS Lett* **583(16):** 2647–2653.
18. Hartl FU, Hlodan R, Langer T. (1994) Molecular chaperones in protein folding: the art of avoiding sticky situations. *Trends Biochem Sci* **19(1):** 20–25.
19. Muchowski PJ, Wacker JL. (2005) Modulation of neurodegeneration by molecular chaperones. *Nat Rev Neurosci* **6(1):** 11–22.
20. Weibezahn J, *et al.* (2005) Novel insights into the mechanism of chaperone-assisted protein disaggregation. *Biol Chem* **386(8):** 739–744.
21. Chang AY, *et al.* (2006) Heat shock protein 60 in rostral ventrolateral medulla reduces cardiovascular fatality during endotoxaemia in the rat. *J Physiol* **574(Pt 2):** 547–564.
22. Lanneau D, *et al.* (2007) Apoptosis versus cell differentiation: role of heat shock proteins HSP90, HSP70 and HSP27. *Prion* **1(1):** 53–60.
23. Kaufmann SH. (1992) Heat shock proteins in health and disease. *Int J Clin Lab Res* **21(3):** 221–226.
24. Richter-Landsberg C, Goldbaum O. (2003) Stress proteins in neural cells: functional roles in health and disease. *Cell Mol Life Sci* **60(2):** 337–349.
25. Soti C, *et al.* (2005) Heat shock proteins as emerging therapeutic targets. *Br J Pharmacol* **146(6):** 769–780.
26. Morimoto RI, Sarge KD, Abravaya K. (1992) Transcriptional regulation of heat shock genes. A paradigm for inducible genomic responses. *J Biol Chem* **267(31):** 21987–21990.
27. Goldberg AL. (2003) Protein degradation and protection against misfolded or damaged proteins. *Nature* **426(6968):** 895–899.
28. Rattan SI. (2008) Increased molecular damage and heterogeneity as the basis of aging. *Biol Chem* **389(3):** 267–272.
29. Aridon P, *et al.* (2011) Protective role of heat shock proteins in Parkinson's disease. *Neurodegener Dis* **8(4):** 155–168.
30. Cohen FE, Kelly JW. (2003) Therapeutic approaches to protein-misfolding diseases. *Nature* **426(6968):** 905–909.
31. Sigler PB, *et al.* (1998) Structure and function in GroEL-mediated protein folding. *Annu Rev Biochem* **67:** 581–608.
32. Xu Z, Horwich AL, Sigler PB. (1997) The crystal structure of the asymmetric GroEL-GroES-(ADP)7 chaperonin complex. *Nature* **388(6644):** 741–750.
33. Martin J, Horwich AL, Hartl FU. (1992) Prevention of protein denaturation under heat stress by the chaperonin Hsp60. *Science* **258(5084):** 995–998.
34. Martin J, *et al.* (1993) The reaction cycle of GroEL and GroES in chaperonin-assisted protein folding. *Nature* **366(6452):** 228–233.
35. Voos W, Rottgers K. (2002) Molecular chaperones as essential mediators of mitochondrial biogenesis. *Biochim Biophys Acta* **1592(1):** 51–62.
36. Dencher NA, *et al.* (2007) Proteome alterations in rat mitochondria caused by aging. *Ann N Y Acad Sci* **1100:** 291–298.

37. Lin MT, Beal MF. (2006) Mitochondrial dysfunction and oxidative stress in neurodegenerative diseases. *Nature* **443(7113):** 787–795.
38. Karbowski M, Neutzner A. (2011) Neurodegeneration as a consequence of failed mitochondrial maintenance. *Acta Neuropathol.*
39. Nunomura A, *et al.* (2001) Oxidative damage is the earliest event in Alzheimer disease. *J Neuropathol Exp Neurol* **60(8):** 759–767.
40. Sullivan PG, Brown MR. (2005) Mitochondrial aging and dysfunction in Alzheimer's disease. *Prog Neuropsychopharmacol Biol Psychiatry* **29(3):** 407–410.
41. Veereshwarayya V, *et al.* (2006) Differential effects of mitochondrial heat shock protein 60 and related molecular chaperones to prevent intracellular beta-amyloid-induced inhibition of complex IV and limit apoptosis. *J Biol Chem* **281(40):** 29468–29478.
42. Lin KM, *et al.* (2001) Combined and individual mitochondrial HSP60 and HSP10 expression in cardiac myocytes protects mitochondrial function and prevents apoptotic cell deaths induced by simulated ischemia-reoxygenation. *Circulation* **103(13):** 1787–1792.
43. Chan JY, *et al.* (2007) Heat shock protein 60 or 70 activates nitric-oxide synthase (NOS) I- and inhibits NOS II-associated signaling and depresses the mitochondrial apoptotic cascade during brain stem death. *J Biol Chem* **282(7):** 4585–4600.
44. Bross P, *et al.* (2008) The Hsp60-(p.V98I) mutation associated with hereditary spastic paraplegia SPG13 compromises chaperonin function both *in vitro* and *in vivo*. *J Biol Chem* **283(23):** 15694–15700.
45. Magen D, *et al.* (2008) Mitochondrial hsp60 chaperonopathy causes an autosomal-recessive neurodegenerative disorder linked to brain hypomyelination and leukodystrophy. *Am J Hum Genet* **83(1):** 30–42.
46. Christensen JH, *et al.* (2010) Inactivation of the hereditary spastic paraplegia-associated Hspd1 gene encoding the Hsp60 chaperone results in early embryonic lethality in mice. *Cell Stress Chaperones* **15(6):** 851–863.
47. Kampinga HH, Craig EA. (2010) The HSP70 chaperone machinery: J proteins as drivers of functional specificity. *Nat Rev Mol Cell Biol* **11(8):** 579–592.
48. Lanneau D, *et al.* (2008) Heat shock proteins: essential proteins for apoptosis regulation. *J Cell Mol Med* **12(3):** 743–761.
49. Frydman J, *et al.* (1994) Folding of nascent polypeptide chains in a high molecular mass assembly with molecular chaperones. *Nature* **370(6485):** 111–117.
50. Beckmann RP, Mizzen LE, Welch WJ. (1990) Interaction of Hsp 70 with newly synthesized proteins: implications for protein folding and assembly. *Science* **248(4957):** 850–854.
51. Eggers DK, Welch WJ, Hansen WJ. (1997) Complexes between nascent polypeptides and their molecular chaperones in the cytosol of mammalian cells. *Mol Biol Cell* **8(8):** 1559–1573.

52. Ben-Zvi AP, Goloubinoff P. (2001) Review: mechanisms of disaggregation and refolding of stable protein aggregates by molecular chaperones. *J Struct Biol* **135(2):** 84–93.
53. Shorter J. (2011) The mammalian disaggregase machinery: Hsp110 synergizes with Hsp70 and Hsp40 to catalyze protein disaggregation and reactivation in a cell-free system. *PLoS One* **6(10):** e26319.
54. Melcher A, *et al.* (1998) Tumor immunogenicity is determined by the mechanism of cell death via induction of heat shock protein expression. *Nat Med* **4(5):** 581–587.
55. Zhu H, *et al.* (2011) Why are hippocampal CA1 neurons vulnerable but motor cortex neurons resistant to transient ischemia? *J Neurochem.*
56. Kose S, *et al.* (2005) The 70-kD heat shock cognate protein (hsc70) facilitates the nuclear export of the import receptors. *J Cell Biol* **171(1):** 19–25.
57. Banski P, *et al.* (2010) Nucleolar targeting of the chaperone hsc70 is regulated by stress, cell signaling, and a composite targeting signal which is controlled by autoinhibition. *J Biol Chem* **285(28):** 21858–21867.
58. Kodiha M, *et al.* (2005) Stress inhibits nucleocytoplasmic shuttling of heat shock protein hsc70. *Am J Physiol Cell Physiol* **289(4):** C1034–C1041.
59. Jana NR, *et al.* (2000) Polyglutamine length-dependent interaction of Hsp40 and Hsp70 family chaperones with truncated N-terminal huntingtin: their role in suppression of aggregation and cellular toxicity. *Hum Mol Genet* **9(13):** 2009–2018.
60. Schmidt T, *et al.* (2002) Protein surveillance machinery in brains with spinocerebellar ataxia type 3: redistribution and differential recruitment of 26S proteasome subunits and chaperones to neuronal intranuclear inclusions. *Ann Neurol* **51(3):** 302–310.
61. Lindersson E, *et al.* (2004) Proteasomal inhibition by alpha-synuclein filaments and oligomers. *J Biol Chem* **279(13):** 12924–12934.
62. Magrane J, *et al.* (2004) Heat shock protein 70 participates in the neuroprotective response to intracellularly expressed beta-amyloid in neurons. *J Neurosci* **24(7):** 1700–1706.
63. Shimura H, *et al.* (2004) CHIP-Hsc70 complex ubiquitinates phosphorylated tau and enhances cell survival. *J Biol Chem* **279(6):** 4869–4876.
64. Petrucelli L, *et al.* (2004) CHIP and Hsp70 regulate tau ubiquitination, degradation and aggregation. *Hum Mol Genet* **13(7):** 703–714.
65. Hoshino T, *et al.* (2011) Suppression of Alzheimer's disease-related phenotypes by expression of heat shock protein 70 in mice. *J Neurosci* **31(14):** 5225–5234.
66. Rumora L, *et al.* (2007) Impairments of heat shock protein expression and MAPK translocation in the central nervous system of follitropin receptor knockout mice. *Exp Gerontol* **42(7):** 619–628.
67. Hall DM, *et al.* (2000) Aging reduces adaptive capacity and stress protein expression in the liver after heat stress. *J Appl Physiol* **89(2):** 749–759.

68. Bernstein SL, *et al.* (2000) Heat shock cognate-70 gene expression declines during normal aging of the primate retina. *Invest Ophthalmol Vis Sci* **41(10):** 2857–2862.
69. Young JC, Moarefi I, Hartl FU. (2001) Hsp90: a specialized but essential protein-folding tool. *J Cell Biol* **154(2):** 267–273.
70. Richter K, Buchner J. (2001) Hsp90: chaperoning signal transduction. *J Cell Physiol* **188(3):** 281–290.
71. Trepel J, *et al.* (2010) Targeting the dynamic HSP90 complex in cancer. *Nat Rev Cancer* **10(8):** 537–549.
72. Shen HY, *et al.* (2005) Geldanamycin induces heat shock protein 70 and protects against MPTP-induced dopaminergic neurotoxicity in mice. *J Biol Chem* **280(48):** 39962–39969.
73. Teiten MH, *et al.* (2009) Induction of heat shock response by curcumin in human leukemia cells. *Cancer Lett* **279(2):** 145–154.
74. Chow AM, Brown IR. (2007) Induction of heat shock proteins in differentiated human and rodent neurons by celastrol. *Cell Stress Chaperones* **12(3):** 237–244.
75. Paris D, *et al.* (2010) Reduction of beta-amyloid pathology by celastrol in a transgenic mouse model of Alzheimer's disease. *J Neuroinflammation* **7:** 17.
76. Garcia-Alloza M, *et al.* (2007) Curcumin labels amyloid pathology *in vivo*, disrupts existing plaques, and partially restores distorted neurites in an Alzheimer mouse model. *J Neurochem* **102(4):** 1095–1104.
77. Baum L, *et al.* (2008) Six-month randomized, placebo-controlled, double-blind, pilot clinical trial of curcumin in patients with Alzheimer disease. *J Clin Psychopharmacol* **28(1):** 110–113.
78. Lee BS, *et al.* (1995) Pharmacological modulation of heat shock factor 1 by antiinflammatory drugs results in protection against stress-induced cellular damage. *Proc Natl Acad Sci USA* **92(16):** 7207–7211.
79. Housby JN, *et al.* (1999) Non-steroidal anti-inflammatory drugs inhibit the expression of cytokines and induce HSP70 in human monocytes. *Cytokine* **11(5):** 347–358.
80. Kalia SK, Kalia LV, McLean PJ. (2010) Molecular chaperones as rational drug targets for Parkinson's disease therapeutics. *CNS Neurol Disord Drug Targets* **9(6):** 741–753.
81. Kurganov BI, Topchieva IN. (1998) Artificial chaperone-assisted refolding of proteins. *Biochemistry (Mosc)* **63(4):** 413–419.
82. Sasaki Y, Akiyoshi K. (2010) Development of an artificial chaperone system based on cyclodextrin. *Curr Pharm Biotechnol* **11(3):** 300–305.
83. Rozema D, Gellman SH. (1996) Artificial chaperone-assisted refolding of carbonic anhydrase B. *J Biol Chem* **271(7):** 3478–3487.
84. Rozema D, Gellman SH. (1996) Artificial chaperone-assisted refolding of denatured-reduced lysozyme: modulation of the competition between renaturation and aggregation. *Biochemistry* **35(49):** 15760–15771.

85. Daugherty DL, *et al.* (1998) Artificial chaperone-assisted refolding of citrate synthase. *J Biol Chem* **273(51):** 33961–33971.
86. Narayan P, *et al.* (2011) The extracellular chaperone clusterin sequesters oligomeric forms of the amyloid-beta(1-40) peptide. *Nat Struct Mol Biol* **19(1):** 79–83.
87. Baig S, *et al.* (2012) Clusterin mRNA and protein in Alzheimer's disease. *J Alzheimers Dis.*
88. Wyatt AR, *et al.* (2011) Clusterin facilitates *in vivo* clearance of extracellular misfolded proteins. *Cell Mol Life Sci* **68(23):** 3919–3931.
89. Maiorano G, *et al.* (2010) Effects of cell culture media on the dynamic formation of protein-nanoparticle complexes and influence on the cellular response. *ACS Nano* **4(12):** 7481–7491.
90. Zhu ZJ, *et al.* (2011) Stability of quantum dots in live cells. *Nat Chem* **3(12):** 963–968.
91. Ruh H, *et al.* (2010) Identification of serum proteins bound to industrial nanomaterials. *Toxicol Lett* **208(1):** 41–50.
92. Zhang H, *et al.* (2011) Quantitative proteomics analysis of adsorbed plasma proteins classifies nanoparticles with different surface properties and size. *Proteomics* **11(23):** 4569–4577.
93. Gilmore JL, *et al.* (2008) Novel nanomaterials for clinical neuroscience. *J Neuroimmune Pharmacol* **3(2):** 83–94.
94. Morimoto N, *et al.* (2006) Nanogel engineered designs for polymeric drug delivery. *ACS Symp Ser* **924:** 88–101.
95. Vinogradov SV, Batrakova EV, Kabanov AV. (2004) Nanogels for oligonucleotide delivery to the brain. *Bioconjug Chem* **15(1):** 50–60.
96. Akiyoshi K, *et al.* (1993) Self-aggregates of hydrophobized polysaccharides in water. Formation and characteristics. *Macromolecules* **26(12):** 3062–3068.
97. Akiyoshi K, *et al.* (1997) Microscopic structure and thermoresponsiveness of a hydrogel nanoparticle by self-assembly of a hydrophobized polysaccharide. *Macromolecules* **30(4):** 857–861.
98. Akiyoshi K, *et al.* (1998) Self-assembled hydrogel nanoparticle of cholesterol-bearing pullulan as a carrier of protein drugs: complexation and stabilization of insulin. *J Control Release* **54(3):** 313–320.
99. Akiyoshi K, Sasaki Y, Sunamoto J. (1999) Molecular chaperone-like activity of hydrogel nanoparticles of hydrophobized pullulan: thermal stabilization with refolding of carbonic anhydrase B. *Bioconjug Chem* **10(3):** 321–324.
100. Nomura Y, *et al.* (2003) Protein refolding assisted by self-assembled nanogels as novel artificial molecular chaperone. *FEBS Lett* **553(3):** 271–276.
101. Asayama W, *et al.* (2008) Comparison of refolding activities between nanogel artificial chaperone and GroEL systems. *Int J Biol Macromol* **42(3):** 241–246.
102. LaFerla FM, Green KN, Oddo S. (2007) Intracellular amyloid-beta in Alzheimer's disease. *Nat Rev Neurosci* **8(7):** 499–509.

103. Haass C, Selkoe DJ. (2007) Soluble protein oligomers in neurodegeneration: lessons from the Alzheimer's amyloid beta-peptide. *Nat Rev Mol Cell Biol* **8(2):** 101–112.
104. Santos AN, *et al.* (2012) Amyloid-beta oligomers in cerebrospinal fluid are associated with cognitive decline in patients with Alzheimer's disease. *J Alzheimers Dis.*
105. Jin M, *et al.* (2011) Soluble amyloid beta-protein dimers isolated from Alzheimer cortex directly induce tau hyperphosphorylation and neuritic degeneration. *Proc Natl Acad Sci USA* **108(14):** 5819–5824.
106. Shankar GM, *et al.* (2008) Amyloid-beta protein dimers isolated directly from Alzheimer's brains impair synaptic plasticity and memory. *Nat Med* **14(8):** 837–842.
107. Sasaki Y, Akiyoshi K. (2010) Nanogel engineering for new nanobiomaterials: from chaperoning engineering to biomedical applications. *The Chemical Record* **10(6):** 366–376.
108. Boridy S, *et al.* (2009) The binding of pullulan modified cholesteryl nanogels to Abeta oligomers and their suppression of cytotoxicity. *Biomaterials* **30(29):** 5583–5591.
109. Dandrea MR, *et al.* (2001) Application of triple immunohistochemistry to characterize amyloid plaque-associated inflammation in brains with Alzheimer's disease. *Biotech Histochem* **76(2):** 97–106.
110. Ribot E, *et al.* (2007) Microglia used as vehicles for both inducible thymidine kinase gene therapy and MRI contrast agents for glioma therapy. *Cancer Gene Ther* **14(8):** 724–737.
111. Madsen SJ, *et al.* (2011) Macrophages as cell-based delivery systems for nanoshells in photothermal therapy. *Ann Biomed Eng.*
112. Simard JM, Szymanski B, Rotello VM. (2005) Reversible regulation of chymotrypsin activity using negatively charged gold nanoparticles featuring malonic acid termini. *Med Chem* **1(2):** 153–157.
113. De M, Rotello VM. (2008) Synthetic "chaperones": nanoparticle-mediated refolding of thermally denatured proteins. *Chem Commun (Camb)* **2008(30):** 3504–3506.
114. Cavalieri F, *et al.* (2004) Study of gelling behavior of poly(vinyl alcohol)-methacrylate for potential utilizations in tissue replacement and drug delivery. *Biomacromolecules* **5(6):** 2439–2446.
115. Martens PJ, Bryant SJ, Anseth KS. (2003) Tailoring the degradation of hydrogels formed from multivinyl poly(ethylene glycol) and poly(vinyl alcohol) macromers for cartilage tissue engineering. *Biomacromolecules* **4(2):** 283–292.
116. Takasu A, *et al.* (2002) Micellar polymerization of amphiphilic poly(vinyl alcohol) macromonomer having a methacrylate end group prepared by

aldol-type group-transfer polymerization. *J Polym Sci A Polym* **40(24):** 4477–4484.

117. Cavalieri F, Chiessi E, Paradossi G. (2007) Chaperone-like activity of nanoparticles of hydrophobized poly(vinyl alcohol). *Soft Matter* **3(6):** 718–724.
118. Lin SC, *et al.* (2000) Enhanced protein renaturation by temperature-responsive polymers. *Biotechnol Bioeng* **67(5):** 505–512.
119. Lu D, *et al.* (2005) The mechanism of PNIPAAm-assisted refolding of lysozyme denatured by urea. *Biochem Eng J* **24(1):** 55–64.
120. Lu D, *et al.* (2006) Dextran-grafted-PNIPAAm as an artificial chaperone for protein refolding. *Biochem Eng J* **27(3):** 336–343.
121. Sharma A, Gupta MN. (2002) Macroaffinity ligand-facilitated three-phase partitioning (MLFTPP) for purification of xylanase. *Biotechnol Bioeng* **80(2):** 228–232.
122. Roy I, Gupta MN. (2003) pH-responsive polymer-assisted refolding of urea- and organic solvent-denatured alpha-chymotrypsin. *Protein Eng* **16(12):** 1153–1157.
123. Wang C, *et al.* (2010) Refolding of denatured/reduced lysozyme at high concentrations by artificial molecular chaperone-ion exchange chromatography. *Biotechnol Prog* **26(4):** 1073–1079.
124. Sasaki Y, *et al.* (2011) Amphiphilic polysaccharide nanogels as artificial chaperones in cell-free protein synthesis. *Macromol Biosci* **11(6):** 814–820.
125. Takahashi H, Sawada S, Akiyoshi K. (2011) Amphiphilic polysaccharide nanoballs: a new building block for nanogel biomedical engineering and artificial chaperones. *ACS Nano* **5(1):** 337–345.
126. Nochi T, *et al.* (2010) Nanogel antigenic protein-delivery system for adjuvant-free intranasal vaccines. *Nat Mater* **9(7):** 572–578.
127. Watanabe K, *et al.* (2011) The use of cationic nanogels to deliver proteins to myeloma cells and primary T lymphocytes that poorly express heparan sulfate. *Biomaterials* **32(25):** 5900–5905.
128. Alles N, *et al.* (2009) Polysaccharide nanogel delivery of a TNF-alpha and RANKL antagonist peptide allows systemic prevention of bone loss. *Eur J Pharm Sci* **37(2):** 83–88.
129. Shimizu T, *et al.* (2008) Nanogel DDS enables sustained release of IL-12 for tumor immunotherapy. *Biochem Biophys Res Commun* **367(2):** 330–335.
130. Kitano S, *et al.* (2006) HER2-specific T-cell immune responses in patients vaccinated with truncated HER2 protein complexed with nanogels of cholesteryl pullulan. *Clin Cancer Res* **12(24):** 7397–7405.
131. Narhi L, *et al.* (1999) Both familial Parkinson's disease mutations accelerate alpha-synuclein aggregation. *J Biol Chem* **274(14):** 9843–9846.
132. Polymeropoulos MH, *et al.* (1997) Mutation in the alpha-synuclein gene identified in families with Parkinson's disease. *Science* **276(5321):** 2045–2047.
133. DiFiglia M, *et al.* (1997) Aggregation of huntingtin in neuronal intranuclear inclusions and dystrophic neurites in brain. *Science* **277(5334):** 1990–1993.

134. Stieber A, Gonatas JO, Gonatas NK. (2000) Aggregation of ubiquitin and a mutant ALS-linked SOD1 protein correlate with disease progression and fragmentation of the Golgi apparatus. *J Neurol Sci* **173(1):** 53–62.
135. Ikeda K, *et al.* (2006) Inhibition of the formation of amyloid beta-protein fibrils using biocompatible nanogels as artificial chaperones. *FEBS Lett* **580(28–29):** 6587–6595.
136. Rekas A, *et al.* (2004) Interaction of the molecular chaperone αB-crystallin with α-synuclein: effects on amyloid fibril formation and chaperone activity. *J Mol Biol* **340(5):** 1167–1183.
137. Narayanan S, *et al.* (2006) alphaB-crystallin competes with Alzheimer's disease beta-amyloid peptide for peptide-peptide interactions and induces oxidation of Abeta-Met35. *FEBS Lett* **580(25):** 5941–5946.
138. Shinder GA, *et al.* (2001) Mutant Cu/Zn-superoxide dismutase proteins have altered solubility and interact with heat shock/stress proteins in models of amyotrophic lateral sclerosis. *J Biol Chem* **276(16):** 12791–12796.
139. Shimura H, Miura-Shimura Y, Kosik KS. (2004) Binding of tau to heat shock protein 27 leads to decreased concentration of hyperphosphorylated tau and enhanced cell survival. *J Biol Chem* **279(17):** 17957–17962.
140. Firdaus WJ, *et al.* (2006) Analysis of oxidative events induced by expanded polyglutamine huntingtin exon 1 that are differentially restored by expression of heat shock proteins or treatment with an antioxidant. *FEBS J* **273(13):** 3076–3093.
141. Bruinsma IB, *et al.* (2011) Inhibition of alpha-synuclein aggregation by small heat shock proteins. *Proteins* **79(10):** 2956–2967.
142. Anthony SG, *et al.* (2003) Stress protein expression in the Alzheimer-diseased choroid plexus. *J Alzheimers Dis* **5(3):** 171–177.
143. Shin Y, *et al.* (2005) The co-chaperone carboxyl terminus of Hsp70-interacting protein (CHIP) mediates alpha-synuclein degradation decisions between proteasomal and lysosomal pathways. *J Biol Chem* **280(25):** 23727–23734.
144. Auluck PK, *et al.* (2002) Chaperone suppression of alpha-synuclein toxicity in a Drosophila model for Parkinson's disease. *Science* **295(5556):** 865–868.
145. Kobayashi Y, Sobue G. (2001) Protective effect of chaperones on polyglutamine diseases. *Brain Res Bull* **56(3–4):** 165–168.
146. Salminen A, *et al.* (2011) Hsp90 regulates tau pathology through co-chaperone complexes in Alzheimer's disease. *Prog Neurobiol* **93(1):** 99–110.
147. Jinwal UK, *et al.* (2011) The Hsp90 kinase co-chaperone Cdc37 regulates tau stability and phosphorylation dynamics. *J Biol Chem* **286(19):** 16976–16983.
148. Cao W, Konsolaki M. (2011) FKBP immunophilins and Alzheimer's disease: a chaperoned affair. *J Biosci* **36(3):** 493–498.
149. Falsone SF, *et al.* (2009) The molecular chaperone Hsp90 modulates intermediate steps of amyloid assembly of the Parkinson-related protein alpha-synuclein. *J Biol Chem* **284(45):** 31190–31199.

150. Akiyama E, *et al.* (2007) Self-assembled nanogels of cholesteryl-modified polysaccharides: effect of the polysaccharide structure on their association characteristics in the dilute and semidilute regimes. *Biomacromolecules* **8(8):** 2366–2373.
151. Ayame H, Morimoto N, Akiyoshi K. (2008) Self-assembled cationic nanogels for intracellular protein delivery. *Bioconjug Chem* **19(4):** 882–890.
152. Morimoto N, *et al.* (2005) Design of hybrid hydrogels with self-assembled nanogels as cross-linkers: interaction with proteins and chaperone-like activity. *Biomacromolecules* **6(4):** 1829–1834.
153. Morimoto N, *et al.* (2005) Hybrid nanogels with physical and chemical cross-linking structures as nanocarriers. *Macromol Biosci* **5(8):** 710–716.

Chapter 20

Bio-Functional Polymer Vesicles for Applications in Nanomedicine

Hans-Peter M. de Hoog*,†, Bo Liedberg* and Madhavan Nallani*,†,‡

Polymersomes are vesicular assemblies formed from amphiphilic block-copolymers. They are reminiscent of liposomes and cell membranes, but because of their higher stability and chemical versatility they may be applied in areas where liposomes, because of their fragile nature, have seen only limited applications. A major area of active research is therefore in the use of polymersomes for application in nanomedicine. In the past decade, it has been demonstrated that polymersomes are well-suited for the integration and immobilization of bio-functional molecules, such as peptides, antibodies and membrane proteins. At the same time it has been shown that the structure of the constituting block copolymers can be tuned to endow the vesicles with stimuli-responsive behavior. In this chapter we will explore to what extent these properties have contributed to the performance of polymersomes in cell targeting and *in vivo* drug delivery.

*Centre for Biomimetic Sensor Science, School of Materials Science and Engineering, Nanyang Technological University, 50 Nanyang Drive, Singapore 637 553.
†Institute for Materials Research and Engineering (IMRE), 3 Research Link, 117602, Singapore.
‡Corresponding author: nallanim@a-star.imre.edu.sg

Introduction

An important aim of nanomedicine is the development of nano-structured materials that can be used to improve the efficacy of administered drugs. Drugs that may benefit from being constricted to a carrier are multiple anti-cancer drugs and drugs of high molecular weight, such as therapeutic proteins and single interference RNAs, which in the past decade have come up as promising therapies for diseases which currently are impossible to cure.[1] When brought into the bloodstream, these drugs are rapidly cleared and only a fraction reaches the intended disease site. What is more, these biomacromolecules can be highly toxic to normal cells and tissue. By sequestering the drug into carriers, the residence time in the bloodstream is prolonged and the toxicity to healthy tissue is reduced. In addition, by functionalization of their surface with cell-specific ligands, such carriers can enhance the accumulation of the drug at the disease site, giving rise to high local concentrations. In the ideal case, the carriers are taken up by the cell where they release their payload into the cytosol to exercise their full effect. Although far away from practical realization, this ideal drug delivery vehicle forms the blueprint for ever more complex drug delivery systems.[2]

The design of these carriers has been greatly improved by developments in nano-technology, which aims to design particles from the bottom-up with properties that can be engineered on demand. In the last decade this has yielded a plethora of different nanoparticles with each system having its own advantages and disadvantages, and possibly it own niche of application. For diagnostic applications, inorganic nanoparticles, such as gold nanoparticles and quantum dots, as well as dendrimers, appear to be especially promising. For drug delivery applications, particles which have been constructed via self-assembly or template-assisted assembly seem particularly successful.[3] This category includes particles which have already passed through the long and complex route of commercialization. A popular example are liposomes: self-assembled spherical compartments (or vesicles) with a lipid bilayer membrane. Because of their simplicity, the inherent biodegradability of lipids and the relatively large amount of drug that can be encapsulated, liposomes have been extensively researched for drug delivery purposes. Still, after 40 years of research, the only commercially available liposomal drug formulation is the drug Doxil, which is a formulation of the anti-cancer drug doxorubicin (DOX) packaged into liposomes. This lack in valorization is partly due to the labile nature of lipid assemblies, which limits the options available to tweak their properties for optimal *in vivo* performance.[4]

A promising alternative to currently used liposomal formulations are polymersomes (Fig. 1). They have a similar vesicular structure with a bilayer membrane, but, instead of lipids, amphiphiles, block copolymers with a hydrophobic and a hydrophilic polymer block are used for self-assembly, leading to a greatly enhanced stability. In analogy to liposomes, their aqueous cores can serve as carriers of complex biological macromolecules such as proteins and enzymes, while the hydrophobic membranes can solvate hydrophobic drugs. The chemistry available to tune the properties of block copolymers and consequently that of the constituting polymers is enormous and is developing rapidly. The same holds for conjugation methodologies to introduce biofunctionality.[5] It may therefore be concluded that polymersomes combine the desirable aspects of stability and tunability

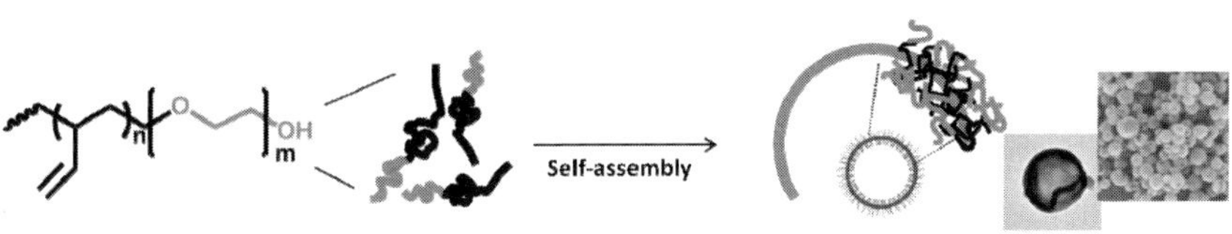

(a) (b)

Fig. 1. (a) Chemical structure of the commonly used amphiphilic block copolymer polybutadiene-poly(ethylene glycol) (PB-PEG), with the hydrophobic polybutadiene polymer block indicated in black and the hydrophilic PEG block indicated in grey. (b) Self-assembly of amphiphilic block copolymers into polymersomes. In water the hydrophobic blocks organize in bilayers to minimize the hydrophobic area which is exposed to the water molecules. At a certain volume fraction of the hydrophilic block (35 ± 10%), the vesicular structure is energetically the most favored, and polymersomes are formed. The possibility of their imaging under the high vacuum conditions of the electron microscope is testimonial to their stability.

into one system. A considerable part of research on polymersomes is focused on the design of the constituting polymer, such that the polymersomes formed can degrade on a certain stimulus, in particular stimuli such as pH, redox state and temperature, which are heavily controlled in biological settings and show cell-type and organelle specific variation.[6–11] In practice, only a small subset of the systems reported has been subjected to the study of performance in cell targeting and *in vivo* drug delivery. This chapter discusses the advances made with this small subset of systems with an emphasis on how these particular polymersomes have been tuned to interact with biological matter.

General Design and Physical Performance of Polymersomes

Polymersomes posses several measurable advantages over liposomes: the foremost being their stability. Their stability *in vitro* has been adequately demonstrated by the groups of Discher and Eisenberg for a series of block copolymers where poly(ethylene glycol) (PEG) was the hydrophilic block and polyethylene or polybutadiene (PB) was the hydrophobic block.[12] When assessing the stability by micropipette aspiration, it was found that the strain needed to rupture the polymersomes was at least 10 times higher than for liposomes. This stability results mainly from the increased molecular weight of amphiphiles and the increased membrane thickness of the polymersomes. The long polymer chains do not form a perfect bilayer but rather exhibit interdigitation (mixing of the separate monolayers).[13] Obviously, the larger the polymer (typical molecular weights are 2000–10,000) the greater the stability, and also the thickness of the membrane.[14] Thus, where phospholipids fall within a regime where stability and fluidity are optimal for biological systems to exert their function, with polymersomes it is possible to escape this strict regime.

The exact stability characteristics depend very much on the polymers used.[15] For hydrophobic blocks, a most important parameter is the melting or glass transition temperature (Tg) of the polymer (Fig. 2). If polystyrene (PS) is used, the membrane is in a glassy state

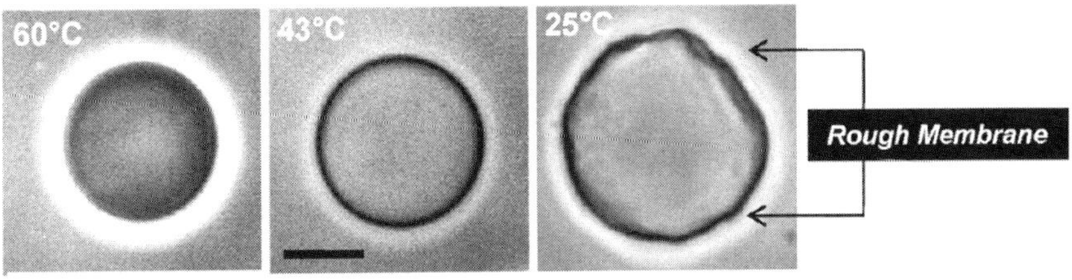

Fig. 2. Effect of temperature on the morphology of polymersomes formed from block copolymers of poly(caprolactone)-poly(ethylene glycol) (PCL-PEG) of high Tg. Around the Tg (~60 °C) the membrane is fluid and a perfect sphere is visible in the phase contrast image. Below Tg the membranes solidify and an irregular membrane with defects results at 25 °C. The high phase contrast at 60 °C is the result of the different inner (glucose) and outer (sucrose) solutions. The disappearance of the contrast upon lowering the Tg indicates that the fluids mix. Scale bar is 10 μm. (Adapted with permission from Ref. 210. Copyright (2010) American Chemical Society.[21])

and stability is high but fluidity is virtually absent. Such structures are kinetically trapped (non-ergodic), and their preparation involves dissolution of the polymer in an organic solvent such as tetrahydrofuran (THF) and adding this to a non-solvent for the hydrophobic block, such as water. Alternatively, the non-solvent may be added to a solution of the polymer. An illustrative example of this kinetic trapping was demonstrated by Kim *et al.*, who used PS-PEG to trap polymersomes in a state where they assumed a bowl-like appearance resembling stomatocytes.[16] For PB and related block copolymers, polymersomes are usually prepared by the film rehydration technique. Here, a solution of the polymer is dried to give a film, which is then rehydrated by adding water and applying some mechanical force, mostly stirring or sonication.

Obviously, a dense polymer membrane obstructs the entrance of most small organic molecules and ions. For the often used systems, such as PB-PEG, PS-poly(acrylic acid) (PAA) and poly(2-methyloxazoline)-b-poly-(dimethylsiloxane)-b-poly(2-methyloxazoline) (ABA), the permeability to water was reported to lie within 1 and 10 $\mu m \cdot s^{-1}$, while for liposomes, typical values are 15–150 $\mu m \cdot s^{-1}$ [17] However, because of the inherent rich structural variety of block copolymers, systems with much higher permeability are frequently reported. For instance, the commercially available block copolymer PEG_5-poly(propylene oxide)$_{68}$-PEG_5 showed virtually no response to osmotic shocks, demonstrating its high permeability to water.[18] An interesting way to control the permeability of polymersomes is by the incorporation of membrane proteins, which has been pioneered by the group of Meier.[19] Here, it is assumed that the flexible polymers in the bilayer membrane adapt their conformation to accommodate the membrane protein, with a minimal mismatch between the hydrophobic regions. Using a variety of membrane proteins, passive or active transport over the polymersome membrane could be achieved. More recently polymersomes were used as a platform for the direct insertion of cell-free synthesized membrane proteins. This direct insertion could greatly enhance the generally poor stability of membrane protein preparations, while at the same time the use of inherently complex cell-based platforms is omitted.[20]

Interaction of Polymersomes with Cells *In Vitro*

Since their introduction, the development of polymersomes for nanomedicine applications has followed two directions: (1) the design/study of architectures that effectively deliver their payload in cells, and (2) systems which are actually introduced into living systems and assessed for their distribution into the body and their effectiveness in curing or diagnosis.

The first point of cellular targeting implies the presence of targeting moieties on the polymersomes, although successful uptake has also been demonstrated for non-functionalized polymersomes (see further). Several methods have been reported for the coupling of targeting moieties to the polymersomes amongst which click chemistry and more recently copper-free click chemistry are of special interest. More traditional methods, such as EDC/NHS coupling and biotin-streptavidin chemistry have however also been applied with success.[22,23] An early example of cellular targeting was the conjugation of biotin to ABA polymers, which after polymersome formation were covered with streptavidin.[24] The streptavidin coated polymersomes were then incubated with biotinylated polyguanylic acid (poly-G). The polymersomes were targeted to macrophages that expressed the SRA1 receptor, which specifically binds to poly-G. Indeed, colocalization of cell fluorescence with that of fluorescently labeled polymersomes was observed while non-functionalized polymersomes did not show overlap. It was observed that at 4 °C the particles ended up sticking to the surface of the cells whereas at 37 °C they ended up in the endosomes, indicating that active SRA1 receptor mediated endocytosis was responsible for uptake. The same conjugation strategy (i.e. via biotin streptavidin interactions) was used by Therien and Hammer to design leuko-polymersomes.[25] These polymersomes carried two different ligands, namely the anti ICAM-1 antibody and sialyl Lewis X, which bind to ICAM-1 and P-selectin, respectively, the latter being receptors that play a synergistic role in recruiting leukocytes to inflamed sites. Adherence to the inflamed site occurs by a rolling (mediated by the Lewis X/P-selectin interaction) and stick mechanism (mediated by the ICAM interaction). By optimizing the ratio between the antibody ICAM-1 and sialyl Lewis X it was possible to selectively arrest leuko-polymersomes under flow onto a monolayer of inflamed cells, whereas no adhesion occurred for "healthy" cells. It would be interesting to see how such leukopolymersomes behave in targeting endothelial cells lining the vasculature *in vivo*.

For targeting selectively to specific cells, peptides can be used to bind specific receptors. A popular targeting moiety is the RGD sequence, which binds to the physiologically important transmembrane glycoprotein family of integrins. Peptides containing this sequence are potent inhibitors of tumor-induced angiogenesis, tumor growth and tumor metastasis. A derivative of this peptide, termed PR_b (sequence: KSSPHSRN$(SG)_5$RGDSP) was used for conjugation to PB-PEG polymersomes that were subsequently tested for their interaction with LNCaP human prostate cancer cells expressing the integrin $\alpha_5\beta_1$ (LNCaP cells being a common cancer line).[26] The take-up by the cancer cells of the PR_b polymersomes, encapsulating the green fluorescent calcein, could be visualized using confocal microscopy. Although individual polymersomes cannot be observed with conventional confocal microscopy, green fluorescent regions were observed that appeared to be localized in the cells' trafficking compartments, i.e. the endosomes and lysosomes. Here, the amount of functionalization was quite high (50% of the surface was theoretically functionalized), but further targeting experiments with polymersomes that carried the apoptosis inducing protein TNF-alpha showed that at a PR_b coverage of 5%, the cytotoxicity of the polymersomes was already 2.6 times higher than for free TNF-alpha. Compared to a similar study with liposomes, it was observed that the effectiveness of the polymersomes and the liposomes was comparable, but that non-functionalized liposomes (i.e. PEG-ylated liposomes containing TNF but without targeting peptide) significant cell death was observed as opposed to non-functionalized polymersomes. This can be attributed to the superior stability of the polymersomes and possibly their resistance to non-specific adhesion. An unanswered question in this particular study was how the obviously robust polymersomes delivered their cargo once in the cell.

Another popular cell targeting peptide is the TAT-peptide, which is derived from HIV and contains multiple cationic arginines, which allow facile membrane translocation. In the broader perspective of being able to track dendritic cell migration, dendritic cells were incubated with TAT functionalized near infra-red (NIR) emitting polymersomes.[27] After 3 hr of incubation, uptake of the TAT-labeled polymersomes was around 10× higher than that of non-functionalized polymersomes, where it was observed that most of the fluorophore emission came from intracellular trafficking compartments. It was furthermore noticed that TAT and polymersomes fluorescence remained associated while in the cell. Interestingly, the authors were able to extract a number for the amount of polymersomes taken up per cell: 70,000. Unfortunately, the cells were not completely inert to the targeting, as a 25% of toxicity was observed and dendritic cell-specific functions were somewhat impaired. Because non-functionalized polymersomes did not show an adverse effect, this was attributed to the TAT-peptide. In an attempt to introduce functional proteins/enzymes into cells as a form of protein therapy, Van Dongen *et al.* studied the delivery of functional horseradish peroxidase (HRP) into HeLa, Jurkat and HEK 293 cells (Fig. 3).[28] The particular polymersomes were prepared from

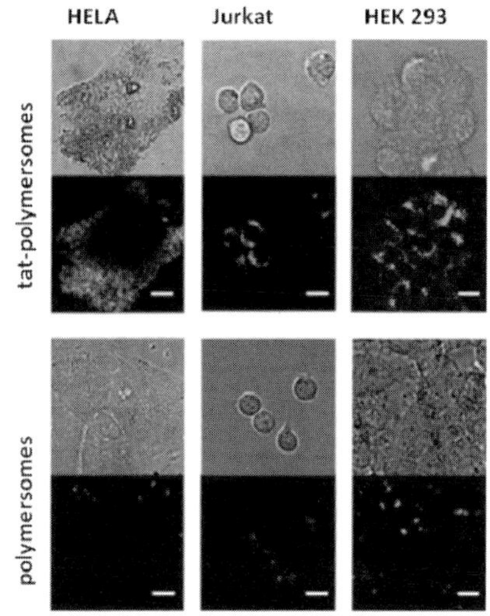

Fig. 3. Cell-uptake of PS-PIAT polymersomes in HeLa, Jurkat and HEK 293 cells. The PS-PIAT polymersomes were functionalized with the cell penetrating TAT peptide and GFP was encapsulated for tracking. Compared to the non-functionalized polymersomes, the uptake (after 4 hr) is considerably higher. Scale bar 10 μm. (Reprinted with permission from Ref. 28. Copyright (2010) Wiley.)

polystyrene-block-polyisocyanoalanine-(2-thiophene-3-yl-ethyl)amide (PS-PIAT) and conjugated with the TAT-peptide. PS-PIAT is a unique block copolymer in the sense that the polymersomes it forms are intrinsically porous, while enzymes can be actively encapsulated.[29,30] Cell uptake experiments suggested that the polymersomes entered the cells via macropinocytosis, and that after 4 hr, the GPF-containing polymersomes resided partly in the lysosomes and partly in a non-specified location, which could either be late endosomes or the cytosol. Interestingly, the enzyme residing in the polymersomes was still accessible to substrates and active in the conversion of a model substrate that had been added externally to the cells. This activity was sustained for over 10 hr.

An important parameter to consider for the practical application of polymersomes is the composition of the block copolymer in terms of cytotoxicity and immunogenity. Apart from biodegradable block copolymers (see *in vivo* section) bio-based materials containing natural polymers are preferred, as they are intrinsically biocompatible.[31] An interesting platform in this respect are vesicles prepared from block copolypeptides. A block copolymer containing the polypeptide poly(benzyl glutamate) (PBLG) as the hydrophobic block and hyaluronan (HA, a polysaccharide) as the hydrophilic block was used to form polymersomes.[32] Here, the hyaluronan, apart from stabilizing the polymersomes, served the function of being a ligand for the CD44 receptor, which is a receptor that is widely expressed in mammalian cells and is implicated in treatment strategies for cancer. The PBLG-HA polymersomes were loaded with DOX and incubated with MCF-7 cell lines expressing a high level of CD44 receptors, and U87 cell lines expressing low amounts of the receptor. Indeed, high uptake of the polymersomes (as measured by the fluorescence of the anticancer drug) was observed for MCF-7 cell lines, while uptake by U87 was considerably lower. The effect observed was however somewhat less pronounced when the cell uptake was analyzed by flow cytometry and was further complicated by the fact that U-87 cells harbor the p-glycoprotein pump that expels drugs from the cell's interior. By following the uptake for 48 hr, a gradual accumulation of the particles into the endosomes was observed, whereas free DOX directly distributed to the cell nucleus, resulting in a fast build-up of fluorescence in the nucleus followed by a gradual decrease over time. Concerning the cytotoxicity, in the long run, the encapsulated cancer drug was more effective than the free drug, most probably because of the higher uptake in the cells combined with a slow release from the polymersomes.

A conceptually comparable example was demonstrated for the uptake of polymersomes based entirely on polypeptides, i.e. poly(arginine)-poly(leucine), where the poly(arginine) formed the hydrophilic block.[33] In this case, the observation that polymersomes based on poly(lysine) were only marginally taken up by cells led the authors to consider arginine-based polypeptides, where arginine confers the positive charge that is essential for the functioning of cell-penetrating peptides such as TAT. Efficient uptake by human Hela cells was observed even though the targeting moiety resided in the membrane of the polymersomes. By treating the cells with various drugs that selectively inhibit one of the cellular internalization pathways, the authors determined the pathway used by the cell for uptake of the polymersomes to be macropinocytosis, as opposed to caveolae-dependent endocytosis, and clathrin-coated pit endocytosis, being more specific internalization pathways. Furthermore, the polymersomes were found to end up in the endosomes where they apparently were trafficked back to the cell membrane because colocalization with a lysosome-specific dye was not observed. Interestingly, upon disruption of the endosomes by chloroquine, the punctuate fluorescence did not spread out over the cells, indicating that the polymersomes remained intact.

As was stressed in the introduction, polymersomes can be readily tuned for specific applications by variation of the chemical structure. The design of stimuli-responsive polymersomes, on which a great deal of research is being directed, generally does not include attempts to target the structures to cells by means of a targeting moiety. At the time of writing (2011), we are only aware of two papers on these particular systems.[34,35] The most convincing report dealt with the delivery of a fluorescently labeled dextran to SKBR3 cancer cells by biodegradable polymersomes of PEG-poly(D,L-lactide) (PEG-PLA). In between the blocks the peptide sequence Gly-Phe-Leu-Gly-Phe was installed, which can be cleaved by Cathepsin B (Cath B).[34] This enzyme resides in the lysosomes, either associated to the membrane or freely dissolved. Polymersomes that end up in the lysosome will be amenable to attack by Cath B, which will cleave the polymer and lead to degradation of the polymersomes and the release of encapsulated drugs. To target the polymersomes to the cells, an antibody to the epidermal growth factor receptor EGFR was (non-specifically) attached. The EGF receptor is over-expressed on surfaces of various human cancers. After three days of incubating the polymersomes with the cells, broad fluorescent spots could be observed, which appeared to reside in the cytosol. For polymersomes without the cleavable linker the fluorescence was present as isolated spots near the nucleus, suggesting their localization into lysosomes.

Finally, with respect to cell targeting, there have been several interesting reports on cell targeting without any targeting ligands present. From the results reported above, one can agree that this should not pose a problem because in the majority of the studies non-functionalized polymersomes are also taken up by cells (albeit slower and to lesser extent). For targeting to cancers, this type of "passive" targeting has proved to be quite efficient because of the enhanced permeability and retention effect in cancerous and inflamed tissue (EPR effect). This effect basically refers to the accumulation of nano-sized particles at the tumor site as a result of the leaky vasculature (as opposed to healthy tissue).[4] As will be discussed later on, several successful studies have used exactly this passive targeting to deliver drugs at tumor sites *in vivo*. Still, with *in vivo* applications in mind, a targeting moiety significantly enhances accumulation and uptake at the disease site while stimulating cell-uptake.

With respect to such "*in vitro* passive" targeting, there are two systems that should not go unmentioned in this chapter. They both concern stimuli-responsive particles. The first report on the feasibility of using stimuli-responsive polymersomes for the delivery of substances into cells dealt with the construction of a polymersome where the blocks of the constituting polymers were linked by a disulfide moiety (Fig. 4).[36] Outside the cells, such structures are expected to be stable

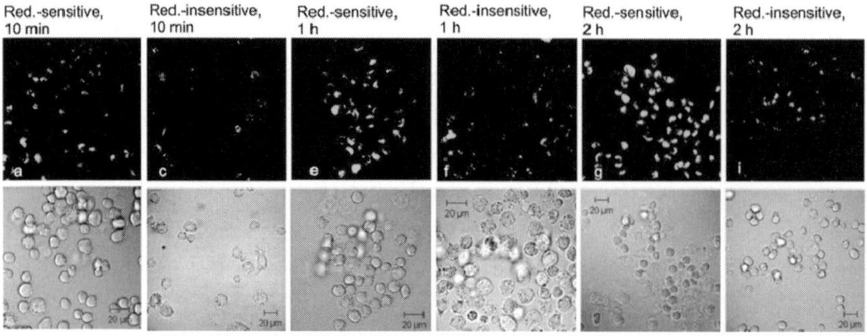

Fig. 4. Cell uptake of polymersomes formed from block copolymers that either contain a reduction sensitive disulfide bond or a normal covalent bond. Once taken up by the cells the reduction sensitive polymersomes release their payload as a result of the degradation of the polymer leading to a fluorescent signal (calcein encapsulated at self-quenching concentrations). The reduction insensitive polymersomes show only a modest increase in fluorescence over the same time frame. Reprinted with permission from Ref. 36. (Copyright (2007) American Chemical Society.)

because of the low concentration of free thiols in the extracellular medium. However, inside cells, a different redox balance exists, and a relatively high millimolar concentration of GSH is present, leading to the reduction of the disulfide bond and eventually the degradation of the polymer constituting the polymersomes. The polymersomes were rapidly taken up by J774A-1 cells: after 10 min punctuate intracellular fluorescence was seen. Because calcein (encapsulated at self-quenching concentrations) was used as a fluorescent dye, the appearance of fluorescence indicated that the polymersomes also readily disintegrated once in the cell. Furthermore, via an unknown mechanism, the reduction of the polymer also led to disruption of the endosome in time, because over a time span of 120 min the fluorescence spread out over the volume of the whole cell. As observed with biodegradable polymersomes (see further), this may result from insertion of the hydrophobic block into the endosomal membrane.

A successful pH-sensitive system was reported by Battaglia and Armes *et al.*[37] As is clear from the above examples, polymersomes are taken up by endocytosis, with the polymersomes ending up in the endosomes and later the lysosomes. During this trafficking the pH decreases to a pH of 5 in the lysosomes. This mechanism was used to design pH-responsive polymersomes from the block copolymer poly (2-(methacryloyloxy)ethyl phosphorylcholine)-(2-(diisopropylamino) ethyl methacrylate), (PMPC-PDPA), of which the hydrophobic PDPA-block exhibits pH-responsive behavior. At physiological pH this block copolymer formed stable vesicles. At pH 6.4 or lower, however, the secondary amino groups on the PDPA become protonated and the polymer dissolves, resulting in a fast dissociation (milli-seconds) of the polymersomes. In cells, a similar fate can be envisioned and, indeed, after 24 hr of incubation an encapsulated rhodamine dye had partitioned over the cytoplasm of the cell, indicating that after take up, the polymersomes dissociated and released their cargo. This was demonstrated for 24 cell types, showing the general applicability of this system. Interestingly, the number of polymersomes taken up by cells after 24 hr was as high as 10 million, which is considerably higher than that found for the number of polymersomes taken up by dendritic cells (70,000).[27]

Interaction of Polymersomes with Organisms

Multiple studies have highlighted the properties needed for nanoparticles to stay for an appreciable time in the bloodstream before being either taken up by macrophages residing in the blood stream or being filtered out of the bloodstream in tissues such as the liver and the spleen. Size is an important parameter here: if smaller than ~5 nm, particles will end up in the kidney. If larger than ~200 nm, they will rapidly be filtered out by the spleen or other systems that contain filter systems, primarily the liver.[38] For polymersomes that can be prepared via film rehydration, sizes can be conveniently cut down by extrusion. For polymersomes with a glassy membrane, the size is harder to control, but generally the desired size can be achieved by manipulation of the experimental conditions during formation. An intensively studied model system in this respect are polymersomes prepared from PS-PAA, where size-control could be achieved by the constitution of the solvent system.[39]

A common strategy to increase the half-life ($t_{1/2}$) of any nanoparticle is its surface functionalization with PEG, which avoids opsonization (the adherence of blood proteins that destine the particles for degradation by macrophages). An important advantage of polymersomes is that block-copolymers can be used that contain a PEG as the hydrophilic block. The inert PEG layer therefore forms part of the membrane, and the resulting polymersomes show comparable or enhanced stealth characteristics to PEG-ylated liposomes. For the commonly used PB_{46}-PEG_{26} polymersomes the $t_{1/2}$ in mice was determined to be 15.8 ± 2.2 hr (assessed from the fluorescence of a hydrophobic dye), whereas for PEG-ylated liposomes a circulation time of 10–20 hr is usually found. With increasing length of the PEG block the circulation time increased to well above 20 hr: for PB_{130}-PEG_{80} a half life of $25.5\,h \pm 2.7$ was found.[40] Another factor of influence in the design of long-circulating nanoparticles is their charge, where it has been found that a slightly negative charge reduces uptake by the spleen and liver.[41] For PB-PEG polymersomes, this was investigated by mixing 10% (w/w) of the negatively charged PS-PAA into the membrane.[42] At this percentage of PS-PAA, a surface charge of

−13.3 mV was measured, which is comparable to the zeta-potential of red blood cells (−15 mV). *In vivo* tracking of the polymersomes by incorporation of a NIR fluorescent lipid dye showed an increased uptake of the polymersomes in the liver but a greatly reduced uptake in the spleen compared to neutral PB-PEG polymersomes. Hennink and Feijen investigated this effect further by monitoring the biodistribution of PEG-PLA polymersomes with varying zeta potentials, i.e. −7.6, −25.3 and −38.7 mV.[43] Here, both the polymers and the encapsulated dextran were radioactively labeled to assure that intact polymersomes were detected (instead of leaked dye). It was found that the difference between the values obtained with either label was less than 5%, indicating that these biodegradable polymersomes were stable *in vivo* even though they were constructed from a biodegradable polymer. The circulation time of the polymersomes increased with decreasing surface charge: the longest $t_{1/2}$ of 47.3 ± 12.7 hr was found for a surface charge of −7.6 mV. As a control, stealth liposomes were also tested, which showed a circulation half time of 10.6 ± 1.8 hr. Importantly, in this case of passive targeting the long circulation time was connected to an increase in accumulation in tumor tissue and a decrease in liver uptake. This observation contrasts with that observed for PB-PEG/PS-PAA polymersomes, where an enhanced liver uptake was found.

The effect of size on the circulation time of polymersomes was investigated more thoroughly for polyion complex polymersomes (PICsomes).[44] These polymersomes are formed from water-soluble block copolymers containing a hydrophilic polyaspartate polypeptide block connected to a PEG, which is mixed with an equimolar solution of the oppositely charged polyaspartate homopolymer (obtained by aminolysis of the polyaspartate) to obtain PLCsomes. Interestingly, the size distribution of such polymersomes can be tuned by varying the concentration of the polyaspartate homopolymer. In this way PICsomes were obtained with diameters that ranged from 102–298 nm, which were then assessed for their circulation time in tumor-bearing mice. The optimal size with respect to half-life in the blood circulation of mice was 158 nm and 197 nm having $t_{1/2}$ of 23.3 hr and 19.6 hr respectively. With respect to the accumulation of the particles in tumor tissue however, a PIC micelle of 38 nm in diameter outperformed the PICsomes and only 102 nm PICsomes showed reasonable accumulation. Instead, the long-circulating PICsomes with diameters larger or equal to 158 nm showed increased absorption by the spleen.

One study reported the actual tuning of the PB-PEG polymersome membrane to enhance its interaction with cells and consequently, the accumulation in tumors (Fig. 5).[45] To this end, the authors mixed the block copolymers with 10–25% of lipid (phosphatidyl choline), and a small (undefined) percentage of a PEG-ylated lipid, which either carried a folate group or an amino group for copper-free click conjugation to an affibody against the HER2/neu receptors. The folate receptor is frequently over-expressed in cancers, and HER2/neu receptors are over-expressed in some breast cancers. Indeed, the incorporation of lipids led to a 5 times higher uptake of ICB cells for the 25% hybrid folate functionalized polymer when

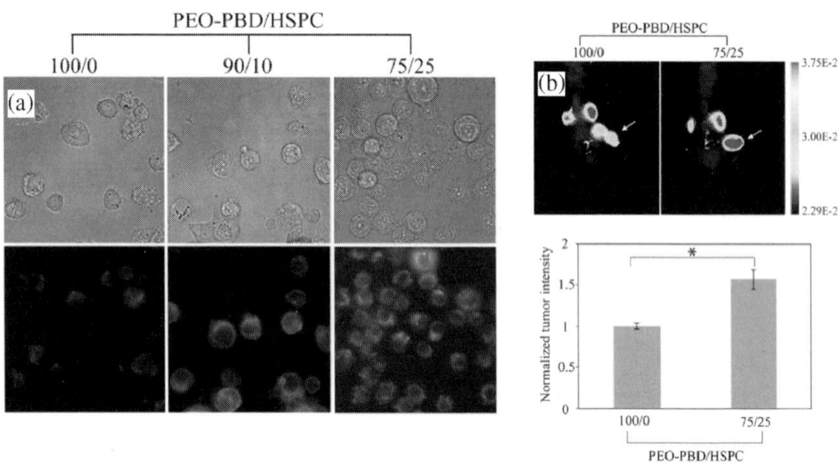

Fig. 5. (a) Incubation of KB cells (expressing folate receptor) by polymersome-lipid hybrids functionalized with folate. With increasing concentration of lipid (0%, 10%, and 25%), an increase in cell uptake of fluorescently labeled polymersomes is seen. (b) *In vivo* tracking of NIR polymersome-lipid polymersomes functionalized with anti HER2/neu affibody after injection into a mouse. An increase in fluorescence, indicating enhanced accumulation of the polymersomes, is seen at the tumor site (arrow) when 25% of lipid is mixed into the polymersome membrane. (Adapted with permission from Ref. 45. Copyright (2011) American Chemical Society.)

compared to functionalized 100% PB-PEG polymersomes. *In vivo* analysis (mouse) by labeling the affibody-functionalized polymersomes with a NIR dye gave a 1.5 fold improvement of the signal at the site of the tumor, indicating the enhanced tumor uptake of the hybrids.

Concerning the actual curing of diseases, most studies employing polymersomes have focused on anti-cancer agents, where the role of the polymersome is to solubilize the drug and to function as tumor targeting delivery vehicle by taking advantage of the EPR effect. This is one of the fields where liposomes, being structurally closest to polymersomes, have been applied with success, even leading to the development of a marketable drug formulation.[46] As has been shown above, their stability in the circulation and the ease by which they can be chemically manipulated may give polymersomes a competitive edge over liposomes. Thus, several studies have reported the delivery of anticancer drugs entrapped in polymersomes to *in vivo* models such as rat and mouse. Usually this involves the use of biodegradable amphiphilic block copolymers, such as PEG-PLA and PEG-PCL, because the hydrolytic degradation provides a release mechanism for the loaded drugs, where otherwise non-degradable polymersomes (such as PB-PEG) would, once in the tumor, presumably fail to deliver the drug once in the tumor.[47] To study the biodistribution and the effectiveness of drug-loaded polymersomes a combination of two regular anti-cancer drugs, DOX and paclitaxel (TAX) were encapsulated in PLA-PEG polymersomes. Here, DOX was loaded in the polymersomes interior and TAX solubilized in the membrane. The release kinetics in water at neutral pH and 37°C showed that in 1 day 80% of the DOX and 60% of the TAX was released for the degradable vesicles, while for the PB-PEG vesicles the percentages were 30% for DOX and 25% for TAX.[48] The fluorescently labeled vesicles were visible for atleast 48 hr in the blood stream of mice. When injected into mouse that carried tumors, the area of the tumor (as a measure of effectiveness) decreased to half its original size in 18 hr, reaching a constant value of 40% of the initial dose (% ID) three days post-injection. The free drug induced a much slower shrinkage of the tumor which started increasing in volume again after 3 days. It appeared that part of this effect derived from the fact that for the polymersomes a higher dose of DOX (3.5 mg/kg) and TAX (7.5 mg/kg) could be used, leading to a 1.5–2 times higher accumulation of drug in the tumor. An effect of the polymersome carrier on the pharmacokinetics was however clearly observed for the drug concentration over time, with the accumulation of free drug peaking in the first 6 hr and the accumulation of encapsulated drug peaking around 20 hr.

The effect of doxorubicin encapsulated in polymersomes on the therapeutic efficacy was also studied for the PBLG-HA polymersomes described above.[32,49] A different tumor system and animal model were used, which makes comparison between the two formulations difficult. Here, upon injection of DOX loaded polymersomes, at a dose of 5.0 mg/kg bodyweight, the tumor volume decreased progressively to 15% after 30 days and no mortality was observed after 60 days. Free drug showed a similar trend in the first 15 days, but the tumor volume stabilized after that period and no rat survived after 58 days. The encapsulated drug also showed a reduced cardiotoxicity (a major side-effect of DOX), which was assessed by measuring the levels of lactate dehydrogenase and serum creatine kinase.

DOX-loaded polymersomes were in addition used to deliver the drug across the blood brain barrier (Fig. 6).[50] Many potential drugs targeting the central nervous system are ineffective because they are not able to cross this barrier, whose function is mainly conveyed by the endothelial cells that line the brain capillaries. By conjugating drugs to ligands that bind to endogenous transporters (i.e. by making use of the process that transports essential molecules into the brain), delivery of substances into the brain is feasible, however.[51] Biodegradable PEG-PCL polymersomes were therefore conjugated to transferrin, which bind to the transferrin receptor. Four hours after being injected into mouse that carried a tumor in the brain (termed glioma), the transferrin conjugated polymersomes indeed showed a 3.2-fold higher accumulation in the tumor tissue as compared to plain polymersomes. As a measure of effectiveness the brain delivery efficiency was determined to be 0.053% ID/g glioma, which compares well to a value of 0.03% ID/g brain tissue for a similar study on liposomes. Furthermore, the glioma volume after 14 days was

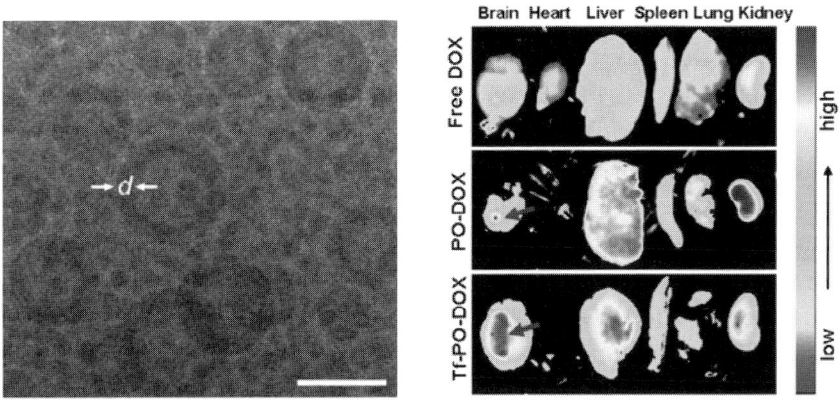

(a) (b)

Fig. 6. (a) Cryo-TEM image of PEG-PCL polymersomes conjugated with transferrin. Bar is 100 nm. (b) Fluorescence imaging of the rat major organs 4 hr post-injection. An increase in fluorescence can be observed at the brain tumor site for transferrin conjugated polymersomes. Higher accumulation (indicated by arrow) is also visible for the liver and the kidneys as a result of the decreased shielding effect of the PEG-ylation by derivatization with the transferrin protein. (Reprinted with permission from Ref. 50. Copyright (2007) American Chemical Society.)

significantly smaller than that observed after administration of the plain polymersomes or the free drug. Although these values are short of the values reported for drugs that are directly conjugated to endogenous transporter ligands, the results may be improved by a more specific immobilization of transferrin to the polymersomes.

Oligonucleotide-based therapies, especially that using small interfering RNA (siRNA), can benefit from *in vivo* delivery by carriers such as polymersomes, first of all because the stability of the free oligonucleotide in the bloodstream is low, and furthermore because the free molecule can illicit an immune response.[52] In addition, the traditional way of packaging the RNA into cationic liposomes may lead to toxic effects from the cationic carrier itself.[53] To explore the use of polymersomes for delivery oligonucleotides, both siRNA (a duplex) and antisense oligonucleotides (AON) were loaded into biodegradable PEG-PCL and PEG-PLA polymersomes.[54] At physiological pH, a gradual release of the oligonucleotides was observed with a $t_{1/2}$ of

3–5 hr, with a higher release rate at pH = 5. It was observed that the release, furthermore, occurred concurrently with the hydrolytic degradation of the particles. This degradation is essential to the effective delivery of the content into the cytoplasm, which was demonstrated by experiments that showed the ability of PEG-PCL and PEG-PLA polymersomes to be able to lyse cells at high polymer concentrations, presumably by intercalation of the hydrophobic blocks into the endosomal bilayer membrane. The polymersomes containing siRNA, which targeted mRNA encoding lamin A/C proteins, were incubated with A594 cancer cells and the expression of lamin was assessed by fluorescence. As a result, the siRNA containing polymersomes reduced the expression of the protein to around 60%, which was comparable to the expression level of the commercially available transfection kit, lipofectamine 2000. AON polymersomes on their turn were tested for their ability to restore the expression of dystrophin in mouse muscles, which was knocked down in this particular mouse model of human muscular dystrophy. Twelve hours after injection of the polymersomes nuclear localization of the AON could be seen in 50% of the inspected muscle cells. Free AON showed less than 10% nuclear localization, with the effect localized to the site of the injection. Three weeks after injection immunostained sections of the treated muscles showed that over the entire muscle, expression was enhanced, showing the successful *in vivo* delivery of oligonucleotides into cells by polymersome carriers.

Conclusion

Since the first reports appeared on polymersomes, they have been suggested to be useful alternatives to their lipid counterparts, i.e. liposomes. This is mainly attributed to their stability and the possibility of introducing responsive polymer moieties and biofunctionalities without compromising their basic structure. Although the main part of the research on polymersomes is constrained to the field of polymer chemistry and physical chemistry, this review testifies that in the past five years, significant steps have been taken to study the interaction of polymersomes with living matter, be it cells or animal models. These studies have followed the steps previously taken by liposomes and have demonstrated that polymersomes perform equally well as the stealth liposomes in terms of drug loading, cell uptake and *in vivo* distribution. Furthermore, the carriers themselves do not seem to confer significant toxicity to the models in which they are studied.

Obviously, the picture until now is somewhat limited because the body of literature available as yet is minimal and no clinical studies on humans have so far been reported. It could, therefore, be concluded that the use of polymersomes for biological applications is in its infancy. So far, the majority of the field is populated by polymer scientists and for the field to really take off, input from biological sciences is needed. At the same time, it has now been demonstrated that through input from polymer science a level of control can be obtained over the vesicular structure that hitherto has been unthinkable. Although living matter imposes severe limitations on the type of chemistry that can be used, it seems only a matter of time before polymersomes will go *in vivo* where liposomes have not gone before.

References

1. Allen TM, Cullis PR. (2004) *Science* **303:** 1818–1822.
2. Yoo JW, Irvine DJ, Discher DE, Mitragotri S. (2011) *Nat Rev Drug Discov* **10:** 521–535.
3. Johnston APR, Such GK, Ng SL, Caruso F. (2011) *Curr Opin Colloid Interface Sci* **16:** 171–181.
4. Zasadzinski JA, Wong B, Forbes N, *et al.* (2011) *Curr Opin Colloid Interface Sci* **16:** 203–214.
5. Li MH, Keller P. (2009) *Soft Matter* **5:** 927–937.
6. Qin S, Geng Y, Discher DE, Yang S. (2006) *Adv Mater* **18:** 2905–2909.
7. Li Y, Lokitz BS, McCormick CL. (2006) *Angew Chem Int Ed* **45:** 5792–5795.
8. Kim H, Jeong SM, Park JW. (2011) *J Am Chem Soc* **133:** 5206–5209.
9. Gaitzsch J, Appelhans D, Gräfe D, *et al.* (2011) *Chem Commun* **47:** 3466–3468.
10. Chen W, Meng F, Cheng R, Zhong Z. (2010) *J Control Release* **142:** 40–46.
11. Sanson C, Diou O, Thévenot J, *et al.* (2011) *ACS Nano* **5:** 1122–1140.
12. Discher BM, Won YY, Ege DS, *et al.* (1999) *Science* **284:** 1143–1146.
13. Battaglia G, Ryan AJ. (2005) *J Am Chem Soc* **127:** 8757–8764.
14. Bermudez H, Brannan AK, Hammer DA, *et al.* (2002) *Macromolecules* **35:** 8203–8208.

15. Kita-Tokarczyk K, Grumelard J, Haefele T, Meier W. (2005) *Polymer* **46:** 3540–3563.
16. Kim KT, Zhu J, Meeuwissen SA, *et al.* (2011) *J Am Chem Soc* **132:** 12522–12524.
17. Carlsen A, Glaser N, Le Meins JF, Lecommandoux S. (2011) *Langmuir* **27:** 4884–4890.
18. Rodríguez-García R, Mell M, López-Montero I, *et al.* (2011) *Soft Matter* 7: 1532–1542.
19. Renggli K, Baumann P, Langowska K, *et al.* (2011) *Adv Funct Mater* **21:** 1241–1259.
20. Nallani M, Andreasson-Ochsner M, Tan CWD, *et al.* (2011) *Biointerphases* **6:** 153–157.
21. Rajagopal K, Mahmud A, Christian DA, *et al.* (2010) *Macromolecules* **43:** 9736–9746.
22. de Hoog HPM, Nallani M, Liedberg B. (2012) *Polymer Chemistry* doi: 10.1039/C1PY00413A.
23. Egli S, Schlaad H, Bruns N, Meier W. (2011) *Polymers* **3:** 252–280.
24. Broz P, Benito SM, Saw C, *et al.* (2005) *J Control Release* **102:** 475–488.
25. Robbins GP, Saunders RL, Haun JB, *et al.* (2010) *Langmuir* **26:** 14089–14096.
26. Demirgöz D, Pangburn TO, Davis KP. (2009) *Soft Matter* **5:** 2011–2019.
27. Christian NA, Milone MC, Ranka SS, *et al.* (2007) *Bioconjug Chem* **18:** 31–40.
28. Van Dongen SFM, Verdurmen WPR, Peters RJRW, *et al.* (2010) *Angew Chem Int Ed* **49:** 7213–7216.
29. Vriezema DM, Hoogboom J, Velonia K, *et al.* (2003) *Angew Chem Int Ed* **42:** 772–776.
30. de Hoog HPM, Nallani M, Cornelissen JJLM, *et al.* (2009) *Org Biomol Chem* **7:** 4604–4610.
31. Van Dongen SFM, de Hoog HPM, Peters RJRW, *et al.* (2009) *Chem Rev* **109:** 6212–6274.
32. Upadhyay KK, Bhatt AN, Mishra AK, *et al.* (2010) *Biomaterials* **31:** 2882–2892.
33. Sun VZ, Li Z, Deming TJ, Kamei DT. (2011) *Biomacromolecules* **12:** 10–13.
34. Lee JS, Groothuis T, Cusan C, *et al.* (2011) *Biomaterials* **32:** 9144–9153.
35. Yang X, Grailer JJ, Rowland IJ, *et al.* (2010) *ACS Nano* **4**: 6805–6817.
36. Cerritelli S, Velluto D, Hubbell JA. (2007) *Biomacromolecules* **8:** 1966–1972.
37. Massignani M, Lopresti C, Blanazs A, *et al.* (2009) *Small* **5:** 2424–2432.
38. Moghimi SM, Hunter AC, Murray JC. (2001) *Pharmacol Rev* **53:** 283–318.
39. Luo L, Eisenberg A. (2001) *Langmuir* **17:** 6804–6811.
40. Photos PJ, Bacakova L, Discher B, *et al.* (2003) *J Control Release* **90:** 323–334.
41. Yamamoto Y, Nagasaki Y, Kato Y, *et al.* (2001) *J Control Release* **77:** 27–38.

42. Christian DA, Garbuzenko OB, Minko T, Discher DE. (2010) *Macromol Rapid Commun* **31:** 135–141.
43. Lee JS, Ankone M, Pieters E, *et al.* (2011) *J Control Release* **155:** 282–288.
44. Anraku Y, Kishimura A, Kobayashi A, *et al.* (2011) *Chem Commun* **47:** 6054–6056.
45. Cheng Z, Elias DR, Kamat NP, *et al.* (2011) *Bioconjug Chem* **22:** 2021–2029.
46. Torchilin VP. (2005) *Nat Rev Drug Discov* **4:** 145–160.
47. Ahmed F, Discher DE. (2004) *J Control Release* **96:** 37–53.
48. Ahmed F, Pakunlu RI, Brannan A, *et al.* (2006) *J Control Release* **116:** 150–158.
49. Sanson C, Schatz C, Le Meins JF, *et al.* (2009) *Langmuir* **26:** 2751–2760.
50. Pang Z, Gao H, Yu Y, *et al.* (2011) **22:** 1171–1180.
51. Pardridge WM. (2007) *Drug Discov Today* **12:** 54–61.
52. Castanotto D, Rossi JJ. (2009) *Nature* **457:** 426–433.
53. Behlke MA. (2006) *Mol Ther* **13:** 644–670.
54. Kim Y, Tewari M, Pajerowski JD, *et al.* (2009) *J Control Release* **134:** 132–140.

Chapter 21

Carbon Nanotubes in Cancer and Stem Cell Therapeutics

Wei Shao, Arghya Paul and Satya Prakash*

1. Introduction

Carbon nanotubes (CNTs) are synthetic nanomaterials discovered by Iijima in 1991. Structurally, CNTs can be viewed as hollow tubes wrapped from graphitic sheets. Based on the numbers of layers of the graphitic sheets, the CNTS can be classified into two categories: single-walled carbon nanotubes (SWNTs) and multi-walled carbon nanotubes (MWNTs).[1] Figure 1 depicts the lattice structure of SWNTs and MWNTs. Owing to their well-ordered molecular structure, CNTs present remarkable physical properties, including ultra-high surface area, high aspect ratio, high tensile strength[2–4]; distinct optical properties (e.g. high absorption in the near-infrared (NIR) range, strong Raman shift, etc.)[5]; and excellent electrical and thermal conductivities.[6–8] Each of the above-mentioned properties of CNTs brought them specific applications. Over the last two decades, CNTs have been advanced in a wide range of industrial applications. For example, CNTs are used as key components for

*Corresponding author: Biomedical Technology and Cell Therapy Research Laboratory, Department of Biomedical Engineering and Artificial Cells and Organs Research Centre, Faculty of Medicine, McGill University, 3775 University Street, Montreal, Quebec, H3A 2B4, Canada. E-mail:Satya.prakash@mcgill.ca

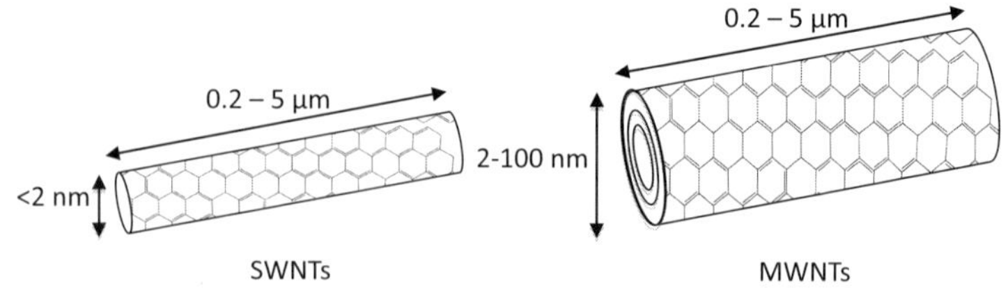

Fig. 1. Typical morphologies and dimensions of SWNTs and MWNTs: CNTs are hollow tubes rolled from sheets of graphene. Single-walled nanotubes (SWNTs) contain one layer of graphene sheet and have a diameter of less than 2 nm with a length 0.2–5 μm. Multi-walled carbon nanotubes (MWNTs) consist of a number of concentric carbon tubes and have a diameter of 2–100 nm.

bicycles and reinforcing additives for composite materials[9,10] due to their strong tensile strength. CNTs are also used to make high-speed transistors[11,12] and electrical sensors[13,14] due to their good electrical conductivity.

Inspired by the unique structural, optical, electrical, biochemical and other properties of CNTs, a variety of the biomedical applications are being investigated. These investigations include biological imaging agents, drug carriers, biosensors, and tissue engineering scaffolds and others. This chapter focuses on the major biomedical applications of CNTs in targeted delivery of therapeutics and stem cells for cancer therapy and other applications.

2. Functionalization of Carbon Nanotubes for Cancer and Other Biomedical Applications

Pristine CNTs, persisting metallic nature, are highly hydrophobic. Therefore, modification of CNT surface, or CNT functionalization, so as to disperse them into aqueous solutions becomes a key step for biomedical application of CNTs. Both covalent and non-covalent methods are applied to modify CNT surface. Covalent modifications form chemical bonding on carbon atoms on CNT surface via chemical reactions, and further conjugation of hydrophilic molecules renders CNTs water-soluble. Non-covalent modifications of CNTs utilize their hydrophobic nature for coating amphiphilic molecules onto CNT sidewalls. These modifications not only generate water-soluble CNTs but also create functional moieties which enable further linking of therapeutics, such as, DNA, siRNA, peptides and small molecule drugs, etc.[15,16]

For covalent modification of CNTs, two types of chemical reactions, oxidation and cycloaddition reaction, are widely applied (Fig. 2). Oxidation reaction is carried out by refluxing CNTs in concentrated nitric acid, which creates carboxyl groups on tips or defect sites of CNTs.[17] Cycloaddition reactions form pyrrolidine-fused rings on CNTs surfaces by reacting with chemical moieties.[18] After the above initial reactions, the CNTs obtain certain solubility due to the functional groups generated on CNTs surface, however much increased solubility is obtained with further conjugation of water-soluble polymers to these functional groups, since polymers provide CNTs with good colloidal stability in aqueous solutions because of their steric structure.[19,20]

The CNTs can be easily functionalized non-covalently by amphiphilic molecules, such as surfactants, amphiphilic polymers, and others using a simple sonication process. Typically, two types anchoring moieties exhibit strong association with CNTs, aliphatic chains[8] and poly aromatic rings[21] (Fig. 2). Aliphatic chains wrap CNTs through the van de Waals force and hydrophobic associations, whereas the poly aromatic rings stick very tightly to CNTs through strong π-stacking; it has been shown that the pyrene derivative can be irreversibly absorbed onto the CNTs.[22] Compared with covalent modifications, the non-covalent methods do not disturb CNTs molecular structure, and therefore the physical properties of CNTs (e.g. optical or electronic properties, etc.) are preserved, which are important for biomedical applications of CNTs in imaging and other applications.

Surfactants are efficient in dispersion of CNTs, however they are useless in biological applications since they cause damage to cell membranes. So, amphiphilic polymers for CNT coating are predominant in biomedical applications. Up to date, a variety of natural and synthetic polymers have been explored for coating on CNTs, including lipid-containing polymers,[8,23,24] proteins,[3,25] DNA,[26] chitosan,[27]

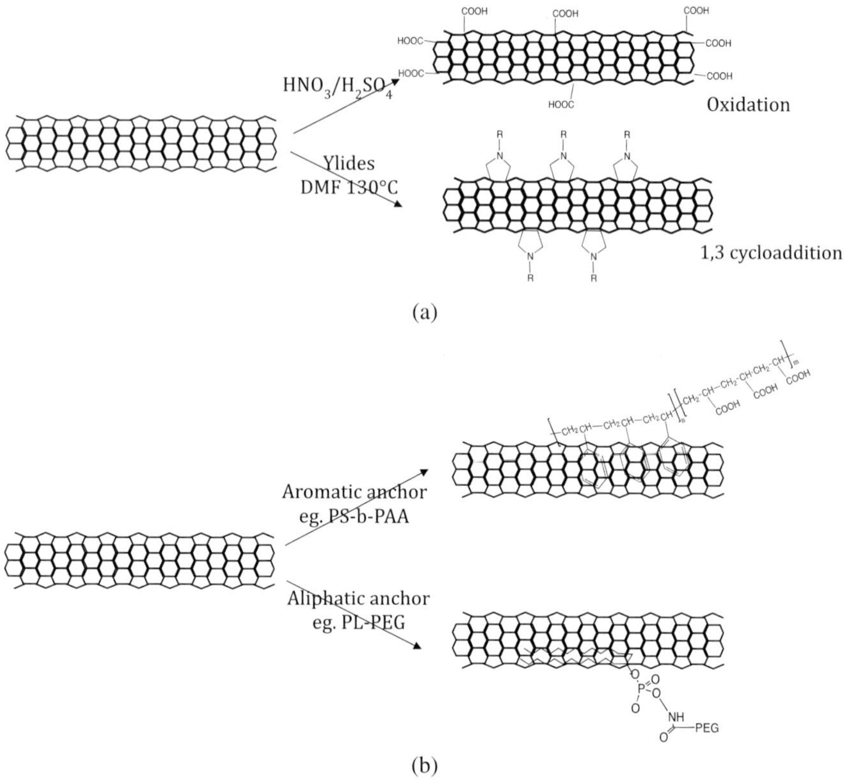

Fig. 2. Schemes of functionalization of carbon nanotubes. (a) CNTs can be covalently modified by oxidation or 1,3 cycloaddition reactions and then conjugated with hydrophilic polymers; (b) CNTs can be coated with amphiphilic molecules containing poly aromatic ring or aliphatic chain structure, e.g. Polystyrene-b-polyacrylic acid (PS-b-PAA), phospholipid polyethylene glycol (PL-PEG).

and aromatic rings containing polymers.[21,22,28,29] Natural polymers, such as lipids, proteins, or DNA coated on CNTs could be degraded by enzymes present in blood in system delivery, and therefore might not be suitable for *in vivo* applications. Among the polymers tested, phospholipid conjugated polymer PEG (PL-PEG) forms stable hybrids with CNTs and provides good stability in a serum-containing environment, which gives rise to a large number of biomedical applications ranging from SiRNA, peptide and drug delivery, as well as biological sensing and imaging.[8]

3. Carbon Nanotube-based Targeted Drug Delivery Systems in Cancer Therapy

The discovery of CNTs opens up new opportunities for cancer drug delivery. The unique structural and optical properties of CNTs allow the design of multifunctional drug delivery systems that contain drugs, targeting moieties and imaging reagents for efficient targeted delivery. The needle-like structure of CNTs provides sufficient surface area for loading large amounts therapeutics along the longitude of tubes without affecting their cell penetration capability. With the adequate loading capacity, the CNTs are able to simultaneously carry multiple therapeutic agents, one or more targeting molecules, contrast agents, endothelial tight-junction permeation enhancer, etc. for tumor targeting. In recent years, a variety of CNT-based tumor targeted delivery systems have been developed to shuttle DNA, siRNA, peptides, or small molecule drugs into cancer cells to exert killing effects.

3.1. *Carbon Nanotube-based Drug Delivery Platform for Tumor Targeting*

Both passive and active tumor-targeting strategies are explored for CNT-based drug delivery systems. Passive targeting utilizes special features of tumor vasculature. Tumors, as fast growing tissues, display enhanced vascular permeability due to their high demand for nutrients and oxygen. The enhanced permeability in tumor vascular architecture can be utilized for delivery of anti-cancer reagents since macromolecules tend to accumulate in tumor tissues, which is termed as the tumor-selective enhanced permeability and retention (EPR) effect. CNTs exhibit macromolecular structure that is advantageous for tumor targeting through EPR effect. The tumor targeting efficiency of CNTs can be further enhanced by surface coating with biocompatible polymers, such as PEG.[30] Studies have shown that SWNTs coated with long or branched PEG generally obtain a prolonged blood circulation, which favors tumor specific targeting via EPR effects.[31,32]

Active targeting relies on recognition of biomarkers on the cancer cell surface. A variety of tumor recognition molecules have been investigated in CNT-based therapeutics delivery systems, including antibodies, peptides and others (summarized in Table 1). As antibodies and peptides contain amino groups, they can be easily conjugated to oxidized CNTs via an esterification process. Antibodies have been proved to have therapeutic effects of their own, and they can also serve as tumor recognition molecules for drug delivery systems in targeted deliveries. Originally, the antibody against CD20 has been applied to treat CD20 positive B cell lymphoma clinically. It has also been conjugated to a variety of drug delivery systems as a tumor targeting moiety. SWNTs were constructed to attach multiple copies of anti-CD20 antibodies to carry metal-ion chelate and fluorescent chromophore moieties for targeting B cell lymphoma.[33] The nano-construct was found to accumulate in tumor sites in a mice model of disseminated human lymphoma. In another study, HER2 antibody conjugated SWNTs for targeting HER positive cancer. *In vitro* study has shown the HER2 conjugated SWNTs were able to efficiently kill HER2-expressing SK-BR-3 cells but not receptor-negative MCF-7 cells by photothermal ablation.[34] Peptide conjugated SWNTs have also shown efficient targeting *in vivo.* Peptides containing the Arginine-Glycine-Aspartate (RGD) triad display a strong affinity to alpha (V) beta (^{3}) receptors overexpressed on many types of tumor cells. RGD tagged CNTs demonstrated effective targeting for delivery of chemotherapy drugs, *in vivo* imaging and photothermal ablation of tumors.[31,35,36] Similar to RGD, another peptide containing the Asn- Gly-Asp (NGR) triad is able to bind to the endothelial cells on neoangiogenic vessels. SWNTs decorated with NGR have been shown to carry chemotherapy drug docetaxel for targeting tumor angiogenesis.[37] Combination of SWNTs-NGR-docetaxel and photothermal ablation showed effective tumor growth inhibition in a murine S180 cancer model.[37] Folic acid (FA) is a small molecule that binds tightly to the folate receptor (FR), which is a well-recognized tumor antigen that captures its ligands from the extracellular milieu and transports them inside the cell via endocytosis.[38] The FA-SWNTs formulated doxorubicin

Table 1. Examples of Cancer Targeting Molecules Applied in CNT Based Delivery Systems

Type of CNTs	Targeting Moieties	Cancer Biomarkers	Types of Cancers and Animal Models	Payloads/ Treatments	References
SWNTs	Antibody against CD20 (Rituximab)	CD20 epitope	B lymphoma (murine model lymphoma)	Radiometal ion fluorescent chromophore	33
SWNTs	Anti-CD133	CD133 epitope	Glioblastoma	photothermal ablation	94
SWNTs	Antibody against CD22	CD22 epitope	B lymphoma cells	photothermal ablation	57, 95
SWNTs	Antibody against HER2	HER2	HER2 positive breast cancer	Toxin RTA	96
SWNTs	Antibody against HER2	HER2	HER2 positive breast cancer	photothermal ablation	34
SWNTs	RGD peptide	integrin $\alpha v\beta 3$	Glioblastoma (U87MG xenograft tumor model)	Raman Spectroscopy imaging	35
SWNTs	RGD peptide	integrin $\alpha v\beta 3$	Breast cancer	Doxorubicin	52
SWNTs	NGR peptide	Endothelial cells	PC3 cell line (Murine S180 cancer model)	Docetaxel combined photothermal ablation	37
SWNTs	Folic acid	Folate receptor	Breast cancer	platinum (IV)	97
SWNTs	Folic acid	Folate receptor	Breast cancer	Doxorubicin	40
SWNTs	Epidermal growth factor	HER1	Pancreatic cancer	Raman spectroscopy imaging	98
SWNTs	Epidermal growth factor	HER1	Squamous cancer	cisplatin	41, 99

Abbreviations: SWNTs, single walled carbon nanotubes; EGF, epidermal growth factor; HER, human epidermal growth factor receptor; RTA, ricin A chain protein; RGD, peptide with arginine-glycine-aspartate sequence; NGR, peptide with asparagine-glycine-arginine sequence.

demonstrated improved cellular uptake and drug efficacy *in vitro* and *in vivo*.[39,40] Epidermal growth factor (EGF) was conjugated to SWNTs for delivery of drug cisplatin to squamous cancer cells that overexpress EGF receptors. SWNTs-cisplatin-EGF could internalize rapidly into squamous cancer cells, but showed limited uptake by the EGF receptor negative cells. The cellular uptake of SWNTs-cisplatin-EGF could be blocked by siRNA knockdown of EGF receptor in squamous cancer cells.[41]

3.2. *Carbon Nanotubes as Gene Delivery Vehicles*

Gene therapy is one of the most promising approaches to treat cancer and other genetic diseases. DNA and siRNA are macromolecules that cannot pass through the cell membrane by themselves. With strong cell penetration capacity, CNTs are highly promising for delivery of DNA and siRNA into cells. Three strategies are commonly applied for loading DNA or siRNA onto CNTs: (1) association of DNA or siRNA with CNTs functionalized by cationic polymers via electrostatic interactions; (2) conjugation of DNA or siRNA to functionalized CNTs via cleavable chemical bonds; (3) wrapping pristine CNTs with DNA/siRNA directly (Fig. 3).

Cationic polymers, such as ammonium containing molecules, polyethylene imine (PEI), can be covalently linked to CNTs modified by oxidation or 1,-3-dipoolar cycloaddition reactions.[42–45] Pantarotto *et al.* applied an amino-terminated oligoethylene glycol attached to CNTs for DNA delivery,[42,43] in which plasmid pCMV-βgal was used as a test gene. They showed that the level of gene expression was 5–10 times higher than that of DNA alone, but was less efficient than that using liposomes. The charge ratio between the ammonium groups at the CNT surface and the phosphate groups of the DNA backbone was an important factor determining the level of gene expression.[43] Similar to DNA delivery, the ammonium terminated CNTs (CNTs-NH_3^+) can also be applied for siRNA delivery.[46,47] CNTs-NH_3^+ mediated delivery of siRNA for silencing cyclin A2 resulted pronounced suppression effect in chronic myelogenous leukemia K562 cells.[47] SWNTs-NH_3^+ carrying Suppressor of Cytokine Signaling

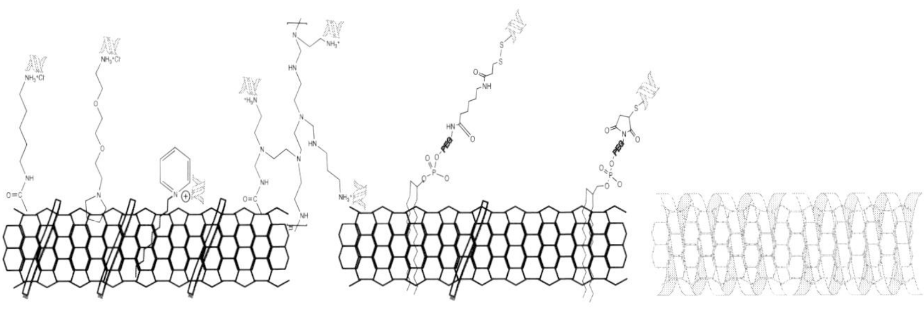

Fig. 3. Strategies for DNA or siRNA delivery by carbon nanotubes (CNTs). (Left) CNTs are firstly covalently modified either by oxidation or 1,3 cycloaddition, then conjugated with cationic polymers, e.g. ammonium end functional group containing molecules, polyethylene imine (PEI). Negative charged DNA or siRNA can be loaded to CNTs through electrostatic association with cationic polymers. (Middle) DNA or siRNA can be covalently conjugated to functionalized CNTs via cleavable chemical bonds. (Right) single-stranded DNA (ssDNA) or siRNA form helical wrapping around CNTs with their bases binding strongly to the hydrophobic sidewalls and the sugar-phosphate groups extending to aqueous environment.

1(SOCS1) siRNA significantly inhibited SOCS1 expression and retarded the tumor growth in an established murine B16 tumor model.[46] In another study, SWNTs-NH_3^+ mediated *in vivo* siRNA delivery for silencing telomerase reverse transcriptase (TERT) gene through intravenous injection route could suppress tumor growth in tumor xenograft model in nude mice.[45] Apart from ammonium terminated CNTs, the polycationic polymer PEI functionalized CNTs have been applied for siRNA delivery.[4,48] PEI is an efficient gene delivery reagent on its own; however, a high amount of PEI is toxic for cells. The PEI-grafted MWNTs showed improved gene expression to the equivalent amounts of PEI polymer alone, but with reduced cytotoxicity.[48]

Alternatively, DNA or siRNA can be covalently conjugated functionalized CNTs.[8,49,50] Incorporation of cleavable chemical bonds (e.g. disulfide bond) allows releasing of DNA or siRNA cargos from

CNTs in a controlled manner. SWNTs were non-covalently functionalized with PL-PEG-NH_3, then thiol-modified DNA or siRNA were covalently linked terminal amine group of PL-PEG[50] (Fig. 3). The release of DNA or siRNA cargo was triggered by the cleavage of disulfide bonds via thiol digesting enzymes upon cellular uptake of the CNTs-gene complexes. Application of the CNTs-PL-PEG-gene delivery system in hard-to transfect human T cells and primary cells lines demonstrated higher gene transfection efficiency compared with liposome-based delivery system. This effect was due to strong cell penetration capability and the high cargo loading capacity of CNTs.[49]

Interestingly, single-stranded DNA (ssDNA) or siRNA can be used to coat pristine SWNTs directly to obtain stable aqueous suspensions.[26] The DNA or siRNA forms a helical wrapping around the SWNTs via the bases binding strongly to the hydrophobic sidewall of SWNTs and the sugar-phosphate groups extending to water phase to impart aqueous solubility. Only a simple ultrasonic process is required to form of the SWNTs/DNA or SWNTs/siRNA complexes. In the complexes, DNA or siRNA serve both to disperse the SWNTs and to act as the cargos. The SWNT/siRNA complexes readily enter cells and retain their biological activity in cell culture.[26] Intratumoral administration of SWNT/siRNA complexes for targeting hypoxia-inducible factor 1 alpha (HIF-1α) significantly inhibited HIF-1α activity *in vivo*.[51]

3.3. *Carbon Nanotubes as Drug Delivery Vehicles*

One limitation of conventional chemotherapy drugs is their system toxicity, since most cancer drugs act to kill all fast dividing cells, including cancer cells and all other types of the cells that divide rapidly under normal circumstances. A CNT-based targeted delivery system could overcome this limitation. The benzene ring structure of CNTs allows tight binding of drugs containing poly-benzene ring structure. For example, Doxorubicin (DOX) is an important chemotherapeutic agent that has been used for treatment of a variety of human cancers. Due to the flat poly-benzene ring structure of DOX, a high amount of DOX can be loaded onto CNTs by π–π stacking.[52,53] *In vivo* study with SWNTs-PL-PEG/DOX showed significantly enhanced therapeutic efficacy in a murine model of breast cancer.[52] Folic acid conjugated chitosan was applied to disperse DOX loaded SWNTs for tumor targeted delivery. The targeted drug delivery system could effectively inhibit growth of cancer cells *in vitro* and *in vivo*.[53]

Alternatively, small molecule drugs can be covalently conjugated to functionalized CNTs to form drug-CNTs prodrugs. The prodrugs are usually inactive. The release of the free active drugs is essential for drug to take effects, and therefore, the linkages in prodrugs have to be cleavable. For example, first line anticancer agent cisplatin was conjugated to oxidized SWNTs via a peptide linkage.[41] The peptide linkage in SWNTs-cisplatin can be selectively cleaved by proteases overexpressed in tumor cells. In this study, for specifically targeting squamous cancer, the SWNTs-cisplatin was tagged with EGF. SWNTs-cisplatin-EGF showed more efficient tumor growth inhibition compared to both free cisplatin and non-targeted SWNTs-cisplatin.[41] Another antitumor drug paclitaxel (PTX) was conjugated to functionalized CNTs for targeted delivery. PTX is one of the most important drugs for metastatic breast cancer. Poor water solubility of this drug causes additional side effects in clinical applications. PTX was covalently linked to SWNTs-PL-PEG via a cleavable ester linkage.[54] The SWNTs-PL-PEG-PTX displayed higher efficacy in tumor growth inhibition and decreased side effects in a murine breast cancer model compared with clinical paclitaxel formulation, Taxol.[54]

For prodrugs, the rational design of linkers is a key factor to ensure the overall efficacy of the drug conjugates. The cleavable linkages of CNTs–drug conjugates have to be stable in circulation and the release of active drugs preferably occurs inside the target cells. To examine the effect of type of linkers to the efficacy of CNT-prodrugs, Samori *et al.* developed two different linkers: peptide and ester linkers for CNTs-Methotrexate (MTX) conjugate.[55] The CNTs-MTX drug with peptide linker showed higher cytotoxicity than that with ester linker, which was probably due to the more efficient cleavage of peptide linker than that of ester linker inside of the cancer cells.[55]

4. Carbon Nanotubes as Photo Thermal Agents in Cancer Therapy

SWNTs emit heat when constantly absorbing energy in the near-infrared (NIR) region. Tissue is transparent in this wavelength; this property therefore can be used for optical stimulation of nanotubes for heating up in living cells. Continuous heating leads to cell death. Selectively thermal ablation of tumor can be achieved by conjugation of targeting moiety to SWNTs. Kam *et al.* showed that folic acid decorated SWNTs can selectively kill cancer cells by thermal ablation without harming neighboring receptor-free normal cells.[56] Monoclonal antibody (mAb) against human CD22 or CD25 was coupled to CNTs for examination of specific targeting and thermal ablation effects of CNTs.[57] Interestingly, anti-CD22 mAb coupled CNTs selectively targeted and killed CD22(+) CD25(−) Daudi cells, whereas anti-CD25 mAb coupled CNTs only target and kill CD22(−) CD25(+) activated peripheral blood mononuclear cells. Both targeting and internalization effects of CNTs are important for effective thermal ablation of tumor cells.[58]

5. Artificial Cell Microencapsulated Carbon Nanotubes in Cancer Therapeutics

Artificial cells have been used to encapsulate CNTs for their targeted delivery applications. In a first feasibility study, they have been encapsulated in artificial cells made from biocompatible polymeric membrane for target specific delivery. Alginate-poly-L-lysine-alginate (APA) artificial cells were used to pack CNT drugs and to protect them from the harsh environments.[59] The optimum degradation condition of the artificial cell microcapsule polymeric membrane allows the release of CNT drugs at the targeted sites in a controlled manner (Fig. 4). This system is ideal for targeting the therapeutics to desired sites for cancer and other targeted delivery applications.

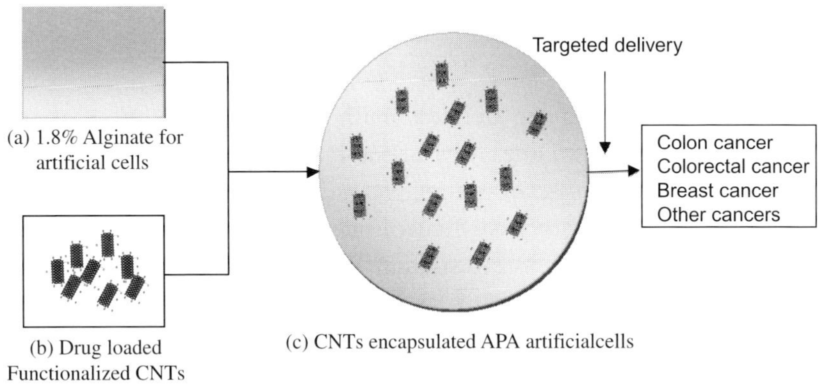

Fig. 4. Carbon nanotubes encapsulated artificial cell for targeted deivery: Functionalized carbon nanotubes are trapped into the core of artificial cells made from alginate-poly-L-lysine-alginate (APA) microcapsules for targeted delivery of therapeutics for a variety of cancers.

6. Carbon Nanotubes in Stem Cell Therapy

The wide spectrum of nano- and biotechnological tools holds immense promise for the study of stem cell biology and the development of new approaches for stem cell expansion, differentiation, and transplantation. Recent focus has been in the development of technologies to modulate stem cell, including both embryonic (ESCs) and mesenchymal (MSCs) stem cell, microenvironments by mimicking the extracellular matrices with nanotubes, nanofibers and other biocompatible nanomaterials.

Stem cells, with the ability to divide and differentiate into diverse specialized cell types, have been considered a potential tool for tissue engineering and regeneration therapy. Over the years, there has been an increasing body of evidence indicating that the extracellular matrix (ECM) plays a critical role in stem cell viability, proliferation and differentiation.[60–62] Designing a microenvironment prepared from polymeric scaffolds which imitate the physical characteristics of natural ECM has been the central strategy in tissue engineering. The emergence of nanomaterials provides opportunities to design scaffolds for hosting and directing stem cell differentiation.

6.1. *Carbon Nanotubes as Nanoscaffolds for Stem Cell Culture and Differentiation*

Embryonic stem cells are derived from the inner cell mass of pre-implantation embryos, and they are able to differentiate into tissue types from all three germ layers. The derivation of neurons from hESCs holds promise to treat neurological pathologies of the central and peripheral nervous system such as Parkinson's disease, spinal cord injury and glaucoma.[63] Recent reports suggest that CNTs may be a promising scaffold material for directing hESCs' differentiation toward neuronal lineage.[64]

This is because CNTs resemble the size of ECM molecules such as collagens and laminins, which have been reported to favor neuron growth.[65] Moreover CNTs are flexible, and have excellent mechanical strength because of which they are able to maintain the scaffolds' structural integrity during cell proliferation. They are also conductive in nature, and this remains unchanged under harsh environments. Because of their conductive nature, scaffolds impregnated with CNTs can induce electrical stimulation and thus can be used in the future to develop biomedical devices which can interact directly with the neuronal system. Sridharan *et al.* reported that a composite of rigid fibril bundles prepared from collagen–carbon nanotube (collagen/CNTs) complex can polarize the growth and differentiation of hESCs, resulting in more than 90% of the cells to the ectodermal lineage.[66] It has been reported that substrates prepared using CNTs are biocompatible and can support stem cell growth and neuron differentiation.[67] It has also been observed that neurons grown on a CNT network have exhibited better signal transmission.[68]

A poly (methacrylic acid)-grafted CNT thin film scaffold, providing nanofibrillar surfaces and mimicking the characteristics of ECM with enhanced cell adhesion and growth factor adsorption, has been shown to encourage stem cell attachment and differentiation.[62] All these newly found unique characteristics of CNTs makes them a very promising material system to provide a supportive environment for stem based tissue engineering, as well as other related medical applications.

6.2. *Synergistic Effect of Carbon Nanotubes and Stem Cells in Cancer Therapy*

There are several lines of evidence showing that CNTs may be used to initiate and maintain immune responses.[69] It has been shown that peptide-functionalized CNTs can act as proficient immunomodulators and consequently generate specific antibody responses.[70] ESCs have also been shown to play an important role as a cellular cancer vaccine that stimulates biological systems to destroy colon cancer cells by eliciting an immune boost vaccination with human pluripotent stem cells which generates a broad spectrum of immunological and clinical responses against colon cancer. Further studies have demonstrated that ESCs, in combination with MWNTs, can induce an enhanced immune boost and provide subsequent anticancer protection in mice with colon cancer by suppressing the proliferation and development of malignant colon tumors (Fig. 5).[71]

6.3. *Carbon Nanotubes in Modulating Stem Cell Behavior*

Bone marrow-derived mesenchymal stem cells (MSCs), also known as stromal stem cells, represent essentially a population of nonhematopoietic progenitor cells, which can be found in the bone marrow stroma.[72] Further studies confirmed the pluripotent potential of these cells.[73] Although the extent of their plasticity is still under investigation, studies within the last few years have demonstrated the capacity of MSCs to differentiate into cells of lineages other than mesenchymal.[74] Because these cells do not have the ethical or tumorigenicity problems of embryonic stem cells, their plasticity has generated much excitement, giving hope to their therapeutic use in a wide range of diseases. Moreover, due to their immunotolerance property, the establishment of MSCs as effective "universal donor cells"[75] could considerably expand their therapeutic potential for cell therapy such as cellular cardiomyoplasty.

Recently it has been reported that human mesenchymal stem cells (hMSCs) grown on CNT networks behave differently to different CNT patterns.[76] Thus it is possible to regulate the growth direction

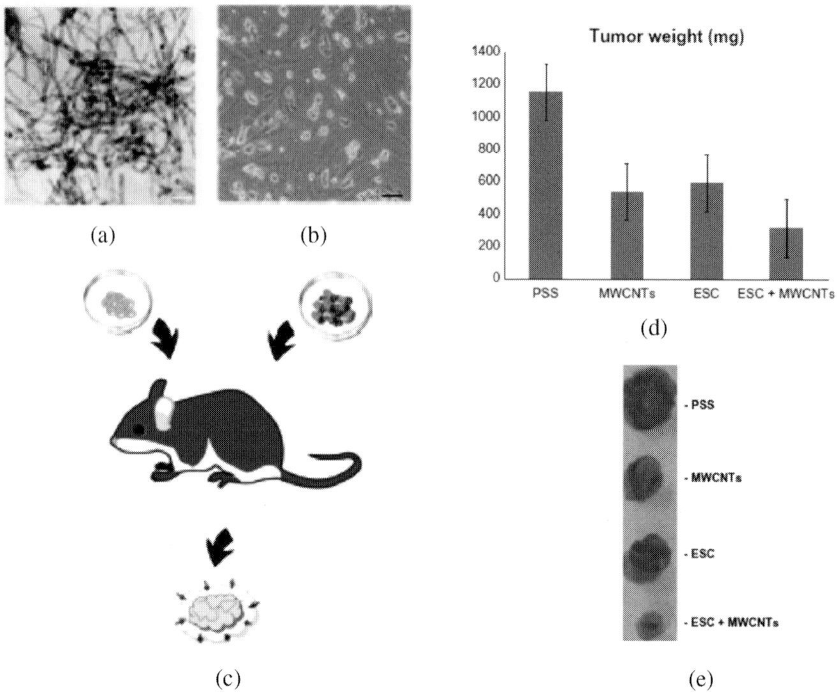

Fig. 5. Embryonic stem cells (ESC) combined with multiwall CNTs (MWCNTs) as anticancer immunization agents: (a) Transmission electron micrographs of MWCNTs used for the immunization. The scale bar represents 100 nm. (b) *In vitro* C57 BL/6 mouse embryonic stem cells used for immunization (scale bar: 50 μm). (c) Schematic illustration of the proposed vaccine containing 1×10^6 ESCs and 300 μL of 0.1 mg/mL MWCNT administered weekly for 6 weeks. (d) The results of the tumor weights (mg) at the end of the study, i.e. 21 days post cancer cell administration. (e) Macroscopic aspect of the harvested tumors at the endpoint.[75]

and differentiation of the hMSCs into different cell lineages such as osteoblasts and neuronal,[76–79] using different patterns of CNTs as the biocompatible nanophase material. Further studies in this direction illustrated that surface modifying CNTs with polymers such as PLGA, poly (carbonate) urethane and biomolecules such as fibronectin can enhance cellular functions, promote cell growth and induce MSC differentiation into osteoblasts.[80,81] It has been proposed that CNTs can have immense other biomedical applications such as stem cell guiding to different target sites using external magnetic fields,[82] cell tracking and labeling, nanosensors, and vehicles for controlled release of drugs.[60]

Addition of MWNTs to electrospun PLA (poly lactic acid) fibers can reduce the fiber diameter, increase the conductivity of scaffold, and potentially provide a functional composite for tissue engineering. This has been illustrated in an interesting study, where MWNTs were coated with poly (lactic acid) to form electrospun nanofiber composite (diameter: 700 nm).[83] Adipose-derived human mesenchymal stem cells (hMSCs) were seeded on the scaffolds containing 1 wt% MWNTs. Tensile testing showed an increase in the tensile modulus for a MWNT loading of 0.25 wt% compared with electrospun nanofibrous mats without MWNT reinforcement. SEM images of cells on PLA with 1% MWNTs at 14 days showed closely packed, longitudinally aligned, confluent layers of cells with fibroblastic morphology. By quantifying the amount of DNA present in each scaffold, we were able to determine a significant difference at day 14 between the PLA scaffolds with MWNTs compared with PLA scaffolds alone (Fig. 6). This difference could be attributed to the increased surface area due to the fibers being considerably smaller.

7. Conclusion

CNTs hold a great potential in biomedical applications on the basis of their interesting physiological and biochemical properties. Many techniques are available to enhance these features towards biomedical application in general and towards cancer therapeutics. The surface functionalization of CNTs, for example, enables covalent conjugation or non-covalent coating of biocompatible polymers for efficiently loading therapeutics or as scaffolds for stem cell culture. In this chapter many novel biomedical devices based on CNTs in cancer application have been elaborated. Although extremely promising, these CNT-based therapeutics require further investigation and development towards clinical applications. The major concern for biomedical application of CNTs relates to their toxicity. Although earlier studies

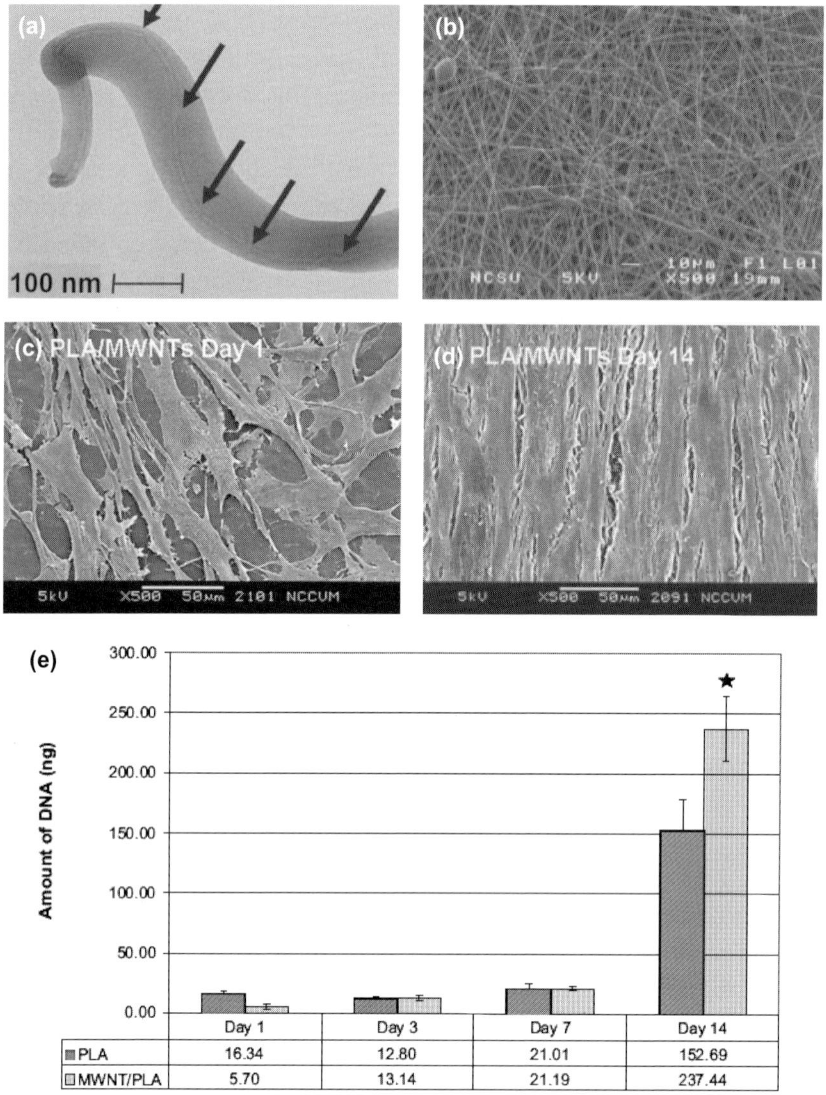

Fig. 6. Scanning electron microscopic (SEM) images of electrospun PLA containing 1% MWNT (a and b). SEM images hMSCs on Day 1 (c) and Day 14 (d), completely confluent on the surface of the electrospun PLA with MWNT. Number of hMSCs (in terms of amount of DNA) present throughout the course of the experiment on the electrospun tissue scaffolds with and without MWNT. ($n = 3$, three scaffolds/time point) (Error bars = Standard Deviation), where star indicates significance p-value <0.05.[87]

reported obvious toxicity with pristine CNTs,[84,85] more studies clarified that properly functionalized CNTs show little or no toxicity.[86–89] For example, PEG coated CNTs demonstrated much decreased cytotoxic than uncoated oxidized CNTs.[90] *In vivo* study examining the survival and clinical parameters (including blood pressure, blood counts, and serum chemistry, etc.) in experimental animal revealed that PEG coated SWNTs did not exert apparent toxicity after 4 months exposure.[91] Our present knowledge on toxicity of CNTs is still inadequate, however. Despite the uncertainty on CNTs toxicity, CNT research has been advanced to clinical trial stage. For example, a clinical trial on SWNT for siRNA delivery in cancer is on the way.[92] And recently, the FDA has approved the use of CNTs to improve colorectal cancer imaging.[93] These advancements encourage biomedical application of CNTs in general and cancer therapies in particular.

Acknowledgments

This work is partially supported by a research grant to SatyaPrakash from the Canadian Institutes of Health Research (CIHR) (MOP 93641). W. Shao acknowledges the *Excellence Award* from the Biomedical Engineering Department, McGill University. A. Paul acknowledges the *Alexander Graham Bell Post Graduate Scholarship-Doctoral* from Natural Sciences and Engineering Research Council of Canada (NSERC).

References

1. Harris PJF. (1999) *Carbon Nanotubes and Related Structures.* Cambridge University Press, 1999.
2. Xie SS, Li WZ, Pan ZW, *et al.* (2000) Mechanical and physical properties on carbon nanotube. *J Phys Chem Solids* **61(7):** 1153–1158.
3. Chen RJ, Zhang Y, Wang D, Dai H. (2001) Noncovalent sidewall functionalization of single-walled carbon nanotubes for protein immobilization. *J Am Chem Soc* **123(16):** 3838–3839. Epub 2001/07/18.
4. Liu Y, Wu DC, Zhang WD, *et al.* (2005) Polyethylenimine-grafted multiwalled carbon nanotubes for secure noncovalent immobilization and efficient delivery of DNA. *Angew Chem Int Ed Engl* **44(30):** 4782–4785. Epub 2005/07/05.

5. Chen Z, Zhang X, Yang R, *et al.* (2011) Single-walled carbon nanotubes as optical materials for biosensing. *Nanoscale* **3(5):** 1949–1956. Epub 2011/03/17.
6. Ostling D, Tomanek D, Rosen A. (1997) Electronic structure of single-wall, multiwall, and filled carbon nanotubes. *Phys Rev B* **55(20):** 13980–13988.
7. Lin MF, Chuu DS, Shung KWK. (1996) Thermal conductance and the Peltier coefficient of carbon nanotubes. *Phys Rev B* **53(16):** 11186–11192.
8. Liu Z, Tabakman SM, Chen Z, Dai H. (2009) Preparation of carbon nanotube bioconjugates for biomedical applications. *Nat Protoc* **4(9):** 1372–1382. Epub 2009/09/05.
9. Zhan GD, Kuntz JD, Wan J, Mukherjee AK. (2003) Single-wall carbon nanotubes as attractive toughening agents in alumina-based nanocomposites. *Nat Mater* **2(1):** 38–42. Epub 2003/03/26.
10. Yadav SK, Bera T, Saxena PS, *et al.* (2010) MWCNTs as reinforcing agent to the hap-gel nanocomposite for artificial bone grafting. *J Biomed Mater Res Part A* **93(3):** 886–896. Epub 2009/08/26.
11. Ozel T, Gaur A, Rogers JA, Shim M. (2005) Polymer electrolyte gating of carbon nanotube network transistors. *Nano Lett* **5(5):** 905–911. Epub 2005/05/12.
12. Weitz RT, Zschieschang U, Forment-Aliaga A, *et al.* (2009) Highly reliable carbon nanotube transistors with patterned gates and molecular gate dielectric. *Nano Lett* **9(4):** 1335–1340. Epub 2009/04/09.
13. Bekyarova E, Itkis ME, Cabrera N, *et al.* (2005) Electronic properties of single-walled carbon nanotube networks. *J Am Chem Soc* **127(16):** 5990–5995. Epub 2005/04/21.
14. Lee CY, Baik S, Zhang J, *et al.* (2006) Charge transfer from metallic single-walled carbon nanotube sensor arrays. *J Phys Chem B* **110(23):** 11055–11061. Epub 2006/06/15.
15. Klumpp C, Kostarelos K, Prato M, Bianco A. (2006) Functionalized carbon nanotubes as emerging nanovectors for the delivery of therapeutics. *Biochim Biophys Acta* **1758(3):** 404–412. Epub 2005/11/26.
16. Liu Z, Tabakman S, Welsher K, Dai H. (2009) Carbon nanotubes in biology and medicine: *in vitro* and *in vivo* detection, imaging and drug delivery. *Nano Res* **2(2):** 85–120. Epub 2010/02/23.
17. Hu H, Zhao B, Itkis ME, Haddon RC. (2003) Nitric acid purification of single-walled carbon nanotubes. *J Phys Chem B* **107(50):** 13838–13842.
18. Tagmatarchis N, Prato M. (2004) Functionalization of carbon nanotubes via 1,3-dipolar cycloadditions. *J Mater Chem* **14(4):** 437–439.
19. Zhao B, Hu H, Yu A, *et al.* (2005) Synthesis and characterization of water soluble single-walled carbon nanotube graft copolymers. *J Am Chem Soc* **127(22):** 8197–8203. Epub 2005/06/02.
20. Prato M, Kostarelos K, Bianco A. (2008) Functionalized carbon nanotubes in drug design and discovery. *Acc Chem Res* **41(1):** 60–68. Epub 2007/09/18.

21. Kang YJ, Taton TA. (2003) Micelle-encapsulated carbon nanotubes: a route to nanotube composites. *J Am Chem Soc* **125(19):** 5650–5651.
22. Chen RJ, Zhang YG, Wang DW, Dai HJ. (2001) Noncovalent sidewall functionalization of single-walled carbon nanotubes for protein immobilization. *J Am Chem Soc* **123(16):** 3838–3839.
23. Richard C, Balavoine F, Schultz P, *et al.* (2003) Supramolecular self-assembly of lipid derivatives on carbon nanotubes. *Science.* **300(5620):** 775–778. Epub 2003/05/06.
24. Park S, Yang HS, Kim D, *et al.* (2008) Rational design of amphiphilic polymers to make carbon nanotubes water-dispersible, anti-biofouling, and functionalizable. *Chem Commun* **(25):** 2876–2878.
25. Wang S, Humphreys ES, Chung SY, *et al.* (2003) Peptides with selective affinity for carbon nanotubes. *Nat Mater* **2(3):** 196–200. Epub 2003/03/04.
26. Zheng M, Jagota A, Semke ED, *et al.* (2003) DNA-assisted dispersion and separation of carbon nanotubes. *Nat Mater* **2(5):** 338–342. Epub 2003/04/15.
27. Jiang L, Liu C, Jiang L, *et al.* (2004) A chitosan-multiwall carbon nanotube modified electrode for simultaneous detection of dopamine and ascorbic acid. *Anal Sci.* **20(7):** 1055–1059. Epub 2004/08/06.
28. Petrov P, Stassin F, Pagnoulle C, Jerome R. (2003) Noncovalent functionalization of multi-walled carbon nanotubes by pyrene containing polymers. *Chem Commun (Camb)* **(23):** 2904–2905. Epub 2003/12/19.
29. Kang Y, Taton TA. (2003) Micelle-encapsulated carbon nanotubes: a route to nanotube composites. *J Am Chem Soc* **125(19):** 5650–5651. Epub 2003/05/08.
30. Liu Z, Davis C, Cai W, *et al.* (2008) Circulation and long-term fate of functionalized, biocompatible single-walled carbon nanotubes in mice probed by Raman spectroscopy. *Proc Natl Acad Sci USA* **105(5):** 1410–1415. Epub 2008/01/31.
31. Liu Z, Cai W, He L, *et al.* (2007) *In vivo* biodistribution and highly efficient tumour targeting of carbon nanotubes in mice. *Nat Nanotechnol* **2(1):** 47–52. Epub 2008/07/26.
32. Yang ST, Fernando KA, Liu JH, *et al.* (2008) Covalently PEGylated carbon nanotubes with stealth character *in vivo. Small* **4(7):** 940–944. Epub 2008/06/25.
33. McDevitt MR, Chattopadhyay D, Kappel BJ, *et al.* (2007) Tumor targeting with antibody-functionalized, radiolabeled carbon nanotubes. *J Nucl Med.* **48(7):** 1180–1189. Epub 2007/07/04.
34. Xiao Y, Gao X, Taratula O, *et al.* (2009) Anti-HER2 IgY antibody-functionalized single-walled carbon nanotubes for detection and selective destruction of breast cancer cells. *BMC Cancer* **9:** 351. Epub 2009/10/06.
35. Zavaleta C, de la Zerda A, Liu Z, *et al.* (2008) Noninvasive Raman spectroscopy in living mice for evaluation of tumor targeting with carbon nanotubes. *Nano Lett* **8(9):** 2800–2805. Epub 2008/08/08.

36. De la Zerda A, Zavaleta C, Keren S, *et al.* (2008) Carbon nanotubes as photoacoustic molecular imaging agents in living mice. *Nat Nanotechnol* **3(9):** 557–562. Epub 2008/09/06.
37. Wang L, Zhang M, Zhang N, *et al.* (2011) Synergistic enhancement of cancer therapy using a combination of docetaxel and photothermal ablation induced by single-walled carbon nanotubes. *Int J Nanomedicine* **6:** 2641–2652. Epub 2011/11/25.
38. Lu YJ, Low PS. (2002) Folate-mediated delivery of macromolecular anticancer therapeutic agents. *Adv Drug Deliv Rev* **54(5):** 675–693.
39. Huang H, Yuan Q, Shah JS, Misra RD. (2011) A new family of folate-decorated and carbon nanotube-mediated drug delivery system: synthesis and drug delivery response. *Adv Drug Deliv Rev* **63(14–15):** 1332–1339. Epub 2011/04/26.
40. Zhang X, Meng L, Lu Q, *et al.* (2009) Targeted delivery and controlled release of doxorubicin to cancer cells using modified single wall carbon nanotubes. *Biomaterials* **30(30):** 6041–6047. Epub 2009/08/01.
41. Bhirde AA, Patel V, Gavard J, *et al.* (2009) Targeted killing of cancer cells *in vivo* and *in vitro* with EGF-directed carbon nanotube-based drug delivery. *ACS Nano* **3(2):** 307–316. Epub 2009/02/25.
42. Pantarotto D, Singh R, McCarthy D, *et al.* (2004) Functionalized carbon nanotubes for plasmid DNA gene delivery. *Angew Chem Int Ed Engl* **43(39):** 5242–5246. Epub 2004/09/30.
43. Singh R, Pantarotto D, McCarthy D, *et al.* (2005) Binding and condensation of plasmid DNA onto functionalized carbon nanotubes: toward the construction of nanotube-based gene delivery vectors. *J Am Chem Soc* **127(12):** 4388–4396. Epub 2005/03/24.
44. Wang X, Song Y, Ren J, Qu X. (2009) Knocking-down cyclin A(2) by siRNA suppresses apoptosis and switches differentiation pathways in K562 cells upon administration with doxorubicin. *PLoS One* **4(8):** e6665. Epub 2009/08/18.
45. Zhang Z, Yang X, Zhang Y, *et al.* (2006) Delivery of telomerase reverse transcriptase small interfering RNA in complex with positively charged single-walled carbon nanotubes suppresses tumor growth. *Clin Cancer Res* **12(16):** 4933–4939. Epub 2006/08/18.
46. Yang R, Yang X, Zhang Z, *et al.* (2006) Single-walled carbon nanotubes-mediated *in vivo* and *in vitro* delivery of siRNA into antigen-presenting cells. *Gene Ther* **13(24):** 1714–1723. Epub 2006/07/14.
47. Wang X, Ren J, Qu X. (2008) Targeted RNA interference of cyclin A2 mediated by functionalized single-walled carbon nanotubes induces proliferation arrest and apoptosis in chronic myelogenous leukemia K562 cells. *ChemMedChem* **3(6):** 940–945. Epub 2008/02/21.
48. Nunes A, Amsharov N, Guo C, *et al.* (2010) Hybrid polymer-grafted multiwalled carbon nanotubes for *in vitro* gene delivery. *Small* **6(20):** 2281–2291. Epub 2010/09/30.

49. Liu Z, Winters M, Holodniy M, Dai H. (2007) siRNA delivery into human T cells and primary cells with carbon-nanotube transporters. *Angew Chem Int Ed Engl* **46(12):** 2023–2027. Epub 2007/02/10.
50. Kam NW, Liu Z, Dai H. (2005) Functionalization of carbon nanotubes via cleavable disulfide bonds for efficient intracellular delivery of siRNA and potent gene silencing. *J Am Chem Soc* **127(36):** 12492–12493. Epub 2005/09/08.
51. Bartholomeusz G, Cherukuri P, Kingston J, *et al.* (2009) *In vivo* therapeutic silencing of hypoxia-inducible factor 1 alpha (IF-1alpha) using single-walled carbon nanotubes noncovalently coated with siRNA. *Nano Res* **2(4):** 279–291. Epub 2010/01/07.
52. Liu Z, Sun X, Nakayama-Ratchford N, Dai H. (2007) Supramolecular chemistry on water-soluble carbon nanotubes for drug loading and delivery. *ACS Nano* **1(1):** 50–56. Epub 2007/08/01.
53. Ji Z, Lin G, Lu Q, *et al.* (2012) Targeted therapy of SMMC-7721 liver cancer *in vitro* and *in vivo* with carbon nanotubes based drug delivery system. *J Colloid Interface Sci* **365(1):** 143–149. Epub 2011/10/07.
54. Liu Z, Chen K, Davis C, *et al.* (2008) Drug delivery with carbon nanotubes for *in vivo* cancer treatment. *Cancer Res* **68(16):** 6652–6660. Epub 2008/08/15.
55. Samori C, Ali-Boucetta H, Sainz R, *et al.* (2010) Enhanced anticancer activity of multi-walled carbon nanotube-methotrexate conjugates using cleavable linkers. *Chem Commun (Camb)* **46(9):** 1494–1496. Epub 2010/02/18.
56. Kam NW, O'Connell M, Wisdom JA, Dai H. (2005) Carbon nanotubes as multifunctional biological transporters and near-infrared agents for selective cancer cell destruction. *Proc Natl Acad Sci USA* **102(33):** 11600–11605. Epub 2005/08/10.
57. Chakravarty P, Marches R, Zimmerman NS, *et al.* (2008) Thermal ablation of tumor cells with antibody-functionalized single-walled carbon nanotubes. *Proc Natl Acad Sci USA* **105(25):** 8697–8702. Epub 2008/06/19.
58. Marches R, Mikoryak C, Wang RH, *et al.* (2011) The importance of cellular internalization of antibody-targeted carbon nanotubes in the photothermal ablation of breast cancer cells. *Nanotechnology* **22(9):** 095101. Epub 2011/01/25.
59. Kulamarva A, Raja PM, Bhathena J, *et al.* (2009) Microcapsule carbon nanotube devices for therapeutic applications. *Nanotechnology* **20(2):** 025612. Epub 2009/05/07.
60. Harrison BS, Atala A. (2007) Carbon nanotube applications for tissue engineering. *Biomaterials* **28(2):** 344–353.
61. Ferreira L, Karp JM, Nobre L, Langer R. (2008) New opportunities: the use of Nanotechnologies to manipulate and track stem cells. *Cell Stem Cell* **3(2):** 136–146.
62. Chao TI, Xiang S, Lipstate JF, *et al.* (2010) Poly(methacrylic acid)-grafted carbon nanotube scaffolds enhance differentiation of hESCs into neuronal cells. *Adv Mater* **22(32):** 3542–3547. Epub 2010/07/24.

63. Levenberg S, Burdick JA, Kraehenbuehl T, Langer R. (2005) Neurotrophin-induced differentiation of human embryonic stem cells on three-dimensional polymeric scaffolds. *Tissue Eng* **11(3–4):** 506–512. Epub 2005/05/05.
64. Jan E, Kotov NA. (2007) Successful differentiation of mouse neural stem cells on layer-by-layer assembled single-walled carbon nanotube composite. *Nano Lett* 7**(5):** 1123–1128. Epub 2007/04/25.
65. Freire E, Gomes FC, Linden R, *et al.* (2002) Structure of laminin substrate modulates cellular signaling for neuritogenesis. *J Cell Sci* **115(Pt 24):** 4867–4876. Epub 2002/11/15.
66. Sridharan I, Kim T, Wang R. (2009) Adapting collagen/CNT matrix in directing hESC differentiation. *Biochem Biophys Res Commun* **381(4):** 508–512. Epub 2009/02/24.
67. Ni Y, Hu H, Malarkey EB, *et al.* (2005) Chemically functionalized water soluble single-walled carbon nanotubes modulate neurite outgrowth. *J Nanosci Nanotechnol* **5(10):** 1707–1712. Epub 2005/10/26.
68. Lovat V, Pantarotto D, Lagostena L, *et al.* (2005) Carbon nanotube substrates boost neuronal electrical signaling. *Nano Lett* **5(6):** 1107–1110. Epub 2005/06/10.
69. Kostarelos K, Bianco A, Prato M. (2009) Promises, facts and challenges for carbon nanotubes in imaging and therapeutics. *Nat Nanotechnol* **4(10):** 627–633. Epub 2009/10/08.
70. Pantarotto D, Partidos CD, Hoebeke J, *et al.* (2003) Immunization with peptide-functionalized carbon nanotubes enhances virus-specific neutralizing antibody responses. *Chem Biol* **10(10):** 961–966. Epub 2003/10/30.
71. Mocan T, Iancu C. (2011) Effective colon cancer prophylaxis in mice using embryonic stem cells and carbon nanotubes. *Int J Nanomedicine* **6:** 1945–1954.
72. Atoui R, Shum-Tim D, Chiu RC. (2008) Myocardial regenerative therapy: immunologic basis for the potential "universal donor cells". *Ann Thorac Surg***86(1):** 327–334. Epub 2008/06/25.
73. Dezawa M, Ishikawa H, Itokazu Y, *et al.* (2005) Bone marrow stromal cells generate muscle cells and repair muscle degeneration. *Science* **309(5732):** 314–317. Epub 2005/07/09.
74. Brazelton TR, Rossi FM, Keshet GI, Blau HM. (2000) From marrow to brain: expression of neuronal phenotypes in adult mice. *Science* **290(5497):** 1775–1779. Epub 2000/12/02.
75. Atoui R, Asenjo JF, Duong M, *et al.* (2008) Marrow stromal cells as universal donor cells for myocardial regenerative therapy: their unique immune tolerance. *Ann Thorac Surg* **85(2):** 571–580.
76. Namgung S, Baik KY, Park J, Hong S. (2011) Controlling the growth and differentiation of human mesenchymal stem cells by the arrangement of individual carbon nanotubes. *ACS Nano* **5(9):** 7383–7390. Epub 2011/08/09.

77. Baik KY, Park SY, Heo K, *et al.* (2011) Carbon nanotube monolayer cues for osteogenesis of mesenchymal stem cells. *Small* **7(6):** 741–745. Epub 2011/03/23.
78. Nho Y, Kim JY, Khang D, *et al.* (2010) Adsorption of mesenchymal stem cells and cortical neural stem cells on carbon nanotube/polycarbonate urethane. *Nanomedicine (Lond)* **5(3):** 409–417. Epub 2010/04/17.
79. Mooney E, Dockery P, Greiser U, *et al.* (2008) Carbon nanotubes and mesenchymal stem cells: biocompatibility, proliferation and differentiation. *Nano Lett* **8(8):** 2137–2143. Epub 2008/07/16.
80. Lin C, Wang Y, Lai Y, *et al.* (2011) Incorporation of carboxylation multiwalled carbon nanotubes into biodegradable poly(lactic-co-glycolic acid) for bone tissue engineering. *Colloids Surf B, Biointerfaces* **83(2):** 367–375. Epub 2011/01/07.
81. Khang D, Kim SY, Liu-Snyder P, *et al.* (2007) Enhanced fibronectin adsorption on carbon nanotube/poly(carbonate) urethane: independent role of surface nano-roughness and associated surface energy. *Biomaterials* **28(32):** 4756–4768.
82. Vittorio O, Quaranta P, Raffa V, *et al.* (2011) Magnetic carbon nanotubes: a new tool for shepherding mesenchymal stem cells by magnetic fields. *Nanomedicine (Lond)* **6(1):** 43–54.
83. McCullen SD, Stevens DR, Roberts WA, *et al.* (2007) Characterization of electrospun nanocomposite scaffolds and biocompatibility with adipose-derived human mesenchymal stem cells. *Int J Nanomedicine* **2(2):** 253–263.
84. Bottini M, Bruckner S, Nika K, *et al.* (2006) Multi-walled carbon nanotubes induce T lymphocyte apoptosis. *Toxicol Lett* **160(2):** 121–126. Epub 2005/08/30.
85. Jos A, Pichardo S, Puerto M, *et al.* (2009) Cytotoxicity of carboxylic acid functionalized single wall carbon nanotubes on the human intestinal cell line Caco-2. *Toxicol In Vitro* **23(8):** 1491–1496. Epub 2009/07/14.
86. Dumortier H, Lacotte S, Pastorin G, *et al.* (2006) Functionalized carbon nanotubes are non-cytotoxic and preserve the functionality of primary immune cells. *Nano Lett* **6(7):** 1522–1528. Epub 2006/07/13.
87. Montes-Fonseca SL, Orrantia-Borunda E, Aguilar-Elguezabal A, *et al.* (2011) Cytotoxicity of functionalized carbon nanotubes in J774A macrophages. *Nanomedicine (Lond).* Epub 2011/10/29.
88. Sayes CM, Liang F, Hudson JL, *et al.* (2006) Functionalization density dependence of single-walled carbon nanotubes cytotoxicity *in vitro. Toxicol Lett* **161(2):** 135–142. Epub 2005/10/19.
89. Ali-Boucetta H, Al-Jamal KT, Muller KH, *et al.* (2011) Cellular uptake and cytotoxic impact of chemically functionalized and polymer-coated carbon nanotubes. *Small* **7(22):** 3230–3238. Epub 2011/09/16.
90. Heister E, Lamprecht C, Neves V, *et al.* (2010) Higher dispersion efficacy of functionalized carbon nanotubes in chemical and biological environments. *ACS Nano* **4(5):** 2615–2626. Epub 2010/04/13.

91. Schipper ML, Nakayama-Ratchford N, Davis CR, *et al.* (2008) A pilot toxicology study of single-walled carbon nanotubes in a small sample of mice. *Nat Nanotechnol* **3(4):** 216–221. Epub 2008/07/26.
92. Nanowerk. (2010) Ensysce receives funding for carbon nanotube therapeutics for siRNA delivery. http://wwwnanowerkcom/news/newsid=16674php
93. NIH. (2010) Nanotechnology in clinical trials. http://nanocancergov/learn/now/clinical-trialsasp
94. Wang CH, Chiou SH, Chou CP, *et al.* (2011) Photothermolysis of glioblastoma stem-like cells targeted by carbon nanotubes conjugated with CD133 monoclonal antibody. *Nanomedicine (Lond)* **7(1):** 69–79. Epub 2010/07/14.
95. Marches R, Chakravarty P, Musselman IH, *et al.* (2009) Specific thermal ablation of tumor cells using single-walled carbon nanotubes targeted by covalently-coupled monoclonal antibodies. *Int J Cancer* **125(12):** 2970–2977. Epub 2009/06/19.
96. Weng X, Wang M, Ge J, *et al.* (2009) Carbon nanotubes as a protein toxin transporter for selective HER2-positive breast cancer cell destruction. *Mol Biosyst* **5(10):** 1224–1231. Epub 2009/09/17.
97. Dhar S, Liu Z, Thomale J, *et al.* (2008) Targeted single-wall carbon nanotube-mediated Pt(IV) prodrug delivery using folate as a homing device. *J Am Chem Soc* **130(34):** 11467–11476. Epub 2008/07/30.
98. Karmakar A, Iancu C, Bartos DM, *et al.* (2011) Raman spectroscopy as a detection and analysis tool for *in vitro* specific targeting of pancreatic cancer cells by EGF-conjugated, single-walled carbon nanotubes. *J Appl Toxicol.* Epub 2011/12/08.
99. Bhirde AA, Patel S, Sousa AA, *et al.* (2010) Distribution and clearance of PEG-single-walled carbon nanotube cancer drug delivery vehicles in mice. *Nanomedicine (Lond)* **5(10):** 1535–1546. Epub 2010/12/15.

Chapter 22

Chitosan-Based Nanocarriers for Efficient and Targeted siRNA Delivery

Surendra Nimesh* and Ramesh Chandra†,‡

Abstract

RNA interference mediated by small interfering RNAs (siRNA) is rapidly emerging as promising therapeutic strategy for the treatment of inherited and acquired diseases, such as cancer. However, the rapid enzymatic degradation and poor cellular uptake of siRNA hampers the clinical realization of this approach, hence there is a need to design a delivery system which can protect and efficiently transport siRNA to the target cells. Polymeric nanocarriers have been proposed as an alternative choice with reduced cytotoxicity and enhanced efficacy. These vectors not only protect siRNA from enzymatic degradation by forming condensed complexes but also lead to tissue and cellular targeting, improve cellular penetration, and release the siRNA in the right intracellular compartment. Nanocarriers prepared from cationic polymers like chitosan have been widely investigated due to ease of

* Department of Biotechnology, Central University of Rajasthan, Bandar Sindri NH-8, Kishangarh, Ajmer (Rajasthan), 305801, India.
Corresponding author email: surendranimesh@curaj.ac.in
† Department of Chemistry, University of Delhi, Delhi – 110007, India.
‡ Dr.B.R. Ambedkar Center for Biomedical Research, University of Delhi, Delhi – 110007, India.

manipulability, stability, non-immunogenicity, biodegradability, low cost and high flexibility regarding the size of transgene delivered. This chapter presents an overview of siRNA delivery strategies employing chitosan-based nanocarriers, with emphasis on targeted, self-assembled polymeric nanoparticles.

1. Introduction

Since the discovery of antisense oligonucleotides for specific inhibition of gene expression, nucleic acid-based therapeutics made rapid progress during past decades.[1,2] Nucleic acids have been widely investigated for the treatment of several inheritable or acquired diseases. As compared to conventional therapeutics, the effect of nucleic acid therapeutics is upstream, at the level of gene expression producing those proteins. RNA interference (RNAi) mediated inhibition of gene expression related to diseases presents a new paradigm shift in nucleic acid-based therapeutics. RNAi is mediated by small sequences of 21 to 23 nucleotides double stranded RNAs, referred to as siRNAs (small interfering RNAs). However, the successful clinical implication of RNAi-based therapeutics is limited due to insufficient delivery to intracellular and extracellular targets and rapid degradation by nucleases.[3,4] Hence, there is a need for a delivery vehicle capable of administering siRNA efficiently, safely, and repeatedly to *in vitro* and *in vivo* milieu. Nanomedicine-based vectors have emerged as systems of choice for effective siRNA delivery.

The term "nanomedicine" is used to collectively mention vectors such as polymeric micelles, quantum dots, liposomes, polymer–drug conjugates, dendrimers, biodegradable nanocarriers, inorganic nanoparticles and other nanomaterials with therapeutic relevance. Of these vectors, polycationic nanocarriers or nanoparticles have found numerous applications in targeted drug and gene delivery. Due to their small size and condensed nature, nanocarriers provide better tissue penetration and targeting.[5] Nanocarriers prepared from polycationic polymers have been largely explored to deliver DNA and siRNA. These nanocarriers can be subcategorized into (1) nanospheres are spherical nanometer size particles where the desired molecules can be either entrapped inside the sphere or adsorbed on the outer surface or both; (2) nanocapsules have a solid polymeric shell and an inner liquid core and the desired molecules can be entrapped (Fig. 1). The fate of nanoparticles within the body is governed by the size, shape, surface charge and nature (hydrophobic or hydrophilic) of the polymer. Several natural and synthetic polycationic polymers including chitosan, polyethylenimine (PEI) have been explored for preparing nanoparticles to deliver siRNA.

Chitosan, an aminoglucopyran, is composed of randomly distributed *N*-acetylglucosamine and glucosamine residues. Owing to its versatile biological activity, biocompatibility, and biodegradability along with low toxicity, chitosan has emerged as a promising candidate of highly sophisticated functions.[6] To harness the unique properties and to tap the full potential of this versatile polysaccharide, enormous attempts have been made to functionalize and derivatize it. Chitin is the second most abundant natural biopolymer extracted from exoskeletons of crustaceans and from cell walls of fungi and insect. Structurally, chitin is a linear cationic heteropolymer of randomly distributed

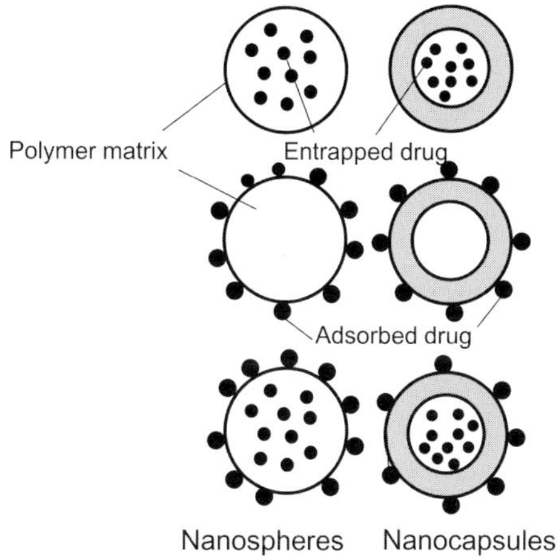

Fig. 1. Nanocarriers types: (a) Nanospheres, where the drug is either embedded in the polymer matrix or adsorbed onto the surface or both. (b) Nanocapsules, where the drug is either entrapped inside the hollow capsule or adsorbed onto the surface or both.

N-acetylglucosamine and glucosamine residues with β-1,4-linkage. Chitosan is produced by *N*-deacetylation of chitin in the presence of alkali (Fig. 2). Controlled derivatization of chitin results in chitosan with degree of deacetylation (DDA) between 40% to 98% and molecular weight (MW) between 5×10^4 Da and 2×10^6 Da.[7] The biological application of chitosan depends on DDA and degree of polymerization (DP), which also decides the MW of the polymer. Chitosan possess reactive hydroxyl and amino groups and is usually less crystalline than chitin. On heating, it degrades prior to melting, thus this polymer has no melting point. Chitosan is readily soluble in dilute acidic solutions below pH 6.0. It can be considered as a strong base as it possesses primary amino groups with a pKa value of 6.3. Due to the presence of amino groups, the charged state and properties of chitosan are dictated by the pH.[8] At low pH, the amino groups get protonated and become positively charged, making chitosan a water-soluble cationic polyelectrolyte. On the other hand, as the pH is increased above 6, chitosan amino groups become deprotonated; the polymer loses its charge and becomes insoluble. This transition between solubility and insolubility occurs at its pKa value between pH 6 and 6.5. Under physiological

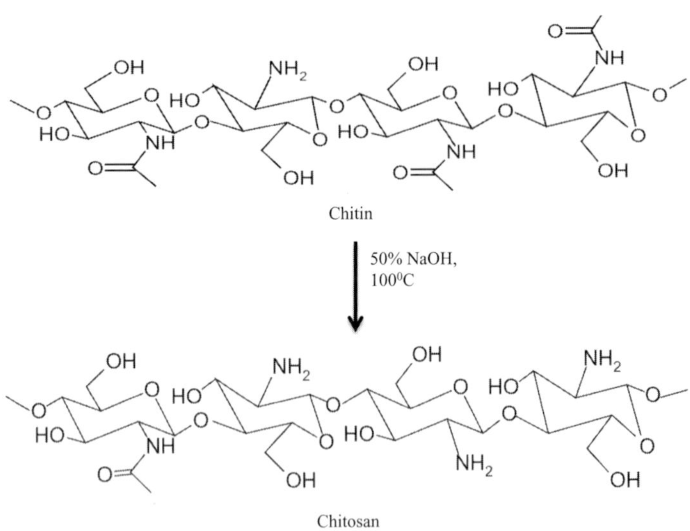

Fig. 2. Preparation of chitosan from chitin.

conditions, chitosan can be easily digested either by lysozymes or by chitinases, which can be produced by the normal flora in the human intestine or exist in the blood.[9–11] Owing to these properties, it has been widely employed for drug delivery both in pharmaceutical research and in industry.[12] Lately, chitosan has been investigated as a safer alternative to other non-viral vectors such as cationic lipids for siRNA delivery.

2. Strategies for Targeted siRNA Delivery

Different tissues or cell types can be specifically targeted using cell type-specific affinity moieties such as peptides or antibodies. Targeted delivery of siRNA to desired site not only increases its efficacy but also prevent possible off target side effects. The major step in targeting is that the cellular receptors should be efficiently internalized after binding to their ligands, followed by rapid re-appearance on the cell surface to allow repeated targeting. Another key parameter is to design ligands for cell surface receptors for siRNA delivery. Tissues at various tumor stages express a variety of cell surface protein, which can be exploited to modulate delivery system for target specificity. The various strategies for targeting siRNA delivery are discussed below in detail.

2.1. *Transferrin*

Appearance of transferrin (TF) is known to increase on the malignant cell surfaces in many types of tumors thereby providing a cell surface receptor that can be targeted. A recent study reported nanoparticles containing a linear cyclodextrin-based polymer (CDP), a human TF protein to bind TF receptors (TFR) on the surface of the cancer cells, PEG and siRNA to knockdown the expression of the M2 subunit of ribonucleotide reductase (RRM2).[13] Tumor biopsies from melanoma treated patients showed the presence of intracellularly localized nanoparticles and a reduction in both the specific mRNA and the protein levels of RRM2. Also, A/J mice bearing subcutaneous Neuro2A tumors treated with i.v. injection of siRNA against RRM2 containing TF conjugated CDP nanoparticles showed decrease in tumor growth.[14]

2.2. *Peptides*

Small peptides, such as a disulfide stabilized Arg-Gly-Asp (RGD) peptide targeting integrin expression, which is upregulated at sites of neovasculature, have been employed due to their specificity.[15] Self-assembling nanoparticles of PEGylated PEI with an RGD peptide ligand attached at the distal end of the PEG were prepared with siRNA against vascular endothelial growth factor receptor-2 (VEGF R2) expression and thereby tumor angiogenesis. RGD peptide conjugated to chitosan nanoparticles (RGD-CH-NP) by thiolation reaction was used to target alphanubeta3 integrin.[16] These nanoparticles were employed to deliver PLXDC1-targeted siRNA into the alphanubeta3 integrin positive tumor endothelial cells in the A2780 tumor-bearing mice.

2.3. *Antibodies*

Antibody-mediated targeted siRNA delivery has attracted much attention due to its high *in vivo* stability and specificity to the target cell surface protein. Chen *et al.* developed a LPH (liposome-polycation-hyaluronic acid) nanoparticle formulation modified with tumor targeting single chain antibody fragment (scFv) for systemic delivery of siRNA and microRNA into experimental lung metastasis of murine B16F10 melanoma. The siRNAs delivered by the scFv targeted nanoparticles efficiently downregulated the target genes (c-Myc/MDM2/VEGF) in the lung metastasis.[17] Monoclonal antibodies generated against oncogenes such as HER2 or VEGF have been widely used for cancer treatments. Recently, Gao *et al.* synthesized PEGylated DOPE immunoliposomes conjugated with the Fab' of recombinant humanized anti-HER2 monoclonal antibody (PIL) for siRNA delivery.[18] The results demonstrated that the lyophilized PIL (LPIL) prepared by the lyophilization/rehydration, containing 2.5% PEG showed the best HER1 gene silencing activity.

2.4. *Targeting Ligands*

Small molecules such as folate, carbohydrates and cholesterol have been reported as targeting moieties in several studies. Folate receptors (FRs) are over-expressed on the cell surfaces in many different types of cancers. To enhance the intracellular delivery of shRNA and reduce cytotoxicity, low MW PEI derivatized to folate-chitosan-graft-PEI (FC-g-PEI) copolymer.[19] Aerosol delivery of FC-g-PEI/Akt1 shRNA complexes suppressed lung tumorigenesis in a urethane induced lung cancer model mouse through the Akt signaling pathway. Kim *et al.* delivered antisense oligodeoxynucleotide (AS-ODN), synthetic siRNA and plasmid siRNA to inhibit green fluorescent protein (GFP) using PEI-PEG-folate conjugates in FRs over-expressing KB cells.[20] Among the three nucleic acids synthetic siRNA exhibited the most dose effective and fastest gene silencing effect.

The asialoglycoprotein receptor (ASGP-R) recognizes many different types of carbohydrates and is highly expressed in liver cells. Nie *et al.* employed galactosylated PEG-graft-PEI (Gal-PEG-PEI) to deliver psiRNA to knockdown HLA-E gene expression in HepG2 cells. Real time PCR analysis and western blot (WB) analysis revealed up to 60% inhibition of HLA-E gene expression, 48 hr post transfection.[21] Cholesterol and its analogues have also been exploited as targeting ligands as these are recognized by cholesterol receptors on hepatocytes. Watanabe *et al.* reported that i.v. delivery of apolipoprotein B specific siRNA with a sixth generation of dendritic poly(L-lysine) (KG6) resulted in siRNA mediated knockdown of ApoB in healthy C57BL/6 mice without hepatotoxicity and observed significant reduction in serum low-density lipoprotein cholesterol in apolipoprotein E deficient mice.[22]

3. Factors Dictating siRNA Delivery Efficiency of Chitosan-based Nanoparticles

Chitosan is one of the most widely investigated non-viral, naturally derived polymeric gene delivery vectors. The transfection efficiency of chitosan/DNA nanoparticles depends on several factors such as the DDA and MW of the chitosan, pH, protein interactions, N/P ratio, cell type, nanoparticle size and interactions with cells.[23] The DNA binding affinity and transfection efficiency have been found to increase with increase in DDA or MW while maximum expression levels are achieved by obtaining an intermediate stability through control of MW and DDA.[24,25] Recently, chitosan has been reported to deliver siRNA. The siRNA delivery efficacy of chitosan is dictated by various structural and experimental parameters. There are two methods of generating siRNAs for gene silencing studies. One involves the preparation of siRNAs via chemical or enzymatic synthesis. The other relies on siRNA expression vectors. Synthetic dsRNA, which mimics siRNA, generated by Dicer, can induce gene silencing processes. However, synthetic siRNA has a short half-life in cells, making its effect transient.

3.1. *Degree of Deacetylation*

The positive charge density of chitosan depends on the DDA when dissolved in acidic conditions. The DDA value represents the percentage of deacetylated primary amino groups along the molecular chain, which further determines the positive charge density of chitosan. The chitosan with higher DDA (DDA usually above 80%) has been shown to possess better siRNA binding capacity.[26] On the other hand, chitosan with low DDA has low charge interaction with siRNA; particles formed are unstable and showed low knockdown in H1299 human lung carcinoma cells.[26]

3.2. *Chitosan Molecular Weight and Concentration*

The physicochemical properties, i.e. size, zeta potential, morphology, complex stability and *in vitro* gene silencing of chitosan/siRNA nanoparticles is influenced by chitosan MW. Stability of nanoparticles is desirable for extracellular siRNA protection, however, decomplexation is essential for siRNA to mediate gene silencing through interaction with intracellular components such as RISC. Hence, a subtle balance between protection and release of siRNA is required for biological activity of siRNA delivered with chitosan.

The particle size of the chitosan/siRNA complex is dependent on chitosan MW. Katas *et al.* showed that a smaller mean particle size of chitosan nanoparticles was obtained when the lower MW of chitosan

(110 kDa) was used compared to the higher MW (270 kDa) chitosan.[27] Liu *et al.* investigated chitosan/siRNA nanoparticles formulated in MW range 8.9–173 kDa.[26] The size of the particles at low MW (8.9 kDa) was 3500 nm, which further reduced to 200 nm when chitosan of high MW (173 kDa) was employed. The chitosan molecules (MW 64.8–170 kDa), 5–10 times the length of the siRNA (MW of 13.36 kDa), could form stable complexes with siRNA through electrostatic forces resulting in high gene silencing efficiency in H1299 human lung carcinoma cells. On the other hand, chitosan with a low MW (10 kDa) could not condense siRNA into stable particles, resulting in the formation of large aggregates and almost negligible knockdown. Also, Ji *et al.* observed very small nanoparticles of 148 nm with low polydispersity index using higher MW chitosan (190–310 kDa).[28] These nanoparticles which incorporated siRNA specific for the *FHL2* gene reduced about 70% *FHL2* gene expression in human colorectal cancer Lovo cells. Hence, MW of chitosan could be considered an important parameter dictating its siRNA-mediated knockdown efficiency.

The chitosan concentration also influences the properties of the complexes. Mean particle size of chitosan-siRNA complexes prepared by simple complexation for both chitosan hydrochloride and glutamate was increased when the concentration of chitosan was increased from 25 to 300 μg/mL in distilled water.[27] Also, the comparative positive value of surface charge (zeta potential) of the chitosan-siRNA complexes increased with the increasing concentration of chitosan at a constant siRNA concentration.

3.3. *N/P Ratio*

For chitosan/siRNA nanoparticles, the N/P ratio is defined as the ratio of amino groups (N) of chitosan to phosphate groups (P) of siRNA. Howard *et al.* studied the influence of N/P ratio on size of chitosan/siRNA nanoparticles and reported higher particle size at lower N/P ratio.[29] For instance, at low concentrations (250 μg/mL) of chitosan, nanoparticles formed at N/P ratio of 7 measured 181.6 nm but increased to 223.6 nm at N/P ratio of 6. Further, Liu *et al.* investigated the influence of N/P ratio on the gene silencing efficacy of chitosan 170 kDa (DDA 84%) in H1299 human lung carcinoma cells and found that the level of enhanced green fluorescent protein (EGFP) silencing increased at higher N/P ratios (50 and 150) in comparison to low N/P ratio (2 and 10) formulations and nanoparticles formed at N/P 150 showed the greatest level (80%) of EGFP expression silencing.[26] The total chitosan used at high N/P ratios does not completely participate in the formation of the nanoparticles, and it was proposed that this proportion of chitosan more likely loosely associates with nanoparticles at high N/P and contributes to the improved stability and increased gene silencing. Moreover, removal of excess chitosan prior to transfection resulted in virtually no cellular knockdown, suggesting the requirement of stabilized particles for cellular entry or a possible role of excess chitosan in cellular permeation.

3.4. *pH of the System*

Although the pH of the system determines the chitosan charge, investigation of the influence of pH on the chitosan-siRNA nanoparticles size revealed that no significant difference in particle size was observed when complexing chitosan with siRNA in acetate buffer at pH 4.5 (0.1 M) compared to in distilled water for all the tested chitosans.[27] However, in a recent study by AFM, the interaction strengths between siRNA and chitosan were pH dependent, and the adhesive interactions decreased as the pH was increased from 4.1 to 6.1, 7.4, and 9.5, exhibiting distinct multimodal distributions of the interaction forces between siRNA and chitosan molecules at acidic pH; only negligible adhesive forces were observed at neutral or high pH. The strong pH dependence of siRNA–chitosan interactions can provide a convincing rationale for chitosan-siRNA complex formation and nanoparticle stability under low acidic conditions.[30] The influence of pH on the gene silencing activity of siRNA in cell culture has not been reported so far, and exhaustive studies are required to further understand this parameter. We have systematically examined the effect of pH on the physicochemical properties and gene delivery efficacy of chitosan-DNA nanoparticles.[31] Nanoparticles prepared with low MW chitosan

(10 kDa) suspended in double distilled water (pH 6.1) gave a uniform hydrodynamic diameter distribution of 243 ± 12 nm, which increased to 911 ± 39.6 nm and 1213 ± 84 nm in PBS pH 6.5 and pH 7.1, respectively. Cells incubated for 24 hr with complexes prepared from rhodamine labeled chitosan revealed the presence of large aggregates at pH 7.4 adhering to cell surfaces while RITC-chitosan fluorescence appeared to be more uniformly distributed on the cells at pH 6.5 (Fig. 3). Transfection efficiency expressed as % of cells expressing EGFP was 26.3% at pH 6.5 and then dropped considerably at higher pH of 7.1

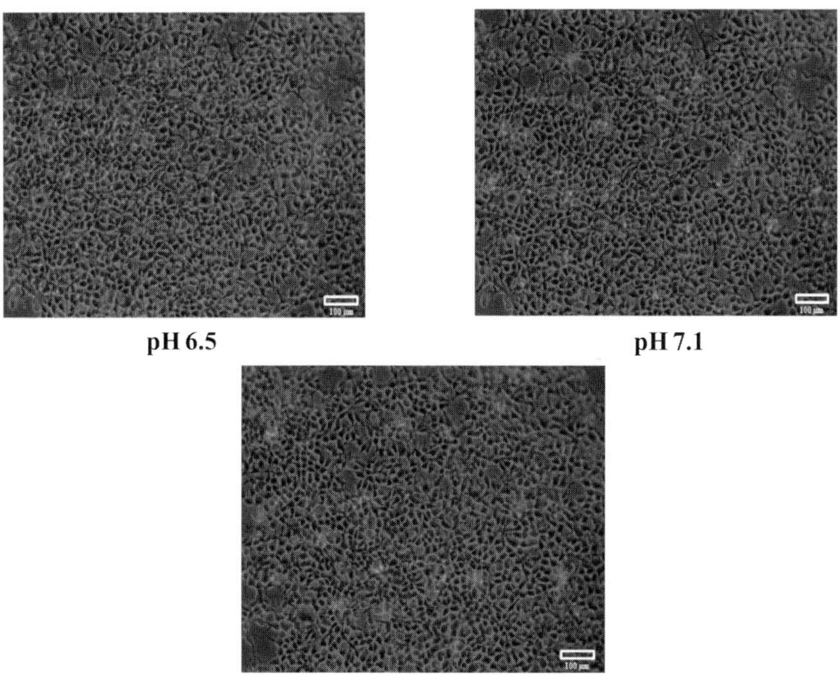

Fig. 3. Fluorescence microscope images of HEK 293 cells transfected at different pH. Cells were exposed to rhodamine labeled chitosan/DNA complexes, and 24 hr post-transfection media was exchanged with PBS and cells with internalized or adsorbed complexes visualized under fluorescent microscope at 10× magnification. Large aggregates present on cell surface at pH 7.4 while complexes were more uniformly distributed at lower pH 6.5.

and 7.4 with 9.2% and 0.2%, respectively.[31] Further, uptake of nanoparticles by HEK 293 cells, as assessed by flow cytometry, was pH dependent with maximum uptake occurring in medium at pH 6.5.

3.5. *Chitosan Salt Form*

The properties of the chitosan-siRNA nanoparticles are also dictated by chitosan salt form. Chitosan glutamate (G213, G113), which has higher MW than chitosan hydrochloride (Cl213, Cl113) produced smaller complexes with siRNA than chitosan hydrochloride.[27] Complete binding of siRNA with the chitosan was observed when the chitosan nanoparticles to siRNA weight ratio approached 100:1 except for lower MW chitosan hydrochloride, Cl113. Also, higher siRNA loading efficiency was observed for siRNA adsorbed onto chitosan glutamate (83%±0.9% for G213 and 90%±0.3% for G113) compared to chitosan hydrochloride (72%±1.1% for Cl213 and 59%±0.8% for Cl113).[27] Further, chitosan glutamate, G213 (470 kDa) showed the highest gene silencing effect at 24 hr post-transfection either by simple complexation (51% of gene knockdown) or ionic gelation (82% and 63% of gene knockdown for siRNA entrapment and siRNA adsorption, respectively) compared to chitosan hydrochloride in CHO K1 cells.

3.6. *Presence of Serum*

Stability against digestion by nucleases is highly desirable for siRNA to achieve maximal knockdown activity in cells. Katas *et al.* investigated the stability of chitosan-siRNA nanoparticles in the presence of serum by incubating free and chitosan complexed siRNA in 5% FBS at 37°C, respectively.[27] It was observed that siRNA was intact only up to 30 min, and it was fully degraded after 48 hr. In contrast, siRNA recovered from chitosan-TPP nanoparticles started to degrade after 24 hr incubation and fully degraded after 72 hr incubation. The experiment was further extended by incubating siRNA as well as chitosan-siRNA nanoparticles in 50% serum concentration. The chitosan-siRNA nanoparticles remarkably protected siRNA from nuclease activity. Complete degradation of siRNA was observed as early as time-point 0, with degradation occurring during the mixing of siRNA with serum and freezing steps. In contrast with unformulated siRNA, the siRNA recovered from chitosan-siRNA nanoparticles was intact up to 7 hr and fully degraded after 48 hr incubation in 50% serum.

3.7. *Methods of siRNA Association*

The association of siRNA with the chitosan has been observed to play an important role in the silencing effect. Katas *et al.* compared three methods of siRNA association: simple complexation, ionic gelation (siRNA entrapment) and adsorption of siRNA onto the surface of pre-formed chitosan nanoparticles for preparation of chitosan based nanoparticles.[27] Owing to their high binding capacity and loading efficiency, chitosan-TPP nanoparticles with entrapped siRNA were shown to be better vectors as siRNA delivery vehicles compared to chitosan-siRNA complexes in two different types of cell lines, CHO K1 and HEK 293. On the other hand, a low inhibition of gene expression was observed for nanoparticles prepared by simple complexation and adsorption of siRNA onto the preformed chitosan-TPP nanoparticles.

4. Therapeutic Applications of Chitosan/siRNA Complexes

From the ongoing research, it is evident that chitosan and chitosan-based vectors are safe and efficient gene delivery systems. But clinical trials are hampered due to their low transfection efficiency. Although, the formation of stable chitosan-siRNA nanoparticles is prerequisite for stability and extra cellular protection of siRNA, efficient disassembly is needed to allow RNA mediated gene silencing. Thus, a proper balance between protection and release plays a vital role in the biological activity of siRNA.[32]

Katas *et al.* seem to be the first group to report the use of chitosan to deliver siRNA *in vitro*.[27] Nanoparticles of chitosan-TPP entrapping siRNA were found to be better vectors as compared to chitosan-siRNA complexes possibly due to their high binding capacity and loading efficiency. On comparing the gene silencing activity at 24 hr and 48 hr post transfection, higher activity was obtained at 24 than 48 hr post transfection for the cells treated with siRNA associated with chitosan nanoparticles or complexes, suggesting the release of siRNA might occur within the first 24 hr for most of the tested chitosans. Howard *et al.* engineered novel chitosan-based siRNA nanoparticles which mediated knockdown of endogenous EGFP in both H1299 human lung carcinoma cells and murine peritoneal macrophages (77.9% and 89.3% reduction).[29] Further, the ability of the nanoparticles to knock down expression of the BCR/ABL-1 oncogene, found in chronic myelogenous leukemia, was tested to demonstrate the therapeutic potential of the chitosan system. The BCR/ABL-1-expressing cell line K562 was transiently transfected with nanoparticles (N/P 57) complexed with either a fusion-specific or a non-specific siRNA. Western blotting analysis using an antibody recognizing the N-terminal part of BCR, demonstrated ~90% *BCR/ABL-1* knockdown in cells transfected with nanoparticles containing BCR/ABL-1-specific siRNA. A detailed study showed that the physicochemical properties (size, zeta potential, morphology and complex stability) as well as *in vitro* gene silencing of chitosan/siRNA nanoparticles are dictated by chitosan MW and DDA.[26] High MW and DDA chitosan resulted in the formation of discrete stable nanoparticles ~200 nm in size. Chitosan/siRNA formulations (N/P 50) prepared with low MW (~10 kDa) showed almost no knockdown of *EGFP* in H1299 cells, whereas those prepared from higher MW (64.8–170 kDa) and DDA (~80%) showed gene silencing ranging between 45% and 65%. The highest gene silencing (80%) was achieved using chitosan/siRNA nanoparticles at N/P 150 using higher MW (114 and 170 kDa) and DDA (84%) that correlated with formation of stable nanoparticles of ~200 nm. In contrast, Rojanarata *et al.* reported maximal silencing (70–73%) of *EGFP* gene by chitosan-thiamine pyrophosphate complexes prepared from low MW chitosan (20 and 45 kDa) at a weight ratio of 80.[33] The observed high efficiency was attributed to increased siRNA binding and improved water solubility of chitosan owing to incorporation of extra amino groups from thiamine pyrophosphate, and salt formation between the phosphate group of thiamine pyrophosphate and the amino group of chitosan.[33]

Anderson *et al.* described specific and efficient knockdown of *EGFP* (approximately 70%) in H1299 human lung carcinoma cells transfected in plates pre-coated with a chitosan/siRNA formulation containing sucrose as lyoprotectant.[34] This methodology alleviates the necessity for siRNA complexation immediately prior to use and addition onto cells. Moreover, chitosan/siRNA formulation displayed silencing activity over the period studied (approximately two months) when stored at room temperature. Delivery of shRNA employing chitosan nanoparticles has also been reported. Potent knockdown of *TGFB1* by shRNA resulting in a decrease in rhabdomyosarcoma (RD) cell growth *in vitro* has been observed.[35] Controlled size chitosan nanoparticles (110–430 nm) were prepared using coacervation method in the presence of polyguluronate encapsulating siRNA.[36] These nanoparticles were not only efficient in delivering siRNA to HEK 293FT and HeLa cells but also showed low cytotoxicity. In another study, chitosan/siRNA nanoparticles were observed as irregular, lamellar and dendritic structures with a hydrodynamic radius size of about 148 nm and net positive charge with zeta potential value of 58.5 mV.[37] The knockdown effect of the chitosan/siRNA nanoparticles showed that FHL2 siRNA formulated within chitosan nanoparticles could knock down about 69.6% *FHL2* gene expression.

Although siRNA delivery employing polymers is quite new, still, several successful *in vivo* studies have been reported. The gene silencing efficacy of siRNA-polymer complexes is influenced by the route of administration. Significant RNA interference was observed after nasal administration of chitosan/siRNA formulations (37% and 43% reduction) in bronchiole epithelial cells of transgenic EGFP mice.[29] Intraperitoneal administration of chitosan/siRNA nanoparticles targeting *TNF*-α expression in systemic macrophages in mice have been observed to downregulate systemic and local inflammation in arthritis.[38] In another study, chitosan-TPP nanoparticles used to deliver shRNA to inhibit *TGFB1* expression reported a decrease in tumorigenicity in nude mice.[35] Glud *et al.* examined the pulmonary gene silencing effect of short interfering locked nucleic acids (siLNAs) targeting *EGFP* in lung bronchoepithelium. Chitosan/siLNA nanoparticles significantly reduced the EGFP protein expression after intranasal delivery.[39] The neuronal uptake and distribution of a chitosan-siRNA targeting specific muscarinic acetylcholine receptors (mAChRs) was investigated by intrathecal injection in rats.[40] The protein levels of mAChR subtypes in the spinal cord were significantly downregulated by siRNA treatment. In an attempt to make a stable and tumor homing vector, glycol chitosan (GC) and PEI were modified with 5 β-cholanic acid and mixed to form self-assembled GC-PEI nanoparticles (GC-PEI NPs), which were further complexed with siRNAs against red fluorescent protein (RFP) expression.[41] These nanoparticles showed considerable inhibition of *RFP* gene expression in RFP/B16F10 bearing mice, due to their higher tumor targeting ability. Tumor targeted chitosan nanoparticles were further prepared by attaching a RGD moiety to chitosan.[16] RGD-chitosan nanoparticles loaded with siRNA significantly increased selective intratumoral delivery in orthotopic animal models of ovarian cancer. In a recent study, aerosolized chitosan/siRNA nanoparticles used to dose transgenic EGFP mice showed significant *EGFP* gene silencing (68% reduction compared to mismatch group).[42]

Chitosan has been shown to improve the transfection efficiency of other vectors in siRNA delivery as a coating material. Pille *et al.* demonstrated that intravenous administration of encapsulated anti-RhoA siRNA in chitosan-coated polyisohexylcyanoacrylate nanoparticles to nude mice with xenografted aggressive breast cancers can significantly inhibit the tumor growth by at least 90%.[43] Further, a chitosan-coated PLGA nanoparticle has been reported to successfully bind the antisense oligonucleotide and to be taken up by A549 cells after 6 hr of incubation.[44]

5. Future Perspectives

Though numerous reports have been published on the development of efficient chitosan-based vectors for gene delivery, only limited reports are available regarding the application of chitosan and its derivatives for siRNA delivery. The delivery of siRNA is a multi-step process that requires bypassing a series of extra- and intracellular barriers. The vector should consist of three different functional moieties: (i) a cationic chitosan polymer component which efficiently condenses siRNA to form complexes, that facilitates endosomal escape of the complexes after cellular uptake, and provides the effective unpacking of the complexes in the cytoplasmic compartment; (ii) a hydrophilic component, such as PEG, which imparts solubility and stability to the complexes in the biological fluid; (iii) a cell- or tissue-specific ligand for enhanced targetability toward target cells or tissues and the efficient cellular uptake by receptor-mediated endocytosis. However, due to the shorter length and linearity of siRNA compared to DNA, it behaves differently. Studies disclosed that siRNA binds to chitosan in a different manner from that observed with pDNA.[29] Recent studies have also indicated that even for the same chitosan vector, different results can be obtained for siRNA delivery compared to that of DNA delivery. Moreover, the extra vulnerability of siRNA to nuclease degradation may offer additional challenges to chitosan-mediated siRNA transfer. Also, more *in vivo* studies need to be done before moving to clinical trials as the majority of reports published till date have only been conducted *in vitro* and in animal models.

6. Conclusion

The chitosan-mediated siRNA delivery has shown enormous potential to evolve as therapeutic modalities, despite the existing problems. The transfection efficiency is dictated by a series of formulation depend parameters, such as the MW of chitosan, DDA, N/P ratio, chitosan salt form, siRNA concentration, pH, serum, additives, preparation techniques of chitosan/nucleic acid particles and routes of administration. Taking into account the combined effects of the formulation parameters, it is expected that the complexes formed with high MW and high DDA would be highly stable and may lead to low or delayed transfection. On the other hand, complexes formed with a low MW and low DDA chitosan would not be sufficiently stable to transfect cells efficiently. Hence, there is a possibility that there is a range of intermediate values of MW and DDA that form complexes of intermediate stability and that transfect efficiently. Besides formulation optimization, the structural modification of chitosan or additives is an effective way to improve the stability of the polyplexes in biological fluids, enhance targeted cell delivery and facilitate endo-lysosomal release of the complex. Overall, the transfection efficiency of chitosan-based delivery systems can be adjusted by playing with formulation related parameters.

References

1. Zamecnik PC, Stephenson ML. (1978) Inhibition of Rous sarcoma virus replication and cell transformation by a specific oligodeoxynucleotide. *Proc Natl Acad Sci USA* **75:** 280–284.
2. Stephenson ML, Zamecnik PC. (1978) Inhibition of Rous sarcoma viral RNA translation by a specific oligodeoxyribonucleotide. *Proc Natl Acad Sci USA* **75:** 285–288.
3. Behlke MA. (2006) Progress towards *in vivo* use of siRNAs. *Mol Ther* **13:** 644–670.
4. Zhang S, Zhao B, Jiang H, *et al.* (2007) Cationic lipids and polymers mediated vectors for delivery of siRNA. *J Control Release* **123**: 1–10.
5. Peer D, Karp JM, Hong S, *et al.* (2007) Nanocarriers as an emerging platform for cancer therapy. *Nat Nano* **2:** 751–760.
6. Rinaudo M. (2006) Chitin and chitosan: Properties and applications. *Prog Polym Sci* **31:** 603–632.
7. Hejazi R, Amiji M. (2003) Chitosan-based gastrointestinal delivery systems. *J Control Release* **89:** 151–165.
8. Yi H, Wu LQ, Bentley WE, *et al.* (2005) Biofabrication with chitosan. *Biomacromolecules* **6:** 2881–2894.
9. Aiba S. (1992) Studies on chitosan: 4. Lysozymic hydrolysis of partially N-acetylated chitosans. *Int J Biol Macromol* **14:** 225–228.
10. Zhang H, Neau SH. (2002) *In vitro* degradation of chitosan by bacterial enzymes from rat cecal and colonic contents. *Biomaterials* **23:** 2761–2766.
11. Escott GM, Adams DJ. (1995) Chitinase activity in human serum and leukocytes. *Infect Immun* **63:** 4770–4773.
12. van der Lubben IM, Verhoef JC, Borchard G, Junginger HE. (2001) Chitosan and its derivatives in mucosal drug and vaccine delivery. *Eur J Pharm Sci* **14:** 201–207.
13. Davis ME, Zuckerman JE, Choi CHJ, *et al.* (2010) Evidence of RNAi in humans from systemically administered siRNA via targeted nanoparticles. *Nature* **464:** 1067–1070.
14. Bartlett DW, Su H, Hildebrandt IJ, *et al.* (2007) Impact of tumor-specific targeting on the biodistribution and efficacy of siRNA nanoparticles measured by multimodality *in vivo* imaging. *Proc Natl Acad Sci USA* **104:** 15549–15554.

15. Schiffelers RM, Ansari A, Xu J, *et al.* (2004) Cancer siRNA therapy by tumor selective delivery with ligand-targeted sterically stabilized nanoparticle. *Nucleic Acids Res* **32:** e149.
16. Han HD, Mangala LS, Lee JW, *et al.* (2010) Targeted gene silencing using RGD-labeled chitosan nanoparticles. *Clin Cancer Res* **16:** 3910–3922.
17. Chen Y, Zhu X, Zhang X, *et al.* (2010) Nanoparticles modified with tumor-targeting scFv deliver siRNA and miRNA for cancer therapy. *Mol Ther* **18:** 1650–1656.
18. Gao J, Sun J, Li H, *et al.* (2010) Lyophilized HER2-specific PEGylated immunoliposomes for active siRNA gene silencing. *Biomaterials* **31:** 2655–2664.
19. Jiang HL, Xu CX, Kim YK, *et al.* (2009) The suppression of lung tumorigenesis by aerosol-delivered folate-chitosan-graft-polyethylenimine/Akt1 shRNA complexes through the Akt signaling pathway. *Biomaterials* **30:** 5844–5852.
20. Kim SH, Mok H, Jeong JH, *et al.* (2006) Comparative evaluation of target-specific GFP gene silencing efficiencies for antisense ODN, synthetic siRNA, and siRNA plasmid complexed with PEI-PEG-FOL conjugate. *Bioconjug Chem* **17:** 241–244.
21. Nie C, Liu C, Chen G, *et al.* (2010) Hepatocyte-targeted psiRNA delivery mediated by galactosylated poly(ethylene glycol)-graft-polyethylenimine *In Vitro*. *J Biomater Appl* **26(3):** 255–275.
22. Watanabe K, Harada-Shiba M, Suzuki A, *et al.* (2009) *In vivo* siRNA delivery with dendritic poly(l-lysine) for the treatment of hypercholesterolemia. *Mol Biosyst* **5:** 1306–1310.
23. Mansouri S, Cuie Y, Winnik F, *et al.* (2006) Characterization of folate-chitosan-DNA nanoparticles for gene therapy. *Biomaterials* **27:** 2060–2065.
24. Lavertu M, Methot S, Tran-Khanh N, Buschmann MD. (2006) High efficiency gene transfer using chitosan/DNA nanoparticles with specific combinations of molecular weight and degree of deacetylation. *Biomaterials* **27:** 4815–4824.
25. Ma PL, Lavertu M, Winnik FM, Buschmann MD. (2009) New insights into chitosan-DNA interactions using isothermal titration microcalorimetry. *Biomacromolecules* **10:** 1490–1499.
26. Liu XD, Howard KA, Dong MD, *et al.* (2007) The influence of polymeric properties on chitosan/siRNA nanoparticle formulation and gene silencing. *Biomaterials* **28:** 1280–1288.
27. Katas H, Alpar HO. (2006) Development and characterisation of chitosan nanoparticles for siRNA delivery. *J Control Release* **115:** 216–225.
28. Ji AM, Su D, Che O, *et al.* (2009) Functional gene silencing mediated by chitosan/siRNA nanocomplexes. *Nanotechnology* **20:** 405103.
29. Howard KA, Rahbek UL, Liu XD, *et al.* (2006) RNA interference in vitro and *in vivo* using a chitosan/siRNA nanoparticle system. *Mol Ther* **14:** 476–484.
30. Xu S, Dong M, Liu X, *et al.* (2007) Direct force measurements between siRNA and chitosan molecules using force spectroscopy. *Biophys J* **93:** 952–959.

31. Nimesh S, Thibault MM, Lavertu M, Buschmann MD. (2010) Enhanced gene delivery mediated by low molecular weight chitosan/DNA complexes: effect of pH and serum. *Mol Biotechnol* **46:** 182–196.
32. Kim TH, Jiang HL, Jere D, *et al.* (2007) Chemical modification of chitosan as a gene carrier *in vitro* and *in vivo*. *Prog Polym Sci* **32:** 726–753.
33. Rojanarata T, Opanasopit P, Techaarpornkul S, *et al.* (2008) Chitosan-thiamine pyrophosphate as a novel carrier for siRNA delivery. *Pharm Res* **25**: 2807–2814.
34. Andersen MØ, Howard KA, Paludan SR, *et al.* (2008) Delivery of siRNA from lyophilized polymeric surfaces. *Biomaterials* **29:** 506–512.
35. Wang SL, Yao HH, Guo LL, *et al.* (2009) Selection of optimal sites for TGFB1 gene silencing by chitosan-TPP nanoparticle-mediated delivery of shRNA. *Cancer Genet Cytogenet* **190:** 8–14.
36. Lee DW, Yun KS, Ban HS, *et al.* (2009) Preparation and characterization of chitosan/polyguluronate nanoparticles for siRNA delivery. *J Control Release* **139:** 146–152.
37. Yu JH, Quan JS, Kwon JT, *et al.* (2009) Fabrication of a novel core-shell gene delivery system based on a brush-like polycation of alpha, beta-poly (L-aspartate-graft-PEI). *Pharm Res* **26:** 2152–2163.
38. Howard KA, Paludan SR, Behlke MA, *et al.* (2009) Chitosan/siRNA nanoparticle-mediated TNF-alpha knockdown in peritoneal macrophages for anti-inflammatory treatment in a murine arthritis model. *Mol Ther* **17:** 162–168.
39. Glud SZ, Bramsen JB, Dagnaes-Hansen F, *et al.* (2009) Naked siLNA-mediated gene silencing of lung bronchoepithelium EGFP expression after intravenous administration. *Oligonucleotides* **19:** 163–168.
40. Cai YQ, Chen SR, Han HD, *et al.* (2009) Role of M2, M3, and M4 muscarinic receptor subtypes in the spinal cholinergic control of nociception revealed using siRNA in rats. *J Neurochem* **111:** 1000–1010.
41. Huh MS, Lee SY, Park S, *et al.* (2010) Tumor-homing glycol chitosan/polyethylenimine nanoparticles for the systemic delivery of siRNA in tumor-bearing mice. *J Control Release* **144:** 134–143.
42. Nielsen EJ, Nielsen JM, Becker D, *et al.* Pulmonary gene silencing in transgenic egfp mice using aerosolised chitosan/siRNA nanoparticles. *Pharm Res* DOI 10.1007/s11095-010-0255-y.
43. Pille JY, Li H, Blot E, *et al.* (2006) Intravenous delivery of anti-RhoA small interfering RNA loaded in nanoparticles of chitosan in mice: safety and efficacy in xenografted aggressive breast cancer. *Hum Gene Ther* **17:** 1019–1026.
44. Nafee N, Taetz S, Schneider M, *et al.* (2007) Chitosan-coated PLGA nanoparticles for DNA/RNA delivery: effect of the formulation parameters on complexation and transfection of antisense oligonucleotides. *Nanomedicine* **3:** 173–183.

Chapter 23

Application of Polyethylenimine-Based Nanoparticles for RNA Therapeutics

Surendra Nimesh* and Ramesh Chandra†,‡

Abstract

RNA interference is a post-transcriptional gene silencing mechanism mediated by double-stranded small interfering RNA (siRNA). The application of siRNA as a therapeutic modality is rapidly emerging as a novel approach for the treatment of several intractable diseases. Although therapeutic use of siRNA evolved at a rapid pace, the systemic application is still ambiguous due to the limitations of siRNA, such as low enzymatic stability, cellular internalization and body distribution after systemic administration. Polymeric nanoparticles have been investigated as an alternative choice with minimal toxicity and high efficiency. These vectors not only protect siRNA from enzymatic degradation by forming condensed complexes but also lead to tissue and cellular targeting, improve cellular penetration, and release of siRNA in the right intracellular compartment. This chapter describes

* Department of Biotechnology, Central University of Rajasthan, Bandar Sindri NH-8, Kishangarh, Ajmer (Rajasthan), 305801, India.
Corresponding author email: surendranimesh@curaj.ac.in
† Department of Chemistry, University of Delhi, Delhi – 110007, India.
‡ Dr.B.R. Ambedkar Center for Biomedical Research, University of Delhi, Delhi – 110007, India.

the recent progress and strategies of siRNA delivery systems using cationic polymer, polyethylenimine-based nanoparticles.

1. Introduction

Successful clinical application of siRNA is limited due to problems such as: (1) rapid enzymatic degradation resulting in a short half-life in the blood; (2) poor cellular uptake; (3) insufficient tissue bioavailability.[1–3] Because of these limitations, native delivery of siRNA to the cells is not effective. The siRNA can be chemically modified to circumvent these problems, but these modifications are also associated with certain disadvantages, such as decreased mRNA hybridization, higher cytotoxicity and increased unspecific effects.[4] Therefore, a delivery system is required which can protect and efficiently transport siRNA to the cytoplasm of the target cells. The majority of approaches tested so far have utilized viral vectors, but recently, the research on non-viral vectors has gained momentum as they offer several advantages, such as easy manipulability, stability, safety, low cost and high flexibility regarding the size of transgene delivered.[5] Among various delivery systems, polycationic polymers have been preferred over other systems due to their ease of preparation, purification and chemical modification along with the long shelf life.[6,7]

Polyethylenimine (PEI), often considered the gold standard of gene transfection, is one of the most widely explored cationic polymers for gene delivery. Since the first successful application by Behr *et al.* of PEI-mediated oligonucleotide delivery, it has been further derivatized to improve the physicochemical and biological properties of polyplexes.[8,9] Its high cation density PEI (a positive charge per 43 Da, which is the monomer's MW) also contributes to the formation of highly condensed particles by interacting with nucleic acids. Depending on the arrangement of the repeating ethylenimine units, PEI occurs in branched and linear morphological isomers. Branched PEI (BPEI) is synthesized by acid-catalyzed polymerization of aziridine, whereas linear PEI (LPEI) is prepared via ring opening polymerization of 2-ethyl-2-oxazoline followed by hydrolysis (Fig. 1).[10,11]

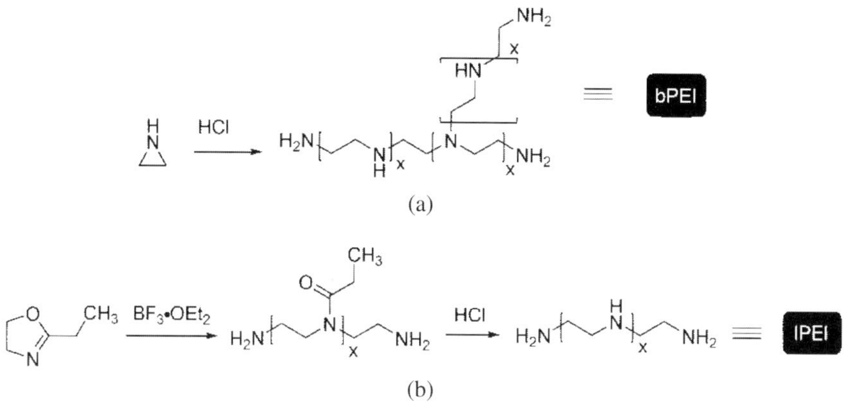

Fig. 1. Synthesis of PEI by (a) Acid-Polymerization of aziridine to yield branched PEI; (b) Ring-Opening polymerization of 2-ethyl-2-oxazoline followed by hydrolysis to yield linear PEI.

LPEI consists of secondary amines in its backbone except the terminal primary groups. On the other hand, BPEI contains primary, secondary and tertiary amino groups at the estimated ratio of 1:2:1.[12] The different types of amine group have different pKa values and could be protonated at different levels at a given pH. This confers PEI with a superior buffering capacity spreading over a wide range of pH. PEI uses the "proton sponge" mechanism to promote the release of endocytosed polyplexes from the endosomes (Fig. 2).[13–16] According to this mechanism, the unprotonated amines with different pKa values confer a buffering effect over a wide range of pH. This buffering may protect the DNA from degradation in the endosomal compartment during the maturation of the early endosomes to late endosomes and their subsequent fusion with the lysosomes. The buffering property also allows the polycation PEI to escape from the endosome: at lower pH, the buffering by PEI causes an influx of chloride ions and water into the endosomes, which eventually burst due to increased osmotic pressure, thus facilitating intracellular release of PEI-DNA polyplexes.

PEIs, ranging from low molecular weight to higher ones, have extensively been exploited as effective gene delivery vehicles.[17–21] They

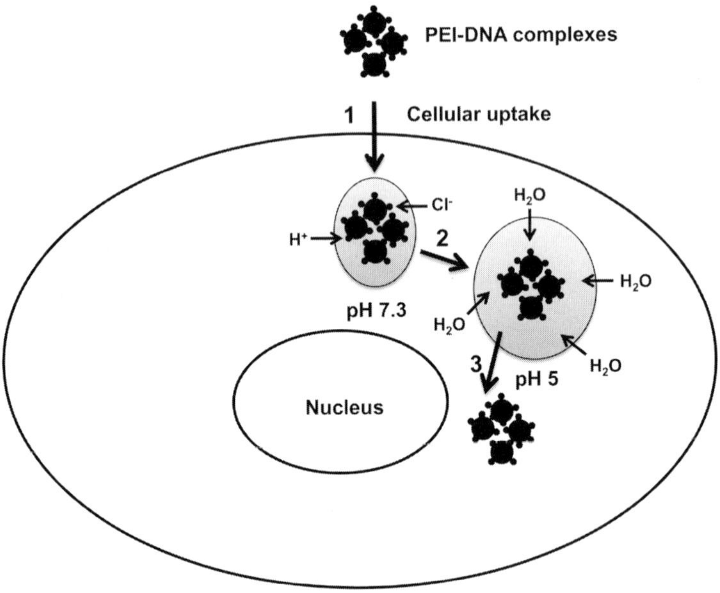

Fig. 2. Schematic representation of the "proton sponge effect": the initial step is endocytosis of the cationic complexes (1), followed by acidic endosome buffering (2), which leads to increased osmotic pressure and finally to lysis (3).

have also been shown to be efficient versatile agents for *in vivo* gene delivery via a number of routes.[22–24] The transfection efficiency of PEI is directly related to size of the polymer and the charge-associated cytotoxicity.[18,25–27] High molecular weight PEI is reported to exhibit superior transfection efficiency compared to low molecular weight PEI. However, the cytotoxicity limits its applications. It has been earlier reported that of linear and branched PEIs, the latter are more toxic and less efficient for transfection, particularly at higher N/P ratios.[28]

Recently, PEIs have been employed for siRNA delivery. Although PEI is highly efficient for DNA delivery, its efficacy in siRNA delivery is considered to be far less pronounced.[29] The reduced efficacy of PEI in the later case arises due to the short length of siRNA. The weaker electrostatic interaction between the negatively charged siRNA and the cationic PEI polymer plays crucial role in dissociation of PEI-siRNA complexes at the anionic cell surface.[30] However, the availability of LPEI and BPEI in a wide range of molecular weights and their implications in various studies to establish the optimal requirements is promising. The most suitable MW of PEI for complexation with DNA is between 5 and 25 kDa, although some groups reported that high (800 kDa) and low (2 kDa) MW PEIs also showed good transfection at their preferred N/P ratios. Hence, appropriate design of cationic vector having PEI not only caters to the cytotoxicity concern, but could also provide sufficient protection against serum or RNase through efficient complexation with siRNA.

2. Basic Principle and Mechanism of RNA Interference

The process of RNAi takes place in the cytoplasm of the cells (Fig. 3). The mechanism of RNAi is a multistep process which initiates with the binding of dsRNA-specific endonuclease (a cytoplasmic ribonuclease III (RNase III)-like protein) called Dicer, which is capable of cleaving long dsRNAs.[31] This binding of Dicer is followed by cleavage of long dsRNA into smaller duplexes with 19 paired nucleotides and two nucleotide overhangs at both 3′-ends, and these small dsRNAs hence generated are called siRNAs.[32] These siRNA molecules further bind to a nuclease containing multi-protein complex called RNA induced silencing complex (RISC).[33] The RISC complex by virtue of its RNA helicase activity unbinds the double stranded siRNA molecule; the sense strand removes which is further degraded by cellular nucleases.[34] The antisense strand of siRNA remains bound to the RISC complex and is directed to the target mRNA sequence, where it anneals complementarily by Watson–Crick base pairing. Finally, the RISC complex cleaves the target mRNA by endonucleolytic activity, thereby preventing its translation into protein.[35] At a later stage, the cleaved products are released and degraded by cellular nucleases, leaving the unengaged RISC complex to further search for more target mRNAs and carry out its activity of interference.[36]

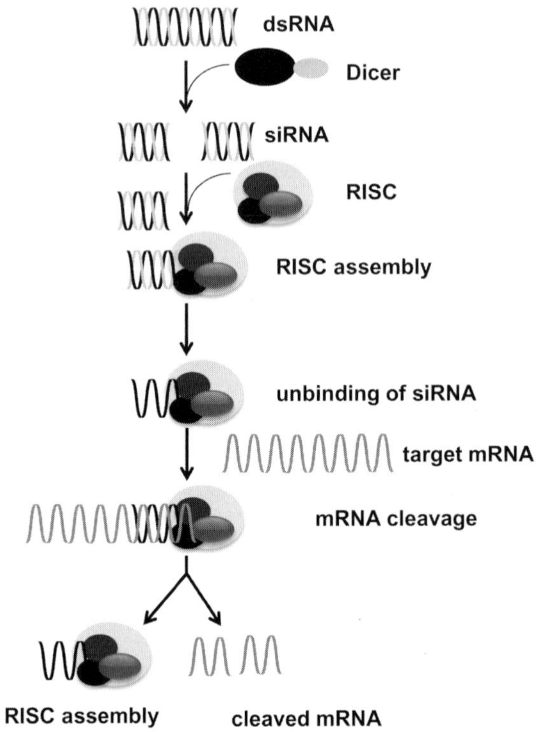

Fig. 3. Mechanism of RNAi. dsRNA molecules are cleaved by Dicer yielding siRNA. siRNAs gets incorporated into RISC assembly, followed by unwinding of the double-stranded molecule by the helicase activity of RISC. The sense strand of siRNA is removed and the antisense strand binds targeted mRNA, which is then cleaved by RISC and subsequently degraded by cellular nucleases.

3. *In Vitro* and *in Vivo* Application of PEI in siRNA delivery

A subtle balance between hydrobhobicity and hydrophilicity is necessary for the vector to cross the plasma membrane barrier of the cells as well as to reduce the cytotoxicity. However, reduction of the net cationic charge density of the polymers should be moderate, and the resulting surface charge must be positive enough to allow the formation of stable polyplexes with siRNA and to exert proton sponge effect to facilitate endosomal release of polyplexes. Hence, there should be an appropriate balance between the introduced hydrophobic character and the reduced positive charge. Oskuee *et al.* grafted alkylcarboxylate residues onto 25 kDa PEI to investigate the structural influence of the different alkyl modifications on the biophysical properties and bioactivity of the formulation, aiming to improve the efficacy of PEI-based carrier systems for siRNA delivery.[37] Studies on various formulations having varied degree of alkylcarboxylation revealed that at low degree of carboxylation (<20%) buffering capacity leading to endosomal escape remained efficient. Also, the increase of the hydrophobic alkyl chain length has been reported to result in improved complex stability with siRNA.[38] The toxicity of modified PEI derivatives reduced significantly in comparison to PEI 25 kDa, owing to the introduction of the additional negative charges lowering the strong interaction potential of PEI with cell surfaces and associated membrane damage. Further, carboxyalkylation of PEI was found to greatly improve siRNA-mediated luciferase gene knockdown.

Further, to overcome instability and low transfection efficiency, PEG (2 kDa)-PEI (25 kDa) was synthesized and investigated as a non-viral carrier of siRNA targeting CD44v6 in gastric carcinoma cells.[39] Physicochemical studies revealed that the size and potential of the complexes below N/P 10 were not suitable for cell transfection because PEG-PEI and siRNA could not form a complex of adequate size and positive potential. Further, complete complexation of siRNA was observed with PEG-PEI at N/P ratios ≥10. Transfection efficiency of PEG-PEI/siRNA was observed to increase at higher N/P ratios, reaching the maximal transfection efficiency at N/P 15. However, no major increase appeared at N/P 30 possibly because over-condensation at high N/P values is not favorable for intracellular release of nucleic acids. Veiseh *et al.* developed a magnetic nanoparticle platform consisting of a superparamagnetic iron oxide (Fe_3O_4) core coated with a cationic copolymer of chitosan-grafted-PEG and PEI for siRNA delivery.[40] To provide site specificity, targeting peptide chlorotoxin was covalently attached to the nanoparticles. This multifunctional magnetic vector was found to deliver siRNA to brain tumor cells through receptor mediated endocytosis and specifically knockdown the transgene expression of GFP in C6/GFP + glioma cells. Creusat *et al.* investigated mechanism of PEI-based siRNA delivery.[41]

Results revealed that the underlining proton sponge property is key to the efficacy of tyrosine-PEI conjugate, as it may allow both endosomal rupture and siRNA liberation via an optimal pH-sensitive dissolution of the PEI self aggregates. In one of our study, we prepared an ionic complex of BPEI (750 kDa) and alginic acid to amalgamate the properties of polycationic PEI with polysaccharide alginate.[42] The content of alginic acid was varied systematically to obtain a series of nanocomposites. As the concentration of alginic acid in nanocomposites was increased, the zeta potential and size of nanocomposites decreased accordingly. Zeta potential profile showed that the excess surface charge on PEI (22 mV) is effectively masked in nanocomposites. We observed maximum transfection efficiency in case of PEI-alginate (6.26%) at w/w ratio 20:1, which was found to be 2–16-fold higher than the native PEI on various cell types. Further, these PEI-alginate (6.26%) nanocomposites efficiently delivered siRNAs into mammalian cells, resulting in 80% suppression of GFP expression (Fig. 4).

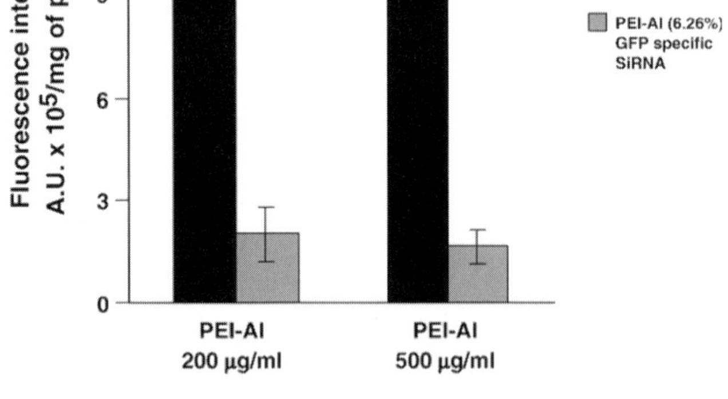

Fig. 4. PEI-alginate (6.26%) as an effective carrier of siRNA. PEI-alginate (6.26%) was tested for its ability to deliver GFP specific siRNA into COS-1 cells. The expression of GFP in cell lysates was reduced by 80% as monitored by measuring fluorescence on spectrofluorometer.

In another published study, we prepared nanoparticles of PEI by acylating PEI with propionic anhydride followed by cross-linking with PEG-bis(phosphate).[43] The nanoparticle size as revealed by DLS studies was found to be ~110 nm, and AFM investigations showed spherical and compact complexes with an average size of 100 nm. The qualitative analysis of level of gene silencing was carried out by observing under an inverted microscope after 48 hr (Fig. 5(a)). *In vitro* siRNA delivery studies in HEK 293 cells with nanoparticles showed up to 85% inhibition of GFP gene expression, which was almost equal to that for siRNA-Lipofectin (81% inhibition) (Fig. 5(a)).

Later, to integrate low cytotoxicity and higher transfection efficiency, Kim *et al.* developed water-soluble lipopolymer (WSLP) by conjugating the cationic head group of low MW BPEI (1.8 kDa) with a hydrophobic lipid anchor, cholesterol chloroformate.[44] The corresponding complex with siRNA was designed to inhibit human VEGF expression and was investigated as a potential siRNA delivery vehicle. It readily formed nano-sized complexes (~100 nm) with siRNA and protected siRNAs from enzymatic degradation in serum-conditioned media. WSLP/siRNA complexes transfected in human prostate cancer (PC-3) cells derived from human prostate adenocarcinomas and inhibited the VEGF production significantly, while complexes of WSLP with control siRNA did not show this inhibitory effect. WSLP/siRNA complexes reduced the VEGF production by 40% when compared with unmodified BPEI. Moreover, WSLP/siRNA complexes reduced tumor volume by 55% at 21 days, and by 65% at 28 days when compared with controls. These results indicate that WSLP has potential as a siRNA delivering agent and can be applied for antiangiogenic tumor therapy.

Further, to develop a multifunctional delivery system which could facilitate target specific treatment of tumor, Biswal *et al.* investigated the FR-mediated delivery of dihydrofolate reductase (DHFR) siRNA to silence the DHFR gene in FR positive KB cells.[45] A DHFR siRNA sequence was cloned into a pSUPER-RNAi vector and complexed with the folate-PEG-PEI (25 kDa) (FOL-PEG-PEI) conjugate. They have carried out the complexation of DHFR siRNA expressing pDNA and FOL-PEG-PEI conjugate, and characterized

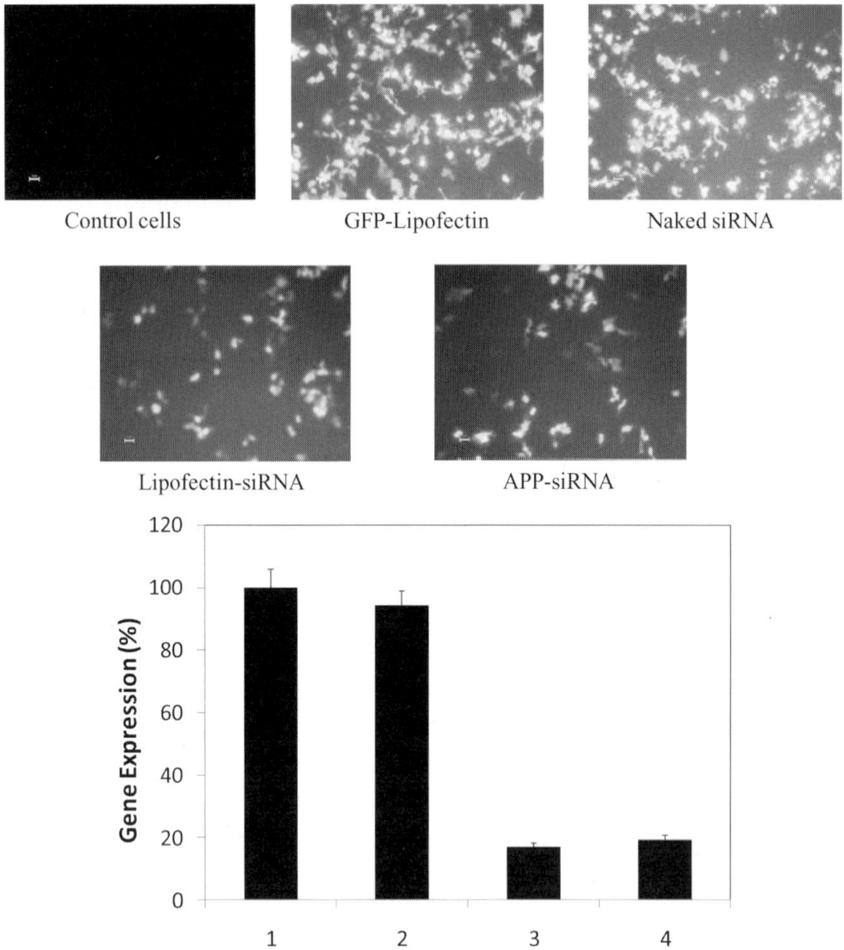

Fig. 5. Comparison of gene silencing efficiency of various siRNA formulations after 48 hr. (a) The GFP expression was observed under fluorescent microscope at 10×magnification. (b) The level of GFP expression was estimated by quantitation of green fluorescence after 48 hr. The data was recorded at optimal inhibition efficiency for APP nanoparticles, i.e. APP nanoparticles: siRNA at 30:1 ratio.

the FOL-PEG-PEI/pSUPER-siDHFR complex by particle size analyzer, gel retardation, and DNase protection assay. The complex was transfected to FRs overexpressing human epidermal carcinoma (KB) and FRs negative human lung carcinoma (A549) cells. The transfection studies by fluorescence microscopy and RT-PCR showed that the complex delivered the siRNA vector and inhibited DHFR gene in KB cells; however, it remained unaffected when applied to A549 cells as control. Therefore, FR-mediated delivery of siDHFR complexed with FOL-PEG-PEI conjugate inhibits the DHFR expression in FR positive cells alone. The target specific delivery was further substantiated by the fact that lipofectamine-mediated transfection of pSUPER-siDHFR delivered the vector and inhibited the DHFR gene in both KB and A549 cells. Target specificity was also confirmed by receptor blocking studies using free folic acid. This strategy can be extended to a wide range of FR-targeted drug delivery and gene silencing therapeutics by siRNA expression pDNA.

Several targeting moieties such as galactose and pullulan have been utilized in liver targeted gene delivery owing to their preferential and high binding ability with ASGP-R. Pullulan is a water-soluble polysaccharide consisting of three α-1,4-linked glucose polymers with different α-1,6-glucosidic linkages. Kang *et al.* introduced pullulan into PEI for liver targeting.[46] They developed a delivery system of pullulan-containing PEI/siRNA complexes for delivery into mice through the tail vein either by a hydrodynamics or nonhydrodynamics-based injection. During systemic injection, the PEI/fluorescein-labeled siRNA complex increased the level of fluorescence in the lung, whereas PEI-pullulan/siRNA complex led to an increased fluorescence level in the liver. Further, an increase in N/P ratio of PEI/siRNA complexes resulted in higher mice mortality, but introduction of pullulan into PEI dramatically reduced mouse death after systemic injection. Therefore, PEI-pullulan polymeric conjugate provides a useful, low toxic approach to efficient delivery of siRNA into the liver.

To engineer a stable and tumor-homing nano-sized carrier, a new nano-sized siRNA carrier system composed of biocompatible/biodegradable glycol chitosan polymer (GC) and PEI was modified with hydrophobic 5β-cholanic acid, and were simply mixed to form self-assembled GC-PEI nanoparticles (GC-PEI).[47] The freshly prepared GC-PEI nanoparticles showed a stable nanoparticle structure (350 nm) and they presented a strongly positive-charged surface (ζ potential = 23.8) that is enough to complex tightly with negatively charged

RFP-siRNAs, designed for inhibiting RFP (red fluorescent protein) expression. The siRNA encapsulated nanoparticles (siRNA-GC-PEI) formed more compact and stable nanoparticle structures (250 nm) at 1:5 weight ratio of siRNA to GC-PEI nanoparticles. Uptake studies done in RFP expressing B16F10 tumor cells (RFP/B16F10) showed that siRNA-GC-PEI nanoparticles showed a rapid time-dependent cellular uptake profile within 1 hr. Moreover, the internalized siRNA-GC-PEI nanoparticles lead to specific mRNA break down. Furthermore, the new formulation of siRNA-GC-PEI nanoparticles presented a significant inhibition of RFP gene expression of RFP/B16F10-bearing mice, due to their higher tumor-targeting ability. These results revealed the promising potential of GC-PEI nanoparticles as a stable and effective nano-sized siRNA delivery system for cancer treatment.

Lee *et al.* suggested reducible polymerized siRNA as a possible solution for low delivery efficiency of siRNA. Dithiol-modified RFP siRNAs at the 5'-ends of both sense and anti-sense strands were disulfide-polymerized. Polymerized siRNA (poly-siRNA) was composed of 30% oligomeric siRNA (50~300 bps) and 66% polymeric siRNA (above ~300 bps) as fractions, and was reducible in reducing solution through disulfide bond cleavage. Poly-siRNA formed more condensed and nano-sized complexes with low MW PEI (1800 Da) by strong electrostatic interaction based on the higher charge density of poly-siRNA, compared with siRNA (mono-siRNA). The compact poly-siRNA/PEI complexes prevented the loss and degradation of siRNA from a polyanion competitor and RNases in serum. Furthermore, poly-siRNA/PEI complexes exhibited superior intracellular uptake by murine melanoma cells (B16F10), and was accompanied with RFP gene silencing efficiency of about 80% compared to untreated cells. These results sufficiently support that strong polyanionic and reducible poly-siRNA can be utilized as a novel powerful therapeutic strategy for human diseases.

Along with tumor models, siRNAs complexed with the linear jetPEI/*in vivo* jetPEI were used for the treatment of other pathologies as well. In a study on the post-exposure protection of guinea pigs against a lethal Ebola virus challenge, the intraperitoneal injection of PEI/siRNA led to targeting of the polymerase (L) gene of the Zaire species of EBOV and protective effects due to a significant reduction in plasma viremia levels.[48] Intraperitoneal application of PEI/siRNA complexes led to the downregulation of hypoxia inducible factor 1α and plasminogen activator inhibitor 1 and to a statistically significant reduction of post-operative abdominal adhesion formation.[49] Ge *et al.* demonstrated that the administration of PEI-complexed siRNAs specific for the conserved regions of influenza virus genes allows the treatment and prevention of lethal influenza infections in the mouse. More specifically, upon administration of the complexes before or after initiation of the virus infection, reduced virus production in the lungs of infected mice was observed.[50] Concomitantly, the lung was identified as one preferential organ of PEI/siRNA complex delivery upon their intravenous injection.[50] This is mainly attributed to the fact that the lung is reached rather quickly after i.v. injection and that the physical properties of the complexes allow efficient interaction with the pulmonary vasculature. Additionally, a certain degree of PEI complex aggregation under physiological conditions may be beneficial for efficient cell transfection in the lung. Intravenous injection was employed for targeting IL-13 as well. PEI-complexed, chemically modified siRNAs were administered to sensitized mice just before airway challenge with allergen, and the treatment significantly reduced airway resistance in the sensitized and challenged mice, suggesting that PEI/siRNA-mediated IL-13 knockdown may prevent the induction of allergen-induced airway dysfunction.[51] Another study showed that systemic delivery of complexes based on fully deacetylated PEI resulted in a marked improvement of siRNA delivery to the lung as well as in reduced toxicity. This was demonstrated by the downregulation of luciferase as a model gene or of the influenza viral nucleocapsid protein gene with a subsequent 94% drop of virus titers in the lungs of influenza-infected animals.[52] In order to decrease toxicity and enhance bioactivity, Werth *et al.* described the purification of a low MW PEI (PEI-F25-LMW) from the commercially available 25 kDa PEI through size-exclusion chromatography.[53] Complexes based on PEI-F25-LMW can be stored lyophilized or frozen, thus avoiding the need to prepare complexes freshly and allowing the preparation of standardized aliquots.[53,54]

4. Degradable PEI for siRNA Delivery

Though high MW PEI (25 kDa) is one of the most potent polymeric vectors because of its high pH-buffering capacity for endosomal escape, it lacks degradable linkages and is too toxic for therapeutic applications. Hence, low MW PEI has been suggested as an alternative to high MW PEI (25 kDa). Also, to overcome the poor transfection efficiency associated with low MW PEI, degradable PEIs consisting of low MW PEIs and degradable cross-linkers for intracellular degradation have been reported. These PEIs displayed high transfection efficiency and low cytotoxicity due to their rapid degradation into small MW water-soluble fragments, which are easily processed and removed by the cells. Tarcha *et al.* seem to be the first group to report of degradable PEI based on chemically condensed low MW PEI containing beta-aminopropionamide for siRNA delivery.[55] The polymer was synthesized by *N*-acylation of degradable PEI prepared by Michael reaction of low MW PEI (800 Da) and hexanediol diacrylate to improve chemical stability relative to ester-containing polymers, but in comparison to PEI (25 kDa), better degradability through the amide linkages. The polymer showed significant *in vitro* knockdown of the luciferase gene, up to 80%, in comparison to non-targeting siRNA in stably transfected HuH7 cells.[55] Breunig *et al.* synthesized degradable PEI by introducing disulfide bonds into the low MW PEI and studied the relationship between cellular uptake and gene silencing activity among linear PEI (5 kDa), cross-linked PEI and branched PEI (25 kDa).[56] The results indicated that the cellular uptake of siRNA was more efficient with increasing branching of the polymer, whereas the siRNA release was promoted by cross-linked PEI, suggesting that a combination of a high branching density and reductively cleavable bonds within the PEI is promising towards improving siRNA delivery.

In another study, degradable PEI based on LMW PEI (423 Da) and PEG diacrylate (258 Da) was evaluated for small interfering/small hairpin (si/sh) RNA delivery in A549 cells.[57] The polymer successfully delivered siRNA targeting EGFP and silenced EGFP expression. The silencing achieved with the polymer was 1.5-fold higher and safer than native PEI (25 kDa). The polymer also exhibited superior protein kinase Akt1 shRNA delivery, and thereby efficiently silenced oncoprotein Akt1. Furthermore, polymer shAkt-mediated Akt1 knockdown hindered cancer cell growth in A549 cells in an Akt1-specific manner due to the degradability of the polymer. Akt (protein kinase B) is an important regulator of cell survival and plays a key role in cancer by stimulating cell proliferation, inhibiting apoptosis, and modulating the protein translation. Later, the same group studied the Akt1 gene silencing after aerosol delivery of degradable PEI/Akt1 siRNA complexes into *K-ras* and urethane-induced lung-cancer-model mice.[58] The aerosol-delivered Akt1 siRNA suppressed the mRNA and protein expression of Akt1 specifically without affecting the Akt2 and Akt3 in the lungs of *K-ras* mice. Also, the number of tumors and the mean tumor diameter were significantly decreased by Akt1 siRNA treatment.

5. Future Perspectives

PEI is one of the most efficient gene and siRNA delivery systems. The availability of PEI in a wide MW range enabled it to find application in several *in vitro* and *in vivo* studies. Although the high MW PEI (25 kDa) is considered as gold standard due to its high transfection efficiency, it also causes high toxicity. The formulations with high transfection efficacies and low cytotoxicity will be the key towards development of siRNA-based clinical therapeutics. Studies employing low MW PEI suffer from low transfection efficiency and those with high MW have high toxicities, hence there is need to find an intermediate MW with pros of both the worlds. PEI modified with biocompatible neutral polymers or with introduction of biodegradable linkages will be preferred for *in vitro* and *in vivo* siRNA delivery.

6. Conclusion

Encouraging results of *in vitro* gene silencing in cell cultures and *in vivo* data in mice models suggest PEIs as promising candidates.

The high cationic charge density infers high transfection efficiency to PEI (25 kDa) along with high toxicity. The problem of cytotoxicity has been addressed by incorporation of neutral moieties such as PEG. A new approach that proposes the use of small MW PEI to prepare high MW PEI of desirable MW is fascinating. Initial *in vitro* and *in vivo* studies suggest that this approach is the upcoming area in PEI-mediated siRNA therapeutics. In a nutshell, the high transfection efficiency of PEI-based delivery vectors can be maintained by reducing the cytotoxicity of the polymer by introduction of biodegradable linkages.

References

1. Elmen J, Thonberg H, Ljungberg K, *et al.* (2005) Locked nucleic acid (LNA) mediated improvements in siRNA stability and functionality. *Nucleic Acids Res* **33:** 439–447.
2. Chiu YL, Ali A, Chu CY, *et al.* (2004) Visualizing a correlation between siRNA localization, cellular uptake, and RNAi in living cells. *Chem Biol* **11:** 1165–1175.
3. Soutschek J, Akinc A, Bramlage B, *et al.* (2004) Therapeutic silencing of an endogenous gene by systemic administration of modified siRNAs. *Nature* **432:** 173–178.
4. Weyermann J, Lochmann D, Georgens C, Zimmer A. (2005) Albumin-protamine-oligonucleotide-nanoparticles as a new antisense delivery system. Part 2: Cellular uptake and effect. *Eur J Pharm Biopharm* **59:** 431–438.
5. Felgner PL. (1997) Nonviral strategies for gene therapy. *Sci Am* **276:** 102–106.
6. Davis ME. (2002) Non-viral gene delivery systems. *Curr Opin Biotechnol* **13:** 128–131.
7. Zhao Z, Wang J, Mao HQ, Leong KW. (2003) Polyphosphoesters in drug and gene delivery. *Adv Drug Deliv Rev* **55:** 483–499.
8. Boussif O, Lezoualc'h F, Zanta MA, *et al.* (1995) A versatile vector for gene and oligonucleotide transfer into cells in culture and *in vivo*: polyethylenimine. *Proc Natl Acad Sci USA* **92:** 7297–7301.
9. Neu M, Fischer D, Kissel T. (2005) Recent advances in rational gene transfer vector design based on poly(ethylene imine) and its derivatives. *J Gene Med* **7:** 992–1009.
10. Jones GD, Langsjoen A, Neumann SMMC, Zomlefer J. (1944) The polymerization of ethylenimine. *J Org Chem* **09:** 125–147.
11. Brissault B, Kichler A, Guis C, *et al.* (2003) Synthesis of linear polyethylenimine derivatives for DNA transfection. *Bioconjug Chem* **14:** 581–587.

12. von Harpe A, Petersen H, Li Y, Kissel T. (2000) Characterization of commercially available and synthesized polyethylenimines for gene delivery. *J Control Release* **69:** 309–322.
13. Boussif O, Lezoualc'h F, Zanta MA, *et al.* (1995) A versatile vector for gene and oligonucleotide transfer into cells in culture and *in vivo*: polyethylenimine. *Proc Natl Acad Sci USA* **92:** 7297–7301.
14. Thomas M, Lu JJ, Ge Q, *et al.* (2005) Full deacylation of polyethylenimine dramatically boosts its gene delivery efficiency and specificity to mouse lung. *Proc Natl Acad Sci USA* **102:** 5679–5684.
15. Thomas M, Klibanov AM. (2002) Enhancing polyethylenimine's delivery of plasmid DNA into mammalian cells. *Proc Natl Acad Sci USA* **99:** 14640–14645.
16. Akinc A, Thomas M, Klibanov AM, Langer R. (2005) Exploring polyethylenimine-mediated DNA transfection and the proton sponge hypothesis. *J Gene Med* **7:** 657–663.
17. Goula D, Benoist C, Mantero S, *et al.* (1998) Polyethylenimine-based intravenous delivery of transgenes to mouse lung. *Gene Ther* **5:** 1291–1295.
18. Boussif O, Lezoualc'h F, Zanta MA, *et al.* (1995) A versatile vector for gene and oligonucleotide transfer into cells in culture and *in vivo* — polyethylenimine. *Proc Natl Acad Sci USA* **92:** 7297–7301.
19. Abdallah B, Hassan A, Benoist C, *et al.* (1996) A powerful nonviral vector for *in vivo* gene transfer into the adult mammalian brain: polyethylenimine. *Hum Gene Ther* **7:** 1947–1954.
20. Erbacher P, Bettinger T, Brion E, *et al.* (2004) Genuine DNA/polyethylenimine (PEI) complexes improve transfection properties and cell survival. *J Drug Target* **12:** 223–236.
21. Kunath K, von Harpe A, Fischer D, *et al.* (2003) Low-molecular-weight polyethylenimine as a non-viral vector for DNA delivery: comparison of physicochemical properties, transfection efficiency and *in vivo* distribution with high-molecular-weight polyethylenimine. *J Control Release* **89:** 113–125.
22. Hong JW, Park JH, Huh KM, *et al.* (2004) PEGylated polyethylenimine for *in vivo* local gene delivery based on lipiodolized emulsion system. *J Control Release* **99:** 167–176.
23. Kichler A, Chillon M, Leborgne C, *et al.* (2002) Intranasal gene delivery with a polyethylenimine-PEG conjugate. *J Control Release* **81:** 379–388.
24. Kichler A, Leborgne C, Coeytaux E, Danos O. (2001) Polyethylenimine-mediated gene delivery: a mechanistic study. *J Gene Med* **3:** 135–144.
25. Fischer D, Bieber T, Li Y, *et al.* (1999) A novel non-viral vector for DNA delivery based on low molecular weight, branched polyethylenimine: effect of molecular weight on transfection efficiency and cytotoxicity. *Pharm Res* **16:** 1273–1279.

26. Godbey WT, Wu KK, Mikos AG. (1999) Size matters: molecular weight affects the efficiency of poly(ethylenimine) as a gene delivery vehicle. *J Biomed Mater Res* **45:** 268–275.
27. Ogris M, Brunner S, Schuller S, *et al.* (1999) PEGylated DNA/transferrin-PEI complexes: Reduced interaction with blood components, extended circulation in blood and potential for systemic gene delivery. *Gene Ther* **6:** 595–605.
28. Wightman L, Kircheis R, Rossler V, *et al.* (2001) Different behavior of branched and linear polyethylenimine for gene delivery *in vitro* and *in vivo*. *J Gene Med* **3:** 362–372.
29. Grayson AC, Doody AM, Putnam D. (2006) Biophysical and structural characterization of polyethylenimine-mediated siRNA delivery *in vitro*. *Pharm Res* **23:** 1868–1876.
30. Spagnou S, Miller AD, Keller M. (2004) Lipidic carriers of siRNA: differences in the formulation, cellular uptake, and delivery with plasmid DNA. *Biochemistry* **43:** 13348–13356.
31. Bernstein E, Caudy AA, Hammond SM, Hannon GJ. (2001) Role for a bidentate ribonuclease in the initiation step of RNA interference. *Nature* **409:** 363–366.
32. Elbashir SM, Harborth J, Lendeckel W, *et al.* (2001) Duplexes of 21-nucleotide RNAs mediate RNA interference in cultured mammalian cells. *Nature* **411:** 494–498.
33. Hammond SM, Bernstein E, Beach D, Hannon GJ. (2000) An RNA-directed nuclease mediates post-transcriptional gene silencing in drosophila cells. *Nature* **404:** 293–296.
34. Nykanen A, Haley B, Zamore PD. (2001) ATP requirements and small interfering RNA structure in the RNA interference pathway. *Cell* **107:** 309–321.
35. Martinez J, Tuschl T. (2004) RISC is a 5' phosphomonoester-producing RNA endonuclease. *Genes Dev* **18:** 975–980.
36. Leung RKM, Whittaker PA. (2005) RNA interference: from gene silencing to gene-specific therapeutics. *Pharmacol Ther* **107:** 222–239.
37. Oskuee RK, Philipp A, Dehshahri A, *et al.* (2010) The impact of carboxyalkylation of branched polyethylenimine on effectiveness in small interfering RNA delivery. *J Gene Med* **12:** 729–738.
38. Philipp A, Zhao X, Tarcha P, *et al.* (2009) Hydrophobically modified oligoethylenimines as highly efficient transfection agents for siRNA delivery. *Bioconjug Chem* **20:** 2055–2061.
39. Wu Y, Wang W, Chen Y, *et al.* (2010) The investigation of polymer-siRNA nanoparticle for gene therapy of gastric cancer *in vitro*. *Int J Nanomedicine* **5:** 129–136.
40. Veiseh O, Kievit FM, Fang C, *et al.* (2010) Chlorotoxin bound magnetic nanovector tailored for cancer cell targeting, imaging, and siRNA delivery. *Biomaterials* **31:** 8032–8042.

41. Creusat G, Rinaldi AS, Weiss E, *et al.* (2010) Proton sponge trick for pH-sensitive disassembly of polyethylenimine-based siRNA delivery systems. *Bioconjug Chem* **21:** 994–1002.
42. Patnaik S, Aggarwal A, Nimesh S, *et al.* (2006) PEI-alginate nanocomposites as efficient *in vitro* gene transfection agents. *J Control Release* **114:** 398–409.
43. Nimesh S, Chandra R. (2009) Polyethylenimine nanoparticles as an efficient *in vitro* siRNA delivery system. *Eur J Pharm Biopharm* **73:** 43–49.
44. Kim WJ, Chang CW, Lee M, Kim SW. (2007) Efficient siRNA delivery using water soluble lipopolymer for anti-angiogenic gene therapy. *J Control Release* **118:** 357–363.
45. Biswal B, Debata N, Verma R. (2010) Development of a targeted siRNA delivery system using FOL-PEG-PEI conjugate. *Mol Biol Rep* **37:** 2919–2926.
46. Kang JH, Tachibana Y, Kamata W, *et al.* (2010) Liver-targeted siRNA delivery by polyethylenimine (PEI)-pullulan carrier. *Bioorg Med Chem* **18:** 3946–3950.
47. Huh MS, Lee SY, Park S, *et al.* (2010) Tumor-homing glycol chitosan/polyethylenimine nanoparticles for the systemic delivery of siRNA in tumor-bearing mice. *J Control Release* **144:** 134–143.
48. Geisbert TW, Hensley LE, Kagan E, *et al.* (2006) Postexposure protection of guinea pigs against a lethal ebola virus challenge is conferred by RNA interference. *J Infect Dis* **193:** 1650–1657.
49. Segura T, Schmokel H, Hubbell JA. (2007) RNA interference targeting hypoxia inducible factor 1alpha reduces post-operative adhesions in rats. *J Surg Res* **141:** 162–170.
50. Ge Q, Filip L, Bai A, *et al.* (2004) Inhibition of influenza virus production in virus-infected mice by RNA interference. *Proc Natl Acad Sci USA* **101:** 8676–8681.
51. Lively TN, Kossen K, Balhorn A, *et al.* (2008) Effect of chemically modified IL-13 short interfering RNA on development of airway hyperresponsiveness in mice. *J Allergy Clin Immunol* **121:** 88–94.
52. Thomas M, Lu JJ, Ge Q, *et al.* (2005) Full deacylation of polyethylenimine dramatically boosts its gene delivery efficiency and specificity to mouse lung. *Proc Natl Acad Sci USA* **102:** 5679–5684.
53. Werth S, Urban-Klein B, Dai L, *et al.* (2006) A low molecular weight fraction of polyethylenimine (PEI) displays increased transfection efficiency of DNA and siRNA in fresh or lyophilized complexes. *J Control Release* **112:** 257–270.
54. Hobel S, Prinz R, Malek A, *et al.* (2008) Polyethylenimine PEI F25-LMW allows the long-term storage of frozen complexes as fully active reagents in siRNA-mediated gene targeting and DNA delivery. *Eur J Pharm Biopharm* **70:** 29–41.

55. Tarcha PJ, Pelisek J, Merdan T, *et al.* (2007) Synthesis and characterization of chemically condensed oligoethylenimine containing beta-aminopropionamide linkages for siRNA delivery. *Biomaterials* **28:** 3731–3740.
56. Breunig M, Hozsa C, Lungwitz U, *et al.* (2008) Mechanistic investigation of poly(ethylene imine)-based siRNA delivery: disulfide bonds boost intracellular release of the cargo. *J Control Release* **130:** 57–63.
57. Jere D, Xu CX, Arote R, *et al.* (2008) Poly(beta-amino ester) as a carrier for si/shRNA delivery in lung cancer cells. *Biomaterials* **29:** 2535–2547.
58. Xu CX, Jere D, Jin H, *et al.* (2008) Poly(ester amine)-mediated, aerosol-delivered Akt1 small interfering RNA suppresses lung tumorigenesis. *Am J Respir Crit Care Med* **178:** 60–73.

Chapter 24

Nanosilver System Technology for Applications in Nanomedicine

Helmut Schmid*

Introduction

Nanotechnology is a particle technology involving active ingredients with a size ranging from 1 to 100 nm (at least in one dimension) — and the desired properties are determined by size and not by chemistry. This technology gives rise to suprising new volume-, surface- and quantum-effects, which are applicable in medicine. Here, contributions in all three classical major areas are possible, namely, basic medical research, diagnosis and therapy as well as in numerous sub-topics. In addition, therapy and diagnostics move together so closely that it makes sense to use the new term, "Thera-Nostics."

Medical examinations, especially in drug development, are carried out at different levels. The impact studies start with *in vitro* investigations, with particular emphasis on human models, followed by *in vivo*

*Fraunhofer-Institute for Chemical Technology, 76327 Pfinztal, Germany
E-mail: sd@ict.fhg.de, Web: www.ict.fhg.de, Phone: +49 721 4640 709.

experiments. This can be accompanied by online measurements with open flow microperfusion.

Among all the analytical procedures — especially during *in vivo* experiments — molecular imaging such as like optical imaging, MRI, PET and SPECT, plays a special role. In optical imaging, quantum dot nanoparticles are used to investigate localization, identification and reaction of components. In diagnostics, nano-coatings are relevant for biosensors and biochips. Nanosilver, for example, enables the fabrication of electrically conductive layers and interconnects. Much broader are applications of nanoparticles in combination with organic molecules in therapy. Here, topics like drug delivery (targeting, controlled drug release), gene delivery, personalized medicine, regenerative medicine, tissue engineering, antimicrobial systems and cancer therapy must be addressed. Here, agents in combination with micells, liposomes, dendrimers, polymer-protein-conjugates and complex-compounds are used. In particular, polymer-protein-conjugates can be coupled with the anti-cancer drug, doxorubicin.

Scientists at the Fraunhofer Institute of Chemical Technology (ICT) carry out research and development work on the following sub-areas, and these are covered in this chapter. Further details are described in the references.[1–14]

— Production, stabilization, characterization and system integration of inorganic nanoparticles, especially transition metals and transition metal oxides. Of importance here are nanosilver with antimicrobial functionality and nanomagnetite used in cancer therapy (thermotherapy).

— Carbon nanotubes for drug delivery.

Tasks and Objectives

The task is to make these nanoparticles available and usable for medicine. The aim is to develop nano-medical products. Although the various nano-systems are addressed, the focus is on nanosilver system technology.

Methodology

Figure 1 shows the procedure to be performed in order to obtain nano products from nanoparticles.

The first step is nanoparticle synthesis which can be done top-down or bottom-up. Derived from experiments, nanoparticles of the order of magnitude around 10 nm show the best effectiveness in nanomedicine. The **synthetic methods** should be optimized in this respect.

For thermodynamic reasons, nanoparticles tend to re-agglomerate, which weakens the intended effects. To prevent this, **nano-stabilization** by chemical additives is required, imprinting resulting surface charges. The selection of suitable additives is carried out with potential calculations based on Poisson-Boltzmann differential equation (1).

$$
\begin{aligned}
\nabla^2 U(\eta,\theta) = {} & \frac{(\cosh\eta - \cos\theta)^2}{B^2} * \frac{\partial^2 U}{\partial \eta^2} + \frac{(\cosh\eta - \cos\theta)^2}{B^2} * \frac{\partial^2 U}{\partial \theta^2} \\
& - \frac{\sinh\eta * (\cos\eta - \cos\theta)}{B^2} * \frac{\partial U}{\partial \eta} \\
& + \left(\frac{(\cosh\eta - \cos\theta)^2}{\tan\theta} - (\cosh\eta - \cos\theta) * \sin\theta \right) \\
& \times * \frac{1}{B^2} * \frac{\partial U}{\partial \theta} = \sinh(U(\eta,\theta)) \qquad (1)
\end{aligned}
$$

U: reduced electrostataial; B: constant in bispheric system of coordinates.

Of importance is the constant B, which correlates with the Hamaker constant of Van-der-Waals interaction. Only such additives are suitable, being capable of minimizing these interactions.

The next step is to **characterize** the particles with selected **nano-analytical-methods,** e.g. scanning electron microscopy, X-ray fluorescence, X-ray diffraction, photon correlation spectroscopy and

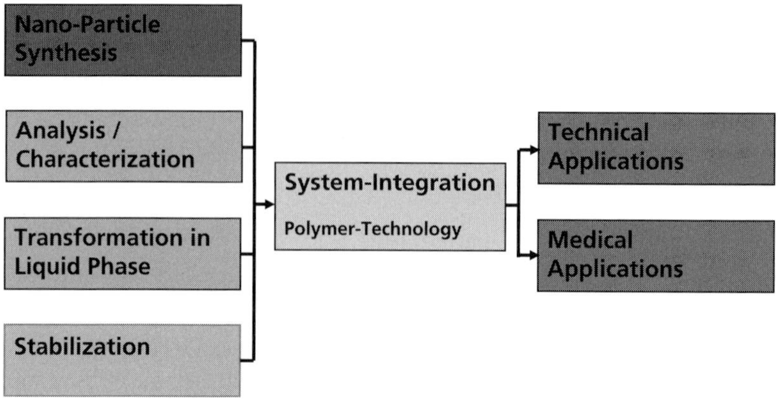

Fig. 1. Overview of the main steps to obtain nano products from nanoparticles.

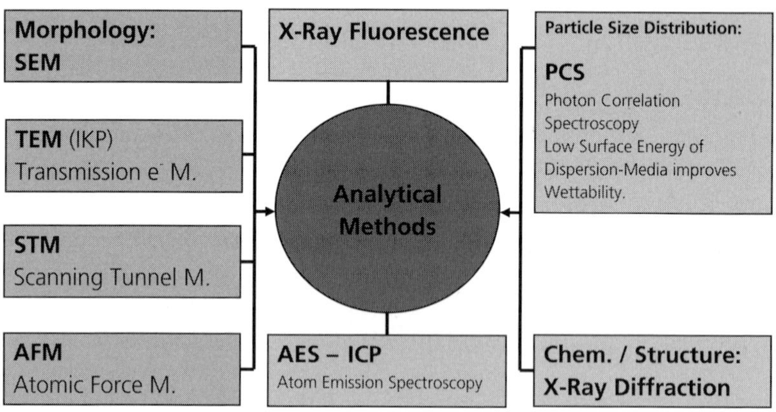

Fig. 2. Overview of the nano-analytical methods used for characterization of nanoparticles.

atomic emission spectroscopy. Figure 2 gives an overview of these methods. The application of the last method is essential to ensure nano-safety.

To provide an **interface to nanomedicine** and to enable **system-integration**, a **combination of nanotechnology and polymertechnology** is needed, with particular emphasis on **biocompatible polymers**. In this way, **stable nano-suspensions** are available and ready for use.

Results and Discussion

As previously mentioned, for nanoparticle synthesis, both top-down and bottom-up methods can be used, leading to intermediates that have a mean particle size of approximately 50 nm. This is demonstrated by the examples of nanosilver (Fig. 3) and nano-magnetite (Fig. 4), where images from scanning electron microscopy show very clearly the particle-size, distribution, morphology and topography. As for medical applications, particles with an average size of about 10 nm show the best effect; in this case post-processing is required.

Optimization of processing parameters leads to a single-step synthesis. These methods of chemical nanotechnology are particularly suitable for the synthesis of inorganic nanoparticles, enabling high production rates and an economic, cost-effective process control. Even more specifically, we prefer direct synthesis in a liquid, because handling of powders is avoided, which means a tremendous safety advantage. Furthermore, the production of nano-suspension, which serves as an interface to polymer combination and sytem integration, is facilitated.

Fig. 3. SEM-record of nano-Ag particles. A stable suspension was applied to an Al-plate and investigated after drying. Particle size approximately 50 nm.

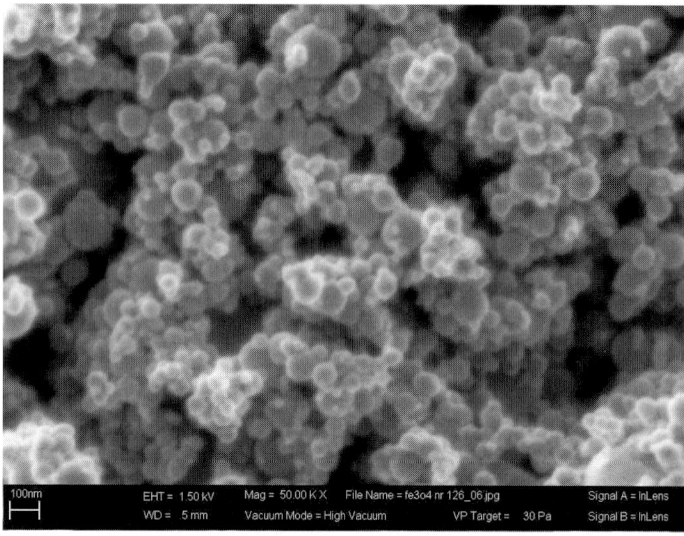

Fig. 4. SEM-record of nano-magnetite particles (Fe_3O_4). A stable suspension was applied to an Al-plate and investigated after drying. Particle size approximately 50 nm.

These nanosilver systems are suitable for producing antimicrobial surfaces that have numerous medical applications, e.g. in hospitals, medical devices, implants, etc. Currently, the commercially most successful products are dispersion paints for hygiene improvement in hospitals to combat dangerous germs like MRSA. Due to the numerous and different reaction mechanisms, the risk of resistance is not increased.

Nano-magnetite (Fig. 4) is another interesting nano drug which can be used in cancer therapy. By excitation of this ferromagnetic material in an alternating magnetic field, a thermo therapy can be applied.

Examples of organic nanoparticles that can be used in medicine for drug delivery, are different types of carbon nanotubes (Fig. 5). Depending on the diameter, an embedded active ingredient can be released according to precise kinetics. For mathematical modeling, we have developed appropriate tools. These principles can be extended by active or passive targeting. For synthesis, physical processes in the form of an arc- or laser-technology are applied.

Fig. 5. SEM-record of carbon nanotubes for drug delivery.

Between the multi-wall and single-wall types, the latter is preferred because high-definition systems can be formulated.

To exclude the risk of toxicity, only biodegradable versions are used. This can be achieved via chemical surface modifications by addition reaction or radical initiated grafting. Oxidation reactions are not suitable because they lead to the destruction of the fibers.

For each drug, release kinetics is defined as a specific target. Thereafter, the diameter of carbon nanotubes is selected.

To describe the kinetics of release, we use a mathematical model that is validated with experimental data. The chosen mathematical approach can be applied very broadly, as both the reaction control and diffusion control and their transitions are taken into account. The solution to the proposed differential equation is an analytical function which can be fitted to the experimental data where the fit parameters are determined. The following examples show a good correlation (Fig. 6).

As already explained, nanoparticle synthesis is followed by analysis with methods described in the chapter on "Methodology."

In the following, selected results are presented.

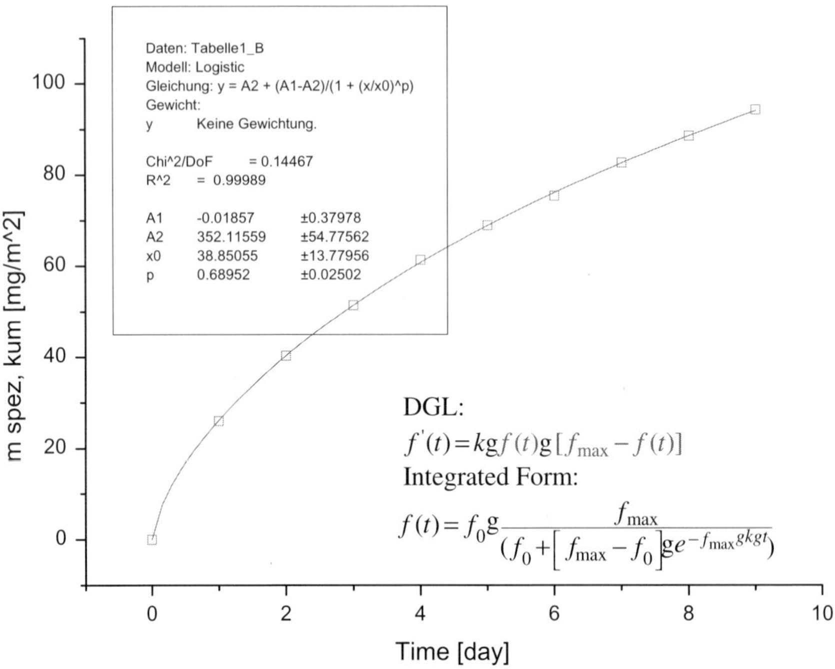

Fig. 6. Example of mathematical modeling of cumulative drug release. A sophisticated correlation between experimental (*squares*) and calculated values (*solid curve*) could be obtained.

In medical applications which require a coating with antimicrobial nano silver, detection of surface morphology by SEM (scanning electron microscopy) is important. This data set is greatly enhanced when a parallel quantitative chemical analysis is performed using EDX (energy dispersive X-ray spectroscopy) (Fig. 7). This is to ensure that the nanoparticles are located at the desired site.

In addition, both on surfaces and in the bulk material, optimal performance requires a homogeneous distribution of the nanoparticles, and this is done by additive stabilization and process adaptation.

As an interface to different applications, nano-suspensions are used. This will create a technology platform that is versatile, cost-effective and sustainable. Also, on this processing level, particle size

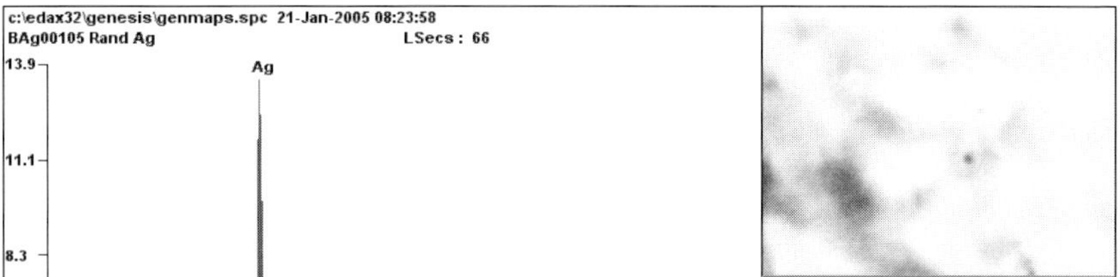

Element	Wt%	At%
C K	41.08	78.57
O K	06.71	09.64
NaK	00.86	00.85
AgL	51.35	10.94

Fig. 7. Quantitative Ag-analysis in sample BAg00105 with X-ray fluorescence.

distribution is controlled by means of photon correlation spectroscopy (PCS) (Fig. 8a and 8b).

In order to detect the primary particle size and to estimate possible agglomeration effects in the solid state, X-ray diffraction is a powerful tool for investigation. By using temperature-resolved X-ray diffraction, agglomeration and grain growth can be analyzed additionally as a function of temperature. The primary crystallite dimension of the prepared nano-Ag is around 10 nm and this correlates with the PCS-data. In addition, starting from 300°C, a grain growth can be detected which leads to particle dimensions of around 130 nm (Fig. 9).

As a next point, it is necessary to describe the principal working mechanism in our coating system. In contrast to the known principles of standard biocides, we focus not on drug release of nanosilver but on surface catalysis. This is advantageous in relation to sustainability, safety and environmental effects. In brief, the partial charge of nano-Ag out of the lattice interacts with the structure proteins of cell walls and kills the microorganism. This dominating reaction mechanism is accompanied by various other working principles, so that the risk of resistance is not increased.

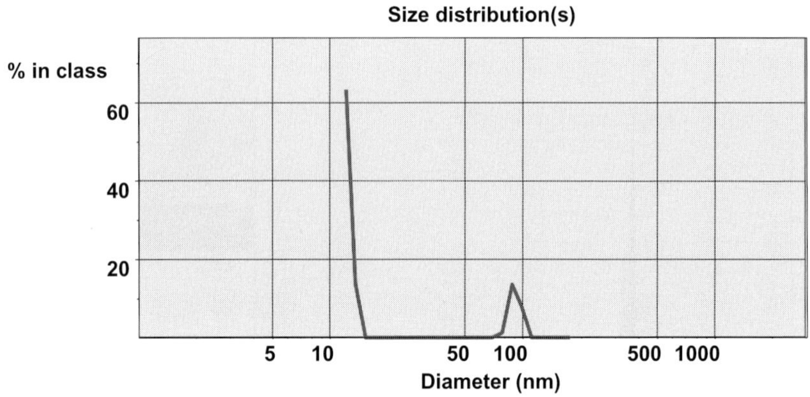

Fig. 8a. Particle size distribution of the used nano-suspension analyzed using photon correlation spectroscopy (PCS).

Size (nm)	% Intensity	Size (nm)	% Intensity	Size (nm)	% Intensity
12.1	63.3	30.8	0.0	78.2	1.2
13.6	13.8	34.6	0.0	87.8	13.6
15.3	0.0	38.9	0.0	98.7	8.1
17.2	0.0	43.7	0.0	110.9	0.0
19.3	0.0	49.0	0.0	124.6	0.0
21.7	0.0	55.1	0.0	140.0	0.0
24.4	0.0	61.9	0.0	157.3	0.0
27.4	0.0	69.6	0.0	176.8	0.0

Fig. 8b. Data set of particle size distribution shown in Fig. 8a.

After sample preparation, the efficiency of antimicrobial effect on the surface must be proved. A small selection of German and International standards is listed below:

- Direktbeschickung, Pharm. Eu. 1997
- (Konservierungs-) Belastungstest
- Quantitativer Suspensionsversuch
- Suspensionstest, EN 1275, 1997

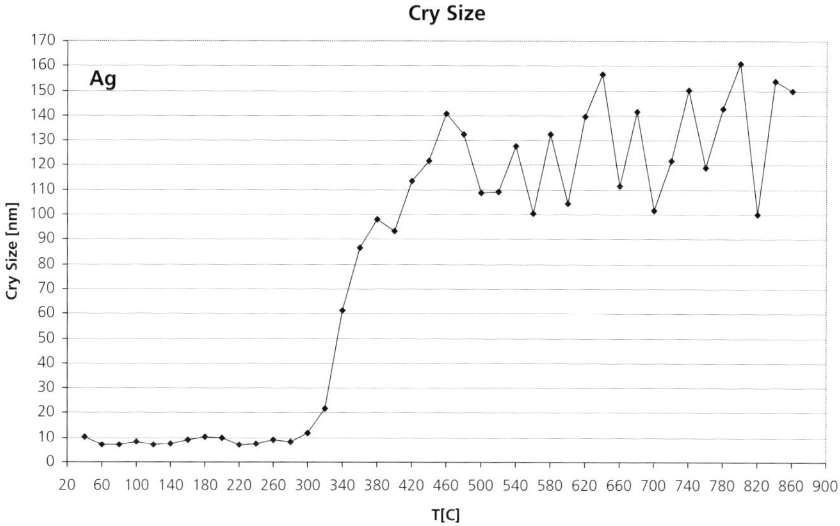

Fig. 9. By means of temperature-resolved X-ray diffraction, the primary crystallite dimension of nano-Ag as a function of temperature is evaluated. Grain growth starts from 300 °C.

- Beschichtungstest
- Test der Oberflächenwirksamkeit, JIS Z 2801
- ASTM E2180–01
- VdL-RL 06/Nov. 1998, Beständigkeit gegen Pilzbefall

Two sets of representative results from these methods are shown in Figs. 10 and 11, clearly demonstrating the effectiveness of antimicrobial activity.

These nanosilver systems are suitable for producing antimicrobial surfaces that have numerous medical applications, e.g. coating of wound-pads (Fig. 12), implants, medical devices, in hospitals, etc.

Currently, the most successful commercially available products are dispersion paints for hygiene improvement in hospitals that can kill even the dangerous germs like MRSA (Figs. 13 and 14).

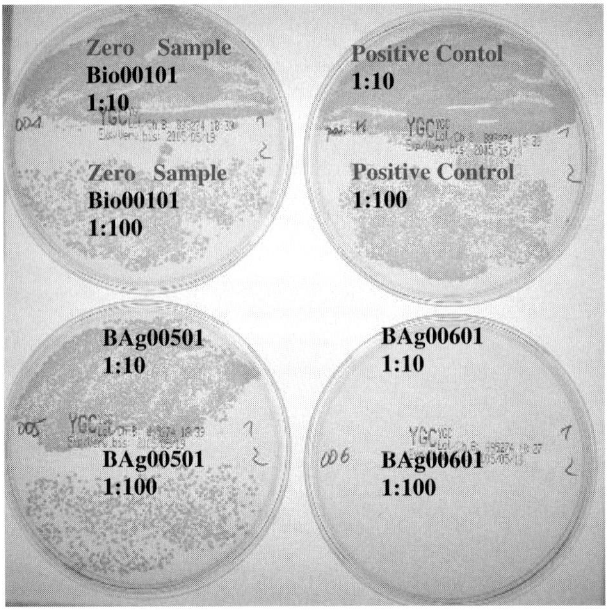

Fig. 10. Result of incubated diluted crop-out of the samples. No fungal attack is found on treated samples BAg00601.

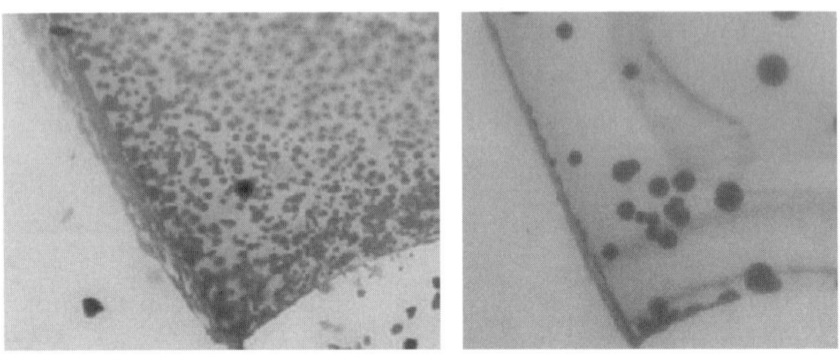

Fig. 11. Nano-Ag in addition is very effective against bacteria. Example of a nano-Ag-coated silicon-polymer (*right*) compared with an untreated sample (*left*) with colonies of *Staphylococcus aureus*. A reduction rate of factor 100 was detected based on the applied nano-Ag-content (Test-setup: Spray-test).
Depending on nano-Ag-content, reduction rates of 99,999% (5 log-levels) are possible.

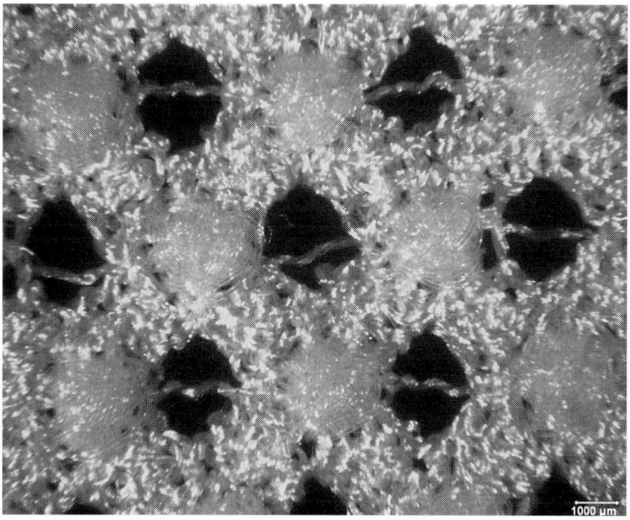

Fig. 12. Wound-dressing that has been coated with nanosilver to accelerate the healing process of burns and reduce the risk of infection.

Fig. 13. Al Wasl Hospital, Dubai, U.A.E., where the nanosilver coating system is applied.

Fig. 14. Al Wasl Hospital (2), Dubai, U.A.E., with nanosilver coatings against MRSA.

Conclusion

For many medical applications, additional functionalizations of surfaces in order to achieve antimicrobial effects are useful and desirable. According to state of the art standard, biocides like thiazolinones are used as active ingredients. The disadvantages are toxicity and lack of sustainability of effect. A new approach, avoiding the mentioned disadvantages, is the use of nanosilver as an active substance.

With this technology, cost-effective solutions are possible, because the required drug concentrations can be kept small. Even the concept of action is fundamentally different: There is no drug-release, but only provision of the active substance on the surface in the extended sense of a catalyst.

In technical applications, mold-formation can be prevented. In medical applications, dangerous bacteria and even multi-resistant germs like MRSA can be killed. This leads to a general improvement of hygiene.

The more numerous the response mechanisms are, the smaller the risk of resistance is. The product safety throughout the life cycle is guaranteed by a combination of nanotechnology and polymer technology. The nanoparticles are chemisorbed in the polymer via dipole-interactions.

Outlook

- Previous investigations of nanosilver system technology in medicine were focused on external applications like antimicrobial coatings for medical devices and surfaces. In the future, we will be looking at internal applications such as implant coatings as well as drug delivery. In this context, we will proceed with efficacy and toxicology studies.
- The expansion of the chemical base leads directly to other nanometals and metal oxides. An outstanding representative is ferromagnetic nano-magnetite as an anti-cancer agent in thermotherapy, where an alternating magnetic field is applied. In this case, combination with biocompatible and bioresorbable polymers is required.
- On the side of organic materials, organically modified carbon nanotubes are functionalized as drug carriers. Synergetic effects together with nanosilver, nano-magnetite, etc. are to be expected.
- In order to improve the dendrimere principle in the sense of thermal controlled drug release systems, we will introduce poly-NIPAM (poly(N-isopropylacrylamide) which can be formulated as reversible thermo-sensitive colloidal particles performing structural changes.

References

1. Schmid H. (2005) "NANO-Technologies, From Basics to Selected Applications in Buildings". ICT-Presentation by Invitation of Ministry of Construction Dubai, September 27, 2005.
2. Schmid H. (2006) "Schutz vor mikrobieller Materialzerstörung durch biozid wirksame Ag-Nano-Partikel". ICT-Presentation by Invitation of DECHEMA

Working Group "Mikrobielle Materialzerstörung und Materialschutz". Frankfurt a.M., January 31, 2006.

3. Schmid H. (2006) "Selected Applications of Nanostructured Materials in Buildings". Materials Science & Technology 2006 Conference & Exhibition, Cincinnati, Convention Center, Ohio, USA, October 15–19, 2006.
4. Schmid H. (2007) "Wirkprinzipien und Sicherheitsaspekte der Nano-Silber-Technologie". ICT-Presentation at German Federal Environmental Agency, Berlin, May 14–16, 2007.
5. Schmid H. (2007) "Nano-Analysis as an Important Step of Nano-Product-Development". Lecture at JETRO and Research Park, Tokyo and Kyoto, Japan, June 9–15, 2007.
6. Schmid H. (2007) "Selected Applications of Nano-Particles for Health, Safety and Convenience". Lecture at International Conference on Nanomedicine, Porto Carras, Greece, September 9–11, 2007.
7. Schmid H. (2007) "Nano-Silber zur Schimmelvermeidung und zur Verbesserung der Innenraumhygiene". Lecture at 1. Internationaler Brenntag "Wassertag", Murten-Meyriez, Switzerland, November 13, 2007.
8. Schmid H. (2007) "Verbesserung der Hygiene mit nicht-humantoxischen Nano-Silber-Systemen". Lecture at Institut für Pharmakologie und Toxikologie der Bundeswehr, München, December 5, 2007.
9. Schmid H. (2007) "Nano-Particles in Plastics to Reach Antimicrobial Functionality and System Improvements". Lecture at International Conference on Biocides and Plastics, Brussels, Belgium, February 26–27, 2008.
10. Schmid H. (2008) "Synthesis and Characterization of Selected Nanoparticles under Consideration of Medical Applications". Lecture at 1st European Conference for Clinical Nanomedicine, Basel, Switzerland, May 19–21, 2008.
11. Schmid H. (2008) "Improvements on Health, Safety and Functionality by Nano-Particle-Applications". Lecture at Symposium "Nano Coating Days", Zürich, Switzerland, May 27, 2008.
12. Schmid H. (2008) "Nano-Silber zur Schimmelvermeidung und zur Verbesserung der Innenraumhygiene". Lecture at Architektenkammer NRW, Gelsenkirchen, November 3–04, 2008.
13. Schmid H. (2008) "Antimicrobial Effects of Nano-Silver: Theory, Analysis, Application, Modelling and Safety Aspects". Lecture at Symposium "4. Moskauer Wirtschaftstage", Düsseldorf, November 5, 2008.
14. Schmid H, *et al.* (2009) "Antimykotische, antibiotische und flammhemmende Nanofunktionalisierung von Mehrschicht-Monofilamenten für die Anwendung im Wachstumsmarkt Sport- und Freizeitrasen sowie Landscaping". Presentation at WING-nano conference, Ulm, April 1–3, 2009.

Index